Selected Problems
in
Theoretical Physics

With Solutions

SELECTED PROBLEMS IN THEORETICAL PHYSICS

(WITH SOLUTIONS)

A. Di Giacomo
G. Paffuti
P. Rossi

Pisa University
Italy

World Scientific
Singapore • New Jersey • London • Hong Kong

Published by

World Scientific Publishing Co. Pte. Ltd.
P O Box 128, Farrer Road, Singapore 9128
USA office: Suite 1B, 1060 Main Street, River Edge, NJ 07661
UK office: 73 Lynton Mead, Totteridge, London N20 8DH

SELECTED PROBLEMS IN THEORETICAL PHYSICS
(with Solutions)

ISBN 981-02-1614-9
      981-02-1615-7 (pbk)

Printed in Singapore by Utopia Press.

Contents

Preface

The collection of problems contained in this book has a common origin: it has been selected from written examinations, given over the years to students of the course entitled "Theoretical Physics", at the University of Pisa. These problems are presented here together with worked-out solutions.

The Appendices establish standard notation and provide some commonly used formulae in relativistic kinematics, group theory, theory of scattering and decay, relativistic wave equations, γ-matrix algebra.

Approximately one half of the problems cover aspects of Elementary Particle Physics, while the rest range over Atomic Physics, Potential Theory, Field Theory and Physics of Nuclei. Some necessary instruments for the solutions are theory of scattering and decay, relativistic quantum mechanics, second quantization and elementary perturbative field theory.

Each examination comprises three problems, of which the candidate must solve one. Statistically, one student out of ten answers correctly all questions, seven answer two questions (from a total of three or four), and the rest turn in insufficient solutions.

The course attaches great importance to recitation sessions, where thorough problem solving becomes a true test of mastery over theoretical background.

Theoretical Physics has been taught for approximately 20 years by one of the authors (A. Di Giacomo). Recitations were held in the past by G. Paffuti and P. Rossi, and more recently by Haris Panagopoulos, for several years.

This book is translated from the original italian edition *Problemi di Fisica Teorica*, A. Di Giacomo, G. Paffuti, P. Rossi, E.T.S. Pisa 1992. The authors are grateful to H. Panagopoulos for his collaboration in preparing the English version.

Problem 1.   Scattering off the potential $V(r) = a\,\delta(r - r_0)$

Consider a spin-0 particle with mass m, scattering off the potential

$$V(r) = a\,\delta(r - r_0)$$

This potential is a thin attractive (a < 0) or repulsive (a > 0) shell of radius r, centered about r_0.

a)   Solve the S-wave Schrödinger equation and calculate the scattering phase as a function of the momentum of the incident particle.

b)   Calculate the Jost functions; study their analytic properties and their zeroes, pointing out their significance.

c)   Determine the scattering length α and the effective range ρ and write down the cross section at low energy.

d)   Study this problem in the Born approximation, comparing the result with the exact solution. Discuss the validity of the approximation.

Solution

a)   The S-wave Schrödinger equation is

$$\chi'' + k^2\chi - 2\,\frac{a\,m}{\hbar^2}\,\delta(r - r_0)\chi = 0 \tag{1}$$

$\chi = r\psi$ is the reduced radial wave function. For $r < r_0$ and $r > r_0$ this equation describes a free particle

$$\chi'' + k^2\chi = 0 \tag{2}$$

Regularity at the origin fixes the interior solution to

$$\chi(r) = \frac{1}{k}\,\sin kr \tag{3}$$

while for $r > r_0$ we write

$$\chi(r) = A\,\sin(kr + \delta_0) \tag{4}$$

Integrating equation (1) over the singularity leads to

$$\chi'_+(r_0) - \chi'_-(r_0) = \frac{2am}{\hbar^2}\,\chi(r_0) \tag{5}$$

where the indices $\pm$ indicate limits from right and left. Using also the condition $\chi_+(r_0) = \chi_-(r_0) = \chi(r_0)$ one obtains

$$A = \frac{1}{k}\,\frac{\sin kr_0}{\sin(kr_0 + \delta_0)}$$

$$\mathrm{cotg}\,\delta_0 = -\frac{\hbar^2 k}{2ma}\,\frac{1}{\sin^2 kr_0} - \mathrm{cotg}\,kr_0 \tag{6}$$

The scattering amplitude is

$$f_0 = \frac{1}{k \cot g\, \delta_0 - i\, k} \tag{7}$$

b)   By definition, the Jost functions are given by

$$A \sin(kr + \delta_0) = \frac{1}{2ik} \left[\phi_+(k)\, e^{ikr} - \phi_-(k) e^{-ikr} \right] \tag{8}$$

whence

$$\phi_\pm(k) = A\, k\, e^{\pm i\delta_0} \tag{9}$$

Substituting the expressions for A and $\cot g\, \delta_0$ we have

$$\phi_\pm(k) = 1 + \frac{2ma}{k\,\hbar^2} \sin kr_0\, e^{\mp ikr_0} \tag{10}$$

These functions are analytic throughout the complex k-plane, except for an essential singularity at infinity. This property stems from the fact that the range of the potential is rigorously finite.

The zeroes of $\phi_-(k)$ are solutions of the equation

$$1 + \frac{\gamma}{x} \left[\sin x \cos x + i \sin^2 x \right] = 0 \tag{11}$$

where the pure numbers γ and x are defined by

$$\gamma = \frac{2mar_0}{\hbar^2} \qquad x = kr_0 \tag{12}$$

We can put this equation in the form

$$1 + \frac{\gamma}{2i\, x} \left\{ e^{2i\,x} - 1 \right\} = 0 \tag{13}$$

γ is positive (negative) for an attractive (repulsive) shell. Setting $-2i\, x = y$, equation (13) becomes

$$e^{-y} = 1 + \frac{y}{\gamma} \tag{14}$$

We can now distinguish the following cases:

- $\gamma > 0$: No real solutions are admitted. The solution $y = 0$ is only possible for $\gamma = -1$, by Eq.(13).

- $-1 < \gamma < 0$: There exists one solution with $y < 0$ ($\mathrm{Im}\, k < 0$), corresponding to an antibound state.

- $\gamma < -1$: There exists one solution with $y > 0$ ($\operatorname{Im} k > 0$), corresponding to a bound state.

- $\gamma = -1$: The pole coincides with the branch point of the cut in the complex energy plane (resonance at zero energy).

For complex values of y, $y = r + i\,s$, Eq.(14) becomes

$$e^{-r}\cos s - 1 = \frac{r}{\gamma} \tag{15a}$$

$$e^{-r}\sin s = -\frac{s}{\gamma} \tag{15b}$$

If $r + i\,s$ is a solution, so is $r - i\,s$, that is, the complex zeroes of ϕ_- come in mirror pairs with respect to the imaginary k-axis. Combining Eqs. (15), we can eliminate γ and obtain

$$\cos s + \frac{r}{s}\sin s = e^{r}$$

This equation has no solution for $r > 0$, since the r.h.s exceeds $(1 + r)$. This implies that the zeroes of ϕ_- must have $\operatorname{Im} k < 0$, and consequently the poles of the resonances lie on the second sheet of the complex energy plane.

c)   The expression for $\cotg \delta_0$, using the small-k expansion

$$k\cotg \delta_0 \simeq -\frac{1}{\alpha} + \frac{1}{2}\rho k^2$$

gives:

$$\alpha = \frac{r_0}{1 + \dfrac{1}{\gamma}} \qquad \rho = \frac{2}{3}r_0\left(1 - \frac{1}{\gamma}\right)$$

d)   In the Born approximation

$$\begin{aligned}
f_0^B &= -\frac{m}{2\pi\hbar^2}\int V(r)\,j_0^2(kr)\,r^2\,dr\,d\Omega \\
&= -\frac{ma}{2\pi\hbar^2}\frac{\sin^2 kr_0}{k^2}\cdot 4\pi = -\frac{2ma}{\hbar^2}\frac{\sin^2 kr_0}{k^2} \tag{16}
\end{aligned}$$

Equation (6), on the other hand, gives

$$\tg \delta_0 = -\frac{\sin^2 kr_0}{\dfrac{k\,\hbar^2}{2ma} + \sin kr_0\cos kr_0} \tag{17}$$

which, as $k \to \infty$, becomes

$$\delta_0 \simeq -\sin^2 kr_0\,\frac{2ma}{k\,\hbar^2}$$

Substituting this in Eq.(7) we have

$$f_0 \simeq -2 \frac{ma}{k^2 \hbar^2} \sin^2 kr_0$$

The above coincides indeed with f_0^B: At high energies the Born approximation is valid, as is well known.

Problem 2.   Reflection and transmission of a one-dimensional barrier

Consider the scattering states of a one-dimensional potential $V(x) = V(-x)$ (even under spatial inversion). Energy and parity are a complete set of observables for this system.

a)   Show that, in a representation in which these observables are diagonal, the S-matrix is also diagonal and its elements have modulus equal to 1.

b)   How rapidly must $V(x)$ tend to zero as $x \to \infty$, in order for S to be well-defined?

c)   Calculate S for the potential

$$\begin{cases} V(x) = V_0 & |x| < a \\ V(x) = 0 & |x| > a \end{cases} \tag{1}$$

for the cases $V_0 > 0$ and $V_0 < 0$.

d)   Define the reflection and transmission coefficients R and T, and show that unitarity implies $R + T = 1$. Express R and T in terms of scattering phases.

e)   Study the solutions $f_\pm(k, P, x)$ with momentum k and parity P, with the boundary conditions

$$f_\pm(k, P, x) \xrightarrow[x \to +\infty]{} e^{\pm i\,kx} \tag{2}$$

Such solutions, and their first derivatives, are typically discontinuous at $x = 0$. By analogy with 3-dimensional scattering, introduce Jost functions and study their analyticity properties, verifying them explicitly for the potential defined in c).

Solution

a)   Energy and parity are a complete set of commuting observables. The S-matrix commutes with both and is unitary; therefore it is diagonal in a representation in which $E = \frac{k^2}{2m}$ and P are diagonal, and its matrix elements have modulus 1.

b)   The Schrödinger equation is formally identical to the radial equation for the reduced S-wave function χ in three dimensions. The existence condition for outgoing solutions, behaving like e^{ikx} as $x \to \infty$, is also the same as for three dimensional S-waves. Thus, the Jost functions, defined below, exist and are analytic for $\mathrm{Im}\, k < 0$ provided

$$x^3 U(x) \xrightarrow[x \to +\infty]{} 0 \tag{3}$$

c) The scattering states of the square well potential are as follows ($k' = \sqrt{k^2 - 2mV_0}$):
Odd solutions

$$|x| < a \qquad \frac{1}{k'} \sin k'x$$

$$x > a \qquad \frac{A_-}{k} \sin(kx + \delta_-) \tag{4}$$

$$x < -a \qquad -\frac{A_-}{k} \sin(-kx + \delta_-)$$

where

$$|A_-|^2 = \cos^2 k'a + \frac{k^2}{k'^2} \sin^2 k'a$$

$$\delta_- = \text{arctg}\left(\frac{k}{k'} \, \text{tg} \, k'a\right) - ka \tag{5}$$

Even solutions

$$|x| < a \qquad \cos k'x$$
$$x > a \qquad A_+ \cos(kx + \delta_+) \tag{6}$$
$$x < -a \qquad A_+ \cos(-kx + \delta_+)$$

$$|A_+|^2 = \sin^2 k'a + \frac{k^2}{k'^2} \cos^2 k'a$$

$$\delta_+ = \text{arctg}\left(\frac{k'}{k} \, \text{tg} \, k'a\right) - ka \tag{7}$$

d) As $x \to \pm\infty$ even solutions behave like

$$\psi_+ \sim \begin{cases} e^{-ikx} + e^{2i\delta_+} e^{ikx} & x \to +\infty \\ e^{ikx} + e^{2i\delta_+} e^{-ikx} & x \to -\infty \end{cases} \tag{8}$$

Similarly, odd solutions have the form

$$\psi_- \sim \begin{cases} e^{-ikx} - e^{2i\delta_-} e^{ikx} & x \to +\infty \\ -e^{ikx} + e^{2i\delta_-} e^{-ikx} & x \to -\infty \end{cases} \tag{9}$$

The following superposition will only give progressive waves for $x \to +\infty$

$$\psi = \text{const.} \left[\frac{A_- e^{-i\delta_-}}{i} \psi_+ + k A_+ e^{-i\delta_+} \psi_-\right] \tag{10}$$

Indeed, for $x \to +\infty$

$$\psi \sim \frac{e^{2i\delta_+} + e^{2i\delta_-}}{2} e^{+ikx} \tag{11}$$

while for $x \to -\infty$

$$\psi \sim e^{ikx} + \frac{e^{2i\delta_+} - e^{2i\delta_-}}{2} e^{-ikx} \tag{12}$$

We thus have

$$T = \frac{1}{4}\left|e^{2i\delta_+} + e^{2i\delta_-}\right|^2$$

$$R = \frac{1}{4}\left|e^{2i\delta_+} - e^{2i\delta_-}\right|^2 \tag{13}$$

Unitarity requires that the phases be real; consequently

$$T \quad + \quad R \quad = \quad 1 \tag{14}$$

e) We denote by $f_{\pm}(E, P, x)$ solutions with parity $P = \pm 1$ obeying boundary condition (2)

$$f_{\pm}(E, P, x) \underset{x \to +\infty}{\sim} e^{\pm ikx}$$

To obtain these solutions one must solve the Schrödinger equation with boundary condition $e^{\pm ikx}$, $x \to +\infty$, on the semiaxis $x > 0$ and then impose the given parity assignment to define them for $x < 0$. In general, solutions will not vanish at $x = 0$, so that antisymmetrization will lead to discontinuous functions at the origin. The regular antisymmetric solution is a linear combination of $f_+(E, P = -1, x)$ and $f_-(E, P = -1, x)$

$$\psi_{REG}(E, P = -1, x) \quad = \quad \frac{1}{2ik}[\phi_-(E, P = -1)\, f_-(E, P = -1, x)$$
$$-\phi_+(E, P = -1)f_+(E, P = -1, x)] \tag{15}$$

The functions $\phi_{\pm}$ are known as Jost functions.

The vanishing of ψ_{REG} at $x = 0$ implies

$$\phi_{\mp}(E, P = -1) = f_{\pm}(E, P = -1, 0) \tag{16}$$

At $x = 0$, one can show that $\psi'_{REG} = 1$ as follows: by Eq.(15), ψ'_{REG} at the origin coincides with the Wronskian of the two solutions f_+ and f_- (aside from the factor $1/2ik$)

$$\psi'_{REG}(E, P = -1, 0) = \frac{1}{2ik} W(f_+, f_-) \tag{17}$$

Since the two solutions have the same energy, the Wronskian is constant and can be evaluated at $x \to \infty$, giving $W = 2ik$. As a result, we have

$$\psi'_{REG}\big|_{x=0} = 1 \tag{18}$$

To obtain $\psi_{REG}(E, P{=}1, x)$ one must impose a zero first derivative at $x = 0$. This leads to

$$\psi_{REG}(E, P = 1, x) \quad = \quad \frac{1}{2ik}[f'_+(E, P = 1, 0)f_-(E, P = 1, x)$$
$$-f'_-(E, P = 1, 0)f_+(E, P = 1, x)] \tag{19}$$

The factor $1/2\mathrm{i}k$ guarantees that $\psi_{REG} = 1$ at $x = 0$. The Jost functions for the odd case, Eq.(15), are formally identical to those in S-wave scattering in three dimensions. For the even case, Eq.(19), they are the derivatives of the functions $f_\pm(E, P{=}1, x)$ at the origin. The analyticity properties in k are the same as for S-wave Jost functions; for a potential with finite range they are entire functions.

For the potential defined in c) we have

$$\phi_-(E, P = -1) = e^{\mathrm{i}ka}\left[\cos k'a - \mathrm{i}\frac{k}{k'}\sin k'a\right] \tag{20a}$$

$$\phi_+(E, P = -1) = e^{-\mathrm{i}ka}\left[\cos k'a + \mathrm{i}\frac{k}{k'}\sin k'a\right] \tag{20b}$$

and

$$f'_+(E, P = 1, 0) = e^{\mathrm{i}ka}\left[k'\sin k'a + \mathrm{i}k\,\cos k'a\right] \tag{21a}$$

$$f'_-(E, P = 1, 0) = e^{-\mathrm{i}ka}\left[k'\sin k'a - \mathrm{i}k\,\cos k'a\right] \tag{21b}$$

Problem 3. The magnetic dipole moment interaction between two spin-1/2 particles

Two spin-1/2 particles with equal mass and normal magnetic moment interact by means of a potential, consisting of a square well term

$$\begin{aligned} V(r) &= V_0 > 0 \qquad r < a \\ V(r) &= 0 \qquad\quad r > a \end{aligned} \tag{1}$$

and a magnetic dipole term

$$H_M = -\boldsymbol{\mu}_1 \cdot \boldsymbol{\mu}_2 \frac{8\pi}{3}\delta^3(\mathbf{r}) + \frac{\mu_1^i \mu_2^j}{r^3}\left\{\delta_{ij} - \frac{3\,r_i\,r_j}{r^2}\right\} \tag{2}$$

which can be treated as a perturbation.
a) In the low energy limit, $ka \ll 1$, calculate the angular distribution, averaged over initial spins and summed over final spins.
b) Calculate the angular distribution with and without helicity flip in a scattering process in which particle 1 has initially helicity $+1/2$, while particle 2 is nonpolarized.
c) What does parity invariance tell us about processes analogous to those described in b), but having initial helicity $-1/2$?

Solution

To first order in perturbation theory. We write

$$H = H_1 + H_M \tag{3}$$

$$H_1 = \frac{p^2}{2m} + V(r) \tag{4}$$

where $V(r)$ and H_M are given in Eqs. (1) and (2), and m is the reduced mass of the system.

By Goldberger theorem, the scattering amplitude f can be written as

$$f = f_0 - \frac{m}{2\pi} \langle f_- | H_M | i_+ \rangle \tag{5}$$

where f_0 is the scattering amplitude in the absence of H_M; $|f_-\rangle$ and $|i_+\rangle$ denote the eigenstates of H_1 whose "out" and "in" boundary condition are the initial and final state.

We now calculate the states $|f_-\rangle$, $|i_+\rangle$ and the scattering amplitude f_0. At low energies only the S-wave is involved and $|f_-\rangle$, $|i_+\rangle$ both coincide with the S state of H_1, $|\psi\rangle$.

In the center of mass of the two particles the problem reduces to a problem of potential scattering, with mass m equal to the reduced mass. The S-wave in the spherical potential well with $E < V_0$ is

$$
\begin{aligned}
\psi(r) &= \frac{\sin(ka + \delta_0)}{\sin \tilde{k}a} \frac{\sinh \tilde{k}r}{kr} \qquad r < a \\
\psi(r) &= \frac{\sin(kr + \delta_0)}{kr} \qquad\qquad r > a
\end{aligned}
\tag{6}
$$

with

$$k^2 = \frac{2mE}{\hbar^2} \qquad \tilde{k}^2 = \frac{2m(V_0 - E)}{\hbar^2} \tag{7}$$

Matching logarithmic derivatives at $r = a$, we obtain

$$
\begin{aligned}
\delta_0 &= \mathrm{arcctg}(\frac{\tilde{k}}{k} \coth \tilde{k}a) - ka \\
\psi(r) &= \frac{\sinh \tilde{k}r}{\tilde{k}r} \frac{1}{\sqrt{1 + \dfrac{V_0}{V_0 - E} \sinh^2 \tilde{k}a}} \qquad r < a
\end{aligned}
\tag{8}
$$

The unperturbed scattering amplitude is

$$f_0 = \frac{1}{k \cotg \delta_0 - ik} \tag{9}$$

Let us now calculate the effect of H_M to first perturbative order. The stationary states of H_1 have the form

$$|E, s\rangle = |\psi\rangle |s_1\rangle |s_2\rangle \tag{10}$$

Since H_M commutes with the total spin

$$\mathbf{s} = \mathbf{s}_1 + \mathbf{s}_2 \tag{11}$$

it is convenient to project onto singlet and triplet states

$$
\begin{aligned}
|\mathbf{s}| &= 0 \quad \frac{1}{\sqrt{2}}\left(|+\rangle_1|-\rangle_2 - |-\rangle_1|+\rangle_2\right) \\
|\mathbf{s}| &= 1 \quad |+\rangle_1|+\rangle_2 \,, \quad \frac{1}{\sqrt{2}}\left(|+\rangle_1|-\rangle_2 + |-\rangle_1|+\rangle_2\right) \,, \quad |-\rangle_1|-\rangle_2
\end{aligned}
\tag{12}
$$

The projectors onto these states are, respectively

$$\Pi_s = \frac{1}{4} - \mathbf{s}_1 \cdot \mathbf{s}_2 \qquad \Pi_t = \frac{3}{4} + \mathbf{s}_1 \cdot \mathbf{s}_2 \tag{13}$$

Further, we have ($\hat{\mathbf{n}} = \mathbf{r}/r$)

$$
\begin{aligned}
\mathbf{s}_1 \cdot \mathbf{s}_2 &= \frac{1}{2}\left(s^2 - \frac{3}{2}\right) \\
(\mathbf{s}_1 \cdot \hat{\mathbf{n}})(\mathbf{s}_2 \cdot \hat{\mathbf{n}}) &= \frac{1}{2}\left[(\mathbf{s} \cdot \hat{\mathbf{n}})^2 - \frac{1}{2}\right] \\
\boldsymbol{\mu}_i &= \mu\, \mathbf{s}_i
\end{aligned}
\tag{14}
$$

H_M now reads

$$H_M = \mu^2 \left\{ \left(\frac{3}{4} - \frac{s^2}{2}\right) \frac{8\pi}{3}\delta^3(\mathbf{r}) + \frac{1}{r^3}\left[\frac{s^2}{2} - \frac{3}{2}(\mathbf{s} \cdot \hat{\mathbf{n}})^2\right] \right\} \tag{15}$$

or, equivalently,

$$H_M = \mu^2 \frac{3\Pi_s - \Pi_t}{4} \frac{8\pi}{3}\delta^3(\mathbf{r}) + \mu^2 \frac{1}{r^3}\Pi_t\left[1 - \frac{3}{2}(\mathbf{s} \cdot \hat{\mathbf{n}})^2\right] \tag{16}$$

The term proportional to $1/r^3$ vanishes in the S-wave and can be neglected at low energies. Thus, the matrix elements of H_M between two eigenstates of H_1 with $\ell = 0$ are

$$
\begin{aligned}
\langle E, s'|H_M|E, s\rangle &= \delta_{ss'}\mu^2 \frac{8\pi}{3}\left[\frac{3}{4}\delta_{s,0} - \frac{1}{4}\delta_{s,1}\right] \frac{\sin^2(ka + \delta_0)}{\sinh^2 \tilde{k}a} \frac{\tilde{k}^2}{k^2} \\
&= \delta_{ss'}\mu^2 \frac{8\pi}{3}\left[\frac{3}{4}\delta_{s,0} - \frac{1}{4}\delta_{s,1}\right] \frac{\tilde{k}^2}{k^2 \sinh^2 \tilde{k}a + \tilde{k}^2 \cosh^2 \tilde{k}a}
\end{aligned}
\tag{17}
$$

Eq.(5) leads to

$$f = f_0 - \frac{4}{3}\mu^2 \left[\frac{3}{4}\delta_{s,0} - \frac{1}{4}\delta_{s,1}\right] \frac{m}{1 + \dfrac{V_0}{V_0 - E}\sinh^2 \tilde{k}a} \tag{18}$$

For $E \ll |V_0|$, $\tilde{k}_0$ becomes

$$\tilde{k}_0 = \sqrt{\frac{2m|V_0|}{\hbar^2}}$$

and Eqs. (8), (9) yield

$$f_0 \simeq \frac{\tanh \tilde{k}_0 a}{\tilde{k}_0} - a \tag{19}$$

Eq.(18) then reads

$$f = \frac{\tanh \tilde{k}_0 a - \tilde{k}_0 a}{\tilde{k}_0} - \frac{4}{3}\left[\frac{3}{4}\delta_{s,0} - \frac{1}{4}\delta_{s,1}\right]\frac{\mu^2 m}{\cosh^2 \tilde{k}_0 a} \tag{20}$$

Expressed as a matrix in spin space, Eq.(20) takes the form

$$f = A + B\boldsymbol{\sigma}_1 \cdot \boldsymbol{\sigma}_2 \tag{21}$$

with

$$A = \frac{\tanh \tilde{k}_0 a - \tilde{k}_0 a}{\tilde{k}_0} \tag{22}$$

$$B = \frac{1}{3}\mu^2 m \left(1 - \tanh^2 \tilde{k}_0 a\right) \tag{23}$$

Averaging over initial spins and summing over final ones, we obtain the cross section

$$\frac{d\sigma}{d\Omega} = \frac{1}{4}\text{Tr}_1\,\text{Tr}_2\,f(\theta)f^\dagger(\theta) = |A|^2 + 3|B|^2 \tag{24}$$

b)   The initial density matrix of particle 1 with positive helicity is

$$\rho_+^{(i)} = \frac{1}{2}(1 + \boldsymbol{\sigma}_1 \cdot \hat{\mathbf{p}}) \tag{25}$$

We are interested in the amplitudes corresponding to final density matrices of the form

$$\rho_\pm^{(f)} = \frac{1}{2}(1 \pm \boldsymbol{\sigma}_1 \cdot \hat{\mathbf{p}}') \tag{26}$$

There follows

$$\frac{d\sigma_\pm}{d\Omega} = \text{Tr}_1\text{Tr}_2\left(f\,\rho_+^{(i)}\,f^\dagger \rho_\pm^{(f)}\right) = |A|^2 + 3|B|^2 \pm \hat{\mathbf{p}}\hat{\mathbf{p}}'(|A|^2 - |B|^2) \tag{27}$$

A plus (minus) sign corresponds to scattering without (with) helicity flip. In particular, we have in the forward direction

$$\frac{d\sigma_+}{d\Omega} = 2(|A|^2 + |B|^2) \qquad \frac{d\sigma_-}{d\Omega} = 4|B|^2 \tag{28}$$

and in the backward direction

$$\frac{\mathrm{d}\sigma_+}{\mathrm{d}\Omega} = 4|B|^2 \qquad \frac{\mathrm{d}\sigma_-}{\mathrm{d}\Omega} = 2(|A|^2 + |B|^2) \tag{29}$$

c)   Under parity the following transformations take place

$$\boldsymbol{\sigma} \to \boldsymbol{\sigma} \,;\, \mathbf{p} \to -\mathbf{p} \,;\, \mathbf{p}' \to -\mathbf{p}' \tag{30}$$

Parity invariance, together with invariance under a rotation by π on the scattering plane, implies the following relation among helicity amplitudes

$$f_{h_1,h_2;h'_1,h'_2} = f_{-h_1,-h_2;-h'_1,-h'_2} \tag{31}$$

Problem 4.   Scattering asymmetry due to $\ell \cdot$ s interaction

The interaction of a spin-1/2 particle in a potential $V(r)$ has the form

$$V(r) + \frac{g}{2m^2}\frac{1}{r}\frac{\mathrm{d}V}{\mathrm{d}r}\, \boldsymbol{\ell} \cdot \mathbf{s} \tag{1}$$

The spin-orbit term has its origin in relativistic corrections. We consider a model in which the complete interaction is given by $U = V + V'$, where

$$V(r) = -V_0\,\theta\,(r_0 - r) \qquad V_0 > 0 \tag{2}$$

and V' is a spin-independent, imaginary interaction simulating inelastic processes

$$V' = -\mathrm{i}\,V_1\,\theta(r_0 - r) \tag{3}$$

a)   Calculate the scattering amplitude in the Born approximation.
b)   Calculate the left-right asymmetry in the scattering of transversely polarized particles, and the induced polarization on nonpolarized particles.

Solution

a)   In the Born approximation we write

$$f(\theta) = -\frac{m}{2\pi}\langle \mathbf{p}'|\,U\,|\mathbf{p}\rangle = -\frac{m}{2\pi}\int \mathrm{e}^{-\mathrm{i}\mathbf{q}\mathbf{r}}\,U(r)\,\mathrm{d}^3\mathbf{r} \tag{4}$$

where $\mathbf{q} = \mathbf{p}' - \mathbf{p}$ is the momentum transfer.

We calculate the spin-independent part making use of the relation

$$\int e^{-i\mathbf{q}\mathbf{r}}\,\theta(r_0 - r)\,\mathrm{d}^3\mathbf{r} = \frac{4\pi r_0}{q^2}\left(\frac{\sin qr_0}{qr_0} - \cos qr_0\right) \tag{5}$$

The spin-orbit part is calculated by means of

$$
\begin{aligned}
\langle \mathbf{p}'|\,\delta(r - r_0)\,\boldsymbol{\ell}\cdot\mathbf{s}\,|\mathbf{p}\rangle
&= \int \mathrm{d}^3\mathbf{r}\, e^{-i\mathbf{p}'\mathbf{r}}\delta(r - r_0)\frac{1}{i}(\mathbf{r}\wedge\boldsymbol{\nabla})\cdot\mathbf{s}\,e^{i\mathbf{p}\mathbf{r}} \\
&= \int \mathrm{d}^3\mathbf{r}\, e^{-i\mathbf{q}\mathbf{r}}\delta(r - r_0)(\mathbf{r}\wedge\mathbf{p})\cdot\mathbf{s} \\
&= i\frac{4\pi r_0^2}{q^2}\left(\frac{\sin qr_0}{qr_0} - \cos qr_0\right)\boldsymbol{\nu}\cdot\mathbf{s}\,p^2\sin\theta
\end{aligned}
\tag{6}
$$

where

$$\boldsymbol{\nu} = \frac{\mathbf{p}\wedge\mathbf{p}'}{|\mathbf{p}\wedge\mathbf{p}'|} \qquad \text{and} \qquad |\mathbf{p}\wedge\mathbf{p}'| = p^2\sin\theta \tag{7}$$

Thus, the complete amplitude reads

$$
\begin{aligned}
f(\theta) &= \left[2m\,(V_0 + iV_1) - i\frac{gV_0}{m}p^2\sin\theta\,\boldsymbol{\nu}\cdot\mathbf{s}\right]\frac{r_0}{q^2}\left(\frac{\sin qr_0}{qr_0} - \cos qr_0\right) \\
&= A + B\boldsymbol{\nu}\cdot\boldsymbol{\sigma}
\end{aligned}
\tag{8}
$$

with

$$
\begin{aligned}
A &= \frac{2mr_0}{q^2}\,(V_0 + iV_1)\left(\frac{\sin qr_0}{qr_0} - \cos qr_0\right) \\
B &= -i\frac{gr_0V_0p^2\sin\theta}{2mq^2}\left(\frac{\sin qr_0}{qr_0} - \cos qr_0\right)
\end{aligned}
\tag{9}
$$

We have set $\mathbf{s} = \boldsymbol{\sigma}/2$, where $\boldsymbol{\sigma}$ are the Pauli matrices.

b) The left-right asymmetry, $\varepsilon(\theta)$, of transversely polarized particles is defined as follows

$$\varepsilon(\theta) = \frac{\mathrm{d}\sigma(\theta) - \mathrm{d}\sigma(-\theta)}{\mathrm{d}\sigma(\theta) + \mathrm{d}\sigma(-\theta)} \tag{10}$$

$\pm\theta$ is the scattering angle on the plane perpendicular to the initial polarization

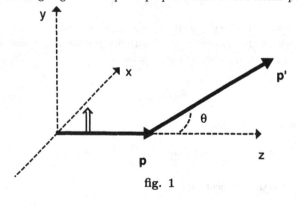

fig. 1

Denoting by $P_\perp$ the transverse polarization of the beam, we have

$$\frac{d\sigma(\theta)}{d\Omega} = \frac{1}{2}\mathrm{Tr}\left\{ f(\theta)\left(1 + \mathbf{P}_\perp \cdot \boldsymbol{\sigma}\right) f^\dagger(\theta)\right\} \tag{11}$$

Substituting the expression (8) for $f(\theta)$ leads to

$$\varepsilon = \frac{2\,\mathrm{Re}\,A(\theta)B^*(\theta)}{|A|^2 + |B|^2} = -\frac{2\,V_0\,V_1\,p^2 g\sin\theta}{4\,m^2\left(V_0^2 + V_1^2\right) + \dfrac{g^2 V_0^2 p^4}{4m^2}\sin^2\theta} \tag{12}$$

The polarization induced on nonpolarized particles and angle θ is obtained through

$$\boldsymbol{P} = \frac{\mathrm{Tr}\left(\boldsymbol{\sigma}f(\theta)f^\dagger(\theta)\right)}{\mathrm{Tr}\left(f(\theta)f^\dagger(\theta)\right)} = \frac{2\,\mathrm{Re}\,AB^*}{|A|^2 + |B|^2}\,\boldsymbol{\nu} = \varepsilon\,\boldsymbol{\nu} \tag{13}$$

We observe that this equals the asymmetry calculated above, as would be expected by time reversal invariance.

Problem 5.   The "static" photoelectric effect

A metal can be represented by a potential well, filled with free electrons. Let n be the number of electrons per unit volume and V_0 the extraction potential. (Typically, $n \approx 10^{23}/cm^3$; $V_0 \approx 2$ Volts.)
a) How deep is this potential well?
b) Suppose now the metal is illuminated by light with intensity I and frequency ν ($h\nu > e\,V_0$). What is the energy distribution of the emitted photoelectrons?
c) If this metal is the negative shield of a capacitor, a static photoelectric effect may take place: the presence of the electric field causes electrons to cross the barrier which binds them to the metal.
    Calculate the current density of these electrons.

Solution

a) The number of electrons in the metal is

$$N = \frac{2V}{h^3}\int_0^{P_F} d^3\mathbf{p} \tag{1}$$

The factor of 2 is due to spin degeneracy and V is the volume of the metal. The corresponding density is

$$n = \frac{8\pi}{3}\frac{1}{h^3}P_F^3 \tag{2}$$

Knowing the density, we can deduce the value of the Fermi momentum P_F.

The energy at the surface of the Fermi sea is

$$- e V_0 = -U + \frac{P_F^2}{2m} \tag{3}$$

Thus the potential well has depth

$$U = e V_0 + \frac{1}{2m} \left(\frac{3 n h^3}{8\pi} \right)^{\frac{2}{3}} \tag{4}$$

With the data provided $U \simeq 9\,\text{eV}$.

b)   The transition amplitude between two momentum eigenstates $|\mathbf{k}\rangle$ and $|\mathbf{k}'\rangle$, induced by a photon of momentum $\mathbf{q}$ and polarization $\boldsymbol{\varepsilon}$, equals

$$
\begin{aligned}
A_{fi} &= e \int \mathrm{d}^3\mathbf{x}\, \langle 0|\mathbf{A}(\mathbf{x})|\mathbf{q}\rangle \, \langle \mathbf{k}'| \frac{\mathbf{P}}{m} | \mathbf{k}\rangle \\
&= e \frac{\mathbf{k}' \cdot \boldsymbol{\varepsilon}}{m} \int \mathrm{d}^3\mathbf{x}\, \langle \mathbf{k}'|e^{i\mathbf{q}\mathbf{x}}| \mathbf{k}\rangle = e \frac{\mathbf{k}' \cdot \boldsymbol{\varepsilon}}{m} (2\pi)^3 \delta^3(\mathbf{k} + \mathbf{q} - \mathbf{k}')
\end{aligned} \tag{5}
$$

The wave functions are normalized to unit density. In the large volume limit, electrons in the initial state can be represented by plane waves. The corresponding cross section is

$$\mathrm{d}\sigma = \frac{1}{2|\mathbf{q}|} \left(\frac{e\,\mathbf{k}'\boldsymbol{\varepsilon}}{m} \right)^2 \frac{\mathrm{d}^3\mathbf{k}'}{(2\pi)^3} (2\pi)^3 \delta^3\left(\mathbf{k} + \mathbf{q} - \mathbf{k}'\right) 2\pi\delta\left(\frac{k^2}{2m} + |\mathbf{q}| - \frac{k'^2}{2m} - U \right) \tag{6}$$

(The initial photon flux is $2q$, while the electron density is 1). To obtain the spectrum of photoelectrons, one must integrate over initial states, with weight

$$2V \frac{\mathrm{d}^3\mathbf{k}}{(2\pi)^3} \tag{7}$$

Doing so, one obtains

$$
\frac{\mathrm{d}N_{el}}{\mathrm{d}^3\mathbf{k}'} =
$$

$$
\begin{aligned}
&2\frac{V}{2|\mathbf{q}|} \int^{P_F} \frac{\mathrm{d}^3\mathbf{k}}{(2\pi)^3} \left(\frac{e\,\mathbf{k}'\boldsymbol{\varepsilon}}{m} \right)^2 (2\pi)\, \delta^3\left(\mathbf{k} + \mathbf{q} - \mathbf{k}'\right) \delta\left(\frac{k^2}{2m} + |\mathbf{q}| - \frac{k'^2}{2m} - U \right) \\
&= \frac{V}{|\mathbf{q}|} \frac{e^2}{m^2} \frac{(\mathbf{k}'\boldsymbol{\varepsilon})^2}{(2\pi)^2} \delta\left(\frac{(\mathbf{k}' - \mathbf{q})^2}{2m} + |\mathbf{q}| - \frac{k'^2}{2m} - U \right) \theta(P_F - |\mathbf{k}' - \mathbf{q}|)
\end{aligned} \tag{8}
$$

N_{el} is the number of photoelectrons per unit photon flux. The θ function enforces the condition $|\mathbf{k}| = |\mathbf{k}' - \mathbf{q}| \leq P_F$; compatibility with energy conservation (the δ function in (8)) requires

$$- |\mathbf{q}| + \frac{k'^2}{2m} + U = \frac{|\mathbf{k}' - \mathbf{q}|^2}{2m} \leq \frac{P_F^2}{2m} \tag{9}$$

which gives

$$|\mathbf{q}| - U + \frac{P_F^2}{2m} \equiv |\mathbf{q}| - e\,V_0 \geq \frac{k'^2}{2m} \tag{10}$$

For a given $|\mathbf{q}|$ this implies $k' \leq k'_{\max} = \sqrt{2m(|\mathbf{q}| - e\,V_0)}$ The δ function requirement

$$\mathbf{k}'\mathbf{q} = m\,|\mathbf{q}| + \frac{\mathbf{q}^2}{2} - U\,m \tag{11}$$

leads to the condition

$$\left| m\,|\mathbf{q}| + \frac{\mathbf{q}^2}{2} - U\,m \right| \leq k'|\mathbf{q}| \tag{12}$$

that is,

$$k' \geq m + \frac{|\mathbf{q}|}{2} - \frac{U\,m}{|\mathbf{q}|} \equiv k'_{\min} \tag{13}$$

Performing also the average over the photon polarizations

$$\langle |\mathbf{k}'\boldsymbol{\varepsilon}|^2 \rangle = \frac{1}{2}\left(k'^2 - \frac{(\mathbf{k}'\mathbf{q})^2}{\mathbf{q}^2} \right) = \frac{1}{2}(k'^2 - k'^2_{\min}) \tag{14}$$

we finally obtain

$$\begin{aligned}
\frac{\mathrm{d}N_{el}}{\mathrm{d}E} &= V\frac{e^2 m}{2\pi\,\mathbf{q}^2}\,(E - E_{\min})\theta(E - E_{\min})\theta(E_{\max} - E) \\
E_{\min} &= \frac{k'^2_{\min}}{2m} \quad E_{\max} = \frac{k'^2_{\max}}{2m} = |\mathbf{q}| - e\,V_0
\end{aligned} \tag{15}$$

Multiplying by the photon flux $\frac{I}{|\mathbf{q}|}$, we find the number of photoelectrons per unit of time, volume and electron energy

$$\frac{1}{V}\frac{\mathrm{d}N_{el}}{\mathrm{d}E} = I\frac{e^2 m}{4\pi}\frac{2}{|\mathbf{q}|^3}(E - E_{\min})\theta(E - E_{\min})\theta(E_{\max} - E) \tag{16}$$

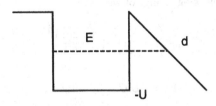

fig. 1

In terms of the electric field $\mathcal{E}$ in the capacitor, the barrier penetration factor for an electron with momentum k_x perpendicular to the surface is given by

$$\Phi = \exp\left[-\frac{2}{\hbar}\int_0^d \sqrt{2m\,e\mathcal{E}\,(d-x)}\,\mathrm{d}x\right] \tag{17}$$

The integration limit is determined by

$$-e\,d\,|\mathcal{E}| = -U + \frac{k_x^2}{2m} \tag{18}$$

(see fig. 1), that is,

$$d = \frac{1}{e|\mathcal{E}|}\left(U - \frac{k_x^2}{2m}\right) \tag{19}$$

Thus the penetration factor equals

$$\Phi = \exp\left[-\frac{2}{\hbar}\frac{\sqrt{2m}}{e\,|\mathcal{E}|}\frac{2}{3}\left(U - \frac{k_x^2}{2m}\right)^{\frac{3}{2}}\right] \tag{20}$$

The current density for a given k_x is

$$\mathrm{d}j = e\,n\,P(k_x)\,\mathrm{d}k_x\frac{k_x}{m}\,\Phi(k_x) \tag{21}$$

where $P(k_x)\,\mathrm{d}k_x$ is the momentum distribution summed over k_y and k_z, and normalized to one.

$$P(k_x) = \int \mathrm{d}k_y\,\mathrm{d}k_z\,\frac{1}{\frac{4\pi}{3}P_F^3} = \frac{\pi(P_F^2 - k_x^2)}{\frac{4\pi}{3}P_F^3} \tag{22}$$

In our case, j equals

$$j = \frac{n\,e}{m}\int_0^{P_F}\frac{\pi(P_F^2 - k_x^2)}{\frac{4\pi}{3}P_F^3}\,k_x\,\mathrm{d}k_x\,\exp\left[-\frac{4\sqrt{2m}}{3\hbar\,e\,|\mathcal{E}|}(U - \frac{k_x^2}{2m})^{\frac{3}{2}}\right] \tag{23}$$

We now perform the following change of variables

$$t = \left(U - \frac{k_x^2}{2m}\right)\left(\frac{4}{3}\frac{\sqrt{2m}}{e\,|\mathcal{E}|\hbar}\right)^{\frac{2}{3}}$$

$$t_1 = U\,C \qquad t_2 = V_0 C$$

$$C = \left(\frac{4}{3}\frac{\sqrt{2m}}{e\,|\mathcal{E}|\,\hbar}\right)^{\frac{2}{3}} \tag{24}$$

Using $n = \frac{8\pi}{3} \frac{P_F^3}{h^3}$, we arrive at

$$j = \frac{m e}{C^2} \frac{4\pi}{h^3} \int_{t_2}^{t_1} dt(t - t_2) e^{-t^{\frac{3}{2}}} \tag{25}$$

A further change of variables, setting $z = t^{3/2}$, leads to

$$dt = \frac{2}{3} \frac{dz}{z^{\frac{1}{3}}}$$

$$j = \frac{8\pi}{3} \frac{m e}{C^2 h^3} \int_{t_2^{\frac{3}{2}}}^{t_1^{\frac{3}{2}}} \frac{dz}{z^{\frac{1}{3}}} [z^{\frac{2}{3}} - z_{\min}^{\frac{2}{3}}] e^{-z} =$$

$$= \frac{8\pi}{3} \frac{m e}{C^2 h^3} e^{-z_{\min}} \int_0^{z_{\max} - z_{\min}} \frac{dx}{[x + z_{\min}]^{\frac{1}{3}}} [(x + z_{\min})^{\frac{2}{3}} - z_{\min}^{\frac{2}{3}}] e^{-x} \tag{26}$$

For typical values of the electric field, together with the data of the problem, we find $t_1 \simeq 4t_2 \gg 1$; this in turn implies that $z_{\max} - z_{\min} \gg 1$ and the limit of integration may be extended to infinity. As an example, for $|\mathcal{E}| \sim 10^4$ Volt/cm we have $C \sim 300\,\mathrm{eV}^{-1}$ and $t_1 \sim 3000$. At the same time, the inequality $z_{\min} \gg 1$ allows us to approximate the integrand as follows

$$\frac{1}{[x + z_{\min}]^{\frac{1}{3}}} [(x + z_{\min})^{\frac{2}{3}} - z_{\min}^{\frac{2}{3}}] \simeq \frac{2}{3} \frac{x}{z_{\min}^{\frac{2}{3}}} \tag{27}$$

Thus,

$$j = \frac{16\pi}{9} \frac{m e}{C^2 h^3} \frac{e^{-z_{\min}}}{z_{\min}^{\frac{2}{3}}} \int_0^\infty x\, e^{-x}\, dx \tag{28}$$

Substituting for C and $z_{\min} = [V_0 C]^{\frac{3}{2}}$ the values given in (24), the current density becomes

$$j = \frac{e^3 |\mathcal{E}|^2}{8\pi\, h\, V_0} \exp\left[-V_0^{\frac{3}{2}} \frac{4}{3} \frac{\sqrt{2m}}{e\, |\mathcal{E}|\hbar} \right] \tag{29}$$

Problem 6. Scattering off a $1/r^2$ potential

A particle of mass m penetrates a center of force described by the potential

$$V(r) = \frac{\hbar^2 \alpha}{2mr^2} \tag{1}$$

α is a real dimensionless parameter.

a)   Write down the equation of motion for the partial wave with angular momentum ℓ, and determine the short distance behaviour of the solution.

b)   Calculate the phase shift.

c)   Discuss the total cross section.

Solution

a)   The radial equation is

$$\chi_\ell'' + \left[k^2 - \frac{\ell(\ell+1) + \alpha}{r^2} \right] \chi_\ell = 0 \tag{2}$$

At short distances the term involving k^2 is negligible and Eq.(2) becomes

$$\chi_\ell'' - \frac{\bar{\ell}(\bar{\ell}+1)}{r^2} \chi_\ell = 0 \tag{3}$$

We have defined

$$\bar{\ell}(\bar{\ell}+1) = \ell(\ell+1) + \alpha \tag{4}$$

implying

$$\bar{\ell} = -\frac{1}{2} + \sqrt{(\ell+\frac{1}{2})^2 + \alpha} \tag{5}$$

At short distances, inserting the functional form $\chi_\ell \sim r^\beta$ in (3) leads to

$$\beta = \bar{\ell} + 1 \quad \text{o} \quad \beta = -\bar{\ell} \tag{6}$$

The physical solution must be regular at the origin and thus behaves as

$$\chi_\ell \simeq c \, r^{\bar{\ell}+1} \tag{7}$$

b)   The phase shift can be written down immediately observing that the exact regular solution of (2) is

$$\chi_\ell(r) = j_{\bar{\ell}}(kr) \tag{8}$$

where $j_{\bar{\ell}}$ is a spherical Bessel function.

At large distances

$$\chi_\ell \sim \sin\left(kr - \bar{\ell}\frac{\pi}{2} \right) \tag{9}$$

Comparing with the free wave function, in which $\bar{\ell} = \ell$, we find for the phase shift

$$\delta_\ell = (\ell - \bar{\ell})\frac{\pi}{2} = \frac{\pi}{2}\left[\left(\ell+\frac{1}{2}\right) - \sqrt{\left(\ell+\frac{1}{2}\right)^2 + \alpha} \right] \tag{10}$$

At large ℓ

$$\delta_\ell \simeq -\frac{\pi\alpha}{4\ell} \tag{11}$$

c)   The total cross section is

$$\sigma = \frac{4\pi}{k^2} \sum_\ell (2\ell + 1) \sin^2 \delta_\ell \tag{12}$$

At large ℓ, by virtue of (11),

$$\sigma \simeq \frac{\pi^3 \alpha^2}{2 k^2} \sum_\ell \frac{1}{\ell} \tag{13}$$

Therefore, σ is infinite.

Problem 7.   Elastic scattering of electrons on atomic hydrogen

A 300 MeV electron beam collides on a hydrogen target; elastically scattered electrons are subsequently detected.

a)   Calculate the scattering amplitude and the differential cross section in the Born approximation. Using the ground state wave function, derive the static charge distribution of the atom. For the proton, one may use the charge distribution $\rho \propto e^{-r/r_0}$, with $r_0 \simeq 2$ fm.

b)   Discuss the angular distribution, identifying the angular regions in which the result is sensitive to the charge distributions of the atomic electron and of the proton.

c)   Parameterize the scattering amplitude due to the proton in terms of an electric form factor and calculate this form factor for the particular charge distribution given above.

Solution

The momentum transfer at a scattering angle θ is

$$|\mathbf{k}| = 2\,|\mathbf{p}|\,\sin\frac{\theta}{2} \tag{1}$$

at energies small with respect to the atomic mass M; $\mathbf{p}$ is the incoming momentum, in our case 300MeV/c. The recoil energy taken up by the atom is

$$k_0 = \frac{|\mathbf{k}|^2}{2M} = |\mathbf{k}|\,\frac{|\mathbf{p}|\,\sin\dfrac{\theta}{2}}{M} \ll |\mathbf{k}| \tag{2}$$

Thus the four-momentum transfer is to all effects purely spatial.

a)   The components $(A_0, \mathbf{A})$ of the electromagnetic potential $A_\mu(k)$ produced by the atom are

$$A_0(\mathbf{k}) = \frac{\rho(\mathbf{k})}{|\mathbf{k}|^2} \tag{3}$$

$$\mathbf{A}(\mathbf{k}) = \frac{\mathbf{j}(\mathbf{k})}{|\mathbf{k}|^2} \tag{4}$$

where $\rho(\mathbf{k})$ $(\mathbf{j}(\mathbf{k}))$ is the Fourier transform of the charge (current) distribution.

Denoting by r_B one half of the Bohr radius, $r_B \simeq 0.25 \cdot 10^{-8}$ cm, the charge distributions in this problem lead to

$$\rho(\mathbf{k}) = e\left\{ \frac{1}{[1 + (|\mathbf{k}|\,r_0)^2]^2} - \frac{1}{[1 + (|\mathbf{k}|\,r_B)^2]^2} \right\} \tag{5}$$

In the Born approximation the scattering amplitude $\mathcal{M}$ equals

$$\mathcal{M} = 2Me\bar{u}(\mathbf{p}')A(\mathbf{k})u(\mathbf{p}) \tag{6}$$

A factor of $2M$ comes from the covariant normalization of the atomic wave function and $\mathbf{k} = \mathbf{p}' - \mathbf{p}$. Making use of the approximation $k_\mu k^\mu = -\mathbf{k}^2$, and noting that $k_\mu A^\mu(\mathbf{k}) = 0$, we find

$$
\begin{aligned}
\overline{|\mathcal{M}|^2} &= \sum_{\text{final spins}} \frac{1}{2} \sum_{\text{init. spins}} |\mathcal{M}|^2 \\
&= \frac{e^2}{2} (2M)^2 \operatorname{Tr} \left\{ (\not{p}' + m) A(\mathbf{k}) (\not{p} + m) A^*(\mathbf{k}) \right\} \\
&= 2e^2 (2M)^2 \left\{ 2|p_\mu A^\mu(\mathbf{k})|^2 - \frac{\mathbf{k}^2}{2} A_\mu(\mathbf{k}) A^\mu(\mathbf{k}) \right\}
\end{aligned}
\tag{7}
$$

Neglecting the mass of the electron (0.5 MeV) as compared to its energy $E = 300$ MeV, we are led to

$$
\overline{|\mathcal{M}|^2} = (2M)^2 \frac{2e^2}{(\mathbf{k}^2)^2} \left(2E^2 - \frac{\mathbf{k}^2}{2} \right) \rho^2(\mathbf{k})
\tag{8}
$$

The cross section is given by

$$
d\sigma = \frac{1}{2M\,2|\mathbf{p}|} \overline{|\mathcal{M}|^2} \, d\Phi
\tag{9}
$$

For small recoil energies we have

$$
\begin{aligned}
d\Phi &= \frac{1}{(2\pi)^2} \delta^4(P_i - P_f) \frac{d^3\mathbf{p}'}{2\,p_0'} \frac{d^3\mathbf{P}}{2\,P_0} \\
&= \frac{1}{(2\pi)^2} \frac{1}{2M} \delta\left(E - E' - \frac{\mathbf{k}^2}{2M} \right) \frac{E'\,dE'}{2} \, d\Omega' \\
&\simeq \frac{E}{4M(2\pi)^2} \, d\Omega'
\end{aligned}
\tag{10}
$$

Finally, using Eq.(1) one obtains

$$
\frac{d\sigma}{d\Omega'} = \frac{e^2}{(2\pi)^2 |\mathbf{k}|^4} E^2 \rho(\mathbf{k})^2 \cos^2 \frac{\theta}{2}
\tag{11}
$$

or equivalently, setting $\rho(\mathbf{k}) = e\tilde{\rho}(\mathbf{k})$, $e^2 = 4\pi\alpha$,

$$
\frac{d\sigma}{d\Omega'} = \frac{\alpha^2}{4E^2 \sin^4 \dfrac{\theta}{2}} \tilde{\rho}(\mathbf{k})^2 \cos^2 \frac{\theta}{2}
\tag{12}
$$

b)   Consider the region $|\mathbf{k}|^2 r_B^2 \leq 1$; in this region $(|\mathbf{k}|r_0)^2 \ll 1$ and we write

$$
\rho(|\mathbf{k}|) \simeq e \left\{ 1 - \frac{1}{[1 + (|\mathbf{k}|r_B)^2]^2} \right\}
\tag{13}
$$

Eq.(12) becomes

$$\frac{d\sigma}{d\Omega'} \simeq 16\alpha^2 E^2 r_B^4 \frac{\left[1 + (|\mathbf{k}|r_B)^2 + \frac{1}{4}(|\mathbf{k}|r_B)^4\right]}{[1 + (E\, r_B \theta)^2]^4} \simeq \frac{16\alpha^2 E^2 r_B^4}{[1 + (E\, r_B \theta)^2]^3} \tag{14}$$

In terms of angles, $(|\mathbf{k}|r_B)^2 \ll 1$ implies

$$\theta^2 E^2 r_B^2 \ll 1 \qquad \theta \leq \frac{1}{E\, r_B} = \frac{\alpha\, m_e}{E} \simeq 10^{-5} \tag{15}$$

This condition identifies the angular region which is sensitive to the charge distribution of the atomic electron.

For $(|\mathbf{k}|\, r_0)^2 \leq 1$, or $\theta \leq 0.3$, and $(|\mathbf{k}|\, r_B)^2 \gg 1$ one obtains instead

$$\frac{d\sigma}{d\Omega'} = \frac{\alpha^2}{4E^2 \sin^4 \frac{\theta}{2}} \frac{1}{[1 + (E\, r_0 \theta)^2]^4} \tag{16}$$

c) The matrix element of the current between proton states is parameterized in terms of form factors as follows

$$J_\mu = \bar{u}(\mathbf{p}') \left\{ G_1(k^2) \frac{1}{2M} (p + p')_\mu + \mathrm{i}\, \frac{\sigma_{\mu\nu} k^\nu}{2M} G_2(k^2) \right\} u(\mathbf{p}) \tag{17}$$

Neglecting the magnetic moment, and comparing with Eq.(3), we find

$$G_1(k^2) \simeq \rho_0(\mathbf{k}^2)$$

where ρ_0 is the Fourier transform of the proton charge distribution, that is, the first term in Eq.(5).

Problem 8.   The Compton effect on bound electrons

At energies small with respect to the electron mass, the invariant Compton amplitude reads

$$A = 2e^2 u_f^\dagger u_i\, \varepsilon_f^* \varepsilon_i \tag{1}$$

where u_i, u_f are Pauli spinors for the initial and final electron states and ε_i, ε_f^* are the initial and final photon polarization vectors.

a)   Consider the outgoing photon in the laboratory frame: Write down its energy ω' as a function of the scattering angle θ, and calculate its angular distribution, summed over final spins and averaged over initial spins.

b)   When X-rays are sent on an atom, a Compton effect takes place, modified by the fact that the initial electron is bound. Derive the angular distribution of the final photon. You may assume an infinite nuclear mass.

Solution

a) Denoting by p (p') and k (k') the four-momenta of the initial (final) electron and photon, respectively, we have

$$p + k = p' + k' \tag{2}$$

which implies

$$(p + k - k')^2 = p'^2 \tag{3}$$

and, since $p^2 = p'^2 = m^2$,

$$p\,k - p\,k' - k\,k' = 0 \tag{4}$$

Thus, in the laboratory frame we have

$$\frac{1}{\omega'} - \frac{1}{\omega} = \frac{1}{m_e}\,(1 - \cos\theta) \tag{5}$$

In the laboratory frame the cross section is given by

$$d\sigma = \frac{1}{4\,m_e\omega}\,|\mathcal{M}_{fi}|^2\,d\Phi^{(2)} \tag{6}$$

with

$$d\Phi^{(2)} = (2\pi)^4\,\delta^{(4)}(p + k - p' + k')\,\frac{d^3\mathbf{p}'}{2E'(2\pi)^3}\,\frac{d^3\mathbf{k}'}{2\omega'(2\pi)^3} \tag{7}$$

Integrating over $d^3\mathbf{p}'$ we find

$$d\Phi^{(2)} = \frac{1}{(2\pi)^2}\,\delta\!\left(m + \omega - \omega' - \sqrt{(\mathbf{k}' - \mathbf{k})^2 + m^2}\right)\frac{d^3\mathbf{k}'}{2\omega'}\,\frac{1}{2E'} \tag{8}$$

Going over to polar coordinates and integrating over $d\omega'$

$$d\Phi^{(2)} = \frac{\omega'}{16\pi^2\,E'}\,\frac{1}{\left|1 + \dfrac{\partial}{\partial\omega'}\sqrt{(\mathbf{k}' - \mathbf{k})^2 + m^2}\right|}\,d\Omega \tag{9}$$

The derivative in the denominator yields

$$E'\left(1 + \frac{\partial}{\partial\omega'}\sqrt{(\mathbf{k}' - \mathbf{k})^2 + m^2}\right) = E' + \omega' - \omega\cos\theta \tag{10}$$

Finally, using (5) to eliminate $\cos\theta$ we obtain

$$d\Phi^{(2)} = \frac{\omega'^2}{16\pi^2\,m_e\omega}\,d\Omega \tag{11}$$

and the differential cross section becomes

$$\frac{d\sigma}{d\Omega} = |\mathcal{M}_{fi}|^2 \frac{1}{64\pi^2 m_e^2} \left(\frac{\omega'}{\omega}\right)^2 \tag{12}$$

Consistently with the approximation leading to Eq. (1), we set $\omega'/\omega \simeq 1$. Summing (averaging) over final (initial) electron spins

$$|\mathcal{M}_{fi}|^2 = \frac{1}{2} \sum_{\text{spin}} \left|2e^2\, u_f^\dagger u_i\, \boldsymbol{\varepsilon}_f' \boldsymbol{\varepsilon}_i\right|^2 = 4e^4 \left|\boldsymbol{\varepsilon}_f' \boldsymbol{\varepsilon}_i\right|^2 \tag{13}$$

Similarly, the sum and average over photon polarizations gives

$$\overline{|\mathcal{M}_{fi}|^2} = 2e^4 \left(1 + \cos^2\theta\right) \tag{14}$$

In conclusion, we find

$$\frac{d\sigma}{d\Omega} = \frac{\alpha^2}{2m_e^2} \left(1 + \cos^2\theta\right) \tag{15}$$

(the Thomson cross section).

b) Wavelengths in the X-ray spectrum ($h\nu \lesssim 1$ KeV) obey the inequality $\lambda > r_B$ (where $r_B = \frac{1}{m_e\alpha}$ is the Bohr radius); indeed, 1 KeV X-rays have $\lambda \simeq 1.2\cdot10^{-7}$ cm, compared to $r_B \simeq 0.5\cdot10^{-8}$ cm. We therefore may use the interaction Hamiltonian in the dipole approximation

$$H_I = -\mathbf{d}\cdot\mathbf{E} \qquad \mathbf{d} = -e\,\mathbf{r} \tag{16}$$

Neglecting atomic recoil, we write for the cross section

$$d\sigma_n = \frac{1}{2\omega}|T_n|^2 \frac{d^3k'}{(2\pi)^3\, 2\omega'}\, 2\pi\delta(E_0 + \omega - E_n - \omega') \tag{17}$$

T_n is the amplitude for this process; to second order in perturbation theory, it is given by

$$T_n = \sum \Big\{ \frac{\langle n; \mathbf{k}'|\mathbf{d}\cdot\mathbf{E}\,|m; 0\rangle\langle m; 0|\mathbf{d}\cdot\mathbf{E}|i; \mathbf{k}\rangle}{E_i + \omega - E_m} \\ + \frac{\langle n; \mathbf{k}'|\mathbf{d}\cdot\mathbf{E}\,|m; \mathbf{k},\mathbf{k}'\rangle\langle m; \mathbf{k},\mathbf{k}'|\mathbf{d}\cdot\mathbf{E}|i; \mathbf{k}\rangle}{E_i + \omega - (E_m + \omega + \omega')} \Big\} \tag{18}$$

Here, $|i\rangle, |m\rangle, |n\rangle$ are initial, intermediate and final atomic bound states, and $|\mathbf{k}\rangle, |\mathbf{k}'\rangle$, $|0\rangle$ are photonic states. Using the relation $\mathbf{E} = -\partial\mathbf{A}/\partial t$, valid in the Coulomb gauge, and setting $\omega_{m,i} = E_m - E_i$, $\omega_{m,n} = E_m - E_n$, we find

$$T_n = \omega\omega'\varepsilon_j\,\varepsilon_i'^* \left[\frac{\langle n|d_i|m\rangle\langle m|d_j|i\rangle}{\omega - \omega_{m,i}} - \frac{\langle n|d_j|m\rangle\langle m|d_i|i\rangle}{\omega_{m,n} + \omega}\right] \tag{19}$$

Thus, in terms of

$$\alpha_{ks}^{ni} = \sum_m \left[\frac{\langle n | r_k | m \rangle \langle m | r_s | i \rangle}{\omega_{mi} - \omega} + \frac{\langle n | r_s | m \rangle \langle m | r_k | i \rangle}{-\omega_{nm} + \omega} \right] \tag{20}$$

the cross section is

$$\frac{d\sigma_n}{d\Omega'} = \alpha^2 \omega \, \omega'^3 |\varepsilon_k \, \varepsilon_j'^* \, \alpha_{kj}^{ni}|^2 \tag{21}$$

For notational simplicity, we set

$$\hat{X}_{ks} = \sum_m \left[r_k \frac{|m\rangle\langle m|}{\omega_{mi} - \omega} r_s + r_s \frac{|m\rangle\langle m|}{-\omega_{nm} + \omega} r_k \right] \tag{22}$$

Then, Eq. (20) becomes

$$\alpha_{ks}^{ni} = \langle n | \hat{X}_{ks} | i \rangle \tag{23}$$

Since both the energy and the projector onto energy eigenstates are rotationally invariant, $\hat{X}_{ks}$ will transform as a two index tensor; hence it can be decomposed into irreducible representations of the rotation group as follows

$$\hat{X}_{ks} = \hat{S}\delta_{ks} + \varepsilon_{ksl}\hat{V}_l + \hat{T}_{ks} \tag{24}$$

where $\hat{T}_{ks}$ is a symmetric traceless tensor. In this notation, the quantity $\alpha_{ks}^{ni}(\alpha_{k's'}^{ni})^*$, averaged over initial polarizations and summed over final ones, takes the form

$$\frac{1}{2J_i + 1} \sum_{M_n, M_i} \alpha_{ks}^{ni} \left(\alpha_{k's'}^{ni} \right)^* = \frac{1}{2J_i + 1} \sum_{M_n, M_i} \langle M_n | \hat{X}_{ks} | M_i \rangle \langle M_i | \hat{X}_{k's'}^\dagger | M_n \rangle \tag{25}$$

Rotational invariance also implies that different terms from the decomposition (24) do not interfere with each other in the product $\hat{X}_{ks}\hat{X}_{k's'}$. With the further notation T_{ks} for matrix elements of the form $\langle n | \hat{T}_{ks} | i \rangle$, Eq. (25) simplifies to

$$\frac{1}{2J_i + 1} \sum_{M_n, M_i} \alpha_{ks}^{ni} \left(\alpha_{k's'}^{ni} \right)^* = \frac{1}{2J_i + 1} \sum_{M_n, M_i} \left\{ |S|^2 \delta_{ks}\delta_{k's'} + V_a V_b^* \varepsilon_{ksa}\varepsilon_{k's'b} + T_{ks}T_{k's'}^* \right\} \tag{26}$$

Once more, rotational invariance gives

$$\frac{1}{2J_i + 1} \sum_{M_n, M_i} V_a V_b^* = \frac{1}{3}\delta_{ab} \frac{1}{2J_i + 1} \sum_{M_n, M_i} V_c V_c^* \tag{27}$$

$$\frac{1}{2J_i + 1} \sum_{M_n, M_i} T_{ks}T_{k's'}^* = \frac{1}{10} \left(\delta_{kk'}\delta_{ss'} + \delta_{ks'}\delta_{sk'} - \frac{2}{3}\delta_{ks}\delta_{k's'} \right) \frac{1}{2J_i + 1} \sum_{M_n, M_i} T_{ab}T_{ab}^*$$

Thus, denoting by a bar the sum over final spin orientations and the average over initial ones

$$\frac{1}{2J_i + 1} \sum_{M_n, M_i} \alpha_{ks}^{ni} \left(\alpha_{k's'}^{ni} \right)^* = \tag{28}$$

$$\overline{|S|^2} \, \delta_{ks}\delta_{k's'} + \frac{1}{3}\overline{V_c V_c^*} \left(\delta_{kk'}\delta_{ss'} - \delta_{ks'}\delta_{k's} \right) + \frac{1}{10}\overline{T_{ks}T_{k's'}^*} \left(\delta_{kk'}\delta_{ss'} + \delta_{ks'}\delta_{sk'} - \frac{2}{3}\delta_{ks}\delta_{k's'} \right)$$

The terms appearing in (28) do not depend on the projection of the angular momentum M_n in the final state; thus, they can be calculated simply by averaging over initial polarizations and multiplying by $(2J_n + 1)$.

Substitution into (21) leads to

$$\frac{d\sigma}{d\Omega'} = \alpha^2 \omega \omega'^3 \left[\overline{|S|^2} |\varepsilon \varepsilon'^*|^2 + \frac{1}{3}\overline{V_c V_c^*}\left(1 - |\varepsilon \varepsilon'|^2\right) + \frac{1}{10}\overline{T_{ks}T_{k's'}^*}\left(1 + |\varepsilon \varepsilon'|^2 - \frac{2}{3}|\varepsilon \varepsilon'^*|^2\right)\right]$$

(29)

The sum (average) over final (initial) photon polarizations gives

$$\frac{d\sigma}{d\Omega'} = \alpha^2 \omega \omega'^3 \left[\frac{1+\cos^2\theta}{2}\overline{|S|^2} + \frac{3-\cos^2\theta}{6}\overline{V_c V_c^*} + \frac{13+\cos^2\theta}{60}\overline{T_{ks}T_{k's'}^*}\right]$$

(30)

We now integrate over directions of the outgoing photon, with the result

$$\sigma_n = \frac{8\pi}{9}\alpha^2 \omega \omega'^3 \left[3\overline{|S|^2} + 2\overline{V_c V_c^*} + \overline{T_{ks}T_{k's'}^*}\right]$$

(31)

The inclusive cross section is

$$\sigma = \sum_n \sigma_n$$

(32)

Let us now consider the high frequency limit of expression (20):

$$\begin{aligned}
\alpha_{kj} &\simeq \frac{1}{\omega^2}\sum_m \left\{-\omega_{m,n}\langle n|r_j|m\rangle\langle m|r_k|i\rangle - \omega_{m,i}\langle n|r_k|m\rangle\langle m|r_j|i\rangle\right\} \\
&= \frac{1}{i\,\omega^2}\sum_m \left\{\langle n|\dot{r}_j|m\rangle\langle m|r_k|i\rangle - \langle n|r_k|m\rangle\langle m|\dot{r}_j|i\rangle\right\} \\
&= \frac{1}{i\,\omega^2}\langle n|[\dot{r}_j, r_k]|i\rangle \\
&= -\delta_{ni}\frac{Z}{m\,\omega^2}\delta_{jk}
\end{aligned}$$

(33)

for a system with Z electrons. In this limit we have elastic (δ_{ni}) and coherent ($A \propto Z$) scattering, and the cross section is

$$\frac{d\sigma}{d\Omega'} = \frac{\alpha^2}{m^2}Z^2|\varepsilon\,\varepsilon'^*|^2$$

(34)

Summing over final spin orientations and averaging over initial ones, we finally obtain

$$\frac{d\sigma}{d\Omega'} = \frac{\alpha^2}{2m^2}Z^2\left(1 + \cos^2\theta\right) = Z^2\left(\frac{d\sigma}{d\Omega'}\right)_{\text{Thomson}}$$

(35)

Compton scattering on individual atomic electrons would result in an incoherent contribution of the form

$$d\sigma \simeq Z\,d\sigma_{\text{Thomson}}$$

(36)

This is clearly the case for $k\,a_B \gg 1$, where the dipole approximation is no longer valid.

Problem 9. Negative ions of noble gases in the semiclassical approximation

Let us consider a rough model for the negative ions of noble gases. In this model, the electron-atom interaction is described by a potential

$$V(r) = -A/r^4 \qquad r > r_0$$
$$V(r) = \infty \qquad r < r_0 \qquad (1)$$

r_0 is a parameter.

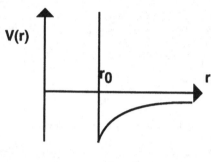

fig. 1

a) Knowing the polarizability χ of the atom, determine the parameter A. For argon, $\chi = 1.63 \cdot 10^{-24} \text{ cm}^3$.

b) Write down the equation obeyed by S-wave bound states, in the semiclassical approximation; derive the bound state spectrum corresponding to $r_0 = 0.8 \chi^{\frac{1}{3}}$

c) Discuss the electron-Argon cross section at low energies.

Some useful formulae:

$$\int_\varepsilon^1 dx \sqrt{\frac{1}{x^4} - 1} = \frac{1}{\varepsilon}\sqrt{1 - \varepsilon^4} - 2\int_\varepsilon^1 dx \frac{x^2}{\sqrt{1-x^4}}$$

$$\int_\varepsilon^1 dx \frac{x^2}{\sqrt{1-x^4}} \underset{\varepsilon \ll 1}{\simeq} \int_0^1 dx \frac{x^2}{\sqrt{1-x^4}} + 0(\varepsilon^3) \simeq \frac{\Gamma(\frac{3}{4})\Gamma(\frac{1}{2})}{\Gamma(\frac{5}{4})} \simeq 0.75 \qquad (2)$$

Solution

a) By definition, polarizability is the proportionality constant between an external field and the induced dipole moment on the atom

$$\mathbf{d} = 4\pi\chi\mathbf{E} \qquad (3)$$

A dipole interacts with electric fields through the potential

$$V = -\mathbf{d} \cdot \mathbf{E} \tag{4}$$

For the electric field produced by an electron

$$\mathbf{E} = -\frac{e}{4\pi r^2} \hat{\mathbf{r}} \tag{5}$$

the potential becomes

$$V(r) = -4\pi \chi \frac{E^2}{2} = -\frac{\chi}{2} \frac{e^2}{4\pi} \frac{1}{r^4} = -\frac{A}{r^4} \tag{6}$$

leading to the relation

$$A = \alpha\chi/2 \tag{7}$$

We note that the factor of $1/2$ is due to the fact that the dipole is induced.

b) The semiclassical condition for bound states is

$$\oint P_r \, dr = \left(n + \frac{1}{2}\right) h \tag{8}$$

In the S-wave this condition reads

$$2 \int_{r_0}^{\bar{r}} \sqrt{\frac{2m_e}{\hbar^2} (E_n - V(r))} \, dr = 2\pi \left(n + \frac{1}{2}\right) \tag{9}$$

where r_0 and $\bar{r}$ are the classical inversion points: r_0 is the radius of the rigid sphere, while $\bar{r}$ is determined from the equation

$$E_n - V(\bar{r}) = 0 \tag{10}$$

The potential in (6) leads to

$$\bar{r} = \left(\frac{A}{|E_n|}\right)^{\frac{1}{4}} \tag{11}$$

and Eq.(9) becomes

$$\int_{r_0}^{\bar{r}} \sqrt{\frac{1}{r^4} - \frac{1}{\bar{r}^4}} \, dr = \pi \sqrt{\frac{\hbar^2}{2m_e A}} \left(n + \frac{1}{2}\right) \tag{12}$$

or, equivalently,

$$\frac{1}{\bar{r}} \int_{r_0/\bar{r}}^{1} \sqrt{\frac{1}{x^4} - 1} \, dx = \pi \sqrt{\frac{\hbar^2}{2m_e A}} \left(n + \frac{1}{2}\right) \tag{13}$$

For $r_0/\bar{r} \ll 1$ we find, using (2),

$$\frac{1}{\bar{r}}\left[\frac{\bar{r}}{r_0}\sqrt{1-(\frac{r_0}{\bar{r}})^4}-1.5\right] \simeq \pi\sqrt{\frac{\hbar^2}{2m_e A}}\left(n+\frac{1}{2}\right) \tag{14}$$

The term $(r_0/\bar{r})^4$ above can be consistently neglected, with the result

$$\frac{1}{r_0}-\frac{1.5}{\bar{r}} = \frac{\pi\hbar}{\sqrt{2m_e A}}\left(n+\frac{1}{2}\right) \tag{15}$$

Finally,

$$E_n = -\frac{A}{(1.5)^4}\left[\frac{1}{r_0}-\frac{\pi\hbar}{\sqrt{2m_e A}}\left(n+\frac{1}{2}\right)\right]^4 \tag{16}$$

Eq.(15) implies that the index n must have a maximum acceptable value, $\bar{n}$, if $\bar{r}$ is to remain positive

$$\bar{n}+\frac{1}{2} = \frac{\sqrt{2m_e A}}{\pi\hbar r_0} \tag{17}$$

Thus, the existence of bound states is subject to the condition

$$\frac{\sqrt{2m_e A}}{\pi\hbar r_0} \geq \frac{1}{2} \tag{18}$$

We define a parameter ρ with dimensions of length

$$\rho = \frac{\sqrt{2m_e A}}{\hbar} = \frac{1}{\hbar}\sqrt{m_e \alpha \chi} \tag{19}$$

The numerical value of ρ in our case is

$$\rho = 1.73\,10^{-8}\text{cm} \tag{20}$$

Eq.(16) now takes the form

$$E_n = -\frac{\hbar^2\rho^2}{2\,m_e(1.5)^4}\left[\frac{1}{r_0}-\frac{\pi(n+\frac{1}{2})}{\rho}\right]^4 \tag{21}$$

provided

$$n+\frac{1}{2} < \frac{\rho}{\pi r_0} \tag{22}$$

The choice $r_0 = 0.8\,\chi^{\frac{1}{3}} \simeq 10^{-8}$ cm gives $\frac{\rho}{\pi\,r_0} \simeq 0.55$; thus, there exists one bound state, corresponding to $n = 0$, whose energy is

$$E_0 = -\frac{\hbar^2 \rho^2}{2m_e (1.5)^4}\left[\frac{1}{r_0} - \frac{\pi}{2\rho}\right]^4 \tag{23}$$

Substituting numerical values

$$E_0 \simeq -1.5 \cdot 10^{-4}\,\text{eV} \tag{24}$$

c) A bound state, whose energy is small compared to the depth of the potential, implies the existence of resonant scattering at low energies. This scattering is dominated by the S-wave, and the corresponding cross section is

$$\sigma = \frac{2\pi\,\hbar^2}{m_e}\,\frac{1}{\varepsilon + |E_0|} \tag{25}$$

where $\varepsilon = k^2/2m_e$ is the energy of incoming electrons. Numerically, we find

$$\sigma \simeq 4 \cdot 10^{-12}\,\text{cm}^2\,\frac{1}{1 + \dfrac{\varepsilon}{|E_0|}} \tag{26}$$

Problem 10.   Physics of the spectral lines $D_1 D_2$ of sodium

Atomic sodium has a spectral line D corresponding to a transition of the external electron from orbital state $3S$ to $3P$. Its wavelength[1] is $\lambda \simeq 5890$ Å. The transition dipole is $e\,r_0$; $r_0 \simeq 10^{-8}$ cm.

a) Calculate, as a function of frequency, the elastic cross section of a photon near resonance scattering off the sodium ground state.

b) Calculate the absorbtion length of sodium gas at temperature T.

c) Calculate the index of refraction, using the forward dispersion relation.

Solution

a) We recall that the interaction Hamiltonian is $H_I = -e\,\mathbf{r} \cdot \mathbf{E}$ and the line width is given by

$$
\begin{aligned}
\mathrm{d}\Gamma_{A^*} &= \frac{\omega}{8\pi^2}|\langle H_I \rangle|^2\,\mathrm{d}\Omega \\
&= \frac{\alpha\omega^3}{2\pi}|\boldsymbol{\varepsilon}\,\mathbf{r}_{fi}|^2\,\mathrm{d}\Omega
\end{aligned}
\tag{1}
$$

---

[1]Actually the line is split in two components D_1 and D_2 with $\lambda_1 = 5896$ Å and $\lambda_2 = 5890$ Å, corresponding to transitions to the fine structure states with $J = 1/2$ and $J = 3/2$. For simplicity, we will consider these two lines superimposed.

In the above $\mathbf{r}_{fi}$ is

$$\mathbf{r}_{fi} = \langle A|\mathbf{r}|A^*\rangle \tag{2}$$

and we have used the following expression for the matrix element between photon states

$$\langle \mathbf{k}, \lambda |\mathbf{E}(0)|0\rangle = -\langle \mathbf{k}, \lambda |\dot{\mathbf{A}}(0)|0\rangle = -i\omega\varepsilon^{(\lambda)*} \tag{3}$$

Performing the angular integration and summing over photon polarizations we obtain

$$\Gamma_{A^*} = \frac{4}{3}\alpha\omega_0^3|\mathbf{r}_{fi}|^2 \tag{4}$$

In (4) we have neglected fine structure effects. Denoting by M the eigenvalue of L_z in the excited state, Eq.(4) can be rewritten more explicitly

$$\Gamma_{A^*} = \frac{4}{3}\alpha\omega_0^3\langle A^*, M|\mathbf{r}|A\rangle\langle A|\mathbf{r}|A^*, M\rangle \tag{5}$$

The rotational invariance of expression (5) ensures that the line width is independent of the angular momentum projection M.

The elastic cross section is obtained from the diagrams of fig.1. The corresponding amplitude is

$$T = -e^2\omega^2\varepsilon_i(\mathbf{k})\varepsilon_j^*(\mathbf{k}')\sum_{A^*} \tag{6}$$

$$\left[\frac{\langle A|r_j|A^*, M\rangle\langle A^*, M|r_i|A\rangle}{\omega + E_A - E_{A^*} + i\Gamma_{A^*}/2} + \frac{\langle A|r_i|A^*, M\rangle\langle A^*, M|r_j|A\rangle}{E_A - \omega - E_{A^*} + i\Gamma_{A^*}/2}\right]$$

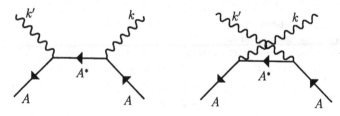

fig. 1

We restrict our attention to the resonant term (the first one in Eq.(6)); indicating by ω_0 the transition frequency $3P \rightarrow 3S$ and by Γ the width, we have

$$T = e^2\omega^2\varepsilon_i(\mathbf{k})\varepsilon_j^*(\mathbf{k}')\sum_M \frac{\langle A|r_j|A^*, M\rangle\langle A^*, M|r_i|A\rangle}{\omega_0 - \omega - i\Gamma_{A^*}/2} \tag{7}$$

Rotational invariance of the initial state (neglecting spin effects) and the sum over intermediate state polarizations allow us to make the replacement $r_i r_j \rightarrow 1/3 \, \delta_{ij} \mathbf{r}^2$.

The factor of 1/3 cancels against the sum over the three values of M (the intermediate state has angular momentum 1) and we can write

$$T = e^2 \omega^2 \varepsilon(\mathbf{k}) \varepsilon^*(\mathbf{k}') \frac{\langle A|\mathbf{r}|A^*, M\rangle \langle A^*, M|\mathbf{r}|A\rangle}{\omega_0 - \omega - i\Gamma_{A^*}/2} \tag{8}$$

In particular the forward amplitude equals

$$T = e^2 \omega^2 \frac{|\mathbf{r}_{fi}|^2}{\omega_0 - \omega - i\Gamma_{A^*}/2} \tag{9}$$

where $\mathbf{r}_{fi}$ is defined in (2). For the elastic cross section we find

$$d\sigma = \frac{1}{2\omega} \sum_{pol} |T|^2 \frac{d^3 k'}{(2\pi)^3 2\omega'} 2\pi \delta(E_A + \omega - E_A - \omega') \tag{10}$$

Integrating out the delta function leads to

$$\frac{d\sigma}{d\Omega} = \frac{1}{16\pi^2} \sum_{pol} |T|^2 = \alpha^2 \omega^4 \sum_{pol} |\varepsilon(\mathbf{k}) \varepsilon^*(\mathbf{k}')|^2 |\frac{\mathbf{r}_{fi}^2}{\omega_0 - \omega - i\Gamma_{A^*}/2}|^2 \tag{11}$$

We now integrate over the direction of the outgoing photon, sum over final polarizations and average over the initial ones, with the result

$$\sigma = \frac{8\pi}{3} \frac{\alpha^2 \omega^4}{(\omega - \omega_0)^2 + \dfrac{\Gamma_{A^*}^2}{4}} |\mathbf{r}_{fi}^2|^2 \tag{12}$$

or, using Eq.(4),

$$\sigma = \frac{3\pi}{2\omega_0^2} \frac{\Gamma_{A^*}^2}{(\omega - \omega_0)^2 + \dfrac{\Gamma_{A^*}^2}{4}} \tag{13}$$

which is the Breit-Wigner formula.

b)   The mean free path of light in a gas is given by

$$d = \frac{1}{\sigma \rho} \tag{14}$$

where ρ is the gas density; in terms of pressure and temperature we write, using the equation of state $\rho = P/kT$,

$$d = \frac{kT}{\sigma P} \tag{15}$$

We are assuming the gas to behave as a perfect gas.

c)   A wave in a medium obeys

$$\mathbf{A}(x) = \boldsymbol{\varepsilon} \, e^{in\mathbf{k}\mathbf{x}} \tag{16}$$

and absorption is measured by

$$2\omega \operatorname{Im} n = \rho \, \sigma \tag{17}$$

By the optical theorem

$$\sigma = \frac{4\pi}{\omega} \operatorname{Im} f(0) \tag{18}$$

where $f(\theta)$ is defined by the equation

$$\frac{d\sigma}{d\Omega} = |f(\theta)|^2 \tag{19}$$

Hence, with our normalization, Eq.(11):

$$f = \frac{T}{4\pi} \tag{20}$$

From Eq.(18) it follows that:

$$\operatorname{Im} f(0) = \frac{\omega^2}{2\pi\rho} \operatorname{Im} n \tag{21}$$

The function $f(0)/\omega^2$ obeys a dispersion relation, as can be verified from expression (9); therefore, the index of refraction, or better $n - 1$, will also obey a dispersion relation

$$n(\omega) = 1 + \frac{1}{\pi} \int \frac{\operatorname{Im} n(\omega')}{\omega' - \omega - i\varepsilon} \, d\omega' \tag{22}$$

Thus, one arrives at

$$\begin{aligned} n(\omega) = 1 + \frac{2\pi\rho}{\omega^2} f(0) &= 1 + \frac{2\pi\rho}{\omega^2} \, \alpha\omega^2 \, \frac{|\mathbf{r}_{fi}|^2}{\omega_0 - \omega - i\Gamma/2} \\ &= 1 + \frac{3\pi\rho}{2\omega^3 c^3} \frac{\Gamma}{\omega_0 - \omega - i\Gamma/2} \end{aligned} \tag{23}$$

In the last step we abandoned natural units, introducing the factor c^3 explicitly.

Problem 11.   On dipole and quadrupole resonances

Consider an electromagnetic transition of an atomic system, $A^* \to A + \gamma$, with frequency ω_0 and width Γ.

a) What is the ω_0-dependence of Γ in the case of an electric dipole transition? How can we determine the dipole from the value of Γ?

b) Answer the above questions for an electric quadrupole transition. Explain why it is difficult to observe such a transition in atoms.

c) In the two cases above, write down the cross section for the process

$$\gamma + A \to \gamma + A \tag{1}$$

at energies close to resonance; derive the angular distribution of scattered photons.

Solution

a) Using a nonrelativistic normalization for atomic states, and neglecting atomic recoil, we have

$$d\Gamma = \sum_f 2\pi \, \delta(E_i - E_f)|H_I|^2 = 2\pi \, \delta(\omega_0 - \omega) \frac{d^3k}{(2\pi)^3} \frac{1}{2\omega}|H_I|^2 \tag{2}$$

which gives

$$d\Gamma = \frac{\omega_0}{8\pi^2}|H_I|^2 \, d\Omega \tag{3}$$

In the dipole approximation $H_I = -e\,\mathbf{r} \cdot \mathbf{E}$ and, in the Coulomb gauge, $\langle \mathbf{k}|\mathbf{E}|0\rangle = -i\,\omega\,\boldsymbol{\varepsilon}^*$. Denoting by M and M^* the angular momentum of the initial and final atomic states, we write

$$d\Gamma = \frac{\alpha}{2\pi}\,\omega_0^3\,d\Omega\,\varepsilon_i\varepsilon_j^* \sum_M \langle A^*, M^*|r_i|A, M\rangle\langle A, M|r_j|A^*, M^*\rangle \tag{4}$$

A sum over polarizations of the outgoing photon,

$$\sum_{pol} \varepsilon_i\,\varepsilon_j^* = \delta_{ij} - n_i\,n_j \tag{5}$$

where $\mathbf{n}$ is the photon direction, leads to

$$d\Gamma = \frac{\alpha}{2\pi}\,\omega_0^3\,d\Omega\,(\delta_{ij} - n_i\,n_j) \sum_M \langle A^*, M^*|r_i|A, M\rangle\langle A, M|r_j|A^*, M^*\rangle \tag{6}$$

Integrating over $\mathbf{n}$ we finally obtain

$$\Gamma_D = \frac{4}{3}\alpha\,\omega_0^3 \sum_M |\langle A, M|\mathbf{r}|A^*, M^*\rangle|^2 \tag{7}$$

It is clear from the above expression that the width does not depend on the angular momentum projection M^* of the initial atomic state. Knowing Γ_D and ω_0, we may determine the dipole matrix element directly through (7).

b)   A quadrupole moment couples to the gradient of the electric field, as follows

$$H_I = -\frac{e}{2}Q_{ab}\,\partial_a E_b$$

$$Q_{ab} = \sum_i \left(x_a^i x_b^i - \frac{1}{3}\mathbf{x}^{i2}\delta_{ab} \right) \tag{8}$$

$\mathbf{E} = \langle \mathbf{k}, \boldsymbol{\varepsilon}|\mathbf{E}|0\rangle$ is the relevant matrix element of the electric field operator.

$$\mathbf{E} = -i\,\omega\,\boldsymbol{\varepsilon} \qquad \partial_a E_b = \omega\,k_a\,\varepsilon_b \tag{9}$$

The corresponding transition width is

$$d\Gamma = \frac{\alpha}{8\pi}\,d\Omega\,\omega_0^5 \sum_M |\langle A, M|Q_{ab}|A^*, M^*\rangle n_a \varepsilon_b^*|^2 \tag{10}$$

A sum over photon polarizations gives

$$\frac{d\Gamma}{d\Omega} = \frac{\alpha}{8\pi}\,\omega_0^5\, n_a n_i (\delta_{bj} - n_b n_j) \sum_M \langle A, M|Q_{ab}|A^*, M^*\rangle \langle A^*, M^*|Q_{ij}|A, M\rangle \tag{11}$$

After angular integration

$$
\begin{aligned}
\Gamma_Q &= \frac{4\pi}{8\pi}\alpha\,\omega_0^5 \sum_M \langle A, M|Q_{ab}|A^*, M^*\rangle \langle A^*, M^*|Q_{ij}|A, M\rangle \\
&\quad \left\{ \frac{1}{3}\delta_{ai}\,\delta_{bj} - \frac{1}{15}\delta_{ai}\delta_{bj} - \frac{1}{15}\delta_{ab}\,\delta_{ij} - \frac{1}{15}\delta_{aj}\,\delta_{bi} \right\} \\
&= \frac{\alpha}{10}\omega_0^5 \sum_M |\langle A, M|Q_{ab}|A^*, M^*\rangle|^2
\end{aligned}
\tag{12}
$$

Using the standard definition of quadrupole moment in spherical coordinates,

$$Q_M^J = r^J \sqrt{\frac{4\pi}{2J+1}} Y_M^J \tag{13}$$

we write

$$Q_M^J Q_M^{J*} = \frac{3}{2} Q_{\alpha\beta} Q_{\alpha\beta}^* \tag{14}$$

$$\Gamma_Q = \frac{\alpha}{15}\omega_0^5 \sum_M |\langle A, M|Q_{J_z}^J|A^*, M^*\rangle|^2 \tag{15}$$

Comparing the dipole width Γ_D of Eq.(7) to the quadrupole width Γ_Q of Eq.(15) we see that

$$\frac{\Gamma_Q}{\Gamma_D} \simeq \omega_0^2\,a^2 \tag{16}$$

where a is the atomic radius.

Since $a \simeq 1$ Å, we find, for wavelengths of the order of 10^4 Å,

$$\frac{\Gamma_Q}{\Gamma_D} \simeq 10^{-8} \tag{17}$$

c) The cross section for $\gamma + A \to \gamma + A$ is, neglecting atomic recoil,

$$d\sigma = \frac{1}{2\omega} |H_{eff}|^2 \, 2\pi \, \delta(\omega + E_A - \omega' - E_{A'}) \frac{d^3 k'}{(2\pi)^3 \, 2\omega'} = \frac{d\Omega}{16\pi^2} |H_{eff}|^2 \tag{18}$$

Dipole and quadrupole transitions have, respectively,

$$H_{eff}^D = -e^2 \omega^2 \varepsilon_i(\mathbf{k}) \varepsilon_j^*(\mathbf{k}') \sum_{A^*, M^*}$$
$$\left[\frac{\langle A|r_j|A^*, M^*\rangle \langle A^*, M^*|r_i|A\rangle}{\omega + E_A - E_{A^*} + i\Gamma_{A^*}/2} + \frac{\langle A|r_i|A^*, M^*\rangle \langle A^*, M^*|r_j|A\rangle}{E_A - \omega - E_{A^*} + i\Gamma_{A^*}/2} \right] \tag{19}$$

$$H_{eff}^Q = -\frac{e^2 \omega^2}{4} \varepsilon_i(\mathbf{k}) \varepsilon_j^*(\mathbf{k}') \sum_{A^*, M^*} \left[\frac{\langle A|Q_{bj}k_b|A^*, M^*\rangle \langle A^*, M^*|Q_{ai}k_a|A\rangle}{\omega + E_A - E_{A^*} + i\Gamma_{A^*}/2} + \right.$$
$$\left. \frac{\langle A|Q_{ai}k_a|A^*, M^*\rangle \langle A^*, M^*|Q_{bj}k_b|A\rangle}{E_A - \omega - E_{A^*} + i\Gamma_{A^*}/2} \right] \tag{20}$$

Let us consider first the dipole interaction. We introduce the (resonant) polarizability matrix

$$\Pi_{ij}^{(M,M')} = \sum_{M^*} \frac{\langle A, M'|r_j|A^*, M^*\rangle \langle A^*, M^*|r_i|A, M\rangle}{\omega_0 - \omega - i\Gamma_{A^*}/2} \tag{21}$$

where $\omega_0 = E_{A^*} - E_A$.

To calculate the total cross section we now sum (average) over the direction and polarization of the outgoing (incoming) photon.

$$\varepsilon_j \varepsilon_k^* \to \frac{1}{2}(\delta_{jk} - n_j n_k) \to \frac{1}{3}\delta_{jk} \qquad \varepsilon_i'^* \varepsilon_\ell' \to \frac{2}{3}\delta_{i\ell} \tag{22}$$

Restricting our attention to the resonant term in (19) we find, using (22),

$$\sigma = \frac{8\pi}{9} \alpha^2 \omega^4 \sum_{M'} \Pi_{ij}^{(M,M')} \Pi_{ij}^{*(M,M')} \tag{23}$$

Making use of rotational invariance, the above tensor structure is simplified as follows

$$\sum_{M'} \Pi_{ij}^{(M,M')} \Pi_{ij}^{*(M,M')} = \frac{\displaystyle\sum_{M', M_1^*, M_2^*} \langle M'|r_j|M_1^*\rangle \langle M_1^*|r_i|M\rangle \langle M|r_i|M_2^*\rangle \langle M_2^*|r_j|M'\rangle}{(\omega_0 - \omega)^2 + \Gamma_{A^*}^2/4}$$

$$= \frac{\displaystyle\sum_{M', M_1^*} |\langle M'|\mathbf{r}|M_1^*\rangle|^2 \, |\langle M|\mathbf{r}|M_1^*\rangle|^2}{(\omega_0 - \omega)^2 + \Gamma_{A^*}^2/4} \tag{24}$$

The quantity $r_j|M'\rangle\langle M'|r_j$ above is rotationally invariant and, by Schur's lemma, proportional to the identity. This implies that $M_2^* = M_1^*$ and that $\sum_{M'} |\langle M'|\mathbf{r}|M_1^*\rangle|^2$ does not depend on M_1^*. Thus the sum over M_1^* acts only on the second absolute value above, which in turn becomes independent of M by Schur's lemma. This allows us to write

$$\sigma = \frac{8\pi}{9}\alpha^2\omega^4 \frac{1}{2J_A+1} \frac{\sum_{M,M',M_1^*} |\langle M'|\mathbf{r}|M_1^*\rangle|^2 \, |\langle M|\mathbf{r}|M_1^*\rangle|^2}{(\omega_0-\omega)^2 + \Gamma_{A^*}^2/4} \tag{25}$$

Use of (7) finally leads to

$$\sigma = \frac{2J_{A^*}+1}{2(2J_A+1)} \frac{\pi}{\omega^2} \frac{\Gamma_{A^*}^2}{(\omega_0-\omega)^2 + \Gamma_{A^*}^2/4} \tag{26}$$

(Breit-Wigner formula).

To calculate the angular distribution, we will assume, for the sake of simplicity, that A has angular momentum 0; consequently, the excited atom has angular momentum 1. The resonant amplitude now is, by rotational invariance,

$$\begin{aligned} H_{eff} &= e^2\omega^2 \, \varepsilon_i(\mathbf{k})\varepsilon_i^*(\mathbf{k}') \sum_{M^*} \frac{1}{3} \frac{|\langle A|\mathbf{r}|A^*, M^*\rangle|^2}{\omega_0 - \omega - i\Gamma_{A^*}/2} \\ &= e^2\omega^2 \, \varepsilon_i(\mathbf{k})\varepsilon_i^*(\mathbf{k}') \frac{|\langle A|\mathbf{r}|A^*, M^* = 0\rangle|^2}{\omega_0 - \omega - i\Gamma_{A^*}/2} \end{aligned} \tag{27}$$

Summing and averaging over polarizations we find for the cross section

$$\frac{d\sigma}{d\Omega} = \frac{\alpha^2\omega^4}{2}(1+\cos^2\theta) \frac{(|\langle A|\mathbf{r}|A^*, M^* = 0\rangle|^2)^2}{(\omega_0-\omega)^2 + \Gamma_{A^*}^2/4} = \frac{3\sigma}{16\pi}(1+\cos^2\theta) \tag{28}$$

The calculation in the quadrupole case goes through in a similar way. The usual sum and average over photon polarizations results in

$$\begin{aligned} \frac{d\sigma}{d\Omega} &= \frac{\alpha^2\omega^4}{16} \frac{1}{(\omega_0-\omega)^2 + \Gamma_{A^*}^2/4} \frac{1}{2} (\delta_{ii'} - n_i n_{i'})(\delta_{jj'} - n_j' n_{j'}') \sum_{M',M_1^*,M_2^*} \\ &\quad \langle M'|Q_{jb}k_b'|M_1^*\rangle\langle M_1^*|Q_{ia}k_a|M\rangle\langle M|Q_{i'a'}k_{a'}|M_2^*\rangle\langle M_2^*|Q_{j'b'}k_{b'}'|M'\rangle \end{aligned} \tag{29}$$

Integrating (averaging) over final (initial) directions, we obtain

$$\begin{aligned} \sigma &= \frac{4\pi}{25} \frac{\alpha^2\omega^8}{32} \frac{1}{(\omega_0-\omega)^2 + \Gamma_{A^*}^2/4} \\ &\quad \cdot \sum_{M',M_1^*,M_2^*} \langle M'|Q_{jb}|M_1^*\rangle\langle M_1^*|Q_{ia}|M\rangle\langle M|Q_{ia}|M_2^*\rangle\langle M_2^*|Q_{jb}|M'\rangle \end{aligned} \tag{30}$$

Just as in the dipole case, rotational invariance leads to

$$
\sigma = \frac{\pi}{2\,(10)^2}\,\frac{\alpha^2\omega^8}{(\omega_0-\omega)^2+\Gamma_{A^*}^2/4}\,\frac{1}{(2J_A+1)}
$$
$$
\cdot \sum_{M,M',M^*} |\langle M|Q_{ia}|M^*\rangle|^2|\langle M'|Q_{jb}|M^*\rangle|^2 \tag{31}
$$

Finally, Eq.(12) gives

$$
\sigma = \frac{\pi}{\omega^2}\,\frac{2J_{A^*}+1}{2(2J_A+1)}\,\frac{\Gamma_{A^*}^2}{(\omega_0-\omega)^2+\Gamma_{A^*}^2/4} \tag{32}
$$

This is again the Breit-Wigner formula.

We now calculate the angular distribution, once again assuming that the initial state $|A\rangle$ has angular momentum 0; the excited state will then have angular momentum 2. By rotational invariance

$$
\langle A|Q_{ab}|A^*,M^*\rangle\langle A^*,M^*|Q_{ij}|A\rangle =
$$
$$
\left(\delta_{ia}\delta_{jb}+\delta_{ib}\delta_{ja}-\frac{2}{3}\delta_{ij}\delta_{ab}\right)\frac{1}{10}\langle A|Q_{mn}|A^*,M^*\rangle\langle A^*,M^*|Q_{mn}|A\rangle =
$$
$$
\frac{1}{2}\left(\delta_{ia}\delta_{jb}+\delta_{ib}\delta_{ja}-\frac{2}{3}\delta_{ij}\delta_{ab}\right)|\langle A|Q_{mn}|A^*\rangle|^2 \tag{33}
$$

Making use of the transversality condition, $\boldsymbol{\varepsilon}\,\mathbf{k}=0$, we write for the scattering amplitude

$$
H_{eff} = \frac{e^2\omega^2}{8}\,\varepsilon_i(\mathbf{k})\varepsilon_a^*(\mathbf{k}')\,k_j k_b'\left(\delta_{ia}\delta_{jb}+\delta_{ib}\delta_{ja}-\frac{2}{3}\delta_{ij}\delta_{ab}\right)\frac{|\langle A|Q_{mn}|A^*\rangle|^2}{\omega_0-\omega-i\Gamma_{A^*}/2} =
$$
$$
\frac{e^2\omega^2}{72}\left[\boldsymbol{\varepsilon}(\mathbf{k})\cdot\boldsymbol{\varepsilon}^*(\mathbf{k}')\,\mathbf{k}\cdot\mathbf{k}'+\boldsymbol{\varepsilon}(\mathbf{k})\cdot\mathbf{k}'\,\boldsymbol{\varepsilon}^*(\mathbf{k}')\cdot\mathbf{k}\right]\frac{|\langle A|Q_{mn}|A^*\rangle|^2}{\omega_0-\omega-i\Gamma_{A^*}/2} \tag{34}
$$

The differential cross section takes the form

$$
\frac{d\sigma}{d\Omega} = \frac{\alpha^2\omega^4}{8^2}\,\frac{\left(|\langle A|Q_{mn}|A^*\rangle|^2\right)^2}{(\omega_0-\omega)^2+\Gamma_{A^*}^2/4}\,\left|\boldsymbol{\varepsilon}(\mathbf{k})\cdot\boldsymbol{\varepsilon}^*(\mathbf{k}')\,\mathbf{k}\cdot\mathbf{k}'+\boldsymbol{\varepsilon}(\mathbf{k})\cdot\mathbf{k}'\,\boldsymbol{\varepsilon}^*(\mathbf{k}')\cdot\mathbf{k}\right|^2 \tag{35}
$$

Summing and averaging over polarizations we find, using (12),

$$
\frac{d\sigma}{d\Omega} = \frac{25}{32}\frac{\Gamma^2}{\omega^2}\,\frac{1}{(\omega_0-\omega)^2+\Gamma_{A^*}^2/4}\left(1-3\cos^2\theta+4\cos^4\theta\right) =
$$
$$
= \frac{5\sigma}{16\pi}\left(1-3\cos^2\theta+4\cos^4\theta\right) \tag{36}
$$

Problem 12.   Quadrupole transitions

A physical system with a quadrupole moment couples to the electromagnetic field by means of the potential

$$V = -\frac{e}{2} Q_{ij} \partial_i E_j \tag{1}$$

where Q_{ij} is the quadrupole momentum operator of the system and $\mathbf{E}(x, t)$ is the operator corresponding to the electric radiation field. For a system of charged particles

$$Q_{ij} = \sum_n q_n \left[r_i^n r_j^n - \frac{1}{3}(\mathbf{r}^n)^2 \delta_{ij} \right] \tag{2}$$

The sum runs over all particles; q_n is the electric charge measured in units of e.

a)   Discuss the selection rules for angular momentum and parity in a decay induced by a quadrupole interaction.

b)   Quadrupole decays and transitions are very difficult to observe in atoms, where one can typically observe only dipole effects, whereas nuclear quadrupole transitions are rather more pronounced. Explain.

c)   Consider a γ transition between a nuclear state with spin 2 and one with spin zero. Determine the decay width and the angular distribution of the emitted γ if the initial state is completely polarized along the z-axis ($J_3 = +2$); repeat for an aligned state (a state having a density matrix whose only nonzero elements are $\rho_{2,2} = \rho_{-2,-2} = 1/2$).

Solution

a)   In the Born approximation, the decay is described by the matrix element of interaction (1), which equals

$$T = \frac{e}{2} \omega \, \varepsilon_j^* \, k_i \langle A'|Q_{ij}|A, M\rangle \tag{3}$$

Here, $\mathbf{k}$ is the momentum of the outgoing photon, $\omega = |\mathbf{k}|$ and $\langle A'|Q_{ij}|A, M\rangle$ is the quadrupole matrix element between the initial and final states of the charged particle system. The quadrupole moment is a symmetric, traceless tensor, transforming under rotations as an irreducible representation of angular momentum 2. Hence the angular momentum selection rules are

$$\Delta J = 0, 1, 2 \tag{4}$$

with the exclusion of $0 \rightleftarrows 0$ and $0 \rightleftarrows 1$ transitions. Furthermore, Q is parity invariant and thus it connects only states with the same parity.

b)   By its definition, Q_{ij} has matrix elements whose order of magnitude is a^2, where a is the typical dimension of the ensemble of charged particles. Thus the matrix element is proportional to $(ka)^2$; this is to be compared to the dipole matrix element, which is determined by the interaction

$$V = -e \sum q_i \, \mathbf{r}_i \cdot \mathbf{E} \tag{5}$$

and therefore is proportional to ka.

Thus in general we have

$$\frac{\Gamma_{quadr}}{\Gamma_{dip}} \sim O(ka)^2 \tag{6}$$

Typical values of k and a for an atom are: $k \simeq 10^{+5}$ cm^{-1} and $a \simeq 10^{-8}$ cm. This implies that $(ka)^2 \simeq 10^{-6}$ and hence quadrupole transitions have a much smaller width than dipole transitions.

c)   If the final state A' has angular momentum 0, the line width is

$$\frac{d\Gamma}{d\Omega} = \frac{\omega}{8\pi^2}\,|T|^2 = \frac{\alpha}{8\pi}\,\omega^5\,\left|\langle A'|Q_{ij}|A,M\rangle\,n_i\,\varepsilon_j^*\right|^2 \tag{7}$$

where $\alpha = \frac{e^2}{4\pi}$. We now sum over the polarizations of the photon and integrate over its directions, with the help of

$$\overline{n_i n_j (\delta_{\ell m} - n_\ell n_n)} = \frac{1}{3}\,\delta_{ij}\,\delta_{\ell m} - \frac{1}{15}(\delta_{ij}\,\delta_{\ell m} + \delta_{i\ell}\,\delta_{jm} + \delta_{im}\,\delta_{j\ell}) \tag{8}$$

obtaining

$$\Gamma = \frac{\alpha}{10}\,\omega^5 \langle A'|Q_{i\ell}|A,M\rangle\langle A,M|Q_{i\ell}^*|A'\rangle \tag{9}$$

Rotational invariance ensures that Γ is independent of the angular momentum projection M.

To obtain the angular distribution, we sum again over photon polarizations; this time however we must take into account the polarization of the atom, described by its density matrix

$$\frac{d\Gamma}{d\Omega} = \frac{\alpha}{8\pi}\omega^5\Big[\langle A'|Q_{i\ell}n_i|A,M\rangle\rho_{M,M'}\langle A,M'|Q_{j\ell}n_j|A'\rangle - $$

$$\langle A'|Q_{i\ell}n_i n_\ell|A,M\rangle\rho_{M,M'}\langle A,M'|Q_{js}n_j n_s|A'\rangle\Big] \tag{10}$$

If the atom is polarized then the only nonzero element of ρ is $\rho_{2,2} = 1$. For an aligned atom the nonzero elements are $\rho_{2,2} = \rho_{-2,-2} = 1/2$. The matrix elements we need to calculate are then

$$\langle A'|Q_{ij}|A,M = \pm 2\rangle \tag{11}$$

Writing the quadrupole tensor in terms of spherical components Q_m^J :

$$Q_{zz} = -\sqrt{\frac{2}{3}}\,Q_0^2$$

$$Q_{xz} \pm i\,Q_{yz} = Q_{\pm 1}^2$$

$$Q_{xx} - Q_{yy} \pm 2i Q_{xy} = Q_{\pm 2}^2$$

$$Q_{xx} + Q_{yy} + Q_{zz} = 0 \tag{12}$$

and using the relation

$$\langle 0|Q_m^2|2,\pm 2\rangle = Q\,\delta_{m,\pm 2} \tag{13}$$

we conclude that the operators Q_{zz}, Q_{xz} and Q_{yz} have zero matrix elements, while

$$\langle A'|Q_{xy}|A, M=\pm 2\rangle = \frac{Q}{2i}(\delta_{M,2} - \delta_{M,-2})$$

$$\langle A'|Q_{xx}|A, M=\pm 2\rangle = \frac{Q}{2}(\delta_{M,2} + \delta_{M,-2})$$

$$\langle A'|Q_{yy}|A, M=\pm 2\rangle = -\frac{Q}{2}(\delta_{M,2} - \delta_{M,-2}) \tag{14}$$

For the two cases of interest, Eq.(10) becomes

$$\frac{d\Gamma}{d\Omega} = \sum_{M=\pm 2} \frac{\alpha}{8\pi}\omega^5 \rho_{M,M}\Big[|\langle A'|Q_{xx}|A, M\rangle n_x + \langle A'|Q_{yx}|A, M\rangle n_y|^2 +$$

$$+|\langle A'|Q_{xy}|A, M\rangle n_x + \langle A'|Q_{yy}|A, M\rangle n_y|^2$$

$$-|\langle A'|Q_{xx}|A, M\rangle n_x^2 + 2\langle A'|Q_{xy}|A, M\rangle n_x\, n_y + \langle A'|Q_{yy}|A, M\rangle n_y^2|^2\Big] \tag{15}$$

Making use of the matrix elements (14) and writing out explicitly the angular dependence of **n**, we finally obtain for both cases

$$\frac{d\Gamma}{d\Omega} = \frac{\alpha}{16\pi}\,\omega^5 Q^2(\sin^2\theta - \frac{1}{2}\sin^4\theta) \tag{16}$$

Problem 13.   The width of the 2P level of hydrogen

a)   Calculate the mean life of the 2P level in the hydrogen atom. The interaction responsible for the decay is

$$H_I = e\,j_\mu(x)\,A^\mu(x)$$

Here, A_μ is the electromagnetic field and j_μ the current operator.
b)   Describe photon scattering off the ground state of the atom, in the vicinity of the 2P resonance.
c)   Describe the absorption of monochromatic photons by hydrogen at temperature T and pressure P, near the resonance.

Solution

a)   In the Coulomb gauge ($A_0 = 0$, $\nabla \cdot \mathbf{A} = 0$), the matrix element of H_I is written as follows

$$\langle f|e^{iH_0 t}e\,\mathbf{j}\mathbf{A}e^{-iH_0 t}|i\rangle = \langle f|e\,\mathbf{v}\mathbf{A}(x)|i\rangle \tag{1}$$

where $|f\rangle = |A^*\rangle|\gamma\rangle$ and $|i\rangle = |A\rangle$. Using the expression

$$\mathbf{A}(x) = \int \left\{ \boldsymbol{\varepsilon}_{(\lambda)}(\mathbf{k}) e^{-ikx} a_{(\lambda)}(\mathbf{k}) + \text{h.c.} \right\} d\Omega_{\mathbf{k}} \tag{2}$$

we obtain

$$\mathcal{M}_{fi}^{(\lambda)} = \langle A | e\, \mathbf{v} \cdot \boldsymbol{\varepsilon}_{(\lambda)}^*(\mathbf{k})\, e^{-i\mathbf{kx}} | A^* \rangle \tag{3}$$

In the dipole approximation the **x**-dependence of the exponential may be ignored. Further, the relation

$$\mathbf{v} = i[H, \mathbf{r}] \tag{4}$$

leads to

$$\begin{aligned} \mathcal{M}_{fi}^{(\lambda)} &= i\, e\, \boldsymbol{\varepsilon}_{(\lambda)}^* \langle A^* | [H, \mathbf{r}] | A \rangle = i\, e\, \boldsymbol{\varepsilon}_{(\lambda)}^* \langle A^* | \mathbf{r} | A \rangle \,\omega_0 \\ \omega_0 &\equiv E_{A^*} - E_A \end{aligned} \tag{5}$$

Thus the decay width takes the form

$$\Gamma(2p \to 1s) = \int \frac{\omega_0}{8\pi^2} |\sum_{(\lambda)} \mathcal{M}_{fi}^{(\lambda)}|^2 \, d\Omega = \frac{4}{3}\alpha\,\omega_0^3 |\langle A^* | \mathbf{r} | A \rangle|^2 \tag{6}$$

Rotational invariance implies that this width does not depend on the angular momentum projection of the 2P state.

The wave functions of the $1S$ and $2P$ state are

$$\begin{aligned} \langle \mathbf{x} | A \rangle &= \frac{2}{\sqrt{a^3}} Y_0^0(\theta, \varphi)\, e^{-r/a} \\ \langle \mathbf{x} | A^* \rangle &= \frac{1}{2\sqrt{6a^3}} Y_1^M(\theta, \varphi)\, \frac{r}{a}\, e^{-r/2a} \end{aligned} \tag{7}$$

where a is the Bohr radius

$$a = 1/m\alpha \tag{8}$$

Carrying out the calculation of the matrix element, we find

$$|\langle A | \mathbf{r} | A^*, M = 0 \rangle|^2 = \frac{2^{15}}{3^{10}} a^2 \tag{9}$$

Since the mean life is independent of M, we have arbitrarily chosen $M = 0$ above. Eqs. (9) and (6) give

$$\Gamma = \frac{2^{17}}{3^{11}} a^2 \alpha \omega_0^3 \tag{10}$$

From the expression for the energy levels

$$E_n = -\frac{1}{2n^2} m\,\alpha^2 \tag{11}$$

we find

$$E_2 - E_1 \equiv \omega_0 = \frac{3}{8}m\,\alpha^2 \tag{12}$$

so that finally

$$\Gamma = \left(\frac{2}{3}\right)^8 m\,\alpha^4 \tag{13}$$

b)   In order to obtain the resonant cross section one must calculate the diagrams

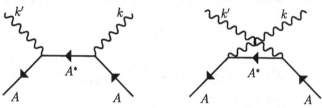

fig. 1

We write

$$d\sigma_{el} = \frac{1}{16\pi^2}|H_{eff}|^2\,d\Omega \tag{14}$$

with

$$H_{eff} = H_I \frac{1}{E - H_0 - PRP} H_I \tag{15}$$

P is a projector onto the state $|A^*\rangle$, while PRP gives the radiative shift and the width due to the decay of $|A^*\rangle$. In the dipole approximation the resonant term in the amplitude equals

$$\langle A|H_{eff}|A\rangle = -e^2\omega^2\varepsilon_i(\mathbf{k})\varepsilon_j'^*(\mathbf{k}')\sum_M \frac{\langle A|r_j|A^*,M\rangle\langle A^*,M|r_i|A\rangle}{\omega + E_A - E_{A^*} + i\Gamma_{A^*}/2} \tag{16}$$

By rotational invariance

$$\sum_M \langle A|r_j|A^*,M\rangle\langle A^*,M|r_i|A\rangle = \sum_M \frac{1}{3}\delta_{ij}|\langle A|\mathbf{r}|A^*,M\rangle|^2 = \delta_{ij}|\langle A|\mathbf{r}|A^*\rangle|^2 \tag{17}$$

The last equality is again a consequence of rotational invariance and of the fact that M can take 3 values in a P state. M-independence allows us to drop M in the final expression. Eq.(16) now reads

$$\langle A|H_{eff}|A\rangle = \frac{e^2\omega^2}{\omega_0 - \omega - i\Gamma/2}\,|\langle A|\mathbf{r}|A^*\rangle|^2\,\varepsilon'^*\varepsilon \tag{18}$$

and the cross section is

$$\frac{d\sigma_{el}}{d\Omega} = \frac{\alpha^2\omega^4}{(\omega - \omega_0)^2 + \Gamma^2/4}\left(|\langle A|\mathbf{r}|A^*\rangle|^2\right)^2|\varepsilon'^*\,\varepsilon|^2 \tag{19}$$

Summing over final polarizations and averaging over initial ones

$$\frac{1}{2}\sum \varepsilon_k \varepsilon_j^* = \frac{1}{2}(\delta_{jk} - n_j n_k) \qquad \sum \varepsilon_i'^* \varepsilon_\ell' = \delta_{i\ell} - n_i' n_\ell' \qquad (20)$$

and integrating over $d\Omega$, we obtain

$$\sigma = \frac{8\pi}{3}\alpha^2 \omega^4 \frac{\left(|\langle A|\mathbf{r}|A^*\rangle|^2\right)^2}{(\omega - \omega_0)^2 + \Gamma^2/4} \qquad (21)$$

A comparison with Eq.(6) gives

$$\sigma = \frac{3\pi}{2\omega^2}\frac{\Gamma^2}{\Delta\omega^2 + \Gamma^2/4} \qquad (22)$$

which is precisely the Breit-Wigner formula.

c) The mean free path of light in a gas is

$$d = \frac{1}{\sigma\rho} \qquad (23)$$

The gas density ρ (particles per unit volume) equals

$$\rho = \frac{P}{kT} \qquad (24)$$

Thus, the temperature and pressure dependence of the mean free path is

$$d = \frac{kT}{\sigma P} \qquad (25)$$

The velocity distribution in a gas widens spectral lines, as a result of the Doppler effect. Calling $\mathbf{v}$ the velocity of the atom and $\mathbf{k}$ the momentum of the incident photon, the "effective" frequency to first order in v/c is $\omega(\mathbf{v}) = \omega(1 - \frac{\mathbf{vk}}{\omega}) = \omega(1 - \mathbf{nv})$. Denoting by $P(\mathbf{v})$ the velocity distribution, normalized to unity,

$$\sigma_{eff} \simeq \frac{3\pi}{2\omega^2}\int d^3\mathbf{v}\, P(\mathbf{v})\frac{\Gamma^2}{(\omega(\mathbf{v}) - \omega_0)^2 + \Gamma^2/4} \qquad (26)$$

The width coming from the Doppler effect is expected to be

$$\Delta\omega = \omega\left\langle \frac{v}{c}\right\rangle \simeq \omega\sqrt{\frac{3KT}{4\,m_A c^2}} \qquad (27)$$

to be compared with the natural width Γ.

If collisions interrupt the absorption process, then their overall effect is to introduce in Eq.(22) an effective width of the order of $\Delta\Gamma$, where

$$\Delta\Gamma \simeq \frac{v}{d} \simeq \frac{v\sigma_{coll}P}{kT} \qquad (28)$$

σ_{coll} is the cross section for interatomic collisions: $\sigma_{coll} \simeq 10^{-16}\,\mathrm{cm}^2$. At normal temperature and pressure, $\Delta\Gamma/\Delta\omega \simeq 10^{-4}$; therefore, the Doppler effect dominates.

Problem 14.   Excitation of atomic levels by electron scattering

a)   An electron scatters off an atom, leaving it in an excited state

$$e^- + A \rightarrow e^- + A^* \tag{1}$$

Using the Born approximation, calculate the angular distribution and the total cross section for a dipole and a quadrupole excitation $A \rightarrow A^*$.

b)   Suppose A has $J = 0$, while A^* has $J = 2$ and the same parity as A; calculate, as a function of the scattering angle, the density matrix for the orientation of A^*, and the polarization of A^*.

c)   Calculate the cross section and the angular distribution for the case in which A is the ground state of the hydrogen atom ($n = 1$) and A^* is the $n = 3$ level, keeping in mind that this level has states with $J = 0, 1, 2$.

Solution

a)   Let us denote by $|\mathbf{k}, A\rangle$ the combined state of an electron with momentum $\mathbf{k}$ and an atom in state A. The scattering amplitude is, in the Born approximation,

$$f = -\frac{m}{2\pi} \langle \mathbf{k}', A^* | H_I | \mathbf{k}, A \rangle \tag{2}$$

In the nonrelativistic approximation, the electron-atom interaction is described by the Coulomb potential generated by the charge distribution $\rho(x)$ of the atom

$$H_I = -\alpha \int \mathrm{d}^3 x' \, \frac{\rho(\mathbf{x}')}{|\mathbf{x} - \mathbf{x}'|} \tag{3}$$

$\mathbf{x}$ is the coordinate of the incoming electron and $\rho(\mathbf{x})$ is, for an atom with atomic number Z,

$$\rho(\mathbf{x}) = Z \, \delta^{(3)}(\mathbf{x}) - \sum_{i=1}^{Z} \delta^3(\mathbf{x} - \mathbf{x}_i) \tag{4}$$

($\mathbf{x}_i$ is the coordinate of the ith atomic electron). Setting $\mathbf{k}' - \mathbf{k} = \mathbf{q}$, (2) becomes

$$f = \frac{2m\alpha}{q^2} \int e^{-i\mathbf{q}\mathbf{x}} \, \mathrm{d}^3 x \left\{ Z \langle A^* | A \rangle \delta^{(3)}(x) - \sum \langle A^* | \delta^{(3)}(\mathbf{x} - \mathbf{x}_i) | A \rangle \right\} \tag{5}$$

If a is a typical atomic dimension and $qa \ll 1$, we may truncate the exponential in (5)

$$e^{-i\mathbf{q}\mathbf{x}} \simeq 1 - i\,\mathbf{q}\mathbf{x} - \frac{1}{2} q^\alpha q^\beta \, x_\alpha x_\beta + \ldots \tag{6}$$

In terms of the matrix elements of the dipole ($\hat{D}_\alpha$) and quadrupole ($\hat{Q}_{\alpha\beta}$) operators

$$D_\alpha = \sum_{i=1}^{Z} \langle A^* | x_\alpha^i | A \rangle \equiv \langle A^* | \hat{D}_\alpha | A \rangle \tag{7}$$

$$Q_{\alpha\beta} = \sum_{i=1}^{Z} \langle A^* | x_\alpha^i x_\beta^i - \frac{1}{3} \mathbf{x}^{i\,2} \delta_{\alpha\beta} | A \rangle \equiv \langle A^* | \hat{Q}_{\alpha\beta} | A \rangle \tag{8}$$

we have

$$f = \frac{2m\alpha}{q^2} \left\{ \sum \frac{\mathbf{q}^2}{6} \langle A^* | \mathbf{x}^i \mathbf{x}^i | A \rangle + i\,q^\alpha D_\alpha + \frac{1}{2} q^\alpha q^\beta Q_{\alpha\beta} + \mathcal{O}[(qa)^3] \right\} \tag{9}$$

The term coming from the nuclear charge is cancelled by the first term in the expansion (6) for the electrons.

The first term in (9) is the electric form factor of the atom, to order q^2; it is absent if we assume, for simplicity, that the angular momenta of states $|A\rangle$ and $|A^*\rangle$ are different. The dipole and quadrupole terms excite states of opposite parity, and therefore there is no interference between them. Thus the differential cross section, averaged over initial spins and summed over final ones, is

$$\left(\frac{d\sigma}{d\Omega} \right)_{Dip} = \frac{k'}{k} \frac{4m^2\alpha^2}{(\mathbf{q}^2)^2} \frac{q^\alpha q^\beta}{2J+1} \sum_{MM'} \langle M' | \hat{D}_\alpha | M \rangle \langle M | \hat{D}_\beta | M' \rangle \tag{10}$$

$$\left(\frac{d\sigma}{d\Omega} \right)_{Quad} = \frac{k'}{k} \frac{m^2\alpha^2}{(\mathbf{q}^2)^2} q^\alpha q^\beta q^{\alpha'} q^{\beta'} \frac{1}{2J+1} \sum_{MM'} \langle M' | \hat{Q}_{\alpha\beta} | M \rangle \langle M | \hat{Q}_{\alpha'\beta'} | M' \rangle \tag{11}$$

M and M' are the eigenvalues of J_3 in the initial and final state, respectively. Rotational invariance now implies

$$q^\alpha q^\beta \sum_{MM'} \langle M' | \hat{D}_\alpha | M \rangle \langle M | \hat{D}_\beta | M' \rangle = \frac{q^\alpha q^\beta}{3} \delta_{\alpha\beta} D^2 = \frac{1}{3} \mathbf{q}^2 D^2$$

$$q^\alpha q^{\alpha'} q^\beta q^{\beta'} \sum_{MM'} \langle M' | \hat{Q}_{\alpha\beta} | M \rangle \langle M | \hat{Q}_{\alpha'\beta'} | M' \rangle = \tag{12}$$

$$\frac{Q^2}{10} (\delta_{\alpha\alpha'} \delta_{\beta\beta'} + \delta_{\alpha\beta'} \delta_{\alpha'\beta} - \frac{2}{3} \delta_{\alpha\beta} \delta_{\alpha'\beta'}) \, q^\alpha q^{\alpha'} \, q^\beta q^{\beta'} = \frac{2}{15} \mathbf{q}^4 Q^2 \tag{13}$$

D and Q are the reduced dipole and quadrupole matrix elements.

One may arrive at the same result using the Wigner-Eckart theorem and the properties of Clebsch-Gordan coefficients; the normalization in Eqs. (13) and (12) is compatible with the one used in connection with this theorem. Denoting indeed by T_ξ^n an irreducible tensor of rank n in q and component ξ, and by Q_ξ^n the corresponding operator, we write

$$f_n \simeq \sum_\xi T_\xi^{n*} \langle M' | Q_\xi^n | M \rangle$$

$$\overline{|f_n|^2} \;=\; \frac{1}{2J+1} \sum_{MM'\xi\eta} T_\xi^{n*} T_\eta^n \langle M'|Q_\xi^n|M\rangle \langle M|Q_\eta^n|M'\rangle^*$$

$$=\; \frac{1}{2J+1} \sum_\xi |T_\xi^n|^2 \sum_{MM'} |\langle M'|Q_\xi^n|M\rangle|^2 = \frac{1}{2J+1} \sum_\xi |T_\xi^n|^2 \frac{1}{2n+1}|Q^n|^2 \quad (14)$$

In particular for the dipole transition

$$\overline{|f_1|^2} \sim \frac{1}{2J+1} q^2 \frac{1}{3} D^2 \tag{15}$$

and for quadrupole transition

$$\overline{|f_2|^2} = \frac{1}{2J+1} \left(q_\alpha q_\beta - \frac{1}{3} q^2 \delta_{\alpha\beta}\right)\left(q^\alpha q^\beta - \frac{1}{3} q^2 \delta^{\alpha\beta}\right) \frac{1}{5} Q^2 = \frac{2q^4 Q^2}{15(2J+1)} \tag{16}$$

The final result for the cross sections is

$$\left(\frac{\mathrm{d}\sigma}{\mathrm{d}\Omega}\right)_{Dip} = \frac{k'}{k} \frac{4m^2}{3q^2} \alpha^2 \frac{D^2}{2J+1} \tag{17}$$

$$\left(\frac{\mathrm{d}\sigma}{\mathrm{d}\Omega}\right)_{Quad} = \frac{k'}{k} \frac{m^2 \alpha^2 Q^2}{2J+1} \cdot \frac{2}{15} \tag{18}$$

In terms of $|\Delta E|$, the difference in energy between states A and A^*, energy conservation gives

$$\frac{k'^2}{2m} = \frac{k^2}{2m} - |\Delta E| \tag{19}$$

where $\mathbf{q}^2 = k'^2 + k^2 - 2k\,k'\cos\theta$.

We now calculate the total cross section, noticing that $2qdq = 2kk'\,\mathrm{d}\cos\theta$; thus

$$\sigma = \frac{2\pi}{k^2} \int_{k-k'}^{k+k'} q\,\mathrm{d}q\,\overline{|f^2(q)|} \tag{20}$$

For fast electrons, $k^2/2m \gg \Delta E$, we may set

$$k' + k \simeq 2k \qquad k - k' \simeq \frac{\Delta E}{v} \tag{21}$$

where v is the velocity of the incoming electron. This leads to

$$\sigma_{Dip} = \frac{8\pi}{3} \frac{1}{2J+1} \frac{m^2 \alpha^2 D^2}{k^2} \ln\frac{2\,k\,v}{\Delta E} \tag{22}$$

$$\sigma_{Quad} = \frac{4\pi}{15} \frac{1}{2J+1} \frac{m^2 \alpha^2 Q^2}{k^2} \int_{\Delta E/v}^{2k} q\,\mathrm{d}q \simeq \frac{8\pi}{15} \frac{m^2 \alpha^2 Q^2}{2J+1} \tag{23}$$

The differential cross section for fast electrons at small angles is, by virtue of $q^2 \simeq (\Delta E/v)^2 + (k\,\theta)^2$,

$$\left(\frac{d\sigma}{d\Omega}\right)_{Dip} = \frac{4}{3}\frac{m^2\alpha^2 D^2}{2J+1}\frac{1}{k^2\theta^2 + (\Delta E/v)^2} \tag{24}$$

This distribution has a forward peak with angular width $\Delta E/vk \ll 1$. The quadrupole cross section, on the other hand, has a uniform angular distribution

$$\left(\frac{d\sigma}{d\Omega}\right)_{Quad} = \frac{2}{15}\frac{m^2\alpha^2 Q^2}{2J+1} \tag{25}$$

b) If the initial state has $J = 0$ and the excited state has $J = 2$, then only the quadrupole matrix element will contribute, by conservation of parity and angular momentum.

$$f_{M'M} \simeq \langle M'|T_\xi^{2*}(q)Q_\xi^2|M\rangle \tag{26}$$

Since $M = 0$, there follows that $\langle M'|Q_\xi^2|0\rangle \propto Q\,\delta_{\xi M'}$, and we find

$$f_{M'0} \propto T_{M'}^{2*}(q) \tag{27}$$

Thus the density matrix $\rho_{M_1 M_2} = (\sum_M f_{M_1 M}f_{M M_2}^*)/(\sum_{MM'} f_{M'M}f_{MM'}^*)$ of the final state takes the form

$$\rho_{M_1 M_2} = \frac{T_{M_1}^{2*}(q)T_{M_2}^2(q)}{\sum_M T_M^{2*}(q)T_M^2(q)} \tag{28}$$

Recalling that $T_M^{(2)}(q) \propto Y_M^{(2)}(\hat{q})$ $(\hat{q} = \mathbf{q}/|\mathbf{q}|)$, we conclude that

$$\rho_{M_1 M_2} = \frac{Y_{M_1}^{2*}(\hat{q})Y_{M_1}^2(\hat{q})}{\frac{5}{4\pi}} = \frac{4\pi}{5}Y_{M_1}^{2*}(\hat{q})\,Y_{M_2}^2(\hat{q}) \tag{29}$$

This corresponds to a pure state; one can immediately verify that $\rho^2 = \rho$.

The polarization is defined by

$$\mathbf{P} = \langle \mathbf{J}\rangle = \sum_{M_1 M_2} \langle M_1|\rho|M_2\rangle\langle M_2|\mathbf{J}|M_1\rangle \tag{30}$$

Choosing the z-axis along $\mathbf{q}$, the only nonzero element of ρ is ρ_{00}. The above equation then gives $\mathbf{P} = \rho_{00}\,\langle 0|\mathbf{J}|0\rangle = 0$: the polarization vanishes.

c) There are three distinct excitations from the $n = 1$ to the $n = 3$ level of the hydrogen atom: $1S \to 3S$, $1S \to 3P$, $1S \to 3D$. The corresponding amplitudes can be read off Eq.(9), using angular momentum conservation

$$f(1S \to 3S) = \frac{m\alpha}{3}\langle 3S|r^2|1S\rangle \tag{31}$$

$$f(1S \to 3P) = \frac{2i\,m\alpha}{q^2}q^\alpha\langle 3P|x_\alpha|1S\rangle \tag{32}$$

$$f(1S \to 3D) = \frac{m\alpha}{q^2}q^\alpha q^\beta\langle 3D|x_\alpha x_\beta - \frac{x^2}{3}\delta_{\alpha\beta}|1S\rangle \tag{33}$$

Rotational invariance gives

$$\sum_m \langle 3P, m|x_\alpha|1S\rangle \langle 1S|x_\beta|3P, m\rangle = \frac{1}{3}\delta_{\alpha\beta}\big|\langle 3P|r|1S\rangle\big|^2 \tag{34}$$

$$\tag{35}$$

$$\sum_m \langle 3D, m|x_\alpha x_\beta - \frac{x^2}{3}\delta_{\alpha\beta}|1S\rangle \langle 1S|x_{\alpha'}x_{\beta'} - \frac{x^2}{3}\delta_{\alpha'\beta'}|3D, m\rangle$$

$$= \frac{1}{15}\big|\langle 3D|r^2|1S\rangle\big|^2 \left(\delta_{\alpha\alpha'}\delta_{\beta\beta'} + \delta_{\alpha\beta'}\delta_{\beta\alpha'} - \frac{2}{3}\delta_{\alpha\beta}\delta_{\alpha'\beta'}\right) \tag{36}$$

where the matrix elements or r are intended between normalized radial wave functions. The corresponding differential cross sections are

$$\left(\frac{d\sigma}{d\Omega}\right)_{1S\to 3S} = \frac{k'}{k}\frac{1}{9}m^2\alpha^2\big|\langle 3S|r^2|1S\rangle\big|^2 \tag{37}$$

$$\left(\frac{d\sigma}{d\Omega}\right)_{1S\to 3P} = \frac{k'}{k}\frac{4}{3}\frac{m^2\alpha^2}{q^2}\big|\langle 3P|r|1S\rangle\big|^2 \tag{38}$$

$$\left(\frac{d\sigma}{d\Omega}\right)_{1S\to 3D} = \frac{k'}{k}\frac{4}{45}m^2\alpha^2\big|\langle 3D|r^2|1S\rangle\big|^2 \tag{39}$$

The explicit form of the wave functions

$$R_{1S} = \frac{2}{\sqrt{a^3}}e^{-r/a} \;;\; R_{3S} = \frac{2}{3\sqrt{3}}\frac{1}{\sqrt{a^3}}e^{-r/3a}\left(1 - \frac{2}{3}\frac{r}{a} + \frac{2}{27}\frac{r^2}{a^2}\right)$$

$$R_{3P} = \frac{8}{27\sqrt{6}}\frac{1}{\sqrt{a^3}}e^{-r/3a}\frac{r}{a}\left(1 - \frac{r}{6a}\right) \;;\; R_{3D} = \frac{4}{81\sqrt{30}}\frac{1}{\sqrt{a^3}}e^{-r/3a}\frac{r^2}{a^2} \tag{40}$$

$(a = 1/m\alpha)$, leads to

$$\langle 3S|r^2|1S\rangle = -\left(\frac{3}{4}\right)^5\frac{8}{\sqrt{3}}a^2$$

$$\langle 3P|r|1S\rangle = \left(\frac{3}{4}\right)^5\frac{16}{3\sqrt{6}}a$$

$$\langle 3D|r^2|1S\rangle = \left(\frac{3}{4}\right)^5\frac{40}{\sqrt{30}}a^2 \tag{41}$$

We make use of $d\Omega = \frac{2\pi q\, dq}{k\, k'}$ to obtain

$$\frac{d\sigma}{dq} = \frac{2\pi q\, m^2\alpha^2}{k^2}\left(\frac{3}{4}\right)^{10}a^4\left[\frac{64}{27} + \frac{512}{81(qa)^2} + \frac{128}{27}\right] \tag{42}$$

Finally, an integral over q results in the total cross section

$$\sigma = 2\pi\left(\frac{3}{4}\right)^{10}a^2\left[\frac{128}{27} + \frac{1}{(ka)^2}\frac{512}{81}\ln\frac{16kv}{3m\alpha^2} + \frac{256}{27}\right] \tag{43}$$

For $k^2/2m \simeq 100\,\text{eV}$, we have $ka \simeq 2.7$ and therefore $\sigma_{Dip}/\sigma_{Quad} \simeq 0.34$.

Problem 15.   Ionization of hydrogen by X-rays

a)   A hydrogen atom is ionized by X-rays with energy $E \sim 1$ KeV. Calculate the amplitude for this process in the Born approximation.

b)   Calculate the cross section as a function of energy.

c)   How does this angular distribution of the electron depend on the polarization (linear or circular) of the incoming photon?

Solution

a)   The electron-photon interaction is described by the Hamiltonian

$$H = \frac{1}{2m_e}\left(\mathbf{p} - e\mathbf{A}\right)^2 \tag{1}$$

We choose to work in the radiation gauge: $A_0 = 0$, $\boldsymbol{\nabla} \cdot \mathbf{A} = 0$. In terms of the photon and electron momenta, $\mathbf{k}$ and $\mathbf{p}$, the cross section is, in the Born approximation,

$$d\sigma = \frac{1}{2\omega}\frac{d^3\mathbf{p}}{(2\pi)^3}\,|\langle \mathbf{p}\,|H_I|\text{Atom} + \gamma\rangle|^2\,2\pi\,\delta(E_A + \omega - \frac{p^2}{2m_e}) \tag{2}$$

where $\omega = |\mathbf{k}|$ and $E_A = -I = -m_e\alpha^2/2$ is the initial energy level. We may neglect nuclear recoil. The amplitude for the process is

$$T = \langle 0|\langle \mathbf{p}\,|H_I|i\rangle|\gamma\rangle \tag{3}$$

Here, $|0\rangle$ is the photon vacuum, $|\gamma\rangle$ is the state of the incoming photon, $|i\rangle$ and $|\mathbf{p}\rangle$ are the initial and final electron states. With our gauge choice the interaction Hamiltonian is

$$H_I = -\frac{e}{m_e}\mathbf{A}\cdot\mathbf{p} \tag{4}$$

Denoting by $\boldsymbol{\varepsilon}$ the polarization of the incoming photon

$$T = -\langle \mathbf{p}|\frac{e}{m_e}e^{i\mathbf{k}\mathbf{x}}\,\mathbf{p}\,|i\rangle\cdot\boldsymbol{\varepsilon} \tag{5}$$

For 1 KeV X-rays one may check that $ka \ll 1$, where $a = 1/m_e\alpha$ is the atomic Bohr radius; this allows us to use the dipole approximation ($e^{i\mathbf{k}\mathbf{x}} \simeq 1$). In the coordinate representation the wave function in the initial state is

$$\langle \mathbf{r}|i\rangle = \frac{1}{\sqrt{\pi a^3}}\,e^{-r/a} \tag{6}$$

This leads to the following expression for the amplitude

$$T = -\frac{e}{m_e}\left(\mathbf{p}\boldsymbol{\varepsilon}\right)\int d^3\mathbf{x}\,e^{-i\mathbf{p}\mathbf{x}}\frac{e^{-r/a}}{\sqrt{\pi a^3}} = -8\sqrt{\pi a^3}\,\frac{e}{m_e}\frac{\mathbf{p}\boldsymbol{\varepsilon}}{\left[1 + (|\mathbf{p}|\,a)^2\right]^2} \tag{7}$$

b) Eq.(2) gives for the cross section

$$d\sigma = \frac{m_e \, |\mathbf{p}|}{8\omega \, \pi^2} \, d\Omega \, |T|^2 \tag{8}$$

Carrying further this calculation, one finds

$$d\sigma = 2^6 a^2 \alpha \, d\Omega \, (\mathbf{n}\varepsilon)^2 \left[\left(1 - \frac{I}{\omega}\right)^{\frac{3}{2}} \left(\frac{I}{\omega}\right)^{\frac{7}{2}} \right] \tag{9}$$

with $\mathbf{n} = \mathbf{p}/|\mathbf{p}|$. The ionization potential $I = m_e \alpha^2/2$ obeys the relation

$$\frac{\mathbf{p}^2}{2m_e} + I = \omega \tag{10}$$

The total cross section for unpolarized photons is

$$\bar{\sigma} = \frac{2^8}{3} \pi \, \alpha \, a^2 \left[\left(1 - \frac{I}{\omega}\right)^{\frac{3}{2}} \left(\frac{I}{\omega}\right)^{\frac{7}{2}} \right] \tag{11}$$

In terms of $\bar{\sigma}$, Eq.(9) can be written as follows

$$\frac{d\sigma}{d\Omega} = \bar{\sigma} \, \frac{3}{4\pi} \, (\mathbf{n}\varepsilon)^2 \tag{12}$$

c) Eq.(12) exhibits the dependence of the angular distribution on the polarization. We will write out this dependence explicitly for the two cases of linear and circular polarization.

Let the incoming photon travel along the z-axis, and let the direction of the outgoing electron be

$$\mathbf{n} = (\sin\theta \, \cos\phi, \sin\theta \, \sin\phi, \cos\theta) \tag{13}$$

The vectors corresponding to linear (L) and circular (C) polarization are

$$\varepsilon_L = (1, 0, 0) \qquad \varepsilon_C = \frac{1}{\sqrt{2}}(1, i, 0) \tag{14}$$

For the two cases we obtain

$$\frac{1}{\bar{\sigma}} \frac{d\sigma_L}{d\Omega} = \frac{3}{4\pi} \sin^2\theta \, \cos^2\phi$$

$$\frac{1}{\bar{\sigma}} \frac{d\sigma_C}{d\Omega} = \frac{3}{8\pi} \sin^2\theta \tag{15}$$

Problem 16.   Atomic scattering of μ at low momentum transfer

a) Write down the electrostatic potential generated by an atom (with infinite mass), in terms of the electronic charge distribution $\rho_e(r)$ in the ground state.

b) Express the amplitude for scattering of relativistic muons in this potential, in the Born approximation; expand this amplitude in powers of v/c up to the lowest order term describing $\mathbf{L} \cdot \mathbf{S}$ coupling.

c) Determine the probability of helicity flip for a forward-polarized μ, as a function of the scattering angle.

d) In the approximation suggested above, a rigid charge distribution has been assumed, neglecting transitions to excited atomic states; discuss the range of validity of this approximation.

Solution

a) The Poisson equation

$$\Delta A_0 = -\rho(r) \tag{1}$$

leads to

$$A_0(\mathbf{q}) = \frac{1}{\mathbf{q}^2}\,\tilde{\rho}(\mathbf{q}) \tag{2}$$

where $\tilde{\rho}(\mathbf{q})$ is the Fourier transform of the charge density.

b) The scattering amplitude is given by

$$\mathcal{M} = e\,A_0(\mathbf{q})\,\bar{u}(\mathbf{p}')\gamma^0 u(\mathbf{p}) \tag{3}$$

where $\mathbf{p}$ and $\mathbf{p}'$ are the initial and final muon momenta and, for elastic scattering, $|\mathbf{p}| = |\mathbf{p}'|$. We express $u(\mathbf{p})$ in terms of the two-component Pauli spinor w

$$u(\mathbf{p}) = \begin{pmatrix} \sqrt{E+m}\,\mathrm{w} \\ \sqrt{E-m}\,\mathbf{n}\cdot\boldsymbol{\sigma}\,\mathrm{w} \end{pmatrix} \tag{4}$$

with $\mathbf{n} = \mathbf{p}/|\mathbf{p}|$. A straightforward expansion in powers of $|\mathbf{p}|/m$ gives

$$\bar{u}(\mathbf{p}')\gamma^0 u(\mathbf{p}) \simeq 2m\,\mathrm{w}^\dagger \left[1 + \frac{|\mathbf{p}|^2}{8m^2} + \frac{(\mathbf{p}\mathbf{p}')}{4m^2} + \frac{\mathbf{p}'^2}{8m^2} - \mathrm{i}\,\frac{(\mathbf{p}\wedge\mathbf{p}')\cdot\boldsymbol{\sigma}}{4m^2} \right]\mathrm{w} \tag{5}$$

That the last term describes $\mathbf{L} \cdot \mathbf{S}$ coupling can be seen as follows

$$-\mathrm{i}\,\frac{(\mathbf{p}\wedge\mathbf{p}')\cdot\boldsymbol{\sigma}}{4m^2}\,A_0 = -\frac{(\mathbf{p}\wedge\boldsymbol{\nabla}A^0)}{4m^2}\cdot\boldsymbol{\sigma} = \frac{1}{r}\frac{\mathrm{d}A_0}{\mathrm{d}r}\frac{\mathbf{L}\cdot\mathbf{S}}{4m^2} \tag{6}$$

c) Let the scattering plane be given by $\varphi = 0$; the scattering angle is denoted θ. The helicity eigenstates for incoming and outgoing particles are

$$w_+ = \begin{pmatrix} 1 \\ 0 \end{pmatrix} \qquad w_- = \begin{pmatrix} 0 \\ 1 \end{pmatrix}$$
$$w'_+ = \begin{pmatrix} \cos\theta/2 \\ \sin\theta/2 \end{pmatrix} \qquad w'_- = \begin{pmatrix} -\sin\theta/2 \\ \cos\theta/2 \end{pmatrix} \tag{7}$$

In terms of the helicity λ, the amplitude takes the form

$$\mathcal{M}_{\lambda'\lambda} = eA_0 \left[(E+m)w'^\dagger_{\lambda'}w_\lambda + \lambda\lambda'(E-m)w'^\dagger_{\lambda'}w_\lambda \right]$$
$$= eA_0\, w'^\dagger_{\lambda'}w_\lambda \left[E(1+\lambda\lambda') + m(1-\lambda\lambda') \right] \tag{8}$$

In the above we have used the equality $\mathbf{n}\cdot\boldsymbol{\sigma}\, w_\lambda = \lambda w_\lambda$. For $E \gg m$ one finds that $\mathcal{M}_{\lambda'\lambda} \sim \delta_{\lambda'\lambda}$, that is, there is no helicity-flip. In general, we have

$$\mathcal{M}_{++} = \mathcal{M}_{--} = 2e\,A_0\,E\cos\frac{\theta}{2}$$
$$\mathcal{M}_{+-} = -\mathcal{M}_{-+} = 2e\,A_0\,m\sin\frac{\theta}{2} \tag{9}$$

This implies

$$\frac{\sigma_{+-}}{\sigma_{++}} = \frac{m^2}{E^2}\,\mathrm{tg}^2\frac{\theta}{2} \tag{10}$$

For $\theta = 0$ one clearly finds $\sigma_{+-} = 0$, by angular momentum conservation along the direction of the μ-beam.

d) Excitations will be suppressed when the excitation frequency ω is large with respect to the frequencies of the electric field produced by μ travelling past the atom; that is, when $\omega > v/r_A$, or $v < \alpha c$. Thus the velocity of muons must be small with respect to that of atomic electrons.

Problem 17. The Frank and Hertz resonances

The experiment of Frank and Hertz involves electrons striking an atom at increasing energies. When the electron energy becomes equal to the energy gap between the ground state and an excited state of the atom, there is a sharp increase in the inelastic cross section.
a) Calculate, in the Born approximation, the amplitude for the process

$$e + A \to e + A^*$$

neglecting atomic recoil; assume that the interaction is due to the Coulomb field generated by the electron on the atom.
b) Calculate the differential and total cross section on a hydrogen atom for energies past the threshold of the second excited level.

c) Give an estimate of the atomic quadrupole contribution; discuss the feasibility of determining the sign of the quadrupole moment from the angular distribution and from the total cross section.

d) Below the threshold for excited states the interaction is of the Van der Waals (polarizability) type

$$V(r) = \varepsilon \left[\left(\frac{r_B}{r} \right)^8 - \left(\frac{r_B}{r} \right)^4 \right] \qquad \varepsilon > 0 \tag{1}$$

Explain the minus sign in front of the fourth power. Explain why at large distances the interaction exhibits a $1/r^4$ behaviour, and determine ε from the structure of the hydrogen atom, knowing that $r_B \simeq 5 \cdot 10^{-9}$ cm.

Solution

a) The electron-atom interaction is

$$V(r) = -\frac{Z\alpha}{r} + \alpha \sum_{i=1}^{Z} \frac{1}{|\mathbf{r} - \mathbf{r}_i|} \tag{2}$$

We denote by $|f\rangle$ and $|e\rangle$ the fundamental and excited states of the atom. In the Born approximation, the matrix element of the scattering operator is

$$\mathcal{M} = \langle e, \mathbf{k}'|V|\mathbf{k}, f\rangle \tag{3}$$

The term coming from the nucleus, being independent of the atomic variables, is diagonal and will not contribute when $|e\rangle \neq |f\rangle$.

Setting $\mathbf{q} = \mathbf{k}' - \mathbf{k}$, we find

$$\langle k'|\frac{1}{|\mathbf{r} - \mathbf{r}_i|}|k\rangle = \int d^3 r \, e^{i\mathbf{q}\mathbf{r}} \frac{1}{|\mathbf{r} - \mathbf{r}_i|} = e^{i\mathbf{q}\mathbf{r}_i} \frac{4\pi}{q^2} \tag{4}$$

Thus the amplitude is

$$M = \frac{4\pi\alpha}{q^2} \langle e|\sum_{i=1}^{Z} e^{i\mathbf{q}\mathbf{r}_i}|f\rangle \tag{5}$$

b) The amplitude for exciting the hydrogen atom to its nth state is

$$M_n = \frac{4\pi\alpha}{q^2} \int d^3 r \, \psi_n^*(\mathbf{r}) \, \psi_{1,s}(\mathbf{r}) \, e^{i\mathbf{q}\mathbf{r}} \tag{6}$$

Up to the second excited state, the hydrogen wave functions are

$$\psi_{1S}(\mathbf{r}) = Y_0^0(\hat{\mathbf{r}}) \frac{2}{(r_B)^{3/2}} e^{-r/r_B}$$

$$\psi_{2S}(\mathbf{r}) = Y_0^0(\hat{\mathbf{r}}) \frac{1}{\sqrt{2}(r_B)^{3/2}} \left(1 - \frac{r}{2r_B} \right) e^{-r/2r_B}$$

$$\psi_{2P,\,\ell_z}(\mathbf{r}) \;=\; Y_1^{\ell_z}(\hat{\mathbf{r}})\,\frac{1}{2\sqrt{6}}\,\frac{r}{(r_B)^{5/2}}\,\mathrm{e}^{-r/2r_B}$$

$$\psi_{3S}(\mathbf{r}) \;=\; Y_0^0(\hat{\mathbf{r}})\,\frac{2}{3\sqrt{3}\,(r_B)^{3/2}}\,\mathrm{e}^{-r/3r_B}\left[1 - \frac{2r}{3r_B} + \frac{2}{3}\left(\frac{r}{3r_B}\right)^2\right]$$

$$\psi_{3P,\,\ell_z}(\mathbf{r}) \;=\; Y_1^{\ell_z}(\hat{\mathbf{r}})\,\frac{8}{27\sqrt{6}}\,\frac{r}{(r_B)^{5/2}}\,\mathrm{e}^{-r/3r_B}\left(1 - \frac{r}{6r_B}\right)$$

$$\psi_{3D,\,\ell_z}(\mathbf{r}) \;=\; Y_2^{\ell_z}(\hat{\mathbf{r}})\,\frac{4}{81\sqrt{30}}\,\frac{r^2}{(r_B)^{7/2}}\,\mathrm{e}^{-r/3r_B} \tag{7}$$

The corresponding integrals have the form

$$\int \mathrm{d}^3\mathbf{r}\, Y_\ell^{\ell_z *}(\hat{\mathbf{r}})\, Y_0^0(\hat{\mathbf{r}})\, R_{n,\ell}(r)\, R_{1,0}(r)\, \mathrm{e}^{\mathbf{i}\mathbf{q}\mathbf{r}} \tag{8}$$

Expanding the exponential in spherical harmonics

$$\mathrm{e}^{\mathbf{i}\mathbf{q}\mathbf{r}} = \sum_{\ell m} \mathrm{i}^\ell\, Y_\ell^{m*}(\hat{\mathbf{q}})\, Y_\ell^m(\hat{\mathbf{r}})\, j_\ell(qr) \tag{9}$$

and setting $R_{n,\ell} = r_B^{-3/2} f_{n,\ell}(r/r_B)$, the integral in (8) becomes

$$\frac{\mathrm{i}^\ell\, Y_\ell^{\ell_z *}(\hat{\mathbf{q}})}{\sqrt{4\pi}} \int_0^\infty \mathrm{d}x\, x^2 f_{1,0}(x)\, f_{n,\ell}(x)\, j_\ell(qr_B x) \tag{10}$$

Define $I_{n,\ell}$ by

$$I_{n,\ell} = \int_0^\infty \mathrm{d}x\, x^2 f_{1,0}(x)\, f_{n,\ell}(x)\, j_\ell(qr_B x) \tag{11}$$

For the different states considered we find $(qr_B \ll 1)$

$$\begin{aligned}
I_{n,0} &= \int_0^\infty x^2\, \mathrm{d}x\, f_{1,0}\, f_{n,0}\left[1 + \frac{x^2(qr_B)^2}{6}\right] \\
&= \delta_{1,n} + \frac{(qr_B)^2}{6}\int_0^\infty x^4 f_{1,0}(x)\, f_{n,0}(x)\, \mathrm{d}x \tag{12a} \\
I_{n,1} &= \frac{qr_B}{3}\int_0^\infty x^3 f_{1,0}(x)\, f_{n,1}(x)\, \mathrm{d}x \tag{12b} \\
I_{n2} &= \frac{(q\,r_B)^2}{15}\int_0^\infty x^4 f_{1,0}(x) f_{n,2}(x)\, \mathrm{d}x \tag{12c}
\end{aligned}$$

The three transitions above, in the given order, correspond to $l = 0,\, 1,\, 2$ (monopole, dipole, quadrupole).

The exponentials inside $I_{n,l}$ provide a cutoff for $x \simeq 1$; this has allowed us to use the approximation, valid for $qr_B \ll 1$,

$$j_\ell(qr_B x) \simeq \frac{(qr_B)^\ell}{(2\ell+1)!!}\, x^\ell \tag{13}$$

Thus $I_{n,l}$ reduces to a multipole matrix element. Transitions to S states are monopole transitions, and cannot be produced by photons. P and D states are reached by dipole and quadrupole transitions. The sign of $I_{n,2}$ is the sign of the quadrupole.

The scattering amplitude for transitions to a state $|n, l\rangle$ is

$$\mathcal{M}_{n,\ell,\ell_z} = i^\ell Y_\ell^{\ell_z*}(\hat{\mathbf{q}}) I_{n,\ell} \frac{4\pi\alpha}{q^2} \tag{14}$$

The cross section is

$$\frac{d\sigma}{d\Omega} = \sum_{n,\ell,\ell_z} \frac{d\sigma(n,\ell,\ell_z)}{d\Omega} = \sum_{n,\ell,\ell_z} \frac{m_e^2 v_n'}{4\pi^2 v} |\mathcal{M}_{n,\ell,\ell_z}|^2 \tag{15}$$

where v and v_n' are the electron velocities before and after the collision.

Summing over the final states of the atom we obtain

$$\begin{aligned} \frac{d\sigma}{d\Omega} &= 4m_e^2 \left(\frac{\alpha}{q^2}\right)^2 \sum_{n,\ell\ell_z} \frac{v_n'}{v} |I_{n,\ell}(qr_B)|^2 Y_\ell^{\ell_z*}(\hat{\mathbf{q}}) Y_\ell^{\ell_z}(\hat{\mathbf{q}}) \\ &= 4m_e^2 \left(\frac{\alpha}{q^2}\right)^2 \sum_{n,\ell} \frac{2\ell+1}{4\pi} \frac{v_n'}{v} |I_{n\,\ell}(qr_B)|^2 \end{aligned} \tag{16}$$

The total cross section is

$$\sigma = \int d\Omega \frac{d\sigma}{d\Omega}(q) \tag{17}$$

c) The quadrupole contribution to the transition $1S \to 3D$ is

$$\left(\frac{d\sigma}{d\Omega}\right)_Q = \frac{5m_e^2}{\pi} (qr_B)^4 \left(\frac{\alpha}{q^2}\right)^2 \frac{v_3'}{v} = \frac{5m_e^2}{\pi} \left(r_B^2\alpha\right)^2 \frac{v_3'}{v} \tag{18}$$

It is not possible to determine the quadrupole sign, since only its square appears in the expression for σ.

d) Let us derive the elastic electron-atom interaction at large distances. The first nonzero contribution in perturbation theory comes from the second order; assuming both particles at rest (adiabatic approximation), and taking $\langle n|n\rangle = 1$,

$$\tilde{V}(r) = \sum_n \frac{\langle 0|V|n\rangle\langle n|V|0\rangle}{E_0 - E_n} \tag{19}$$

where n runs over atomic states. $\tilde{V}(r)$ is negative, since $E_0 < E_n$ and the numerator is a square modulus. At large distances $\mathbf{r}$ between the electron and the nucleus, we have

$$V \simeq \alpha \left(\sum_i \frac{\mathbf{r}_i \mathbf{r}}{r^3} + \dots\right) = \frac{e}{4\pi} \frac{d\mathbf{r}}{r^3} + \dots \tag{20}$$

$\mathbf{r}_i$ are the coordinates of the ith electron with respect to the nucleus. At large distances, V goes like $1/r^2$; thus, equation (19) gives $\tilde{V}(r) \simeq -C/r^4$, where C is

$$C = \alpha^2 \sum \frac{\langle 0| \sum \mathbf{r}_i \hat{\mathbf{r}} |n\rangle \langle n| \sum \mathbf{r}_i \hat{\mathbf{r}} |0\rangle}{|E_0 - E_n|} \tag{21}$$

For each value of E_n the operator $\sum_n |n\rangle\langle n|$ has spherical symmetry; therefore, only the spherically symmetric part of the tensor $\hat{r}^\alpha \hat{r}^\beta$ contributes to (21), giving

$$C = \frac{\alpha^2}{3} \sum_n \frac{|\langle 0| \sum \mathbf{r}_i |n\rangle|^2}{E_n - E_0} = \sum_n \frac{\alpha}{12\pi} \frac{|\langle 0|\mathbf{d}|n\rangle|^2}{E_n - E_0} \tag{22}$$

An order-of-magnitude estimate for the parameter ε in Eq.(1) is then

$$\varepsilon = \frac{C}{r_B^4} \simeq \frac{\alpha^2}{3 r_B^4} \frac{r_B^2}{m\alpha^2} = \frac{1}{3} \frac{\alpha}{r_B} \simeq 9\,\mathrm{eV} \tag{23}$$

For the hydrogen atom, the sum (21) can be evaluated exactly, leading to

$$C = \frac{9}{4} \alpha r_B^3 \tag{24}$$

Thus, in this case, ε equals

$$\varepsilon = \frac{9}{4} \frac{\alpha}{r_B} \simeq 62\,\mathrm{eV} \tag{25}$$

Problem 18. $p\,\bar{p}$ annihilation at low energies

$p\bar{p}$ scattering is highly inelastic (the majority of events are annihilations into pions); it is dominated by the S-wave up to antiproton momenta k of the order of 2 GeV/c in the laboratory frame.

a)   Parameterize the process by a complex scattering phase in the S-wave. In terms of this phase, describe the elastic angular distribution, the elastic and inelastic cross sections, and the total cross section.

b)   State the upper limit imposed by unitarity on the total cross section.

c)   Assuming that $\sigma_{\mathrm{tot}} \simeq c/(k^2 + a^2)$ ($a \simeq 100$ MeV/c), and using an unsubtracted dispersion relation for the forward scattering amplitude, calculate the real part of the elastic forward scattering amplitude.

d)   Study the interference with Coulomb scattering at small angles, as a function of energy.

Solution

a)   The elastic scattering amplitude and cross section are

$$
\begin{aligned}
f_{el} &= \frac{1}{2ik}\left(e^{i\delta - \eta} - 1\right) \qquad \eta \geq 0 \\
\sigma_{el} &= 4\pi |f_{el}|^2 = \frac{\pi}{k^2}\left[1 + e^{-2\eta} - 2e^{-\eta}\cos\delta\right]
\end{aligned}
\tag{1}
$$

By the optical theorem

$$\sigma_T = \frac{4\pi}{k}\operatorname{Im} f_{el} = \frac{4\pi}{k}\frac{\left(1-e^{-\eta}\cos\delta\right)}{2k} = \frac{2\pi}{k^2}\left(1-e^{-\eta}\cos\delta\right)$$

$$\sigma_I = \sigma_T - \sigma_{el} = \frac{2\pi}{k^2}\left(1-e^{-2\eta}\right) \tag{2}$$

b,c)   Eq.(2) gives the following limit for the cross section

$$\sigma_T \le \frac{2\pi}{k^2}\left(1+e^{-\eta}\right) \le \frac{4\pi}{k^2} \tag{3}$$

For $\sigma_T = c/(k^2+a^2)$, we have $\operatorname{Im} f_{el}(0) = k/4\pi\, c/(k^2+a^2)$, and the unsubtracted forward dispersion relation reads

$$f_{el}(E) = \frac{1}{\pi}\int_0^\infty \frac{\operatorname{Im} f(E')}{E'-E-i\varepsilon}\,dE' \tag{4}$$

or

$$\operatorname{Re} f_{el}(E) = \frac{1}{\pi}\int_0^\infty \frac{\operatorname{Im} f(E')}{E'-E}\,dE' \tag{5}$$

leading to

$$f_{el}(E) = \frac{1}{\pi}\int_0^\infty \frac{k'}{4\pi}\frac{c}{k'^2+a^2}\frac{1}{k'^2-k^2-i\varepsilon}\,dk'^2 \tag{6}$$

Integration yields

$$f_{el}(E) = \frac{c}{4\pi}\frac{1}{a-ik} \tag{7}$$

The relation $\sigma_T \le 4\pi/k^2$ implies $c \le 4\pi$.

d)   Typical Coulomb energies are of the order of $E_c = m\alpha^2/2 \simeq 27$ KeV. For $E \gg 27$ KeV we may use the Born approximation

$$f_c(\theta) = -\frac{\mu}{2\pi}\int V_{\text{Coul}}\,e^{-i\mathbf{q}\mathbf{r}}\,d^3\mathbf{r} = \frac{2\mu\alpha}{\mathbf{q}^2} = \frac{\mu\alpha}{2k^2}\frac{1}{\sin^2(\theta/2)} \tag{8}$$

where $\mathbf{q}$ is the momentum transfer and $\mu = m/2$ is the reduced mass.

The total amplitude thus becomes

$$f = \left[\frac{c}{4\pi}\frac{1}{a-ik} + \frac{\mu\alpha}{2k^2\sin^2(\theta/2)}\right] \tag{9}$$

leading to the differential cross section

$$\frac{d\sigma}{d\Omega} = \frac{c^2}{(4\pi)^2}\frac{1}{a^2+k^2}\left\{1 + \frac{1}{c}\frac{4\pi\alpha\,\mu a}{k^2\sin^2(\theta/2)} + \frac{4\pi^2\alpha^2}{c^2}\frac{\mu^2(a^2+k^2)}{k^4\sin^4(\theta/2)}\right\} \tag{10}$$

Problem 19. Partial waves in π - N scattering

$\pi - N$ scattering at low energies may be described in terms of the scattering phases $\delta^{\ell}_{2T,2J}$. The relevant phases, up to 300 MeV, are $\delta^{s}_{1,1}$, $\delta^{s}_{3,1}$, $\delta^{p}_{3,1}$, which are small, and $\delta^{p}_{3,3}$ which has a Breit-Wigner shape with mass $M \simeq 1232$ MeV and width $\Gamma \simeq 115$ MeV.
a)   In the above energy range calculate the total cross section, using the optical theorem.
b)   Calculate the angular distribution, averaged over initial spins and summed over final spins, for the processes

$$\pi^+ p \rightarrow \pi^+ p$$
$$\pi^+ n \rightarrow \pi^+ n$$
$$\pi^+ n \rightarrow \pi^0 p \tag{1}$$

c)   Calculate the asymmetry in $\pi^+ - p$ scattering, if the protons are polarized trasversely to the incident pion momentum.

Solution

a)   The description of scattering in terms of phases $\delta^{\ell}_{2T,2J}$ is made possible by virtue of the invariance under rotations, parity and isospin.

At low enough energies, only partial waves with $\ell = 0$, 1 will contribute. For $\ell = 0$ there is only one state, having $J = 1/2$, while for $\ell = 1$ there are states with $J = 1/2$ and $J = 3/2$. Viewed as a matrix in spin space, the amplitude has the general form

$$f = \sum_{\ell} (2\ell + 1) (a_{\ell} + b_{\ell}\, \boldsymbol{\ell} \cdot \mathbf{s}) P_{\ell}(\cos \theta) \tag{2}$$

$$\boldsymbol{\ell} \cdot \mathbf{s} = \ell_z s_z + \frac{1}{2}(\ell_+ s_- + \ell_- s_+) \tag{3}$$

The following relations

$$\ell_{\pm} P_{\ell}(\cos \theta) = \pm e^{\pm i\varphi} \frac{\partial}{\partial \theta} P_{\ell}(\cos \theta) = \mp e^{\pm i\varphi} \sin \theta \frac{\mathrm{d}}{\mathrm{d}\cos \theta} P_{\ell}(\cos \theta) \tag{4}$$

$$\ell_z P_{\ell} = 0 \quad ; \quad s_- e^{i\varphi} - s_+ e^{-i\varphi} = -2i\, \mathbf{s} \cdot \boldsymbol{\nu} \tag{5}$$

where

$$\boldsymbol{\nu} = \frac{\mathbf{n} \wedge \mathbf{n}'}{|\mathbf{n} \wedge \mathbf{n}'|} \tag{6}$$

is the unit normal to the scattering plane (x-z), allow us to write

$$f(\theta) = \sum_{\ell} (2\ell + 1) \left[P_{\ell}(\cos \theta) \cdot a_{\ell} + i\, b_{\ell}\, \boldsymbol{\nu} \cdot \mathbf{s}\, P^1_{\ell}(\cos \theta) \right] \tag{7}$$

$$P^1_{\ell}(\cos \theta) = -\frac{\mathrm{d}}{\mathrm{d}\theta} P_{\ell}(\cos \theta) \tag{8}$$

Using

$$\boldsymbol{\ell} \cdot \mathbf{s} = \frac{j^2 - \ell^2 - s^2}{2} = \begin{cases} \ell/2 & j = \ell + 1/2 \\ -(\ell + 1)/2 & j = \ell - 1/2 \end{cases} \tag{9}$$

we may express a_ℓ and b_ℓ in terms of the scattering phases

$$\begin{aligned}
a_\ell + \frac{\ell}{2} b_\ell &= \frac{1}{2ik} \left(e^{2i\delta_{2\ell+1}^\ell} - 1 \right) \\
a_\ell - \frac{\ell + 1}{2} b_\ell &= \frac{1}{2ik} \left(e^{2i\delta_{2\ell-1}^\ell} - 1 \right)
\end{aligned} \tag{10}$$

The amplitude now becomes

$$\begin{aligned}
f(\theta) &= A + 2B\boldsymbol{\nu} \cdot \mathbf{s} \\
A &= \frac{1}{2ik} \sum_\ell \left\{ (\ell + 1) \left(e^{2i\delta_{2\ell+1}^\ell} - 1 \right) + \ell \left(e^{2i\delta_{2\ell-1}^\ell} - 1 \right) \right\} P_\ell(\cos\theta) \\
B &= \frac{1}{2k} \sum_\ell \left\{ \left(e^{2i\delta_{2\ell+1}^\ell} - 1 \right) - \left(e^{2i\delta_{2\ell-1}^\ell} - 1 \right) \right\} P_\ell^1(\cos\theta)
\end{aligned} \tag{11}$$

So far, isospin indices were left aside; we now introduce them defining

$$f_{2T} = A_{2T} + B_{2T}\,\boldsymbol{\nu} \cdot \boldsymbol{\sigma} \tag{12}$$

Indicating by $\boldsymbol{\tau}$, $\mathbf{t}$ and $\mathbf{T}$ the nucleon, pion and total isospin, respectively, we write, in terms of Clebsch-Gordan coefficients,

$$f_{\tau_3', t_3'; \tau_3 t_3} = \sum_T C_{\tau_3 t_3 T_3}^{\frac{1}{2} 1 T} C_{\tau_3' t_3' T_3}^{\frac{1}{2} 1 T} f_{2T} \tag{13}$$

In particular,

$$\begin{aligned}
f_{\pi^+ p \to \pi^+ p} &= f_3 \\
f_{\pi^- p \to \pi^- p} &= \frac{1}{3} f_3 + \frac{2}{3} f_1 \\
f_{\pi^+ n \to \pi^+ n} &= \frac{1}{3} f_3 + \frac{2}{3} f_1 \\
f_{\pi^- p \to \pi^0 n} &= f_{\pi^+ n \to \pi^0 p} = \frac{\sqrt{2}}{3} (f_3 - f_1)
\end{aligned} \tag{14}$$

The optical theorem

$$\sigma_{tot} = \frac{4\pi}{k} \operatorname{Im} f(0) \tag{15}$$

then leads to

$$\begin{aligned}
\sigma_{tot}(\pi^+ p) &= \frac{4\pi}{k} \operatorname{Im} f_3(0) \\
\sigma_{tot}(\pi^+ n) &= \sigma_{tot}(\pi^- p) = \frac{4\pi}{k} \operatorname{Im} \left\{ \frac{1}{3} f_3(0) + \frac{2}{3} f_1(0) \right\}
\end{aligned} \tag{16}$$

In terms of scattering phases, Eq.(11) gives the following forward ($\theta = 0$) amplitudes

$$\operatorname{Im} A_{2T}(0) = \frac{1}{k} \sum_{\ell} \left\{ (\ell + 1) \sin^2 \delta^\ell_{2T, 2\ell+1} + \ell \sin^2 \delta^\ell_{2T, 2\ell-1} \right\}$$

$$B_{2T}(0) = 0 \tag{17}$$

Thus, at low enough energies, where higher partial waves may be neglected, we find

$$\sigma_{tot}(\pi^+ p) \simeq \frac{4\pi}{k^2} \left\{ \sin^2 \delta^0_{3,1} + 2 \sin^2 \delta^1_{3,3} + \sin^2 \delta^1_{3,1} \right\}$$

$$\sigma_{tot}(\pi^+ n) = \sigma_{tot}(\pi^- p) = \frac{4\pi}{3k^2} \left\{ \sin^2 \delta^0_{3,1} + 2 \sin^2 \delta^1_{3,3} + \sin^2 \delta^1_{3,1} \right.$$

$$\left. + 2(\sin^2 \delta^0_{1,1} + 2 \sin^2 \delta^1_{1,3} + \sin^2 \delta^1_{1,1}) \right\} \tag{18}$$

and

$$A_3 = \frac{1}{2ik} \left[\left(e^{2i\delta^0_{3,1}} - 1 \right) + \cos\theta \left(e^{2i\delta^1_{3,1}} - 1 \right) + 2\cos\theta \left(e^{2i\delta^1_{3,3}} - 1 \right) \right]$$

$$B_3 = \frac{1}{2k} \left[\left(e^{2i\delta^1_{3,3}} - 1 \right) - \left(e^{2i\delta^1_{3,1}} - 1 \right) \right] \sin\theta$$

$$A_1 = \frac{1}{2ik} \left(e^{2i\delta^1_{1,1}} - 1 \right)$$

$$B_1 = 0 \tag{19}$$

b)   The angular distribution, averaged over initial spins and summed over final spins, may be evaluated in terms of the amplitude $f = A + B\,\boldsymbol{\sigma} \cdot \boldsymbol{\nu}$ and the density matrices

$$\rho_i = \frac{1}{2} \qquad \rho_f = 1 \tag{20}$$

as follows

$$\frac{d\sigma}{d\Omega} = \operatorname{Tr}(\rho_f\, f\, \rho_i\, f^+) \tag{21}$$

This, in turn, reduces to

$$\frac{1}{2} \operatorname{Tr}\, f f^+ = AA^* + BB^* \tag{22}$$

For the processes considered,

$$\frac{d\sigma}{d\Omega}(\pi^+ p \to \pi^+ p) = |A_3|^2 + |B_3|^2$$

$$\frac{d\sigma}{d\Omega}(\pi^+ n \to \pi^+ n) = \frac{1}{9}|A_3 + 2A_1|^2 + \frac{1}{9}|B_3 + 2B_1|^2$$

$$= \frac{1}{9}(|A_3|^2 + |B_3|^2 + 4|A_1|^2 + 4\operatorname{Re} A_3 A_1^*)$$

$$\frac{d\sigma}{d\Omega}(\pi^+ n \to \pi^0 p) = \frac{2}{9}|A_3 - A_1|^2 + \frac{2}{9}|B_3 - B_1|^2$$

$$= \frac{2}{9}(|A_3|^2 + |B_3|^2 + |A_1|^2 - 2\operatorname{Re} A_3 A_1^*) \tag{23}$$

Let us now turn to the process $\pi^+ p$; eliminating the index 3, for simplicity, we write

$$f = A + B\boldsymbol{\sigma} \cdot \boldsymbol{\nu} \tag{24}$$

The initial density matrix for protons with transverse polarization $\mathbf{P}$ is

$$\rho_i = \frac{1}{2}(1 + \boldsymbol{\sigma} \cdot \mathbf{P}) \tag{25}$$

and the resulting differential cross section is

$$
\begin{aligned}
\frac{d\sigma}{d\Omega} &= \text{Tr}[(A + B\boldsymbol{\sigma} \cdot \boldsymbol{\nu})\frac{1}{2}(1 + \boldsymbol{\sigma} \cdot \mathbf{P})(A^* + B^*\boldsymbol{\sigma} \cdot \boldsymbol{\nu})] \\
&= AA^* + BB^* + 2\text{Re}\,(AB^*)\boldsymbol{\nu}\mathbf{P}
\end{aligned}
\tag{26}
$$

By definition, $\boldsymbol{\nu}\mathbf{p} = \boldsymbol{\nu}\mathbf{p}' = 0$, so that

$$\boldsymbol{\nu}\mathbf{P} = \boldsymbol{\nu}\mathbf{P}_\perp \tag{27}$$

$\mathbf{P}_\perp$ is the transverse polarization with respect to the scattering plane.

Choosing $\mathbf{P} \| \boldsymbol{\nu}$ we find on the scattering plane

$$
\begin{aligned}
\frac{d\sigma}{d\Omega}(+\theta) &= |A|^2 + |B|^2 + 2\text{Re}\,(AB^*)\,|\mathbf{P}| \\
\frac{d\sigma}{d\Omega}(-\theta) &= |A|^2 + |B|^2 - 2\text{Re}\,(AB^*)\,|\mathbf{P}|
\end{aligned}
\tag{28}
$$

The quantity

$$\frac{\frac{d\sigma}{d\Omega}(+\theta) - \frac{d\sigma}{d\Omega}(-\theta)}{\frac{d\sigma}{d\Omega}(+\theta) + \frac{d\sigma}{d\Omega}(-\theta)} = \frac{2\,\text{Re}\,(AB^*)\,|\mathbf{P}|}{|A|^2 + |B|^2} \tag{29}$$

is the scattering asymmetry.

Problem 20.   Measuring proton polarization by the asymmetry in p-C_{12} scattering

To measure the transverse polarization of protons one measures the asymmetry in the scattering

$$p + C_{12} \to p + C_{12} \tag{1}$$

The carbon nucleus has spin zero. By asymmetry here one means a left-right asymmetry of the angular distribution with respect to the $\mathbf{p} - \boldsymbol{\zeta}$ oriented plane ($\mathbf{p}$: incident proton momentum, $\boldsymbol{\zeta}$: polarization), as described by the formula

$$\frac{d\sigma}{d\Omega} = f(\theta) + g(\theta)\,\boldsymbol{\nu}\boldsymbol{\zeta} \tag{2}$$

$\nu = \dfrac{\mathbf{p} \wedge \mathbf{p}'}{|\mathbf{p} \wedge \mathbf{p}'|}$; $\mathbf{p}'$ is the momentum of the outgoing photon.

a) Using time reversal invariance, derive the polarization of the outgoing proton in terms of f and g, if the initial proton is nonpolarized.

b) Using the result of question a), discuss a way of calibrating a polarimeter based on reaction (1), by observing double scattering of nonpolarized proton beams.

c) Now suppose that the $p - C_{12}$ interaction is given by a Coulomb potential plus a complex square well potential described by

$$\begin{array}{ll} V = 0 \quad r > R & V = -(V_R + i V_I) \quad r < R \\ V_R = 20 \,\mathrm{MeV} & V_I = 2 \,\mathrm{MeV} \end{array} \tag{3}$$

The spin-orbit relativistic correction is also present. Calculate, in the Born approximation, the functions $f(\theta)$ and $g(\theta)$.

d) Consider a bubble chamber with angular resolution of about $2°$, filled with C_{12} atoms. With what precision can we measure the polarization of proton samples, having kinetic energies $E = 100 \,\mathrm{MeV}$ and $E = 500 \,\mathrm{MeV}$?

Solution

a) The parameterization (2) indicates an asymmetry in the cross section of $p - C_{12}$ scattering. For definiteness, we take incoming protons to lie on the z-axis with polarization along y, and consider scattering events on the x-z plane. Denoting by $\sigma^{(+)}$ and $\sigma^{(-)}$ the cross section for $\mathbf{p} \wedge \mathbf{p}'$ parallel and antiparallel to $\boldsymbol{\zeta}$, we have

$$\frac{\mathrm{d}\sigma^{(+)}}{\mathrm{d}\Omega} = f(\theta) + g(\theta) \tag{4a}$$

$$\frac{\mathrm{d}\sigma^{(-)}}{\mathrm{d}\Omega} = f(\theta) - g(\theta) \tag{4b}$$

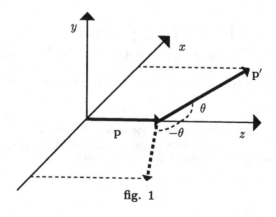

fig. 1

The left-right asymmetry is then

$$a(\theta) = \frac{\dfrac{d\sigma^{(+)}}{d\Omega} - \dfrac{d\sigma^{(-)}}{d\Omega}}{\dfrac{d\sigma^{(+)}}{d\Omega} + \dfrac{d\sigma^{(-)}}{d\Omega}} = \frac{g(\theta)}{f(\theta)} \tag{5}$$

The density matrix in the initial state is $\Pi(\mathbf{p})\frac{1}{2}(1 + \boldsymbol{\zeta}\cdot\boldsymbol{\sigma})$, where $\Pi(\mathbf{p})$ projects onto a state with momentum $\mathbf{p}$. In terms of this density matrix and the S-matrix we write

$$f(\theta) \quad \propto \quad \mathrm{Tr}\left\{S\Pi(\mathbf{p})S^{\dagger}\Pi(\mathbf{p}')\right\}$$

$$g(\theta)\boldsymbol{\nu} \quad \propto \quad \mathrm{Tr}\left\{S\Pi(\mathbf{p})\boldsymbol{\sigma}S^{\dagger}\Pi(\mathbf{p}')\right\} \tag{6}$$

If the incoming protons are not polarized, their density matrix will be $\rho_i = \frac{1}{2}\Pi(\mathbf{p})$, and the polarization of outgoing protons will be given by

$$\zeta'(\theta) = \frac{\mathrm{Tr}\left\{S\,\Pi(\mathbf{p})\,S^{\dagger}\boldsymbol{\sigma}\,\Pi(\mathbf{p}')\right\}}{\mathrm{Tr}\left\{S\,\Pi(\mathbf{p})\,S^{\dagger}\,\Pi(\mathbf{p}')\right\}} \tag{7}$$

Invariance under time reversal T implies $T^{\dagger}ST = S^{\dagger}$. Thus

$$\mathrm{Tr}\left\{S\,\Pi(\mathbf{p})S^{\dagger}\,\Pi(\mathbf{p}')\right\} = \mathrm{Tr}\left\{S^{\dagger}\,\Pi(-\mathbf{p})S\,\Pi(-\mathbf{p}')\right\} \tag{8a}$$

$$= \mathrm{Tr}\left\{S\,\Pi(-\mathbf{p}')S^{\dagger}\Pi(-\mathbf{p})\right\}$$

$$\mathrm{Tr}\left\{S\,\Pi(\mathbf{p})S^{\dagger}\boldsymbol{\sigma}\,\Pi(\mathbf{p}')\right\} = \mathrm{Tr}\left\{S^{\dagger}\,\Pi(-\mathbf{p})S(-\boldsymbol{\sigma})\Pi(-\mathbf{p}')\right\} \tag{8b}$$

$$= -\mathrm{Tr}\left\{S\boldsymbol{\sigma}\Pi(-\mathbf{p}')S^{\dagger}\Pi(-\mathbf{p})\right\}$$

Comparing now the process in which $\mathbf{p} \to \mathbf{p}'$ to the one with $(-\mathbf{p}') \to (-\mathbf{p})$, (see fig. 2)

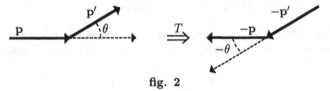

fig. 2

one sees that the scattering angle θ is the same, while $\boldsymbol{\zeta}\boldsymbol{\nu}$ changes sign. Thus, Eqs. (6), (7) and (8) lead to

$$\zeta'(\theta) = \frac{g(\theta)}{f(\theta)}\boldsymbol{\nu} = a(\theta)\boldsymbol{\nu} \tag{9}$$

b)   In order to extract $\boldsymbol{\zeta}$ from a measurement of the differential cross section

$$\frac{d\sigma}{d\Omega} = f(\theta) + g(\theta)\,\boldsymbol{\nu}\boldsymbol{\zeta} \tag{10}$$

we must know the ratio $g(\theta)/f(\theta)$. We can measure this ratio in a double scattering experiment involving nonpolarized protons

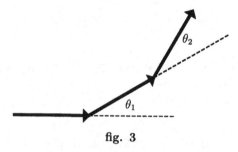

fig. 3

The first scattering generates a transverse polarization

$$\zeta(\theta_1) = \frac{g(\theta_1)}{f(\theta_1)} \tag{11}$$

The asymmetry in the second scattering is

$$a(\theta_2) = \frac{g(\theta_2)}{f(\theta_2)} \zeta(\theta_1) \tag{12}$$

Measuring this asymmetry for $\theta_1 = \theta_2$ one obtains

$$a(\theta_2 = \theta_1) = \frac{g^2(\theta_1)}{f^2(\theta_1)} \tag{13}$$

c)   The properties discussed above can be made explicit in the suggested model

$$
\begin{array}{ll}
V = 0 \quad r > R & V = -(V_R + \mathrm{i}\,V_I) \quad r < R \\
V_R = 20\,\mathrm{MeV} & V_I = 2\,\mathrm{MeV}
\end{array}
$$

The imaginary part of the potential serves as an effective interaction to describe inelastic processes. The spin-orbit correction is given by the usual expression

$$H_{LS} = \frac{g}{4m^2} \frac{1}{r} V'(r)\,\boldsymbol{\sigma} \cdot \mathbf{L} \tag{14}$$

The parameter g depends on the fundamental interaction[1] which gives rise to the potential V, and V' is the derivative of the real part of V.

The Coulomb potential is

$$V_C(r) = \frac{Z\alpha}{r} \tag{15}$$

---

[1]If this interaction is the exchange of a scalar or vector field, taken in the nonrelativistic limit, $g = \pm 1$.

Thus, the total spin-orbit correction is

$$H_{LS} = \frac{g}{4m^2} \left\{ \frac{V_R}{R} \delta(r-R) - \frac{Z\alpha}{r^3} \right\} \boldsymbol{\sigma} \cdot \mathbf{L} \tag{16}$$

In terms of the above, we find for the total interaction Hamiltonian

$$H_I = V(r) + \frac{Z\alpha}{r} + H_{LS} \tag{17}$$

In the Born approximation we write

$$f_B(\theta) = -\frac{m}{2\pi} \langle \mathbf{k}' | H | \mathbf{k} \rangle \tag{18}$$

It is straightforward to carry out the integrations, with the result

$$
\begin{aligned}
f_B(\theta) &= -\frac{m}{2\pi} \left\{ \frac{4\pi Z\alpha}{q^2} - (V_R + \mathrm{i} V_I) \frac{4\pi R}{q^2} \left(\frac{\sin qR}{qR} - \cos qR \right) \right. \\
&\left. + \mathrm{i}\boldsymbol{\nu} \cdot \boldsymbol{\sigma} \frac{g}{4m^2} \frac{4\pi p^2 \sin\theta}{q^2} \left[RV_R \left(\frac{\sin qR}{qR} - \cos qR \right) - Z\alpha \right] \right\}
\end{aligned} \tag{19}
$$

$\mathbf{q} = \mathbf{p}' - \mathbf{p}$ is the momentum transfer.

The Born approximation is valid for

$$|V_R| \ll \frac{p}{mR} \qquad \Longleftrightarrow \qquad p \gg mRV_R \tag{20}$$

For proton-carbon scattering we have m = reduced mass $\simeq$ 866 MeV, $R = 1.3A^{\frac{1}{3}}$fm = 0.015 MeV^{-1}; the Born approximation will therefore be valid for $p \gg 260$ MeV/c, that is $E \gg 40$ MeV.

The Coulomb correction is important at angles satisfying

$$Z\alpha > RV_R \left(\frac{\sin qR}{qR} - \cos qR \right) \tag{21}$$

that is, for

$$\theta^2 \leq \frac{3Z\alpha}{R^3 p^2} \frac{1}{V_R} \simeq \frac{1.12}{E(\mathrm{MeV})}, \qquad |\theta| \leq \frac{6°}{\sqrt{\dfrac{E}{100\,\mathrm{MeV}}}} \tag{22}$$

For $E > 100$ MeV and $\theta > 10°$ the Coulomb correction is negligible. In that case,

$$f_B(\theta) = A(\theta) + B(\theta)\boldsymbol{\sigma} \cdot \boldsymbol{\nu} \tag{23}$$

with

$$
\begin{aligned}
A(\theta) &= \frac{2mR}{q^2}(V_R + \mathrm{i} V_I) \left(\frac{\sin qR}{qR} - \cos qR \right) \\
B(\theta) &= -\mathrm{i} \frac{gRV_R E}{q^2} \left(\frac{\sin qR}{qR} - \cos qR \right) \sin\theta
\end{aligned} \tag{24}
$$

If the initial density matrix has the form $\rho_i = \frac{1}{2}(1 + \boldsymbol{\zeta} \cdot \boldsymbol{\sigma})$, the cross section will be

$$\frac{d\sigma}{d\Omega} = |A|^2 + |B|^2 + 2\mathrm{Re}\, A^* B\, \boldsymbol{\nu}\boldsymbol{\zeta} \tag{25}$$

This results in the following expressions for $f(\theta)$ and $g(\theta)$

$$\begin{aligned} f(\theta) &= |A|^2 + |B|^2 \\ g(\theta) &= 2\mathrm{Re}\, A^* B \end{aligned} \tag{26}$$

Finally, the scattering asymmetry is

$$a(\theta) = -\frac{4g\dfrac{E}{m}V_R V_I \sin\theta}{4\left(V_R^2 + V_I^2\right) + g^2\dfrac{E^2}{m^2}V_R^2 \sin^2\theta} \tag{27}$$

It is clear from Eqs. (25) and (26) that the absorption term is necessary to exhibit asymmetry in the Born approximation.

d) Expression (27) can be made more explicit by substituting the numerical values given in the text; setting also $g = 1$ and $\epsilon = E/100$ MeV, we have

$$a(\theta) \simeq -\frac{\epsilon \sin\theta}{86.46 + 0.292\,\epsilon^2 \sin^2\theta} \tag{28}$$

For $\epsilon \lesssim 1$

$$a(\theta) \simeq \frac{\epsilon \sin\theta}{86.46} \quad \text{and} \quad \frac{\Delta a}{a} \simeq \frac{\Delta\theta}{\theta} \tag{29}$$

For $\epsilon = 5$ ($E = 500$ MeV) and $\delta\theta = 2°$, we list in the following table the asymmetry $a(\theta)$ and the precision $\Delta a/a$ with which the polarization can be measured.

θ	$a(\theta)$	$\dfrac{\Delta a}{a}$
10°	0.0100	0.197
20°	0.0196	0.094
30°	0.0283	0.058
40°	0.0359	0.038
50°	0.0422	0.026

For $\epsilon = 1$ ($E = 100$ MeV) the table becomes

θ	$a(\theta)$	$\dfrac{\Delta a}{a}$
10°	0.0020	0.197
20°	0.0039	0.096
30°	0.0058	0.060
40°	0.0074	0.041
50°	0.0088	0.029

Problem 21. π - C_{12} scattering at low energy

In order to determine the scattering length of the $\pi^{\pm} C$ interaction, low energy pions ($k \simeq 50$ MeV/c) are sent on a carbon target, and the total cross section is measured.

a)   What is the relationship between π^+ and π^- scattering amplitudes, knowing that the C^{12} nucleus has zero isospin?

b)   What is the cross section in the absence of electromagnetic interactions? With what precision can it be approximated in terms of a scattering length?

c)   What is the electromagnetic amplitude, knowing that the charge radius of C^{12} is 2 fm?

d)   How can the magnitude and sign of the scattering length be determined from experimental data?

Solution

a)   If the nucleus has zero isospin, the scattering amplitude coming from the strong interaction will be the same for π^+ and π^-. The corresponding electromagnetic amplitudes, on the other hand, will have the same modulus but opposite sign.

b)   The strong interaction amplitude for low energy pions is

$$f \simeq \frac{e^{2i\delta_0} - 1}{2ik} = \frac{1}{k(\cot \delta_0 - i)} \tag{1}$$

$$\sigma \simeq \frac{4\pi}{k^2} \sin^2 \delta_0 \tag{2}$$

In the "effective range" approximation

$$k \cot g\, \delta_0 = -\frac{1}{a} + \frac{1}{2} r_0 k^2 \tag{3}$$

The validity of this approximation is subject to the condition

$$(ak)^2 + a r_0 k^2 \ll 1 \tag{4}$$

For $k \simeq 50$ MeV/c this implies

$$
\begin{aligned}
a &\ll 4\,\text{fm} \\
r_0 &\ll 4\,\text{fm}
\end{aligned}
\tag{5}
$$

c)   The electromagnetic amplitude is

$$f = -\frac{m}{2\pi} A_0(\mathbf{q})(\pm e) \tag{6}$$

where $A_0(\mathbf{q})$ is the Fourier transform of the electromagnetic potential generated by the nucleus.

Denoting by $\rho(q)$ the Fourier transform of the charge density, we have

$$A_0(\mathbf{q}) = \frac{\rho(\mathbf{q})}{\mathbf{q}^2} \tag{7}$$

At low momentum transfer

$$\rho(\mathbf{q}) \simeq Ze \left[1 - \frac{1}{2} r_c^2 q^2 \right] \tag{8}$$

where r_c is the charge radius. Thus

$$f_{em} = \mp \frac{2\alpha\, mZ}{q^2} \left[1 - \frac{1}{2} r_c^2 q^2 \right] \tag{9}$$

d)   The total amplitude is, for the two values of the pion charge,

$$f_{\pm} \simeq -a \mp \frac{2\alpha\, mZ}{q^2} \left[1 - \frac{r_c^2 q^2}{2} \right] \tag{10}$$

Measuring the total cross section

$$\sigma_{\pm} = 4\pi \left| f_{\pm} \right|^2 \tag{11}$$

we can determine both the sign and the magnitude of the scattering length.

Problem 22.   On the process $n + p \to d + \gamma$ and its inverse

Consider the process

$$n + p \to d + \gamma \tag{1}$$

initiated by slow neutrons.

a)   Knowing that the state of the deuteron is almost completely an S-wave with $J = 1$, show that process (1) proceeds through a channel with $\ell = 0$, $s = 0$. Show that the transition is a magnetic dipole transition.

b)   Calculate the transition probability, knowing that the binding energy of deuterium is 2.3 MeV, and that the $p - n$ scattering length in the spin singlet channel is $a = -23.7\,\mathrm{fm}$.

c)   Calculate the cross section near the threshold of the inverse process

$$\gamma + d \to n + p \tag{2}$$

Solution

a,b) At low energies, the process of radiative capture $n + p \to d + \gamma$ takes place via electric and magnetic dipole transitions. In the usual notation $^{2S+1}L_J$, the deuteron is a $^3S_1$ state, and the possible transitions are

$$^3S_1 \to {}^3S_1 \quad (\mu) \qquad ^3P_{0,1,2} \to {}^3S_1 \quad (d) \qquad ^1S_0 \to {}^3S_1 \quad (\mu) \tag{3}$$

(μ = magnetic dipole, d= electric dipole). If the energy is low enough, the nucleons are in an S-wave and therefore only the magnetic dipole transition is possible.

In the S-wave the magnetic moment operator acts only on spin variables. This implies that the transition $^3S_1 \to {}^3S_1$ will be forbidden: indeed, the scalar product between two wave functions with the same spin orientation is zero if these correspond to different energies, it reduces to a scalar product between orbital states (in this case orthogonal orbitals, one from the continuum and one from the discrete spectrum). On the other hand, given the strong spin-dependence of the $n - p$ potential, singlet and triplet orbitals are different; in particular, there is no reason to expect that orbital wave functions corresponding to different spin and energy will be orthogonal. In conclusion, the only allowed transition is $^1S_1 \to {}^3S_1$.

The cross section for this process is

$$\mathrm{d}\sigma = \frac{1}{v_R}|H_{fi}^I|^2 \frac{\mathrm{d}^3 k}{2\omega(2\pi)^3} 2\pi \,\delta(E_p + E_n - E_d - \omega) \tag{4}$$

Here, $\mathbf{k}$ is the photon momentum, $\omega = |\mathbf{k}|$, $E_d = m_p + m_n - E_3$, E_3 is the triplet binding energy $E_3 = 2.3\,\mathrm{MeV}$, and

$$H^I = -\boldsymbol{\mu} \cdot \mathbf{H} = -\boldsymbol{\mu} \cdot (\mathbf{k} \wedge \boldsymbol{\varepsilon}) \tag{5}$$

Nonrelativistic normalization has been used for the nucleon states in (4).

Integrating over the final photon direction and summing over its polarizations, we obtain

$$\sigma = \frac{4}{3}\frac{1}{v_R}\omega^3 |\boldsymbol{\mu}_{fi}|^2 \tag{6}$$

with

$$\boldsymbol{\mu} = \mu_p\,\boldsymbol{\sigma}_p + \mu_n\,\boldsymbol{\sigma}_n = 2\mu_n \mathbf{s} + 2(\mu_{p} - \mu_n)\mathbf{s}_p \tag{7}$$

μ_p and μ_n are the magnetic moments of the proton and neutron.

The density matrix in the initial state is

$$\rho_{in} = \frac{1}{4}(\Pi_S + \Pi_T) \tag{8}$$

As discussed above, only the singlet part contributes. Summing over deuteron polarizations we find

$$\begin{aligned}
|\boldsymbol{\mu}_{fi}|^2 &= \frac{1}{4}\operatorname{Tr}\Big(\Pi_S\,2(\mu_p - \mu_n)\,\mathbf{s}_p\,\Pi_T\,2(\mu_p - \mu_n)\,\mathbf{s}_p\Big) \\
&= \frac{3}{4}(\mu_p - \mu_n)^2 \int \mathrm{d}^3 x\,\psi_f^*(\mathbf{x})\,\psi_i(\mathbf{x})
\end{aligned} \tag{9}$$

ψ_f corresponds to a state with binding energy E_3, and ψ_i corresponds to the initial state. Since the radius of the bound state wave function, and the wavelength of the initial state are much larger than the range of the potential, r_0, we will approximate them with their asymptotic form, for large r ($r \gg r_0$)

$$\psi_f \simeq \sqrt{\frac{A}{2\pi}}\,\frac{e^{-Ar}}{r}, \qquad A = \sqrt{ME_3}, \qquad M = \frac{m_p + m_n}{2} \tag{10}$$

The initial wave function is

$$\psi_i = e^{-i\delta_1}\,\frac{1}{pr}\sin(pr + \delta_1) \tag{11}$$

where δ_1 is the scattering phase for the spin singlet, and the normalization is compatible with the definition (4) of σ.

At low energies, δ_1 is given by $\delta_1 = -ak$, where a is the scattering length given in the text. To simplify notation, we set

$$\frac{1}{a} = -\sqrt{ME_1} \tag{12}$$

E_1 is the energy corresponding to a pole of the amplitude f in the nonphysical sheet (antibound state). In our case

$$E_1 = 0.067\,\text{MeV} \tag{13}$$

Finally, using expression (10) for the deuteron wave function, we obtain

$$
\begin{aligned}
|\mu_{fi}|^2 &= \frac{3}{4}(\mu_p - \mu_n)^2\,\frac{A}{2\pi}\left|\int d^3\mathbf{r}\,\frac{e^{-Ar}}{r}\,\frac{\sin(pr + \delta_1)}{pr}\right|^2 \\
&= \frac{3}{4}(\mu_p - \mu_n)^2\,\frac{A}{2\pi}\,16\pi^2\left|\text{Im}\frac{e^{i\delta_1}}{A - ip}\,\frac{1}{p}\right|^2
\end{aligned}
\tag{14}
$$

The relations

$$p^2 = M(\omega - E_3), \qquad \cotg\delta_1 = \frac{\sqrt{M E_1}}{p}, \qquad A = \sqrt{M E_3} \tag{15}$$

lead further to

$$
\begin{aligned}
|\mu_{fi}|^2 &= 6\pi(\mu_p - \mu_n)^2\sqrt{M E_3}\,\frac{1}{p^2}\left|\frac{\sqrt{\omega - E_3}}{\sqrt{M}\,\omega}\,\frac{\sqrt{E_1} + \sqrt{E_3}}{\sqrt{\omega - E_3 + E_1}}\right|^2 \\
&= 6\pi(\mu_p - \mu_n)^2\sqrt{M E_3}\,\frac{\omega - E_3}{M\,\omega^2}\,\frac{(\sqrt{E_1} + \sqrt{E_3})^2}{|\omega - E_3 + E_1|}\,\frac{1}{p^2}
\end{aligned}
\tag{16}
$$

Near threshold, $v_r \simeq 2p/M = 2\sqrt{\omega - E_3}/\sqrt{M}$, and the total cross section becomes

$$\sigma = (\mu_p - \mu_n)^2 \omega \sqrt{E_3} \frac{(\sqrt{E_3} + \sqrt{E_1})^2}{|\omega - E_3 + E_1|} \frac{1}{M\sqrt{\omega - E_3}} \tag{17}$$

A typical value for $\omega \simeq 1\,\text{MeV}$ is

$$\sigma \simeq 20\,\text{barn} = 2\cdot 10^{-23}\,\text{cm}^2 \tag{18}$$

(We recall that $\mu_p = 2.793\,\mu$, $\mu_n = -1.913\,\mu$, $\mu = e\hbar/2m_p$.)
c)   The process $\gamma + d \to n + p$ is related to its inverse process by the principle of detailed balance. The cross section, averaged over initial spins (polarizations) and summed over final spins, is

$$\sigma(\gamma + d \to n + p) = \frac{(2s_p + 1)(2s_n + 1)}{(2s_\gamma)(2s_d + 1)} \frac{p^2}{\omega^2} \sigma(n + p \to \gamma + d) =$$

$$= \frac{2}{3} \frac{p^2}{\omega^2} \sigma(n + p \to \gamma + d) \tag{19}$$

Problem 23.   A model of $\pi - N$ interaction

Suppose that pions interact with nucleons only through the exchange of the vector meson ρ $(m_\rho = 750\,\text{MeV}, T = 1)$, according to the interaction Lagrangian

$$\mathcal{L}_I = g\,\bar{\psi}_N \gamma_\mu \frac{\boldsymbol{\tau}}{2} \psi_N\, \boldsymbol{\rho}^\mu + i\,g(\boldsymbol{\pi} \wedge \partial_\mu \boldsymbol{\pi}) \cdot \boldsymbol{\rho}^\mu \tag{1}$$

$\psi_N = (\psi_p,\ \psi_n)$ is the nucleon isospin doublet and $\boldsymbol{\rho}^\mu$ is the field corresponding to ρ (an isospin vector); the triple product in the second term is performed in isospin space.
a)   Determine g knowing the width $\Gamma_{\rho\pi\pi} = 150\,\text{MeV}$.
b)   Show that the Lagrangian (1) conserves isospin; derive the corresponding conserved current.
c)   Deduce the intrinsic parity of ρ and the charge conjugation of ρ^0 from the requirement that $\mathcal{L}_I$ be P- and C-conserving.
d)   Using the Born approximation, calculate the $\pi - N$ cross section in the channels with isospin $1/2$ and $3/2$.

Solution

a)   The interaction

$$\mathcal{L}_I = i\,g(\boldsymbol{\pi} \wedge \partial_\mu \boldsymbol{\pi})\rho^\mu \tag{2}$$

gives the following amplitude for ρ^0 decay

$$\mathcal{M} = g\,\varepsilon_\mu (k_{\pi^-} - k_{\pi^+})^\mu \tag{3}$$

This in turn leads to the decay width

$$
\begin{aligned}
\Gamma &= \frac{1}{2\,m_\rho}|\mathcal{M}|^2 \left(1 - \frac{4\,m_\pi^2}{m_\rho^2}\right)^{\frac{1}{2}} \frac{1}{8\pi} \\
&= \frac{g^2}{48\pi} m_\rho \left(1 - \frac{4\,m_\pi^2}{m_\rho^2}\right)^{\frac{3}{2}}
\end{aligned}
\tag{4}
$$

Substituting known numerical values, we find $g^2 = 3.0$.

The decay widths of ρ^+ and ρ^- are equal to that of ρ^0, as can be directly shown by isospin invariance (see question b).

b) Under an isospin transformation

$$
\psi \to e^{i\,\boldsymbol{\tau}\cdot\boldsymbol{\alpha}/2}\,\psi \qquad \boldsymbol{\pi} \to e^{i\,\mathbf{T}\cdot\boldsymbol{\alpha}}\,\boldsymbol{\pi} \qquad \boldsymbol{\rho}_\mu \to e^{i\,\mathbf{T}\cdot\boldsymbol{\alpha}}\,\boldsymbol{\rho}_\mu
\tag{5}
$$

the Lagrangian rests invariant. Indeed, the quantities $\bar{\psi}\boldsymbol{\tau}\psi$, $\boldsymbol{\rho}_\mu$ and $\boldsymbol{\pi}\wedge\partial_\mu\boldsymbol{\pi}$ transform as vectors under isospin rotations, and the Lagrangian is a made of scalar products between such terms.

The conserved current, as given by Noether's theorem, is

$$
\sum_i \frac{\partial\mathcal{L}}{\partial\left(\partial_\mu\varphi_i\right)}\delta\varphi_i
\tag{6}
$$

where $\delta\varphi_i$ is the infinitesimal variation of the fields under trasformation (5) and the sum runs over all fields. Written in terms of pion and nucleon fields, the current derived from (1) is

$$
\mathbf{j}_\mu = \bar{\psi}_N\,\gamma_\mu\frac{\boldsymbol{\tau}}{2}\psi_N + \boldsymbol{\pi}\wedge\partial_\mu\boldsymbol{\pi}
\tag{7}
$$

c) In the interaction Lagrangian (1) the two quantities coupling to ρ are parts of the isospin current and they are both polar vectors; thus, if ρ is defined to transform under parity as an polar vector, parity invariance is maintained.

Under charge conjugation the bilinear $\bar{\psi}_N\,\gamma^\mu\tau^3/2\,\psi_N$ changes sign. The same is true for the current $\pi^+\partial_\mu\pi^- - \pi^-\partial_\mu\pi^+$. Thus if ρ^0 has charge conjugation eigenvalue -1, C-invariance is also maintained.

d) In the Born approximation, $\pi - N$ scattering is simply described by an exchange of ρ between the two currents

$$
\begin{aligned}
j_a^\mu(\pi) &= (k_\pi + k_\pi')^\mu\,\varepsilon_{abc}e_b\,e_c' \\
j_a^\mu(N) &= u_N'\,\gamma^\mu\frac{\tau_a}{2}\,u_N
\end{aligned}
\tag{8}
$$

where e_a is the isospin wave function of π. The amplitude for this exchange is

$$
\mathcal{M} = g^2\,\bar{u}_N'\,\gamma^\mu\frac{\tau^a}{2}\,u_N\,\frac{1}{(p'-p)^2 - m_\rho^2}\left(g^{\mu\nu} - \frac{q^\mu q^\nu}{m_\rho^2}\right)(k_\pi + k_\pi')_\mu\,\varepsilon_{abc}\,e_b\,e_c'^{*}
\tag{9}
$$

p and p' stand for the initial and final four-momentum of the nucleon and $q^\mu = (p' - p)^\mu$. Using the equations of motion, one can show that the above currents are conserved if the mass differences are neglected; therefore

$$\mathcal{M} = g^2 \bar{u}'_N \, \slashed{k}_\pi \frac{\boldsymbol{\tau} \cdot (\mathbf{e} \wedge \mathbf{e}'^*)}{q^2 - m_\rho^2} \, u_N \tag{10}$$

It is convenient to introduce density matrices which encompass also isospin space; summing over final states, we write

$$\begin{aligned}
\rho_i(N) &= u_N \bar{u}_N = (\slashed{p} + m_N) \frac{1}{2}[1 + \boldsymbol{\tau} \cdot \langle \boldsymbol{\tau}_N \rangle] \\
\rho_f(N) &= \sum u'_N \bar{u}'_N = \slashed{p}' + m_N \\
\rho_f(\pi) &= \sum e'_a e'^*_b = \delta_{ab}
\end{aligned} \tag{11}$$

$\langle \boldsymbol{\tau}^{(N)} \rangle$ is defined as follows

$$\langle \tau_i^{(N)} \rangle = \begin{cases} \delta_{i3} & \text{(proton)} \\ -\delta_{i3} & \text{(neutron)} \end{cases} \tag{12}$$

Substituting in (10) we obtain

$$\begin{aligned}
|\mathcal{M}|^2 &= \left(\frac{g^2}{q^2 - m_\rho^2} \right) \mathrm{Tr}\left\{ \slashed{k}_\pi \frac{\slashed{p} + m_N}{2} \slashed{k}_\pi (\slashed{p}' + m_N) \right\} \\
&\quad \mathrm{Tr}\left\{ \boldsymbol{\tau}(\mathbf{e} \wedge \mathbf{e}'^*) \frac{1}{2}(1 + \boldsymbol{\tau} \cdot \langle \boldsymbol{\tau}_N \rangle) \boldsymbol{\tau} \cdot (\mathbf{e}^* \wedge \mathbf{e}') \right\} \\
&= 2 \left(\frac{g^2}{q^2 - m_\rho^2} \right) [(s - m_\pi^2 - m_N^2) + q^2(s - m_N^2)] \left[1 - \frac{i}{2} \langle \boldsymbol{\tau}_N \rangle (\mathbf{e} \wedge \mathbf{e}^*)\right] \tag{13}
\end{aligned}$$

In the center-of-mass frame the cross section for elastic scattering is

$$\frac{\mathrm{d}\sigma}{\mathrm{d}\Omega} = \frac{1}{64\pi^2} \frac{|\mathcal{M}|^2}{s} \tag{14}$$

We note that all isospin dependence is included in the factor

$$1 - \frac{i}{2} \langle \boldsymbol{\tau}_N \rangle \cdot (\mathbf{e} \wedge \mathbf{e}^*) \tag{15}$$

Further, we have

$$i(\mathbf{e} \wedge \mathbf{e}^*) = \langle \mathbf{T}_\pi \rangle \tag{16}$$

as can be easily verified on pure isospin states $\pi^\pm$, π^0; thus the isospin factor becomes

$$1 - \frac{\langle \boldsymbol{\tau}_N \rangle}{2} \cdot \langle \mathbf{T}_\pi \rangle \tag{17}$$

In terms of the total isospin, $\mathbf{T} = \boldsymbol{\tau}_N/2 + \mathbf{T}_\pi = \mathbf{T}_N + \mathbf{T}_\pi$, we write

$$1 - \langle \mathbf{T}_N \rangle \langle \mathbf{T}_\pi \rangle = \frac{1}{2} \left(-\mathbf{T}^2 + \frac{19}{4} \right) = \frac{1}{2} \left[\frac{19}{4} - T(T+1) \right] \tag{18}$$

with $\mathbf{T}^2 = T(T+1)$.

Finally, defining

$$\begin{aligned}
\frac{d\sigma}{d\Omega} &= \frac{1}{32\pi^2 s} \left(\frac{g^2}{q^2 - m_\rho^2} \right)^2 [(s - m_\pi^2 - m_N^2) + q^2(s - m_N^2)](1 - \langle \mathbf{T}_N \rangle \cdot \langle \mathbf{T}_\pi \rangle) \\
&\equiv \left(\frac{d\sigma}{d\Omega} \right)_0 (1 - \langle \mathbf{T}_N \rangle \cdot \langle \mathbf{T}_\pi \rangle)
\end{aligned} \tag{19}$$

we have

$$\left(\frac{d\sigma}{d\Omega} \right)_{\frac{1}{2}} = 2 \left(\frac{d\sigma}{d\Omega} \right)_0 \ ; \ \left(\frac{d\sigma}{d\Omega} \right)_{\frac{3}{2}} = \frac{1}{2} \left(\frac{d\sigma}{d\Omega} \right)_0 \tag{20}$$

Problem 24.   The N-N interaction through π, ρ and ω exchange

We propose to calculate the nucleon-nucleon interaction due to a pion exchange, in the nonrelativistic limit. The pion-nucleon interaction has the form

$$\mathcal{L}_I = \frac{1}{2} g \, \bar{\psi}(x) \tau_i \gamma_5 \psi(x) \cdot \pi_i \tag{1}$$

$\psi = (\psi_p, \ \psi_n)$ is the nucleon isospin doublet; τ_i are isospin Pauli matrices and π_i ($i = 1, 2, 3$) are the pion fields.

a)   Show that $\mathcal{L}_I$ is invariant under isospin rotations, charge conjugation and parity.

b)   Calculate, in the nonrelativistic limit, the nucleon-nucleon interaction due to single pion exchange and point out its dependence on the total isospin of the two-nucleon state. Show that the scattering length resulting from Lagrangian (1) vanishes to order g^2.

c)   In a similar way, discuss the force due to ρ and ω exchange

$$\mathcal{L}_I^1 = g_\rho \, \bar{\psi} \tau_i \gamma_\mu \psi \, \rho_i^\mu + g_\omega \, \bar{\psi} \gamma_\mu \psi \, \omega^\mu \tag{2}$$

Calculate the scattering length for S-wave singlet and triplet states.

Solution

a)   If pions are assigned a negative intrinsic parity

$$\begin{aligned}
P \, \bar{\psi}(x^0, \mathbf{x}) \gamma_5 \psi(x^0, \mathbf{x}) \, P^\dagger &= -\bar{\psi}(x^0, -\mathbf{x}) \gamma_5 \psi(x^0, -\mathbf{x}) \\
P \, \pi_i(x^0, \mathbf{x}) \, P^\dagger &= -\pi_i(x_0, -\mathbf{x})
\end{aligned} \tag{3}$$

then $\int d^4x \, \mathcal{L}_I$ is parity invariant. In terms of the field redefinitions $\Phi_0 = \pi_3$, $\Phi = \frac{1}{\sqrt{2}}(\pi_1 - i\pi_2)$ (Φ annihilates a π^+), the Lagrangian becomes

$$\mathcal{L}_I = g\frac{1}{2}\Phi_0 \left(: \bar{p}\gamma^5 p - \bar{n}\gamma^5 n :\right) + \frac{g}{\sqrt{2}} : \bar{p}\gamma^5 n : \Phi + \frac{g}{\sqrt{2}} : \bar{n}\gamma^5 p : \Phi^\dagger \tag{4}$$

Under charge conjugation

$$C : \bar{\psi}_1 \gamma^5 \psi_2 : C^{-1} = : \bar{\psi}_2 \gamma^5 \psi_1 : \tag{5}$$

Defining the following transformation of the pion fields under C

$$C\Phi_0 C^{-1} = \Phi_0 \qquad C\Phi C^{-1} = \Phi^\dagger \tag{6}$$

renders $\mathcal{L}_I$ charge conjugation invariant.

The group of isospin rotations is $SU(2)$. Under such a rotation, the fields ψ and π_i transform according to the representations of dimension 2 and 3, respectively. The bilinear $\bar{\psi}\boldsymbol{\tau}\psi$ transforms also as a vector, implying that $\mathcal{L}_I$ stays invariant.

b) The interaction Hamiltonian is the Fourier transform of the scattering amplitude, divided by the normalizations of the states. Two diagrams contribute to the amplitude, shown in fig. 1. (p_1, p_2 (p_1', p_2') are the initial (final) four-momenta; α, β (α', β') are the corresponding isospin indices.)

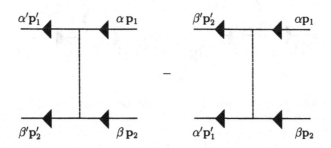

fig. 1

The first diagram gives

$$-i\mathcal{M}^{(1)} = -i\frac{g^2}{4}\bar{u}(\mathbf{p}_1')\gamma^5 \boldsymbol{\tau} u(\mathbf{p}_1)\frac{1}{q^2 - m_\pi^2}\bar{u}(\mathbf{p}_2')\gamma_5 \boldsymbol{\tau} u(\mathbf{p}_2) \tag{7}$$

In the nonrelativistic limit, denoting by φ the Pauli spinors and setting

$$\begin{aligned} p_1 &= (M, \mathbf{p}) & p_1' &= (M, \mathbf{p} + \mathbf{q}) \\ p_2 &= (M, -\mathbf{p}) & p_2' &= (M, -\mathbf{p} - \mathbf{q}) \end{aligned} \tag{8}$$

we find

$$\bar{u}(\mathbf{p}_1')\gamma^5 u(\mathbf{p}_1) \simeq -\varphi_1'^{\dagger}(\boldsymbol{\sigma}\cdot\mathbf{q})\,\varphi_1 \qquad \bar{u}(\mathbf{p}_2')\gamma^5 u(\mathbf{p}_2) \simeq \varphi_2'^{\dagger}(\boldsymbol{\sigma}\cdot\mathbf{q})\,\varphi_2$$

$$\mathcal{M}^{(1)} = \frac{g^2}{4}\frac{1}{\mathbf{q}^2 + m_\pi^2}\,\varphi_1'^{\dagger}(\boldsymbol{\sigma}\cdot\mathbf{q})\,\tau_1\varphi_1\,\varphi_2'^{\dagger}(\boldsymbol{\sigma}\cdot\mathbf{q})\tau_2\varphi_2 \tag{9}$$

Thus, indicating by $\boldsymbol{\sigma}_1$ and $\boldsymbol{\sigma}_2$ the Pauli matrices acting on the first and second nucleon, H_I is

$$H_I = \frac{g^2}{4}\frac{1}{4M^2}\,(\boldsymbol{\sigma}_1\cdot\mathbf{q})(\boldsymbol{\sigma}_2\cdot\mathbf{q})\,\boldsymbol{\tau}_1\cdot\boldsymbol{\tau}_2\,\frac{1}{\mathbf{q}^2 + m_\pi^2} \tag{10}$$

H_I is an operator acting on the spin and isospin subspaces. The second diagram has the two final nucleons exchanged; it gives the same contribution to H_I.

This amplitude vanishes for $q \to 0$, and therefore the S-wave scattering length is zero.

c)   The diagrams involving a single ρ exchange are shown in fig. 2.

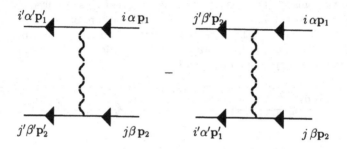

fig. 2

and similarly for ω exchange. Here i, i', j, j' are spin indices.

The vector meson propagator is

$$D_{\mu\nu}(h) = -\mathrm{i}\left(g_{\mu\nu} - \frac{h_\mu h_\nu}{m^2}\right)\frac{1}{h^2 - m^2 + \mathrm{i}\varepsilon} \tag{11}$$

The term with $h_\mu h_\nu$ does not contribute to the amplitude, by current conservation. Thus, at low momentum transfer

$$D_{\mu\nu}(h) \simeq \mathrm{i}g_{\mu\nu}\frac{1}{m^2} \tag{12}$$

In the limit of small velocities one has

$$\bar{u}_{i'}\gamma^\mu u_i \simeq \delta_{ii'}\,2M\delta_{\mu 0} \tag{13}$$

The amplitude then reads

$$-\frac{i\mathcal{M}}{4M^2} = -\frac{i}{m_\rho^2} g_\rho^2 \left\{ \boldsymbol{\tau}_{\alpha'\alpha} \boldsymbol{\tau}_{\beta'\beta} \delta_{ii'} \delta_{jj'} - \boldsymbol{\tau}_{\beta'\alpha} \boldsymbol{\tau}_{\alpha'\beta} \delta_{i'j} \delta_{j'i} \right\}$$

$$-\frac{i}{m_\omega^2} g_\omega^2 \left\{ \delta_{\alpha\alpha'} \delta_{\beta\beta'} \delta_{ii'} \delta_{jj'} - \delta_{\alpha\beta'} \delta_{\beta\alpha'} \delta_{ij'} \delta_{i'j} \right\} \tag{14}$$

If we now use the identity

$$\boldsymbol{\tau}_{\beta'\alpha} \boldsymbol{\tau}_{\alpha'\beta} = -\frac{1}{2} \boldsymbol{\tau}_{\alpha'\alpha} \boldsymbol{\tau}_{\beta'\beta} + \frac{3}{2} \delta_{\alpha\alpha'} \delta_{\beta\beta'} \tag{15}$$

and define singlet and triplet projection operators for spin and isospin

$$\Pi_T^{(I)} = \frac{3 + \boldsymbol{\tau}_1 \cdot \boldsymbol{\tau}_2}{4} \qquad \Pi_T^{(SP)} = \frac{3 + \boldsymbol{\sigma}_1 \cdot \boldsymbol{\sigma}_2}{4}$$

$$\Pi_S^{(I)} = \frac{1 - \boldsymbol{\tau}_1 \cdot \boldsymbol{\tau}_2}{4} \qquad \Pi_S^{(SP)} = \frac{1 - \boldsymbol{\sigma}_1 \cdot \boldsymbol{\sigma}_2}{4} \tag{16}$$

we obtain

$$\delta_{\alpha'\alpha} \delta_{\beta'\beta} = \Pi_T^{(I)} + \Pi_S^{(I)}$$

$$\delta_{\alpha'\beta} \delta_{\beta'\alpha} = \Pi_T^{(I)} - \Pi_S^{(I)}$$

$$\boldsymbol{\tau}_{\alpha'\alpha} \boldsymbol{\tau}_{\beta'\beta} = \Pi_T^{(I)} - 3\Pi_S^{(I)}$$

$$\boldsymbol{\tau}_{\alpha'\beta} \boldsymbol{\tau}_{\beta'\alpha} = \Pi_T^{(I)} + 3\Pi_S^{(I)} \tag{17}$$

(similarly for spin). Substituting in (14) we obtain

$$\frac{\mathcal{M}}{4M^2} = \frac{g_\rho^2}{m_\rho^2} \left\{ [\Pi_T^{(I)} - 3\Pi_S^{(I)}][\Pi_T^{(SP)} + \Pi_S^{(SP)}] - [\Pi_T^{(I)} + 3\Pi_S^{(I)}][\Pi_T^{(SP)} - \Pi_S^{(SP)}] \right\}$$

$$+ \frac{g_\omega^2}{m_\omega^2} \left\{ [\Pi_T^{(I)} + \Pi_S^{(I)}][\Pi_T^{(SP)} + \Pi_S^{(SP)}] - [\Pi_T^{(I)} - \Pi_S^{(I)}][\Pi_T^{(SP)} - \Pi_S^{(SP)}] \right\} \tag{18}$$

or, equivalently,

$$\frac{\mathcal{M}}{4M^2} = \left\{ \left(\frac{2g_\rho^2}{m_\rho^2} + \frac{2g_\omega^2}{m_\omega^2} \right) \Pi_T^{(I)} \Pi_S^{(SP)} + \Pi_S^{(I)} \Pi_T^{(SP)} \left(\frac{2g_\omega^2}{m_\omega^2} - \frac{6g_\rho^2}{m_\rho^2} \right) \right\} \tag{19}$$

We now calculate the scattering length, in terms of the amplitude f

$$f = -\frac{M_R}{2\pi} \mathcal{M} \qquad M_R = \frac{M}{2} \tag{20}$$

Recalling that $a = -\lim_{k \to 0} f$, we find from Eq.(19)

$$a(T = 1, S = 0) = \frac{M}{4\pi} \left(\frac{2g_\rho^2}{m_\rho^2} + \frac{2g_\omega^2}{m_\omega^2} \right) \tag{21a}$$

$$a(T = 0, S = 1) = \frac{M}{4\pi} \left(\frac{2g_\omega^2}{m_\omega^2} - \frac{6g_\rho^2}{m_\rho^2} \right) \tag{21b}$$

Problem 25.   Interaction in the heavy $q\bar{q}$ system

Consider a system consisting of a fermion and its antiparticle, bound by a Coulomb-type potential

$$V(r) = -\frac{\alpha}{r} \qquad \alpha = 0.2 \tag{1}$$

The degeneracy of the corresponding hydrogenoid states is lifted (among other effects) by a perturbation $H_I^{(spin)}$ of the form

$$\begin{aligned}
H_I^{(spin)} &= \frac{3}{2m^2}\frac{1}{r}\frac{dV}{dr}\mathbf{L}\cdot\mathbf{S} + \frac{7}{3m^2}\mathbf{S}_1\cdot\mathbf{S}_2\,\nabla^2 V(r) \\
&\quad - \frac{1}{m^2}\left(\frac{d^2V}{dr^2} - \frac{1}{r}\frac{dV}{dr}\right)\left\{-\frac{\mathbf{S}_1\cdot\mathbf{S}_2}{3} + \frac{(\mathbf{S}_1\cdot\mathbf{r})(\mathbf{S}_2\cdot\mathbf{r})}{r^2}\right\}
\end{aligned} \tag{2}$$

a)   Show that the potential V is the nonrelativistic approximation stemming from the interaction

$$H = g:\bar{\psi}(x)\gamma^\mu\psi(x)\,V_\mu(x): \tag{3}$$

where V_μ is a massless vector meson (gluon). Show that $H_I^{(spin)}$ is the corresponding spin-dependent correction of order $(v/c)^2$.

b)   Calculate the energy of the states $1\,^3S_1$, $1\,^1S_0$, $2\,^3S_1$, $2\,^1S_0$. How do these states transform under charge conjugation?

Solution

a)   The interaction Hamiltonian of a system is given, in momentum representation, by the scattering amplitude in the Born approximation. Denoting by w_1, w_2 the Pauli spinors in the initial state, and by w'_1, w'_2 those in the final state, we write

$$w_1'^\dagger w_2'^\dagger H_I w_1 w_2 = \frac{1}{\sqrt{2E_1\,2E_2\,2E_1'\,2E_2'}}\langle\mathbf{p}_1',\mathbf{p}_2'\,|T|\,\mathbf{p}_1,\mathbf{p}_2\rangle \tag{4}$$

The factors of $\sqrt{2E}$ in Eq.(4) take into account the relativistic normalization of the states.

The scattering amplitude produced by interaction (3) corresponds to the diagrams shown in fig.1; its value is

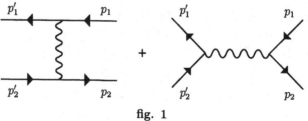

fig. 1

$$\mathcal{M} = g^2 \left[\bar{u}(p_1')\gamma^\mu u(p_1) \, D_{\mu\nu}(p_1' - p_1) \, \bar{v}(p_2)\gamma^\nu v(p_2') \right]$$
$$+ g^2 \left[\bar{v}(p_2)\gamma^\mu u(p_1) \, D_{\mu\nu}(p_1 + p_2) \, \bar{u}(p_1')\gamma^\nu v(p_2') \right] \tag{5}$$

The first term comes from the t-channel diagram (diffusion), and the second one from the s-channel (annihilation). In the t-channel it is convenient to use the Coulomb gauge

$$D_{00} = -\frac{1}{\mathbf{q}^2} \qquad D_{0i} = 0 \qquad D_{ik} = \frac{1}{\mathbf{q}^2 - q_0^2}(g_{ik} - \hat{q}_i\hat{q}_k) \tag{6}$$

The corresponding amplitude $\mathcal{M}_d$ becomes

$$\mathcal{M}_d = g^2 \left[\left(\bar{u}(p_1')\gamma^0 u(p_1) \right) \left(\bar{v}(p_2)\gamma^0 v(p_2') \right) D_{00} \right.$$
$$\left. + \left(\bar{u}(p_1')\gamma^i u(p_1) \right) \left(\bar{v}(p_2)\gamma^j v(p_2') \right) D_{ij} \right] \tag{7}$$

To lowest order in v/c the second term in (7) vanishes, while the first one gives

$$\mathcal{M}_d = -4m^2 (\mathbf{w}_1'^\dagger \mathbf{w}_1)(\mathbf{w}_2'^\dagger \mathbf{w}_2) \frac{g^2}{\mathbf{q}^2} \tag{8}$$

There follows, setting $\alpha = g^2/4\pi$,

$$H_I^{(0)} = -\frac{4\pi\alpha}{\mathbf{q}^2} \tag{9}$$

In coordinate representation this is simply the Coulomb potential

$$V(r) = -\frac{\alpha}{r} \tag{10}$$

Consider now contributions from the next order in v/c: First of all, the free Hamiltonian expanded in powers of v/c becomes

$$E = \sqrt{\mathbf{p}^2 + m_2} \simeq \frac{\mathbf{p}^2}{2m} - \frac{\mathbf{p}^4}{8m^3} \tag{11}$$

This gives rise to the correction term

$$-\frac{\mathbf{p}_1^4}{8m^3} - \frac{\mathbf{p}_2^4}{8m^3} \tag{12}$$

which we incorporate into H_I.

To this order in v/c the quark spinor is

$$u = \sqrt{2E} \begin{pmatrix} (1 - \dfrac{\mathbf{p}^2}{8m^2})\mathbf{w} \\[2mm] \dfrac{\sigma \mathbf{p}}{2m}\mathbf{w} \end{pmatrix} \tag{13}$$

The factor of $\sqrt{2E}$ will cancel against the normalization in Eq.(4). The current matrix elements appearing in (7) are now written as follows

$$\bar{u}(p_1')\gamma^0 u(p_1) = \sqrt{2E_1\,2E_1'}\; w_1'^\dagger \left\{ 1 - \frac{\mathbf{q}^2}{8m^2} + \frac{i\,\boldsymbol{\sigma}\cdot(\mathbf{q}\wedge\mathbf{p}_1)}{4m^2} \right\} w_1 \qquad (14)$$

$$\bar{u}(p_1')\gamma u(p_1) = \frac{\sqrt{2E_1\,2E_1'}}{2m}\, w_1'^\dagger \left\{ i(\boldsymbol{\sigma}\wedge\mathbf{q}) + 2\mathbf{p}_1 + \mathbf{q} \right\} w_1 \qquad (15)$$

where $\mathbf{q} = \mathbf{p}_1' - \mathbf{p}_1$.

Similarly, for the antiquark spinor we have

$$v = \sqrt{2E} \begin{pmatrix} \dfrac{\boldsymbol{\sigma}\mathbf{p}}{2m}\,\tilde{w} \\[2mm] \left(1 - \dfrac{\mathbf{p}^2}{8m^2}\right)\tilde{w} \end{pmatrix} \qquad (16)$$

and

$$\bar{v}(p_2)\gamma^0 v(p_2') = \sqrt{2E_2\,2E_2'}\; \tilde{w}_2^\dagger \left\{ 1 - \frac{\mathbf{q}^2}{8m^2} + \frac{i\,\boldsymbol{\sigma}\cdot(\mathbf{q}\wedge\mathbf{p}_2)}{4m^2} \right\} \tilde{w}_2'$$

$$\bar{v}(p_2)\gamma v(p_2') = \frac{\sqrt{2E_2\,2E_2'}}{2m}\, \tilde{w}_2^\dagger \left\{ i(\boldsymbol{\sigma}\wedge\mathbf{q}) + 2\mathbf{p}_2 - \mathbf{q} \right\} \tilde{w}_2' \qquad (17)$$

The spinors $\tilde{w}$ are related to the Pauli spinors in the standard basis by charge conjugation

$$\tilde{w}_\alpha = \eta\,(\sigma_y)_{\alpha\beta}\, w_\beta^* \qquad (18)$$

where η is a phase. This relation leads to

$$\bar{v}(p_2)\gamma^0 v(p_2') = \sqrt{2E_2\,2E_2'}\; w_2'^\dagger \left\{ 1 - \frac{\mathbf{q}^2}{8m^2} - \frac{i\,\boldsymbol{\sigma}_2\cdot(\mathbf{q}\wedge\mathbf{p}_2)}{4m^2} \right\} w_2$$

$$\frac{\sqrt{2E_2\,2E_2'}}{2m}\,\bar{v}(p_2)\gamma v(p_2') = \frac{1}{c}\, w_2'^\dagger \left\{ -i(\boldsymbol{\sigma}_2\wedge\mathbf{q}) + 2\mathbf{p}_2 - \mathbf{q} \right\} w_2 \qquad (19)$$

Substituting into Eq.(7) and using definition (4), we obtain the contribution of the t-channel amplitude to the interaction Hamiltonian

$$H_I^{(d)}(\mathbf{p}_1,\mathbf{p}_2,\mathbf{q}) = -g^2 \left\{ \frac{1}{\mathbf{q}^2} - \frac{1}{4m^2} + \frac{(\mathbf{q}\mathbf{p}_1)(\mathbf{q}\mathbf{p}_2)}{m^2\mathbf{q}^4} - \frac{(\mathbf{p}_1\mathbf{p}_2)}{m^2\mathbf{q}^2} \right.$$

$$+ \frac{i\,\boldsymbol{\sigma}_1\cdot(\mathbf{q}\wedge\mathbf{p}_1)}{4m^2\mathbf{q}^2} - \frac{i\,\boldsymbol{\sigma}_1\cdot(\mathbf{q}\wedge\mathbf{p}_2)}{2m^2\mathbf{q}^2} - \frac{i\,\boldsymbol{\sigma}_2\cdot(\mathbf{q}\wedge\mathbf{p}_2)}{4m^2\mathbf{q}^2} + \frac{i\,\boldsymbol{\sigma}_2\cdot(\mathbf{q}\wedge\mathbf{p}_1)}{2m^2\mathbf{q}^2}$$

$$\left. + \frac{(\boldsymbol{\sigma}_1\cdot\mathbf{q})(\boldsymbol{\sigma}_2\cdot\mathbf{q})}{4m^2\mathbf{q}^2} - \frac{\boldsymbol{\sigma}_1\cdot\boldsymbol{\sigma}_2}{4m^2} \right\} \qquad (20)$$

In order to cast this operator in the coordinate representation, we must first calculate the Fourier transform

$$\int e^{i\mathbf{q}\mathbf{r}} H_I^{(d)}(\mathbf{p}_1, \mathbf{p}_2, \mathbf{q}) \frac{d^3q}{(2\pi)^3} \tag{21}$$

and then substitute $\mathbf{p}_1$, $\mathbf{p}_2$ by the operators $-i\boldsymbol{\nabla}_1$, $-i\boldsymbol{\nabla}_2$, placed to the right of all other factors. We make use of the formulae

$$\int e^{i\mathbf{q}\mathbf{r}} \frac{\mathbf{q}}{\mathbf{q}^2} \frac{d^3q}{(2\pi)^3} = \frac{i\mathbf{r}}{4\pi r^3}$$

$$\int \frac{(\mathbf{a}\mathbf{q})(\mathbf{b}\mathbf{q})}{\mathbf{q}^2} e^{i\mathbf{q}\mathbf{r}} \frac{d^3q}{(2\pi)^3} = \frac{1}{4\pi r^3}\left\{ (\mathbf{a}\mathbf{b}) - 3\frac{(\mathbf{a}\mathbf{r})(\mathbf{b}\mathbf{r})}{r^2} \right\} + \frac{1}{3}(\mathbf{a}\mathbf{b})\delta^3(\mathbf{r}) \tag{22}$$

with the result

$$H_I^{(d)}(\mathbf{p}_1, \mathbf{p}_2, \mathbf{r}) = -\frac{\alpha}{r} + \frac{\pi\alpha}{m^2}\delta^3(\mathbf{r}) + \frac{\alpha}{2m^2 r}\left(\mathbf{p}_1\mathbf{p}_2 + \frac{1}{r^2}\mathbf{r}(\mathbf{r}\mathbf{p}_1)\mathbf{p}_2 \right)$$

$$-\frac{\alpha}{4m^2 r^3}\left\{ -(\boldsymbol{\sigma}_1 + 2\boldsymbol{\sigma}_2)\cdot(\mathbf{r}\wedge\mathbf{p}_1) + (\boldsymbol{\sigma}_2 + 2\boldsymbol{\sigma}_1)\cdot(\mathbf{r}\wedge\mathbf{p}_2) \right\}$$

$$-\frac{\alpha}{4m^2}\left\{ \frac{\boldsymbol{\sigma}_1\cdot\boldsymbol{\sigma}_2}{r^3} - \frac{3(\boldsymbol{\sigma}_1\cdot\mathbf{r})(\boldsymbol{\sigma}_2\cdot\mathbf{r})}{r^5} - \frac{8\pi}{3}\boldsymbol{\sigma}_1\cdot\boldsymbol{\sigma}_2\delta^3(\mathbf{r}) \right\} \tag{23}$$

We may simplify further, using a set of expressions derived from (10)

$$\frac{dV}{dr} = \frac{\alpha}{r^2} \qquad \frac{d^2V}{dr^2} = -\frac{2\alpha}{r^3} \qquad \nabla^2 V(r) = 4\pi\alpha\delta^3(\mathbf{r}) \tag{24}$$

as well as the following relations, valid in the center-of-mass frame,

$$\mathbf{p}_1 = -\mathbf{p}_2 = \mathbf{p} \qquad \mathbf{r}\wedge\mathbf{p} = \mathbf{L} \tag{25}$$

We obtain

$$H_I^{(d)} = -\frac{\alpha}{r} + \frac{3}{4m^2}\frac{1}{r}\frac{dV}{dr}\mathbf{L}\cdot(\boldsymbol{\sigma}_1 + \boldsymbol{\sigma}_2) + \frac{1}{6m^2}\boldsymbol{\sigma}_1\cdot\boldsymbol{\sigma}_2\nabla^2 V(r)$$

$$+\frac{1}{4m^2}\left(\frac{d^2V}{dr^2} - \frac{1}{r}\frac{dV}{dr} \right)\left[\frac{1}{3}\boldsymbol{\sigma}_1\cdot\boldsymbol{\sigma}_2 - \frac{1}{r^2}(\boldsymbol{\sigma}_1\cdot\mathbf{r})(\boldsymbol{\sigma}_2\cdot\mathbf{r}) \right]$$

$$+\frac{1}{4m^2}\nabla^2 V - \frac{\alpha}{2m^2 r}\left[\mathbf{p}^2 + \frac{1}{r^2}\mathbf{r}(\mathbf{r}\mathbf{p})\mathbf{p} \right] \tag{26}$$

The annihilation diagram is gauge invariant on its own; it is convenient to calculate it in the Feynman gauge

$$D_{\mu\nu} = \frac{1}{q^2}g_{\mu\nu} \simeq g_{\mu\nu}\frac{1}{4m^2} \tag{27}$$

The expansion in v/c leads to

$$\mathcal{M}_a = 4m^2\frac{\pi}{m^2}\alpha\,\tilde{w}_2^*\boldsymbol{\sigma}w_1\,w_1'^*\boldsymbol{\sigma}\tilde{w}_2 \tag{28}$$

In terms of this amplitude, with the help of charge conjugation and the relation

$$\boldsymbol{\sigma}_{ab} \cdot \boldsymbol{\sigma}_{ij} = \frac{3}{2} \delta_{bi} \delta_{aj} - \frac{1}{2} \boldsymbol{\sigma}_{ib} \cdot \boldsymbol{\sigma}_{aj} \tag{29}$$

Eq.(4) gives

$$H_I^{(a)} = \frac{\alpha\pi}{2m^2} (3 + \boldsymbol{\sigma}_1 \cdot \boldsymbol{\sigma}_2) \tag{30}$$

Summing up (26) and (30) one obtains a spin-dependent term which is indeed the one of Eq.(2).

b) The sum of the contributions from Eqs. (12), (26) and (30) can be written as follows

$$
\begin{aligned}
H &= \frac{p^2}{m} - \frac{\alpha}{r} + V_1 + V_2 + V_3 \\
V_1 &= -\frac{p^4}{4m^3} + \frac{\pi\alpha}{m^2} \delta^3(\mathbf{r}) - \frac{\alpha}{2m^2 r} \left\{ p^2 + \frac{\mathbf{r}(\mathbf{rp})\mathbf{p}}{r^2} \right\} \\
V_2 &= \frac{3}{2} \frac{\alpha}{m^2} \frac{1}{r^3} \mathbf{L} \cdot \mathbf{S} \\
V_3 &= \frac{3}{2} \frac{\alpha}{m^2} \frac{1}{r^3} \left\{ \frac{(\mathbf{S} \cdot \mathbf{r})(\mathbf{S} \cdot \mathbf{r})}{r^2} - \frac{1}{3} S^2 \right\} + \frac{\pi\alpha}{m^2} \left(\frac{7}{3} S^2 - 2 \right) \delta^3(\mathbf{r})
\end{aligned} \tag{31}
$$

with $\mathbf{S} = (\boldsymbol{\sigma}_1 + \boldsymbol{\sigma}_2)/2$. In S-states, total angular momentum and spin coincide, so that we may write

$$\langle nS|V_1|nS\rangle = \langle nS| \left\{ -\frac{p^4}{4m^3} + \frac{\pi\alpha}{m^2}\delta^3(\mathbf{r}) - \frac{\alpha}{2mr} \frac{p^2}{m} - \frac{\alpha}{2mr^3} \mathbf{r}(\mathbf{rp})\mathbf{p} \right\} |nS\rangle \tag{32}$$

$$\langle nS|V_2|nS\rangle = 0 \tag{33}$$

$$\langle nS|V_3|nS\rangle = \frac{\pi\alpha}{m^2} \left[\frac{7}{3} S(S+1) - 2 \right] |\psi_{nS}(0)|^2 \tag{34}$$

The first and third terms in Eq.(32) can be calculated by the following relation, valid on wave functions,

$$\frac{p^2}{m} = \left(E_n + \frac{\alpha}{r} \right) \tag{35}$$

For the fourth term we note the following property of S-waves

$$\mathbf{r}(\mathbf{rp})\mathbf{p}\,\psi_S(r) = -r^2 \psi_S''(r) \tag{36}$$

As a result,

$$
\begin{aligned}
\langle nS|V_1|nS\rangle &= -\frac{1}{4m^2} \langle nS| \left(E_n + \frac{\alpha}{r} \right)^2 |nS\rangle \\
&+ \frac{\pi\alpha}{m^2} |\psi_{nS}(0)|^2 - \frac{\alpha^2}{2m} \langle nS| \frac{1}{r} \left(E_n + \frac{\alpha}{r} \right) |nS\rangle \\
&+ \frac{\alpha}{2m} \int \psi_{nS}(r) \frac{1}{r} \psi_{nS}''(r)
\end{aligned} \tag{37}
$$

Introducing now the explicit form of the wave functions

$$\psi_{1S}(r) = \frac{1}{\sqrt{\pi a^3}} e^{-r/a}$$

$$\psi_{2S}(r) = \frac{1}{\sqrt{8\pi a^3}} e^{-r/2a} \left(1 - \frac{r}{2a}\right) \tag{38}$$

where $a = 2/m\alpha$ is the Bohr radius of the system, we find

$$\langle nS|V_1|nS\rangle = \frac{m\alpha^4}{n^4}\left(\frac{11}{64} - \frac{n}{4}\right)$$

$$\langle nS_J|V_3|nS_J\rangle = \frac{m\alpha^4}{8n^3}\left(\frac{7}{3}J(J+1) - 2\right) \tag{39}$$

Finally, the states mentioned in the text have energies

$$E(1\,{}^1S_0) = -\frac{m\alpha^2}{4} - \frac{21}{64}m\alpha^4$$

$$E(1\,{}^3S_1) = -\frac{m\alpha^2}{4} + \frac{49}{192}m\alpha^4$$

$$E(2\,{}^1S_0) = -\frac{m\alpha^2}{16} - \frac{53}{128}\frac{m\alpha^4}{8}$$

$$E(2\,{}^3S_1) = -\frac{m\alpha^2}{16} + \frac{65}{384}\frac{m\alpha^4}{8} \tag{40}$$

Their charge conjugation eigenvalue is

$$C = (-1)^{L+S} \tag{41}$$

Problem 26.   Strong interaction effects on the levels of π-mesic hydrogen

A π-mesic atom consists of a proton and a π^- meson ($m_{\pi^-} = 139.6$ MeV, $m_p = 938.27$ MeV). The force between the two particles is produced, as a first approximation, by single photon exchange.

a)   Formulate the relativistic Hamiltonian of this system and its expansion in powers of v/c. Calculate the energy levels in the nonrelativistic limit including $\mathcal{O}(v^2/c^2)$ corrections.

b)   The π-nucleon scattering lengths in the S-wave are

$$a_{\frac{1}{2}} = 0.19 \text{ fm} \qquad a_{\frac{3}{2}} = 0.08 \text{ fm} \tag{1}$$

The indices 1/2, 3/2 stand for isotopic spin.

The $\pi - N$ strong interaction may be represented by a potential well with radius $r_0 = 1.2$ fm and an isospin-dependent depth. Determine the shift of the energy levels 1S and 2S due to this interaction.

c)   Calculate the width of the $1S$ and $2S$ levels. The following processes contribute to the decay of these levels: i) π-decay ($\tau = 2.6 \cdot 10^{-8}$ sec), ii) The inelastic process $\pi^- p \to \pi^0 n$.

Note: The following formulae may be useful

$$\langle n\ell | \frac{1}{r} | n\ell \rangle = \frac{1}{n^2 r_B}, \qquad \langle n\ell | \frac{1}{r^2} | n\ell \rangle = \frac{1}{r_B^2} \frac{1}{n^3 \left(\ell + \frac{1}{2} \right)} \tag{2}$$

r_B is the Bohr radius of the atom.

Solution

a)   To lowest order in v/c the Hamiltonian is

$$H = \frac{p^2}{2\mu} + \frac{e_1 e_2}{4\pi r} \tag{3}$$

Here, p is the center-of-mass momentum, e_1 and e_2 the two particle charges, r their relative distance and $\mu = m_\pi m_N / (m_\pi + m_N)$ is the reduced mass of the system. In general one can write

$$H = H_\pi^0 + H_N^0 - m_\pi - m_N + H_I(\mathbf{p}_\pi, \mathbf{p}_N, r) \tag{4}$$

$H^{(0)}$ is the free Hamiltonian

$$H^0 = \sqrt{p^2 + m^2} \simeq m + \frac{p^2}{2m} - \frac{1}{8} \frac{p^4}{m^3} + \dots \tag{5}$$

To order v^2/c^2:

$$H = \frac{p_\pi^2}{2m_\pi} + \frac{p_N^2}{2m_N} - \frac{1}{8} \left(\frac{p_\pi^4}{m_\pi^3} + \frac{p_N^4}{m_N^3} \right) + H_{Int} \tag{6}$$

In the center of mass the first two terms reduce to the kinetic term of Eq.(3).

H_I is determined through the scattering amplitude; denoting by w and w′ the initial and final proton spinors, one has

$$w'^\dagger H_I w = \frac{1}{\sqrt{2E_N \, 2E_\pi \, 2E_N' \, 2E_\pi'}} \langle \mathbf{p}_N', \mathbf{p}_\pi' | T | \mathbf{p}_N, \mathbf{p}_\pi \rangle \tag{7}$$

The factors of $\sqrt{2E}$ in (7) take into account the relativistic normalization of the states.

The interaction is described by the following diagram

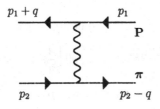

fig. 1

The corresponding amplitude is

$$\mathcal{M} = -e_\pi e_N \, J^\mu_{(p)} \, D_{\mu\nu}(q) \, J^\nu_{(\pi)} \tag{8}$$

In our case, $e_n e_\pi = -4\pi\alpha$. Using the Coulomb gauge, we have

$$D_{00} = -\frac{1}{\mathbf{q}^2}, \quad D_{0i} = 0, \quad D_{ij} = \frac{1}{\mathbf{q}^2 - q_0^2}\left(\delta_{ij} - \frac{q_i q_j}{\mathbf{q}^2}\right) \simeq \frac{1}{\mathbf{q}^2}\left(\delta_{ij} - \frac{q_i q_j}{\mathbf{q}^2}\right) \tag{9}$$

The "retarded" term q_0^2 has been left out in D_{ij}: this term is of order $\mathcal{O}(v^2/c^2)$ and couples to the spatial part of the currents which are already proportional to v/c. Thus, the amplitude takes the form

$$\mathcal{M} = 4\pi\alpha \left[J^0_N \, D^{00} \, J^0_\pi + J^i_N \, D_{ij} \, J^j_\pi \right] \tag{10}$$

The pion current can be read off the minimal coupling term $|(i\,\partial_\mu - eA_\mu)\phi)|^2$; it equals

$$J^\mu_\pi = (2\,p^\mu_\pi - q^\mu) \tag{11}$$

The electromagnetic current of the fermion is

$$J_\mu = \bar{u}(\mathbf{p}')\gamma_\mu u(\mathbf{p}) \tag{12}$$

To second order in v/c, the spinor $u(\mathbf{p})$ is

$$u = \sqrt{2E}\begin{pmatrix}(1 - \dfrac{\mathbf{p}^2}{8m^2})\mathrm{w}\\[2mm] \dfrac{\sigma\cdot\mathbf{p}}{2m}\mathrm{w}\end{pmatrix} \tag{13}$$

so that the current (12) may be approximated by

$$\bar{u}(p'_N)\gamma^0 u(p_N) = \sqrt{2E_N 2E'_N}\, \mathrm{w}'^\dagger_N\left\{1 - \frac{\mathbf{q}^2}{8m^2} + \frac{i\boldsymbol{\sigma}\cdot(\mathbf{q}\wedge\mathbf{p}_N)}{4m^2}\right\}\mathrm{w}_N \tag{14}$$

$$\bar{u}(p'_N)\gamma u(p_N) = \frac{\sqrt{2E_N 2E'_N}}{2m}\, \mathrm{w}'^\dagger_N\left\{i(\boldsymbol{\sigma}\wedge\mathbf{q}) + 2\mathbf{p}_N + \mathbf{q}\right\}\mathrm{w}_N \tag{15}$$

where $\mathbf{q} = \mathbf{p}_N - \mathbf{p}_N$.

Substituting Eqs. (11), (14) and (15) in (8), we find[2]

$$
\begin{aligned}
\mathcal{M} \;=\; & -4\pi\alpha\,\sqrt{2E_N\,2E'_N\,2E_\pi\,2E'_\pi}\,\mathrm{w}'^{\dagger}\Bigg\{\frac{1}{\mathbf{q}^2}-\frac{1}{8m_N^2}+\frac{i\boldsymbol{\sigma}}{4m_N^2\mathbf{q}^2}(\mathbf{q}\wedge\mathbf{p}_N)\\
& -\frac{2}{4m_N m_\pi \mathbf{q}^2}\left[2\,\mathbf{p}_N\mathbf{p}_\pi + i\boldsymbol{\sigma}\cdot(\mathbf{q}\wedge\mathbf{p}_\pi)-\frac{2(\mathbf{q}\mathbf{p}_N)(\mathbf{q}\mathbf{p}_\pi)}{\mathbf{q}^2}\right]\Bigg\}\,\mathrm{w}
\end{aligned}
\tag{16}
$$

We now insert this amplitude into expression (7) and perform a Fourier transform, finding

$$
\begin{aligned}
H_I \;=\; & -\alpha\Bigg\{\frac{1}{r}-\frac{\pi}{2m_N^2}\delta^{(3)}(r)-\frac{1}{4m_N^2}\boldsymbol{\sigma}\cdot\left(\frac{\mathbf{r}}{r^3}\wedge\mathbf{p}_N\right)-\frac{\mathbf{p}_N\,\mathbf{p}_\pi}{m_N\,m_\pi\,r}\\
& +\frac{1}{2\,m_\pi m_N}\boldsymbol{\sigma}\cdot\left(\frac{\mathbf{r}}{r^3}\wedge\mathbf{p}_\pi\right)+\frac{1}{m_\pi m_N}\frac{1}{2r}\left(\mathbf{p}_N\mathbf{p}_\pi-\frac{(\mathbf{p}_N\mathbf{r})(\mathbf{p}_\pi\mathbf{r})}{r^2}\right)\Bigg\}
\end{aligned}
\tag{17}
$$

In the above we used the following Fourier transforms, denoted by $\mathcal{F}$

$$
\mathcal{F}\left(\frac{4\pi}{\mathbf{q}^2}\right)=\frac{1}{r},\qquad \mathcal{F}(1)=\delta^{(3)}(\mathbf{r}),\qquad \mathcal{F}\left(\frac{4\pi\mathbf{q}}{\mathbf{q}^2}\right)=\frac{i\mathbf{r}}{r^3},
$$

$$
\mathcal{F}\left(\frac{4\pi}{\mathbf{q}^4}(\mathbf{A}\mathbf{q})(\mathbf{B}\mathbf{q})\right)=\frac{1}{2r}\left(\mathbf{A}\mathbf{B}-\frac{(\mathbf{A}\mathbf{r})(\mathbf{B}\mathbf{r})}{r^2}\right)
\tag{18}
$$

In the center-of-mass frame, setting $\mathbf{p}=\mathbf{p}_N=-\mathbf{p}_\pi$, and using Eqs. (6) and (17), we obtain

$$
\begin{aligned}
H \;=\; & \frac{\mathbf{p}^2}{2m_\pi}+\frac{\mathbf{p}^2}{2m_N}-\frac{1}{8}\,|\mathbf{p}|^4\left(\frac{1}{m_\pi^3}+\frac{1}{m_N^3}\right)-\frac{\alpha}{r}\\
& +\alpha\Bigg\{\frac{\pi}{2m_N^2}\delta^{(3)}(\mathbf{r})+\frac{\boldsymbol{\sigma}}{2m_N\,r^3}\cdot(\mathbf{r}\wedge\mathbf{p})\left(\frac{1}{2m_N}+\frac{1}{m_\pi}\right)\\
& -\frac{\mathbf{p}^2}{2m_N m_\pi\, r}-\frac{(\mathbf{r}\mathbf{p})^2}{2m_N m_\pi\, r^3}\Bigg\}
\end{aligned}
\tag{19}
$$

In the notation used above, $\mathbf{p}$ are operators acting on the wave functions. Eq.(19) can also be cast in the form

$$
H \;=\; \frac{\mathbf{p}^2}{2\mu}-\frac{\alpha}{r}-\frac{1}{8}\left(\frac{1}{m_\pi^3}+\frac{1}{m_N^3}\right)|\mathbf{p}|^4
$$

---

[2]The time component of the pion current enters the formula with a factor

$$
\frac{(E_\pi+E'_\pi)}{\sqrt{4\,E_\pi E'_\pi}}\simeq\frac{1}{\sqrt{1-\dfrac{(E_2-E_1)^2}{(E_1+E_2)^2}}}
$$

where: $E_\pi=\sqrt{\mathbf{p}_\pi^2+m_\pi^2}$, $E'_\pi=\sqrt{(\mathbf{p}_\pi+\mathbf{q})^2+m_\pi^2}$, $E_2-E_1\simeq\frac{\mathbf{p}_\pi\mathbf{q}}{m}$. Since $p/m\simeq v/c$, this expression equals 1 up to terms of order $(v/c)^4$.

$$+\alpha \left\{ \frac{\pi}{2m_N^2} \delta^{(3)}(\mathbf{r}) + \left(\frac{1}{2m_N} + \frac{1}{m_\pi}\right) \frac{\boldsymbol{\sigma} \cdot \mathbf{L}}{2m_N \, r^3} - \frac{1}{2m_\pi} \frac{1}{m_N \, r} \mathbf{p}^2 \right.$$
$$\left. - \frac{1}{2r^3 \, m_\pi m_N} \mathbf{r}(\mathbf{rp})\mathbf{p} \right\} \tag{20}$$

We note that, in the limit $m_N \to \infty$, H is the Hamiltonian of a Klein-Gordon particle in a Coulomb external field

$$H = eA_0 + \sqrt{m^2 + \mathbf{p}^2} \tag{21}$$

The level shifts can be obtained immediately from Eq.(20). Using the unperturbed Schrödinger equation

$$\mathbf{p}^2 |n\rangle = 2\mu \left(E_n + \frac{\alpha}{r}\right) |n\rangle \tag{22}$$

we find

$$\langle n||\mathbf{p}|^4|n\rangle = (2\mu)^2 \langle n|\left(E_n + \frac{\alpha}{r}\right)^2 |n\rangle$$
$$\langle n|\frac{1}{r^3}\mathbf{r}(\mathbf{rp})\mathbf{p}|n\rangle = 2\mu \langle n|\frac{1}{r}\left(E_n + \frac{\alpha}{r}\right)|n\rangle$$
$$-4\pi|\psi_n(0)|^2 - \ell(\ell+1)\langle n|\frac{1}{r^3}|n\rangle \tag{23}$$

In particular, for S-states,

$$\Delta E_{n,s} = \frac{\mu\alpha^4}{n^3} \left\{ \left(\frac{\mu^3}{m_\pi^3} + \frac{\mu^3}{m_N^3}\right)\left(\frac{3}{8\pi} - 1\right) + \frac{\mu^2}{2m_N^2} + \frac{2\mu^2}{m_\pi m_N} + \frac{\mu^2}{m_\pi m_N}\left(\frac{1}{n} - 4\right) \right\}$$
$$= \frac{\mu\alpha^4}{n^3} \left\{ \left(\frac{\mu^3}{m_\pi^3} + \frac{\mu^3}{m_N^3}\right)\left(\frac{3}{8\pi} - 1\right) + \frac{\mu^2}{m_N m_\pi}\left(\frac{1}{n} - 2\right) + \frac{\mu^2}{2m_N^2} \right\} \tag{24}$$

b) The pairs $\pi^- p$ and $\pi^0 n$ can be decomposed into isospin eigenstates $|T, T_3\rangle$ as follows

$$|\pi^- p\rangle = \sqrt{\frac{1}{3}}|\frac{3}{2}, -\frac{1}{2}\rangle - \sqrt{\frac{2}{3}}|\frac{1}{2}, -\frac{1}{2}\rangle$$
$$|\pi^0 n\rangle = \sqrt{\frac{2}{3}}|\frac{3}{2}, -\frac{1}{2}\rangle + \sqrt{\frac{1}{3}}|\frac{1}{2}, -\frac{1}{2}\rangle \tag{25}$$

The depth of the potential wells $V_{\frac{1}{2}}$ and $V_{\frac{3}{2}}$ can be expressed in terms of the scattering lengths $a_{\frac{1}{2}}$ and $a_{\frac{3}{2}}$ through the relations

$$a_T = \frac{kr_0 - \mathrm{tg}(kr_0)}{k} \qquad k = \sqrt{2\mu V_T} \tag{26}$$

These lead to the numerical values $k_{\frac{1}{2}} r_0 = 4.451$, $k_{\frac{3}{2}} r_0 = 4.478$, whence

$$V_{\frac{1}{2}} = 2.20\,\text{GeV} \qquad\qquad V_{\frac{3}{2}} = 2.23\,\text{GeV} \qquad\qquad (27)$$

To first order in perturbation theory, the shift induced on the energy levels is

$$\begin{aligned}
\langle ns|H_{strong}|ns\rangle &= \langle ns| \left(\frac{2}{3} V_{\frac{1}{2}} + \frac{1}{3} V_{\frac{3}{2}}\right) \theta(r_0 - r)|ns\rangle \\
&\simeq \frac{4}{3n^3} (r_0 \mu \alpha)^3 \left(\frac{2}{3} V_{\frac{1}{2}} + \frac{1}{3} V_{\frac{3}{2}}\right)
\end{aligned} \qquad (28)$$

In the last step we used the fact that $r_0 \ll 1/\alpha\mu$, in order to approximate the wave function for $r \leq r_0$ with $\psi(0)$ (up to $\mathcal{O}(r_0\mu\alpha)$).

Substituting numerical values, we find for the first two energy levels

$$\begin{aligned}
E_{1s} &= -(3234.94 + 0.2367 + 461.78)\,\text{eV} \\
E_{2s} &= -(808.735 + 0.03202 + 57.72)\,\text{eV}
\end{aligned} \qquad (29)$$

In the above, the first correction is electromagnetic, while the second one is induced by H_{strong}.

c)   The two processes contribute to the width as follows

i) The pion decay time τ obviously induces the width

$$\Gamma_\pi = \frac{1}{\tau} = 3.8{\cdot}10^7\,\text{sec}^{-1} \qquad\qquad (30)$$

ii) Since the strong interaction is pointlike as compared to the mesic atom, we may write

$$\Gamma_{nS}^{strong} = v_{Rel}\,|\psi_{nS}(0)|^2\,\sigma(\pi^- p \to \pi^0 n) \simeq \frac{(\mu\alpha)^3 \alpha}{n^4}\,\sigma(\pi^- p \to \pi^0 n) \qquad (31)$$

The decomposition of Eq.(25) leads directly to

$$\begin{aligned}
\sigma(\pi^- p \to \pi^0 n) &= 4\pi \left(\frac{2}{3} a_{\frac{1}{2}}^2 + \frac{2}{3} a_{\frac{3}{2}}^2\right) = \frac{8\pi}{3} \left(a_{\frac{1}{2}}^2 + a_{\frac{3}{2}}^2\right) \\
\Gamma_{nS}^{strong} &= \frac{8\pi}{3} \frac{\mu^3 \alpha^4}{n^4} \left(a_{\frac{1}{2}}^2 + a_{\frac{3}{2}}^2\right)
\end{aligned} \qquad (32)$$

Numerically

$$\Gamma_{1S}^{strong} = 7.07{\cdot}10^{13}\,\text{sec}^{-1} \qquad\qquad (33)$$

$$\Gamma_{2S}^{strong} = 4.42{\cdot}10^{12}\,\text{sec}^{-1} \qquad\qquad (34)$$

Problem 27.   The secant law in the angular distribution of cosmic muons

Primary cosmic rays are made almost exclusively out of nucleons. As they penetrate the atmosphere they produce pions. The total nucleon-nucleon cross section, σ_T, equals approximately 40 mb at high energies; the multiplicity M of charged pions is about 20 at incident energies of 20,000 GeV. The pion distribution is an exponential of the transverse momentum $p_\perp$

$$e^{-p_\perp^2/K^2} d^2 p_\perp \qquad K \simeq 200\,\mathrm{MeV/c} \tag{1}$$

and it is uniform in rapidity.

a)   Consider a flux Φ of incident protons at 20,000 GeV, in a nitrogen atmosphere having 1/50 of the normal density. Calculate the number of pions produced per unit volume.

b)   Calculate the energy spectrum of the pions and their average energy.

c)   The pions which are produced can undergo two competing processes: They may either decay or interact with other nuclei and produce yet more pions. The cross section for pion-nucleon scattering is approximately 20 mb.

We regard the atmosphere roughly as a 10 km-thick layer at 1/50 of normal density, above a layer at normal density. Calculate the angular distribution (with respect to the zenith) of muons produced in the decay of pions with energy greater than 100 GeV.

Some useful numbers:
   Pion half-life: $\tau_\pi = 2.6 \cdot 10^{-8}$ sec
   Avogadro number: $N_A = 6.02 \cdot 10^{23}/$ mole
   Molar volume: $V_m = 2.2 \cdot 10^4\,\mathrm{cm}^3/$ mole
   Molecular weight of nitrogen: 28

Solution

a)   The number of collisions per unit volume and time is

$$\frac{d\nu}{dV\,dt} = \Phi\, N_A\, \sigma_T\, \rho \tag{2}$$

ρ is the mass density of the atmosphere

$$\rho = \frac{1}{50} \frac{\text{Molecular weight}}{V_m} \tag{3}$$

With the given data $\rho = 2.5 \cdot 10^{-5}\,\mathrm{gr/cm}^3$.

   The number of pions produced per unit volume and time is obtained by multiplying the number of collisions (2) by the average pion multiplicity M

$$\frac{dN_\pi}{dV\,dt} = M\, \Phi\, \sigma_T\, N_A\, \rho \tag{4}$$

The corresponding numerical value is

$$\frac{\mathrm{d}N_\pi}{\mathrm{d}V\,\mathrm{d}t} = 1.2 \cdot 10^{-5}\,\mathrm{cm}^{-1}\,\Phi\,(\mathrm{protons/cm^2\,sec}) \tag{5}$$

b)   The pion distribution has the form

$$\mathrm{d}N_\pi \sim \mathrm{e}^{-p_\perp^2/K^2}\,\mathrm{d}^2\mathbf{p}_\perp\,\mathrm{d}y \tag{6}$$

where

$$y = \frac{1}{2}\ln\frac{E + p_\parallel}{E - p_\parallel} \tag{7}$$

$$E = \sqrt{(p_\perp^2 + p_\parallel^2 + m_\pi^2)} \tag{8}$$

We may convert this into an energy distribution noting that

$$E = \sqrt{p_\perp^2 + m_\pi^2}\,\cosh y \tag{9}$$

and

$$\frac{\partial E}{\partial y} = \sqrt{p_\perp^2 + m_\pi^2}\,\sinh y = p_\parallel \tag{10}$$

We thus write

$$\mathrm{d}N_\pi \simeq \mathrm{e}^{-p_\perp^2/K^2}\,\mathrm{d}^2 p_\perp\,\frac{\mathrm{d}E}{\sqrt{E^2 - m_\pi^2 - p_\perp^2}} \tag{11}$$

Carrying out the integration over $p_\perp$ we finally obtain

$$\mathrm{d}N_\pi \sim \frac{\mathrm{d}E}{\sqrt{E^2 - m_\pi^2 - K^2}} \tag{12}$$

To derive the above, the Gaussian integration range was taken to be infinite and numerical factors were dropped. (12) is then valid for $E \gg K$.

The average energy of pions is

$$\langle E \rangle = \frac{\int_0^{E_{max}} E\,\mathrm{d}N_\pi}{\int \mathrm{d}N_\pi} \simeq \frac{E_{max}}{\mathrm{arc}\cosh(E_{max}/\sqrt{m_\pi^2 + K^2})} \tag{13}$$

Numerically, $\langle E \rangle \simeq 2000\,\mathrm{GeV}$.

c)   The pion decay probability per unit distance covered is

$$\frac{\mathrm{d}P}{\mathrm{d}x} = \frac{m_\pi}{\langle E \rangle v \tau_\pi} = \frac{m_\pi}{\langle E \rangle}\,1.3 \cdot 10^{-3}\,\mathrm{cm}^{-1} \tag{14}$$

At the same time, the probability of interaction per unit distance (that is, the inverse mean free path) is

$$\frac{1}{\ell} = \sigma_{\pi N} \rho\, N_A \simeq 3{\cdot}10^{-7}\,\mathrm{cm}^{-1} \qquad \text{if} \qquad \rho = \frac{1}{50}\rho_m \tag{15}$$

Thus we find a probability of decaying in μ

$$P_{(\mu)} = \frac{\mathrm{d}P}{\mathrm{d}x}\,\ell \simeq \frac{m_\pi}{\langle E \rangle}\,0.43{\cdot}10^4 \simeq 0.3 \tag{16}$$

In the rarefied atmospheric layer the decay rate is comparable to the interaction rate. In the dense layer, collisions with nucleons dominate completely. Thus, the number of muons produced per pion is

$$n_{(\mu)} = \frac{\mathrm{d}P}{\mathrm{d}x}\,\Delta x \tag{17}$$

where Δx is the length of the path in the low density layer. Calling θ the angle with respect to the zenith

$$\Delta x = \frac{L}{\cos\theta} \tag{18}$$

$L \simeq 10\,\mathrm{km}$ is the thickness of the low density layer. There follows

$$n_{(\mu)} = \frac{\mathrm{d}P}{\mathrm{d}x}\,\frac{L}{\cos\theta} \tag{19}$$

Eq.(19) is known as the secant law.

Problem 28.   Structure functions for a particle in a box

An electron is bound in a rigid cubic box of side a, in the ground state. A charged projectile with spin $1/2$ is sent into the system, and its momentum variation and energy loss are subsequently measured. The energies involved are nonrelativistic.

a)   What is the vector potential generated by the projectile in the process of scattering off the electron, in the Born approximation?

b)   Write the cross section for the process in which the final state of the electron is not observed; show that this is proportional to the spacetime correlation function of the electromagnetic four-current density in the ground state electron.

c)   Show that, for momentum and energy transfers much greater than $\frac{\hbar}{a}$ and $\frac{\hbar^2}{2m\,a^2}$ respectively, the amplitude becomes independent of a (as if the electron were free).

Solution

a) The Fourier transform of the vector potential generated by a spin-1/2 particle is, to lowest order in e,

$$A_\mu(q) = -\frac{e}{q^2} \bar{u}(\mathbf{p}')\gamma_\mu u(\mathbf{p}) \tag{1}$$

where $q = p' - p$ is the four-momentum transfer.

b) The scattering matrix element can be written as

$$2\pi\,\delta(E_i + q_0 - E_f)\mathcal{M} = e\,A^\mu(q)\int d^4x\,e^{-iqx}\langle f|j_\mu(x)|i\rangle \tag{2}$$

The δ-function is a result of time translation invariance.

$$\langle f|j_\mu(x)|i\rangle = e^{i(E_f - E_i)t}\langle f|j_\mu(\mathbf{x},0)|i\rangle$$

The states of the electron in a box are normalized nonrelativistically: $\langle i|i\rangle = \langle f|f\rangle = 1$. $\mathcal{M}$ is given by

$$\mathcal{M} = e\,A^\mu(q)\int d^3x\,e^{i\mathbf{qx}}\langle f|j_\mu(\mathbf{x},0)|i\rangle \tag{3}$$

To obtain the inclusive cross section we sum over all electron states (they remain unobserved)

$$d\sigma = \frac{1}{2|\mathbf{p}|}\sum_f |\mathcal{M}|^2\,2\pi\,\delta(E_i + q_0 - E_f)\frac{d^3p'}{2p_0'\,(2\pi)^3} \tag{4}$$

Using the equality

$$2\pi\delta(E_i + q_0 - E_f) = \int dt\,e^{i(E_f - q_0 - E_i)t} \tag{5}$$

we have

$$\sum_f |\mathcal{M}|^2 2\pi\,\delta(E_i + q_0 - E_f) =$$
$$= e^2 \int dt\,d^3x\,d^3y\,\langle i|j^\mu(\mathbf{x},0)|f\rangle\langle f|j^\nu(\mathbf{y},0)|i\rangle\,e^{i(E_f - q_0 - E_i)t}\,A_\mu(q)A_\nu(q)$$
$$= e^2 A_\mu(q)A_\nu(q)T^{\mu\nu} \tag{6}$$

where

$$T_{\mu\nu} = \int dt\,d^3x\,d^3y\,e^{i\mathbf{q}(\mathbf{x}-\mathbf{y})-iq_0t}\langle i|j_\nu(\mathbf{x},t)j_\mu(\mathbf{y},0)|i\rangle \tag{7}$$

$$j_\mu(\mathbf{x},t) = e^{iH_0t}j_\mu(\mathbf{x},0)e^{-iH_0t} \tag{8}$$

H_0 is the Hamiltonian describing the electron in the box. Thus the cross section becomes

$$d\sigma = \frac{1}{2|\mathbf{p}|}e\,A_\mu(q)A_\nu^*(q)T^{\mu\nu}\frac{d^3p'}{2p_0'\,(2\pi)^3} \tag{9}$$

$T_{\mu\nu}$ is the correlation function of the electromagnetic current density in the ground state of the target; Eq.(9) shows manifestly the proportionality between $T_{\mu\nu}$ and the cross section.

c)   In the case of a free electron, $T_{\mu\nu}$ assumes the value

$$T_{\mu\nu}^{\text{free}} = \frac{j_\mu(q)j_\nu(q)}{2k_0\, 2k_0'}\, 2\pi\, \delta(k_0' - k_0 - q_0) \tag{10}$$

$$j_\mu(q) = \overline{u}(\mathbf{k}')\gamma_\mu\, u(\mathbf{k}) \tag{11}$$

where
$$q = k' - k \qquad\qquad k = (m, \mathbf{0}) \tag{12}$$

The energy factors in the denominator of (10) ensure that normalizations are set to 1.

In the nonrelativistic limit the only relevant term is

$$T_{00}^{\text{free}} = \frac{|u^*(\mathbf{k}')u(\mathbf{k})|^2}{2m\, 2(m + q_0)} 2\pi\, \delta(k_0' - m - q_0) \simeq |w'^* w|^2\, 2\pi\, \delta(k_0' - m - q_0) \tag{13}$$

where w is a Pauli spinor.

We will then calculate for our system only the term T_{00}, which is the dominant one at low energies

$$T_{00} = \int dt \int d^3\mathbf{x} \int d^3\mathbf{y}\, e^{i\mathbf{q}(\mathbf{x}-\mathbf{y})} \sum_{\mathbf{n}} e^{i(q_0 + E_0 - E_{\mathbf{n}})\, t}\psi_0^*(\mathbf{x})\psi_{\mathbf{n}}(\mathbf{x})\psi_{\mathbf{n}}^*(\mathbf{y})\psi_0(\mathbf{y}) \tag{14}$$

We recall that the eigenfunctions of a particle in a box are

$$\psi_{\mathbf{n}}(\mathbf{x}) = \left(\frac{2}{a}\right)^{\frac{3}{2}} \sin\frac{\pi\, n_1 x_1}{a}\, \sin\frac{\pi\, n_2 x_2}{a}\, \sin\frac{\pi\, n_3 x_3}{a} \tag{15}$$

where $\mathbf{x} = (x_1, x_2, x_3)$ and $\mathbf{n} = (n_1, n_2, n_3)$ with n_1, n_2, n_3 integers.

The corresponding energy eigenvalues are

$$E_{\mathbf{n}} = \frac{\hbar^2}{2m}\left(\frac{\pi}{a}\right)^2 \mathbf{n}^2 \tag{16}$$

In particular, for the ground state we have

$$\psi_0(\mathbf{x}) = \left(\frac{2}{a}\right)^{\frac{3}{2}} \sin\frac{\pi x_1}{a}\, \sin\frac{\pi x_2}{a}\, \sin\frac{\pi x_3}{a} \tag{17}$$

$$E_0 = 3\frac{\hbar^2}{2m}\left(\frac{\pi}{a}\right)^2 \tag{18}$$

The expression for T_{00} becomes

$$T_{00} = \int dt\, e^{i\,q_0 t} \left(\frac{2}{a}\right)^6 \prod_1^3 \int_0^a dx_i dy_i\, e^{i\,q_i(x_i - y_i)}$$

$$\sum_{n_i=1}^\infty e^{\frac{i\hbar\pi^2}{2ma^2}(1-n_i^2)t} \sin\frac{\pi\,x_i}{a} \sin\frac{\pi\,y_i}{a} \sin\frac{\pi\,n_i\,x_i}{a} \sin\frac{\pi\,n_i\,y_i}{a}$$

$$= \int dt\, e^{i\,q_0 t} \prod_1^3 \int_0^a \frac{dx_i}{a}\frac{dy_i}{a}\, e^{i\,q_i(x_i-y_i)} \sum_{n_i=1}^\infty e^{\frac{i\pi^2\hbar}{2ma^2}(1-n_i^2)t}$$

$$\left[\cos\pi\frac{(x_i-y_i)}{a} - \cos\pi\frac{(x_i+y_i)}{a}\right]\left[\cos\pi n_i\frac{(x_i-y_i)}{a} - \cos\pi n_i\frac{(x_i+y_i)}{a}\right] \quad (19)$$

Introducing the variables

$$\xi_i = \frac{x_i}{a} \qquad\qquad \eta_i = \frac{y_i}{a} \tag{20}$$

$$T_{00} = \int dt\, e^{i\,q_0 t} \prod_1^3 \int_0^\pi d\xi_i d\eta_i\, e^{i\,q_i a(\xi_i-\eta_i)} \sum e^{\frac{i\pi^2\hbar}{2ma^2}(1-n_i^2)t}$$

$$[\cos\pi(\xi_i-\eta_i) - \cos\pi(\xi_i+\eta_i)][\cos\pi n_i(\xi_i-\eta_i) - \cos\pi n_i(\xi_i+\eta_i)] \quad (21)$$

Let us now consider the limit

$$q_0 \gg \frac{\pi^2\hbar}{2m\,a^2} \qquad\qquad q_i a \gg \pi \tag{22}$$

We note that terms depending on $\xi_i + \eta_i$ will be suppressed in the evaluation of the integrals, by virtue of oscillations in the integrand which cannot be compensated by oscillations in the term $e^{i\,q_i a(\xi_i-\eta_i)}$. Thus, within the approximation considered,

$$T_{00} \simeq \int dt\, e^{i\,q_0 t} \prod_{i=1}^3 \int_0^1 d\xi_i\, d\eta_i\, e^{i\,q_i a(\xi_i-\eta_i)}$$

$$\sum_{n_i=1}^\infty e^{\frac{i\pi^2\hbar}{2ma^2}(1-n_i^2)t} [\cos\pi(\xi_i-\eta_i)\cos\pi n_i(\xi_i-\eta_i)]$$

$$= \int dt\, e^{i\,q_0 t} \prod_{i=1}^3 \int_0^1 \frac{d(\xi_i+\eta_i)}{2}\, d(\xi_i-\eta_i) \sum_{n_i=-\infty}^{+\infty} e^{i(\xi_i-\eta_i)[q_i a+\pi(n_i+1)]}\, e^{\frac{i\pi^2\hbar}{2ma^2}(1-n_i^2)t} \quad (23)$$

Integration over $(\xi_i-\eta_i)$ produces a Kronecker delta

$$\delta_{q_i a+\pi(n_i+1),\,0} \tag{24}$$

while integration over $(\xi_i + \eta_i)/2$ results in a volume factor equal to 1. Therefore, in this limit,

$$T_{00} \simeq \int dt\, e^{i\,q_0 t} \sum_{n_i=-\infty}^{+\infty} \delta_{q_i a+\pi(n_i+1),\,0}\, e^{\frac{i\pi^2\hbar}{2ma^2}(1-n_i^2)t}$$

$$\simeq \int dt\, e^{i\,q_0 t}\, e^{-i\hbar\frac{q_i^2 t}{2m}} = 2\pi\,\delta\!\left(q_0 - \frac{q_i^2}{2m}\right) \tag{25}$$

corresponding to the nonrelativistic limit of a free particle.

Problem 29.   Inclusive pion distributions in $p-p$ scattering

Pions produced in proton-proton collisions at high energy have the following inclusive distribution in the center of mass

$$d\sigma^{(1)} = e^{-KP_{\perp}^2} f(x,E) \frac{dx}{2x^0} d^2 P_{\perp} \tag{1}$$

where $K = (200\,\text{MeV})^{-2}$. The component $P_{\perp}$ ($P_{\parallel}$) of the pion momentum is perpendicular (parallel) to the incident beam, E is the energy of each incoming proton, and x, x_0 are given by

$$x = P_{\parallel}/E, \qquad x_0 = \sqrt{x^2 + \frac{m_{\perp}^2}{E^2}}, \qquad m_{\perp}^2 = m^2 + P_{\perp}^2 \tag{2}$$

For $E \geq 10$ GeV, the function $f(x,E)$ depends only on x ($f(x,E) \simeq \phi(x)$); this property is known as Feynman scaling. $\phi(x)$ is a finite, continuous function in the interval $-1 \leq x \leq 1$.
a)   How does pion multiplicity depend on energy, if Feynman scaling holds?
b)   Derive the pion distribution with respect to rapidity y

$$y = \text{arc cosh} \frac{E_{\pi}}{\sqrt{P_{\perp}^2 + m^2}} \tag{3}$$

c)   Which is the $P_{\parallel}$ and $P_{\perp}$ distribution in the laboratory frame?
d)   If the double inclusive distribution has the form

$$d\sigma^{(2)} = e^{-K(P_{1\perp}^2 + P_{2\perp}^2)} d^2 \mathbf{P}_{1\perp} d^2 \mathbf{P}_{2\perp} f(x_1, x_2) \frac{dx_1}{2x_1^0} \frac{dx_2}{2x_2^0} \tag{4}$$

calculate $\int d\sigma^{(2)}$ and $\int x^0 d\sigma^{(1)}$.

Solution

a)   The inclusive cross section counts separately every pion which is produced. Therefore

$$\int d\sigma^{(1)} = \sum n\,\sigma(n) = \bar{n}\,\sigma_{Tot} \tag{5}$$

where $\bar{n}$ is the average pion multiplicity. Assuming Feynman scaling

$$
\begin{aligned}
\int d\sigma^{(1)} &= \int d^2 \mathbf{P}_{\perp} e^{-KP_{\perp}^2} \int \frac{\phi(x)\,dx}{2\sqrt{x^2 + \frac{m_{\perp}^2}{E^2}}} \\
&= \int d^2 \mathbf{P}_{\perp} e^{-KP_{\perp}^2} \int \frac{dx}{2\sqrt{x^2 + \frac{m_{\perp}^2}{E^2}}} \{[\phi(x) - \phi(0)] + \phi(0)\}
\end{aligned} \tag{6}
$$

With this separation, the first term vanishes for $x \to 0$; the corresponding integral may be evaluated at high energies by setting $m_\perp^2/E^2 = 0$. The second term is singular for $m_\perp^2/E^2 \to 0$ and gives the dominant contribution at high energies

$$
\int d\sigma^{(1)} \underset{E \gg m_\perp^2}{\simeq} \phi(0) \int d^2\mathbf{P}_\perp e^{-KP_\perp^2} \int_{-1}^{1} \frac{dx}{2\sqrt{x^2 + \dfrac{m_\perp^2}{E^2}}}
$$

$$
\simeq \quad \phi(0) \int d^2\mathbf{P}_\perp e^{-KP_\perp^2} \operatorname{arc\,sinh} \frac{E}{m_\perp} \tag{7}
$$

The integrand in (7) has its maximum at the origin; thus we take $m_\perp \simeq m_\pi$, obtaining

$$
\bar{n} = \frac{\int d\sigma^{(1)}}{\sigma_{Tot}} \simeq \frac{\pi}{K} \frac{\phi(0)}{\sigma_{Tot}} \ln \frac{2E}{m_\pi} \tag{8}
$$

b)   By the definition of rapidity we have

$$
x = \frac{m_\perp}{E} \sinh y\,, \qquad dx = \frac{m_\perp}{E} \cosh y\, dy \tag{9}
$$

Together with the definition of x_0

$$
x_0 = \frac{E_\pi}{E} = \frac{m_\perp}{E} \cosh y \tag{10}
$$

we obtain

$$
\frac{dx}{2x^0} = \frac{dy}{2} \tag{11}
$$

To find the rapidity distribution one must integrate (1) over $P_\perp$. Assuming again Feynman scaling

$$
\frac{d\sigma^{(1)}(y)}{dy} = \int d^2\mathbf{P}_\perp e^{-KP_\perp^2} \frac{1}{2}\phi\left(\frac{m_\perp}{E} \sinh y\right)
$$

$$
\simeq \frac{\pi}{2K} \phi\left(\frac{m_\pi}{E} \sinh y\right) \tag{12}
$$

c)   Collinear Lorentz transformations bring about a mere shift in rapidity

$$
y' = y + \operatorname{arc\,tanh} \beta
$$
$$
dy' = dy \tag{13}
$$

Consequently

$$
\frac{dx'}{2x^{0'}} = \frac{dx}{2x^0} \tag{14}
$$

and

$$
\sinh y' = \gamma(\sinh y + \beta \cosh y) \tag{15}
$$

Noting also that $P'_\perp = P_\perp$ and $m'_\perp = m_\perp$, we are led to

$$d\sigma^{(1)\prime} = e^{-KP_\perp^2}\, d^2\mathbf{p}_\perp\, \phi(\gamma(x' - \beta x^{0\prime}))\, \frac{dx'}{2x^{0\prime}} \tag{16}$$

d) The double inclusive distribution counts each pion pair a number of times equal to the number of pairs which can be formed out of n pions. Therefore

$$\int d\sigma^{(2)} = \sum \frac{n(n-1)}{2}\, \sigma(n) = \frac{\overline{n^2} - \overline{n}}{2}\, \sigma_{Tot} \tag{17}$$

where

$$\overline{n^k} = \frac{1}{\sigma_{Tot}} \sum n^k \sigma(n) \tag{18}$$

The integral $\int x^0\, d\sigma^{(1)}$ may also be written as the following sum

$$\sum \frac{E_\pi(n)}{E}\, \sigma(n) = \frac{\langle E_\pi \rangle}{E} \sigma_{Tot} \tag{19}$$

E_π/E is the fraction of energy taken up by pions with multiplicity n and

$$\langle E_\pi \rangle = \frac{1}{\sigma_{Tot}} \sum E_\pi(n)\sigma(n) \tag{20}$$

Problem 30. The structure functions of a particle bound in a harmonic oscillator potential

A spin-1/2 particle is bound to a compound system; at distances relevant to the ground state, the binding interaction has the form of a harmonic oscillator in the rest frame of the compound (CM).

a) Write down the wave function in momentum representation, in the center of mass.

b) Consider a reference frame in which the compound moves with momentum $\mathbf{P}$ along the z-axis. Express the same wave function in this frame.

c) For $\mathbf{P} \to \infty$ calculate the probability of finding the particle in the interval $d^2p_\perp\, dx$, where $p_\perp$ is the momentum perpendicular to $\mathbf{P}$ and $x = p_z/P$.

d) The particle is made to interact electromagnetically with a high energy electron; subsequently, the energy and direction of the final state electron are measured. Write the contribution of the bound particle to the differential cross section, for a given energy and direction of the scattered electron.

Show that this differential cross section provides a measure of the probability calculated in part c). If the electron energy is much greater than $\langle p_\perp \rangle$, show that the particle momentum can be considered collinear with P for all practical purposes.

Solution

a)   The ground state wave function of a particle with mass m in a harmonic oscillator with frequency ω is, in momentum space,

$$\varphi(\mathbf{p}) = \left(\frac{4\pi}{m\omega}\right)^{\frac{3}{4}} \exp\left[-\frac{\mathbf{p}^2}{2m\omega}\right]$$

$$\int |\varphi(\mathbf{p})|^2 \frac{\mathrm{d}^3\mathbf{p}}{(2\pi)^3} = 1 \tag{1}$$

For simplicity, we will assume $m \ll M$ (M: total mass of the system); then, the relative momentum $\mathbf{p}$ coincides in the center of mass with the particle momentum. Indicating by B all remaining variables, we can write for the complete state of the system

$$|\psi(\mathbf{P})\rangle = \sqrt{\frac{2M}{2m\,2M_B}} \int \frac{\mathrm{d}^3\mathbf{p}}{(2\pi)^3}\, \varphi(\mathbf{p})\,|\mathbf{p}\rangle\,|\mathbf{P}-\mathbf{p}\rangle_B \tag{2}$$

The states have the usual relativistic normalization. Recalling that all momenta are nonrelativistic, we can easily calculate the normalization

$$\langle \psi(\mathbf{P})\,|\,\psi(\mathbf{P}')\rangle = 2M(2\pi)^3\delta(\mathbf{P}-\mathbf{P}') \int \frac{\mathrm{d}^3\mathbf{p}}{(2\pi)^3}\,|\varphi(\mathbf{p})|^2 = 2M(2\pi)^3\delta(\mathbf{P}-\mathbf{P}') \tag{3}$$

Eq.(3) tells us that the normalization of the state is given by the nonrelativistic limit of the usual invariant normalization; we also see that the probability density in $\mathbf{p}$, in the center of mass, has the usual form

$$\frac{\mathrm{d}^3\mathbf{p}}{(2\pi)^3}\,|\varphi(\mathbf{p})|^2 \tag{4}$$

b)   Consider a Lorentz transformation Λ, which boosts the center of mass to a velocity $\boldsymbol{\beta} = \mathbf{P}/E$, where $\mathbf{P}$ and E are the total momentum and energy of the system. In the new frame of reference, state (2) becomes[3]

$$|\psi'\rangle = U(\Lambda)|\psi\rangle = \sqrt{\frac{2M}{2m\,2M_B}} \int \frac{\mathrm{d}^3\mathbf{p}}{(2\pi)^3}\, \varphi(\mathbf{p})|\Lambda\cdot\mathbf{p}\rangle|\Lambda(\mathbf{P}-\mathbf{p})\rangle_B \tag{5}$$

The transformed four-momentum, $p' = \Lambda \cdot p$, is given explicitly by

$$p_\parallel = \gamma\left(p'_\parallel - \beta\varepsilon'\right) \simeq \frac{1}{2\gamma}\varepsilon' \tag{6a}$$

$$p_\perp = p'_\perp \tag{6b}$$

$$\varepsilon = \gamma\left(\varepsilon' - \beta p'_\parallel\right) \simeq \frac{1}{2\gamma}\varepsilon' \tag{6c}$$

---

[3]To keep notation simple, we neglect the spin of the constituent particle; this does not affect what follows.

In the above, $\gamma = E/M$, and $\beta = P/E$. The approximate equalities apply in the limiting case $P \to \infty$. Using Eq.(6), the wave function $\varphi(\mathbf{p})$ can be written in terms of $\mathbf{p}'$ as follows

$$\varphi'(\mathbf{p}') = \varphi(\mathbf{p}) = \left(\frac{4\pi}{m\omega}\right)^{\frac{3}{4}} e^{-\frac{1}{2m\omega}[p_\perp'^2 + \gamma^2(p_\parallel' - \beta\varepsilon')^2]} \tag{7}$$

In the limit $E \to \infty$

$$\begin{aligned} P &= \sqrt{E^2 - M^2} \simeq E - \frac{M^2}{2E} \\ \varepsilon' &= \sqrt{p_\parallel'^2 + m^2 + p_\perp^2} \simeq p_\parallel' + \frac{m^2 + p_\perp^2}{2p_\parallel'} \end{aligned} \tag{8}$$

the quantity appearing in the exponent of (7) becomes

$$\begin{aligned} \gamma(p_\parallel' - \beta\varepsilon') &= \frac{E}{M}\left(p_\parallel' - \frac{P}{E}\varepsilon'\right) \\ &\simeq \frac{M}{2E}p_\parallel' - \frac{E(m^2 + p_\perp^2)}{2Mp_\parallel'} = \frac{M}{2}x - \frac{m^2 + p_\perp^2}{2Mx} \end{aligned} \tag{9}$$

In the above,

$$x = \frac{p_\parallel'}{E} \tag{10}$$

We thus arrive at

$$\varphi'(\mathbf{p}') = \left(\frac{4\pi}{m\omega}\right)^{\frac{3}{4}} e^{-\frac{p_\perp^2}{2m\omega} - \frac{M^2}{8m\omega}\left(x - \frac{m^2 + p_\perp^2}{M^2 x}\right)^2} \tag{11}$$

At the same time, the limit $P \to \infty$ also gives

$$\begin{aligned} \frac{d^3\mathbf{p}}{(2\pi)^3} &= \frac{1}{2\gamma}\frac{d^3\mathbf{p}'}{(2\pi)^3} \\ \frac{\varepsilon'}{2\gamma} &= m \qquad \frac{E_B'}{2\gamma} = M_B \end{aligned} \tag{12}$$

so that (5) can be written as

$$|\psi'\rangle = \int \sqrt{\frac{2M}{2\varepsilon' \, 2E_B'}} \frac{d^3\mathbf{p}'}{(2\pi)^3} \varphi'(\mathbf{p}') |\mathbf{p}'\rangle |\mathbf{P}' - \mathbf{p}'\rangle \tag{13}$$

c)   The probability of finding the particle in the interval d^3p' can be derived from (13). Let us consider the scalar product $\langle \psi'(\mathbf{P''})|\psi(\mathbf{P})\rangle$:

$$
\begin{aligned}
\langle \psi'(\mathbf{P''})|\psi(\mathbf{P})\rangle &= \int \frac{d^3p'}{(2\pi)^3}\frac{d^3p''}{(2\pi)^3}\sqrt{\frac{2M}{2\varepsilon'\,2E_B'}}\sqrt{\frac{2M}{2\varepsilon''\,2E_B''}}\,\varphi'^*(\mathbf{p''})\varphi'(\mathbf{p'})\cdot \\
&\quad \cdot 2\varepsilon'(2\pi)^3\delta^{(3)}(\mathbf{p'}-\mathbf{p''})\,2E_B'(2\pi)^3\delta^{(3)}(\mathbf{P'}-\mathbf{P''}) \\
&= 2M(2\pi)^3\delta^{(3)}(\mathbf{P'}-\mathbf{P''})\int \frac{d^3p'}{(2\pi)^3}|\varphi'(\mathbf{p'})|^2 \\
&= 2E(2\pi)^3\delta^{(3)}(\mathbf{P'}-\mathbf{P''})\frac{M}{E}\int \frac{d^3p'}{(2\pi)^3}|\varphi'(\mathbf{p'})|^2
\end{aligned}
\tag{14}
$$

Having factorized the normalization $2E(2\pi)^3\delta^{(3)}(\mathbf{P'}-\mathbf{P''})$, the rest is interpreted as the probability distribution as a function of the momentum $\mathbf{p'}$:

$$
\frac{d\Pi(\mathbf{p'})}{d^3p'} = \frac{M}{E}\frac{1}{(2\pi)^3}|\varphi'(\mathbf{p'})|^2
\tag{15}
$$

In terms of x and $p_\perp$ the probability density is

$$
\frac{d\Pi(x,p_\perp)}{dx\,d^2p_\perp} = \frac{M}{(2\pi)^3}|\varphi'(\mathbf{p'})|^2
\tag{16}
$$

Integrating out $p_\perp$ we obtain the probability density in x:

$$
\frac{d\Pi(x)}{dx} = f(x) = M\int \frac{d^2p'_\perp}{(2\pi)^3}|\varphi'(\mathbf{p'})|^2
\tag{17}
$$

More explicitly, using the wave function (11), we obtain

$$
f(x) = \int d^2p_\perp \frac{M}{2(\pi m\omega)^{3/2}}e^{-\frac{p_\perp^2}{m\omega}-\frac{M^2}{4\,m\omega}\left(x-\frac{m^2+p_\perp^2}{M^2x}\right)^2}
\tag{18}
$$

d)   The kinematics of the process is shown in fig. 1. The incoming (outgoing) electron has momentum k (k'). The four-momentum transfer is

$$
q = k - k'
\tag{19}
$$

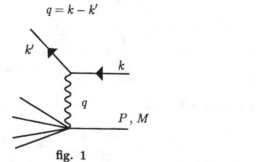

fig. 1

The invariants characterizing the process are q^2 and the energy loss in the lab frame $\nu = Pq/M$. Calling θ the scattering angle we have, for relativistic electrons

$$q^2 = -4k_0 k_0' \sin^2 \frac{\theta}{2} \tag{20}$$

In the following we will make use of the Bjorken variable

$$\bar{x} = -\frac{q^2}{2Pq} = -\frac{q^2}{2M\nu} \tag{21}$$

The kinematic constraint

$$(P+q)^2 \geq P^2 \tag{22}$$

due to the fact that the composite is originally in its ground state, implies

$$0 \leq \bar{x} \leq 1 \tag{23}$$

Let us choose a reference frame in which $|q_z| = q_0$. Such a frame will always exist, since q is spacelike, and it can be obtained by a mere rotation of the laboratory axes (up to further rotations about z). In this frame, $\mathbf{q}_\perp^2 = -q^2$. We now perform a Lorentz transformation leading to $P \to \infty$ in the z-direction. The following relations will hold

$$
\begin{aligned}
P &\simeq (E, \mathbf{0}, E) \\
p &= (\varepsilon, \mathbf{p}_\perp, p_\parallel) \simeq (\varepsilon, \mathbf{p}_\perp, xE) \\
q &\simeq \left(\frac{qP}{2E}, \mathbf{q}_\perp, \frac{qP}{2E} \right) = \left(\frac{M\nu}{2E}, \mathbf{q}_\perp, -\frac{M\nu}{2E} \right)
\end{aligned}
\tag{24}
$$

where we have used the approximate relationship

$$\gamma(q_0 - \beta q_3) = \gamma(1 - \beta)q_0 \simeq \frac{M}{2E}\left(\frac{qP}{M}\right) = \frac{M}{2E}\nu \tag{25}$$

The scattering amplitude for an electron with momentum $\mathbf{k}$ and a target with momentum $\mathbf{P}$ is

$$T = -\frac{i}{q^2}e^2 \bar{u}(\mathbf{k}')\,\gamma_\mu\, u(\mathbf{k})\langle f|J^\mu|\mathbf{P}\rangle \tag{26}$$

The cross section for relativistic electrons, averaged over initial spins and summed over final spins, is

$$
\begin{aligned}
d\sigma(P) &= \frac{1}{2P_0\,2k_0} \frac{d^3\mathbf{k}'}{(2\pi)^3 2k_0'} \frac{e^4}{q^4} \frac{1}{2} \mathrm{Tr}\,(\gamma^\mu \slashed{k}\gamma^\nu \slashed{k}') \\
&\quad \sum_f (2\pi)^4 \delta^4\,(P_f - P - q)\,\langle \mathbf{P}|J_\mu|f\rangle\langle f|J_\nu|\mathbf{P}\rangle
\end{aligned}
\tag{27}
$$

$|f\rangle$ stands for a generic final state. We now introduce the leptonic and hadronic tensors, $L_{\mu\nu}$ and $W_{\mu\nu}$; ignoring mass terms, these can be written as

$$L_{\mu\nu} = \frac{1}{2} \text{Tr}\{\gamma_\mu \not{k} \gamma_\nu \not{k}'\} = r_\mu r_\nu + q^2 g_{\mu\nu} - q_\mu q_\nu \tag{28}$$

$$W_{\mu\nu} = \frac{1}{2\pi} \overline{\sum_f} (2\pi)^4 \delta^4 (P_f - P - q) \langle \mathbf{P}|J_\mu|f\rangle \langle f|J_\nu|\mathbf{P}\rangle \tag{29}$$

with

$$r_\mu = k_\mu + k'_\mu \tag{30}$$

The cross section (27) takes the form

$$\frac{\mathrm{d}\sigma}{\mathrm{d}k'_0 \, \mathrm{d}\Omega} = \frac{\alpha^2}{2q^4} \frac{k'_0}{P_0 k_0} L_{\mu\nu} W^{\mu\nu} \tag{31}$$

Using expression (13) for the state of the system as $P \to \infty$, and keeping in mind that the current couples only to the spin-1/2 particle, we find

$$\begin{aligned}
W_{\mu\nu} &= \frac{1}{2\pi} \overline{\sum_f} (2\pi)^4 \delta^4 (P_f - P - q) \langle \psi(P)|J_\mu|f\rangle \langle f|J_\nu|\psi(P)\rangle = \\
&= \int \frac{\mathrm{d}^3\mathbf{p}}{(2\pi)^3} |\varphi(\mathbf{p})|^2 \frac{2M}{2\varepsilon} \frac{1}{2\pi} \overline{\sum_f} (2\pi)^4 \delta^4 (P_f - p - q) \langle \mathbf{p}|J_\mu|f\rangle \langle f|J_\nu|\mathbf{p}\rangle \\
&= \int \mathrm{d}x \, \mathrm{d}^2\mathbf{p}_\perp \frac{f(x, \mathbf{P}_\perp)}{x} W_{\mu\nu}^{(el)}(p)
\end{aligned} \tag{32}$$

$W_{\mu\nu}^{(el)}(p)$ is the structure function of the elementary constituent; for the case at hand, it reads (cf. (29))

$$W_{\mu\nu}^{(el)} = \frac{1}{2\pi} \int \frac{\mathrm{d}^3\mathbf{p}'}{(2\pi)^3 2p'_0} (2\pi)^4 \delta^4 (p' - p - q) \frac{1}{2} \text{Tr}\{\gamma_\mu \not{p} \gamma_\nu \not{p}'\} \tag{33}$$

Transverse components of the momenta are exponentially cut off and may be neglected, leading to

$$p + p' \simeq 2xP + q \tag{34}$$

The trace appearing in Eq.(33) is rewritten as follows

$$\begin{aligned}
\frac{1}{2} \text{Tr}\{\gamma_\mu \not{p} \gamma_\nu \not{p}'\} &= (p + p')_\mu (p + p')_\nu + q^2 g_{\mu\nu} - q_\mu q_\nu = \\
&q^2 \left[\left(g_{\mu\nu} - \frac{q_\mu q_\nu}{q^2} \right) + \frac{4x^2}{q^2} \left(P_\mu + \frac{q_\mu}{2x} \right) \left(P_\nu + \frac{q_\nu}{2x} \right) \right]
\end{aligned} \tag{35}$$

The momentum delta function in (33) can be integrated immediately. As for the energy delta function, using the kinematic relations (24) we have

$$\frac{2\pi}{2p_0} \delta \left(\sqrt{(xE - \frac{\nu}{2E})^2 + q_\perp^2} - q_0 - xE \right) \simeq \frac{\pi}{M\nu} \delta(x - \bar{x}) = \frac{\pi x}{q^2} \delta(x - \bar{x}) \tag{36}$$

Eq.(36) shows that the Bjorken variable $\bar{x}$ in this system, defined by (21), coincides with the fraction of momentum carried by the particle.

Substituting in (33) we find

$$W_{\mu\nu}^{(el)} = \pi x \delta(x - \bar{x}) \left[\left(g_{\mu\nu} - \frac{q_\mu q_\nu}{q^2} \right) + \frac{4x^2}{q^2} \left(P_\mu + \frac{q_\mu}{2x} \right) \left(P_\nu + \frac{q_\nu}{2x} \right) \right] \qquad (37)$$

This in turn gives for $W_{\mu\nu}$, integrating over $\mathbf{p}_\perp$,

$$\begin{aligned} W_{\mu\nu} &\equiv W_1 \left(g_{\mu\nu} - \frac{q_\mu q_\nu}{q^2} \right) + \frac{W_2}{M^2} \left(P_\mu + \frac{q_\mu}{2\bar{x}} \right) \left(P_\nu + \frac{q_\nu}{2\bar{x}} \right) \\ &= \frac{1}{2} f(\bar{x}) \left[\left(g_{\mu\nu} - \frac{q_\mu q_\nu}{q^2} \right) + \frac{4x^2}{q^2} \left(P_\mu + \frac{q_\mu}{2\bar{x}} \right) \left(P_\nu + \frac{q_\nu}{2\bar{x}} \right) \right] \end{aligned} \qquad (38)$$

The invariant form factors W_1, W_2 equal

$$\begin{aligned} W_1 &= \frac{1}{2} f(\bar{x}) \\ W_2 &= M^2 \frac{2x^2}{q^2} f(\bar{x}) \end{aligned} \qquad (39)$$

and satisfy the scaling laws

$$\begin{aligned} \frac{\nu W_2}{M} &= F_2(\bar{x}) = \bar{x} f(\bar{x}) \\ W_1 &= F_1(\bar{x}) = \frac{f(\bar{x})}{2} \end{aligned} \qquad (40)$$

with $f(x)$ given in (18). Finally, we have for the cross section (27)

$$\frac{d\sigma}{dk_0' \, d\Omega} = \frac{\alpha^2}{4k_0^2 \sin^4 \theta/2} \left(W_2 \cos^2 \frac{\theta}{2} + 2W_1 \sin^2 \frac{\theta}{2} \right) \qquad (41)$$

The approximations which we have made are only valid in the limit in which transverse momentum is negligible compared to the longitudinal momentum xP. This excludes the kinematical region $x \simeq 0$. Since x coincides with the Bjorken variable (21), the kinematically accessible region at large momentum transfer ($q^2 \to \infty$) obeys

$$\lim_{q^2 \to \infty} \frac{q^2}{2M\nu} = \text{const.} \qquad (42)$$

This is the so-called deep inelastic scattering region.

The system considered in this problem is a realization of Feynman's parton model. The main idea in this model is that the cross section (27), in the limit $P \to \infty$, equals the cross section on the elementary constituent with momentum xP , weighted by the probability of having this momentum

$$\sigma(P) = \int f(x)\, dx\, \sigma^{(el)}(xP) \tag{43}$$

For the hadronic tensor this implies

$$\frac{1}{2P_0}W_{\mu\nu} = \int f(x)dx\, \frac{1}{2xP_0}\, W_{\mu\nu}^{(el)} \tag{44}$$

which gives

$$W_{\mu\nu} = \int dx\, \frac{f(x)}{x}\, W_{\mu\nu}^{(el)} \tag{45}$$

This relation was indeed obtained in Eq.(38).

An essential ingredient in the derivation was the exponential damping of the particle's transverse momentum.

Problem 31. The Fermi statistical model for π production

A simple description of relativistic scattering with production of many particles (e.g. $p+\bar{p} \to n\,\pi$) is provided by a model of Fermi. In this model, the matrix element $|\mathcal{M}|^2$ entering the formula

$$d\sigma = \frac{1}{\text{Density} \times \text{Flux}}\, |\mathcal{M}|^2\, d\Phi^{(n)} \tag{1}$$

is taken to be constant (except for a possible dependence on the total energy). The differential

$$d\Phi^{(n)} = (2\pi)^4 \delta^4(P_{\text{in}} - P_{\text{fin}}) \prod_{j=1}^{n} \left[\frac{d^3\mathbf{p}_j}{2p_j^0 (2\pi)^3} \right] \tag{2}$$

is an element of phase space.

a)   What is the length dimension (d_n) of $|\mathcal{M}_n|^2$ for given n, in natural units ($\hbar = c = 1$) ?

b)   Following Fermi, we parameterize the system in terms of a fundamental length $\ell \simeq 10^{-13}$ cm and set $|\mathcal{M}_n|^2 = C\ell^{d_n}$, taking for simplicity the pion mass to be zero. How does the total cross section σ_n depend on the energy?

c)   Calculate the average multiplicity $\langle n \rangle$ as a function of energy.

Solution

a)   Using covariant normalization at high energy we have

$$\text{Density} \cdot \text{Flux} = 2E_{cm}^2 \tag{3}$$

Since $[E_{cm}^2] = [\ell^{-2}]$, $|\mathcal{M}|^2 d\Phi$ must be dimensionless. The mass dimension of $d\Phi$ is $2n - 4$, and therefore $|\mathcal{M}|^2$ has dimension $2n - 4$ in length

$$|\mathcal{M}|^2 = C\ell^{2n-4} \tag{4}$$

b) Let us calculate $\Phi^{(n)}$. For massless particles

$$\Phi^{(2)} = \frac{1}{8\pi} \tag{5}$$

We now use the recursive formula (see Appendix A)

$$\Phi^{(n)} = \int_0^{E_{cm}^2} \frac{dM^2}{2\pi} \, \Phi^{(n-1)}(M) \, \Phi^{(2)}(E_{cm}, M) \tag{6}$$

where

$$\Phi^{(2)}(E_{cm}, M) = \frac{1}{8\pi} v(E_{cm}, M) \tag{7}$$

is the phase space for two particles, having masses zero and M, with total center-of-mass energy E_{cm}

$$v(E_{cm}, M) = 1 - \frac{M^2}{E_{cm}^2} \tag{8}$$

Given that

$$\Phi^{(k)}(M) = c_k M^{2k-4} \tag{9}$$

Eq.(6) becomes

$$\Phi^{(n)}(E_{cm}) = \frac{c_{n-1}}{16\pi^2} \int_0^{E_{cm}^2} M^{2n-6} \left(1 - \frac{M^2}{E_{cm}^2}\right) dM^2 \tag{10}$$

Carrying out the integral we obtain

$$\Phi^{(n)}(E_{cm}) = \frac{1}{(n-1)(n-2)} \frac{c_{n-1}}{16\pi^2} E_{cm}^{2n-4} \tag{11}$$

which results in

$$c_n = c_{n-1} \frac{1}{16\pi^2 (n-1)(n-2)} \tag{12}$$

Iterating the above, with the initial condition (5), leads to

$$c_n = \frac{1}{8\pi} \frac{1}{(4\pi)^{2(n-2)}} \frac{1}{(n-1)!(n-2)!} \tag{13}$$

Thus the cross section is found to be

$$\sigma_n = \frac{1}{2E_{cm}^2} C \left(\frac{\ell E_{cm}}{4\pi}\right)^{2n-4} \frac{1}{8\pi(n-1)!(n-2)!} \tag{14}$$

c) Setting

$$x = \left(\frac{\ell \, E_{cm}}{4\pi}\right)^2 \tag{15}$$

we write

$$\langle n \rangle = \frac{\sum n \, \sigma_n}{\sum \sigma_n} = \frac{\sum \dfrac{n x^{n-2}}{(n-1)!(n-2)!}}{\sum \dfrac{x^{n-2}}{(n-1)!(n-2)!}} \tag{16}$$

The above can also be expressed more compactly as follows

$$\langle n \rangle - 2 \;=\; \frac{\partial \ln \sigma(x)}{\partial \ln x} \tag{17}$$

$$\sigma(x) \;=\; \sum_n \sigma_n(x) \tag{18}$$

Eq.(18) is related to a modified Bessel function

$$\sum_{n=0}^{\infty} \frac{x^k}{n!(n+1)!} = \frac{I_1(2\sqrt{x})}{\sqrt{x}} \tag{19}$$

The asymptotic behaviour of this function

$$I_1(z) \underset{z\to\infty}{\sim} \frac{e^z}{\sqrt{2\pi z}} \tag{20}$$

gives, for large values of x:

$$\ln \sigma(x) \underset{x\to\infty}{\sim} 2\sqrt{x} - \frac{3}{4}\ln x \tag{21}$$

Substituting in Eq.(17) we find

$$\langle n \rangle - 2 \underset{x\to\infty}{\simeq} \sqrt{x} = \frac{\ell E_{cm}}{4\pi} \tag{22}$$

An alternative derivation is based on the observation that, at high energies, (18) is dominated by large values of n. Using Stirling's formula, $n! \sim n^n e^{-n}\sqrt{2\pi n}$, we write

$$\ln \sigma(x) \underset{x\to\infty}{\simeq} \ln \sum x^m (m+1)^{-(m+1)} e^{m+1} m^{-m} e^m \tag{23}$$

This sum can be approximated by an integral, which in turn is evaluated by means of the stationary phase method

$$\int_0^\infty e^{z\ln x + 2z - z\ln z - (z+1)\ln(z+1)}\, \mathrm{d}z \simeq \int_0^\infty e^{z\ln x + 2z - 2z\ln z - \ln z}\, \mathrm{d}z = K e^{\bar{f}} \tag{24}$$

where $\bar{f}$ is the exponent calculated at the stationary point

$$0 = \frac{\partial}{\partial z}[z \ln x + 2z - 2z \ln z - \ln z] \tag{25}$$

that is, at

$$z \simeq \sqrt{x} - \frac{1}{2} \tag{26}$$

Thus

$$\bar{f} = 2\sqrt{x} - \frac{1}{2} \ln x \tag{27}$$

The constant K is proportional to the inverse of

$$\left. \frac{\partial^2 f}{\partial z^2} \right|_{z=\sqrt{x}-\frac{1}{2}} = \frac{1}{x^{1/4}} \tag{28}$$

And finally,

$$\frac{\partial}{\partial \ln x} \ln \sigma(x) \underset{x \to \infty}{\simeq} \frac{\partial}{\partial \ln x} \left[2\sqrt{x} - \frac{3}{4} \ln x \right]$$

This again leads to Eqs. (21) and (22). In conclusion, the multiplicity at high energy is proportional to the total center-of-mass energy.

Problem 32.   A model of $p - \bar{p}$ annihilation

The proton-antiproton interaction can be described by an attractive potential well of radius $r_0 \simeq= 1.6\,\mathrm{fm}$ and depth $V_0 = 1600$ MeV, together with a totally absorbing (bottomless) well of radius $r_1 = r_0/20$, which represents the decay of $p\bar{p}$ into pions.
a) Write down the radial potential for states with $\ell > 0$. Derive the equation for the energy of bound states and resonances in the well, using the semiclassical approximation

$$\oint p_r dr = \left(n + \frac{1}{2} \right) h \tag{1}$$

b) Calculate the elastic width of the resonances, and their width of decay into pions.
c) Calculate the contribution of resonances to the cross section for elastic scattering and for annihilation into pions.

Solution

The radial potential in a state with angular momentum ℓ is

$$V_{\mathrm{eff}} = V + \frac{\hbar^2 \ell(\ell+1)}{2m\,r^2} \tag{2}$$

(m is the reduced mass). The form of V_{eff} is illustrated in fig. 1.

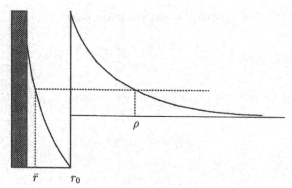

fig. 1

Inside the well we have

$$p_r = \sqrt{2m\left(E - V_{\text{eff}}(r)\right)}$$

$$= \sqrt{2m\left(E + V_0 - \frac{\hbar^2 \ell(\ell+1)}{2m\,r^2}\right)} \tag{3}$$

The semiclassical condition (1) can also be written as

$$\frac{2}{\hbar}\int_{\bar{r}}^{r_0} p_r\,\mathrm{d}r = n + \frac{1}{2} \tag{4}$$

where $\bar{r}$ is the inversion point on the left. Substituting (3), we find

$$2\int_{\bar{r}}^{r_0}\sqrt{\frac{2m(E_n + V_0)}{\hbar^2} - \frac{\ell(\ell+1)}{r^2}}\,dr = 2\pi\left(n + \frac{1}{2}\right) \tag{5}$$

The argument of the square root above is the classical velocity, which must vanish at $\bar{r}$. Thus,

$$\bar{r} = \hbar\left[\frac{\ell(\ell+1)}{2m(E_n + V_0)}\right]^{\frac{1}{2}} \tag{6}$$

Carrying out the integral in (5) we obtain

$$2\sqrt{\ell(\ell+1)}\left\{\sqrt{\left(\frac{r_0}{\bar{r}}\right)^2 - 1} - \arctan\sqrt{\left(\frac{r_0}{\bar{r}}\right)^2 - 1}\right\} = 2\pi\left(n + \frac{1}{2}\right) \tag{7}$$

From the values of $\bar{r}$ satisfying this equation we deduce, by (6), the energy eigenvalues E_n of the states in the well. Those eigenvalues satisfying

$$\frac{\hbar^2 \ell(\ell+1)}{2m\,r_0^2} \geq E_n \geq 0 \tag{8}$$

correspond to resonances, while eigenvalues $E_n < 0$ correspond to bound states.

Defining

$$x = \sqrt{\left(\frac{r_0}{\bar{r}}\right)^2 - 1} \tag{9}$$

equation (7) takes the form

$$x - \arctan x = \frac{\pi}{2} \frac{2n + 1}{\sqrt{\ell(\ell + 1)}} \tag{10}$$

The relations (6) and (9) allow us to reexpress condition (8) in terms of x as follows

$$\frac{2m V_0 r_0^2}{\hbar^2 \ell(\ell + 1)} \geq x^2 \geq \frac{2m V_0 r_0^2}{\hbar^2 \ell(\ell + 1)} - 1 \tag{11}$$

A solution of (4) will thus be a resonance if it satisfies condition (11), or a bound state ($E_n < 0$) if

$$x^2 < \frac{2m V_0 r_0^2}{\hbar^2 \ell(\ell + 1)} - 1 \tag{12}$$

With the numerical values in the problem, $2m V_0 r_0^2 / \hbar^2 \simeq 100$, so that conditions (11) and (12) read

$$\frac{100}{\ell(\ell + 1)} \geq x^2 \geq \frac{100}{\ell(\ell + 1)} - 1 \tag{13}$$

$$x^2 < \frac{100}{\ell(\ell + 1)} - 1 \tag{14}$$

Equation (10) admits a finite number of solutions compatible with conditions (11) and (12); these can be found numerically and are listed in Table 1.

n \ ℓ	1	2	3	4	5	6	7	8	9	10
0	2.266	1.673	1.406	1.246	1.135	1.054	0.991	0.939	0.896	0.860
1	4.693	3.191	2.559	2.197						
2	6.982									

Table 1

The corresponding energies are listed in Table 2.

n \ ℓ	1	2	3	4	5	6	7	8	9	10
0	-877	-772	-643	-490	-313	-113	109	355	623	914
1	-540	-330	-94	165						
2	-50									

E (MeV)

Table 2

b)   Elastic widths are given by

$$\Gamma = \phi\, P \tag{15}$$

where ϕ is the outward flux of particles trapped inside the well and P is the barrier penetration factor

$$P = e^{-\frac{2}{\hbar}\left|\int_{r_0}^{\rho}\sqrt{2m[V_{\text{eff}}(r) - E]}\,dr\right|} \tag{16}$$

In Eq.(16) V_{eff} is

$$V_{\text{eff}} = \frac{\hbar^2 \ell(\ell+1)}{2\,m\,r^2} \tag{17}$$

and ρ is the exterior inversion point, defined by the equation

$$\frac{\hbar^2 \ell(\ell+1)}{2m\,\rho^2} = E \tag{18}$$

The integration is straightforward and gives

$$\int_{r_0}^{\rho}\sqrt{2m[V_{\text{eff}} - E]}\,dr = \sqrt{\ell(\ell+1)}\left\{\operatorname{arcth}\sqrt{1 - \frac{r_0^2}{\rho^2}} - \sqrt{1 - \frac{r_0^2}{\rho^2}}\right\} \tag{19}$$

The flux equals $\phi = 1/\tau$, where τ is the period of one classical oscillation inside the well. For the potential

$$V_{\text{eff}}(r) = \frac{\hbar^2 \ell(\ell+1)}{2m\,r^2} - V_0 \tag{20}$$

we find

$$\tau = \frac{\hbar\sqrt{\ell(\ell+1)}}{E + V_0}\sqrt{\left(\frac{r_0}{\bar{r}}\right)^2 - 1} \tag{21}$$

In conclusion

$$\Gamma_{\text{el}} = \frac{E + V_0}{\sqrt{\ell(\ell+1)}}\,\frac{1}{\sqrt{\left(\frac{r_0}{\bar{r}}\right)^2 - 1}}\,e^{-2\sqrt{\ell(\ell+1)}\left\{\operatorname{arcth}\sqrt{1 - \frac{r_0^2}{\rho^2}} - \sqrt{1 - \frac{r_0^2}{\rho^2}}\right\}} \tag{22}$$

Table 3 contains the value of Γ_{el} for the resonances of Table 1.

n \ ℓ	1	2	3	4	5	6	7	8	9	10
0	–	–	–	–	–	–	0.034	8.1	50	123
1	–	–	–	93.3						
2	–									

Γ_{el} (MeV)   Table 3

Pion annihilation corresponds to crossing the barrier separating the particles from the internal well. The expression for the width can be obtained from Eq.(22), with the substitution $r_0/\rho \rightarrow r_1/\bar{r}$ in the penetration factor. Recalling the definitions

r_0 :   radius of external well r_1 :   radius of internal well
$\bar{r}$:   interior inversion point ρ :   exterior inversion point

we write

$$\Gamma_{in} = \frac{E + V_0}{\sqrt{\ell(\ell+1)}} \frac{1}{\sqrt{\left(\frac{r_0}{\bar{r}}\right)^2 - 1}} e^{-2\sqrt{\ell(\ell+1)}\left\{\text{arcth}\sqrt{1 - \frac{r_1^2}{\bar{r}^2}} - \sqrt{1 - \frac{r_1^2}{\bar{r}^2}}\right\}} \qquad (23)$$

The corresponding numerical values are shown in Table 4. (Negligible widths are left out.)

n \ ℓ	1	2	3	4	5	6	7
0	0.24	$3\cdot10^{-3}$	$2\cdot10^{-5}$	$2\cdot10^{-7}$	$2\cdot10^{-9}$	10^{-11}	10^{-13}
1	2.8	$5.8\cdot10^{-2}$	$8.9\cdot10^{-4}$	10^{-5}			
2	11.5						

Γ_{in} (MeV)   Table 4

c)   The elastic cross section is given by the Breit-Wigner formula

$$\sigma_{el} = \frac{\pi}{k^2} \sum_{\ell,n}(2\ell + 1)\frac{\Gamma_{el}^2(n,\ell)}{(E - E_{n,\ell})^2 + \frac{\Gamma^2(n,\ell)}{4}} \qquad (24)$$

with

$$\Gamma(n,\ell) = \Gamma_{el}(n,\ell) + \Gamma_{in}(n,\ell) \qquad (25)$$

The sum in (24) runs over all resonances.

For the annihilation cross section we write

$$\sigma_{in} = \frac{\pi}{k^2} \sum_{\ell,n}(2\ell + 1)\frac{\Gamma_{el}(n,\ell)\Gamma_{in}(n,\ell)}{(E - E_{n,\ell})^2 + \frac{\Gamma^2(n,\ell)}{4}} \qquad (26)$$

Problem 33.   The decay $\mu^- \rightarrow e^- \gamma$

Experimental data on muon decay provide the following limit

$$BR\left(\frac{\mu^- \rightarrow e^- \gamma}{\mu^- \rightarrow \text{anything}}\right) < 4.9 \cdot 10^{-11} \qquad 90\% \,\text{C.L.} \tag{1}$$

a)   Show that the transition $\mu^- \rightarrow e^- \gamma$, if it exists, is a magnetic dipole transition.

b)   Knowing that the muon mean life is $\tau_\mu = 2.197 \cdot 10^{-6}$sec, find the upper limit imposed by the experimental value (1) on the magnetic moment of the transition.

c)   What is the upper limit set by (1) on $BR\left(\dfrac{\mu^- \rightarrow e^- e^+ e^-}{\mu^- \rightarrow \text{anything}}\right)$?

Solution

a)   Under the assumption that the photon couples to conserved vector currents (as required by gauge and parity invariance), the matrix element for the process $\mu^- \rightarrow e^- \gamma$ takes the following most general form

$$\mathcal{M} = \varepsilon_\mu^* J^\mu$$

$$J_\mu = \bar{u}\left(\mathbf{p}_{e^-}\right)\left\{F_1\left[q^2 P_\mu - (Pq)\,q_\mu\right] + \mathrm{i}\,\frac{e}{2m_\mu}\,F_2\sigma_{\mu\nu}q^\nu\right\}u\left(\mathbf{p}_{\mu^-}\right) \tag{2}$$

where $P^\mu = (p_{\mu^-} + p_{e^-})^\mu$ and $q^\mu = (p_{\mu^-} - p_{e^-})^\mu$. Physical photons have $q^2 = 0$, $\mathbf{q}\varepsilon^* = 0$, so that the first term in the current vanishes. As a result,

$$\mathcal{M} = \mathrm{i}\,\frac{e}{2m_\mu}\,F_2\,\bar{u}\left(\mathbf{p}_{e^-}\right)\sigma_{\mu\nu}q^\nu u\left(\mathbf{p}_{\mu^-}\right)\varepsilon^{*\mu} \tag{3}$$

corresponding to a magnetic dipole transition. F_2 is the transition moment in units of the muon magnetic moment.

b)   The differential width is given by

$$d\Gamma = \frac{|\mathcal{M}|^2}{2m_\mu}\,d\Phi^{(2)} \tag{4}$$

Summing over final polarizations and averaging over initial ones leads to

$$\begin{aligned}
|\mathcal{M}|^2 &= \frac{|F_2|^2}{m_\mu^2}\frac{e^2}{8}\,\mathrm{Tr}\left\{\sigma_{\mu\nu}q^\nu\left(\not{p}_{e^-} + m_e\right)\sigma^{\mu\rho}q_\rho\left(\not{p}_{\mu^-} + m_\mu\right)\right\} \\
&= \frac{|F_2|^2}{m_\mu^2}\frac{e^2}{8}\,\mathrm{Tr}\left\{\left[4\left(qp_{e^-}\right)\not{q} - q^2\not{p}_{e^-}\right]\left(\not{p}_{\mu^-} + m_\mu\right) - 3q^2 m_e\left(\not{p}_{\mu^-} + m_\mu\right)\right\} \\
&= \frac{|F_2|^2}{m_\mu^2}\frac{e^2}{2}\left(m_\mu^2 - m_e^2\right)^2
\end{aligned} \tag{5}$$

The width for this process is then

$$\Gamma = \frac{|F_2|^2}{32\pi} e^2 m_\mu \left(1 - \frac{m_e^2}{m_\mu^2}\right)^3 \tag{6}$$

Knowing the mean life of μ, we find its total width to be $\Gamma_{tot} = 3 \cdot 10^{-16}\,\text{MeV}$. The experimental limit (1) implies

$$\Gamma \le 1.5 \cdot 10^{-26}\,\text{MeV} \tag{7}$$

hence,

$$|F_2| \le 4 \cdot 10^{-13} \tag{8}$$

c) To calculate this branching ratio, we make use of the relation

$$BR\left(\frac{\mu^- \rightarrow e^- e^+ e^-}{\mu^- \rightarrow \text{anything}}\right) = BR\left(\frac{\mu^- \rightarrow e^- e^+ e^-}{\mu^- \rightarrow e^- \gamma}\right) \cdot BR\left(\frac{\mu^- \rightarrow e^- \gamma}{\mu^- \rightarrow \text{anything}}\right) \tag{9}$$

It then suffices to calculate

$$BR\left(\frac{\mu^- \rightarrow e^- e^+ e^-}{\mu^- \rightarrow e^- \gamma}\right) = \frac{\int |\mathcal{M}_3|^2\, d\Phi^{(3)}}{\int |\mathcal{M}_2|^2\, d\Phi^{(2)}} \tag{10}$$

where

$$\mathcal{M}_2 = \mathrm{i}\,\frac{e}{2m_\mu}\,F_2\,\bar{u}\,(p_{e^-})\,\sigma_{\mu\nu}q^\nu u\,(p_{\mu^-})\,\varepsilon^{*\mu} \tag{11}$$

$$\mathcal{M}_3 = \mathrm{i}\,\frac{e}{2m_\mu}\,F_2\,\Big\{ \bar{u}\,(\mathbf{p}_{e^-})\,\sigma_{\mu\nu}q^\nu u\,(\mathbf{p}_{\mu^-})\,\frac{\mathrm{i}}{q^2}\,\bar{u}'\,(\mathbf{p}_{e^{-\prime}})\,\gamma^\mu v\,(\mathbf{p}_{e^+})$$
$$+\, \bar{u}\,(\mathbf{p}_{e^{-\prime}})\,\sigma_{\mu\nu}q'^\nu u\,(\mathbf{p}_{\mu^-})\,\frac{\mathrm{i}}{q'^2}\,\bar{u}\,(\mathbf{p}_{e^-})\,\gamma^\mu v\,(\mathbf{p}_{e^+}) \Big\} \tag{12}$$

and

$$q^\mu = (p_{\mu^-} - p_{e^-})^\mu \qquad q'^\mu = (p_{\mu^-} - p_{e^{-\prime}})^\mu \tag{13}$$

The two terms contained in $\mathcal{M}_3$ correspond to the Feynman diagrams shown in fig.1. The following formulae simplify the calculation

$$\bar{u}\,(\mathbf{p}_{e^-})\,\mathrm{i}\,\sigma_{\mu\nu}q^\nu u\,(\mathbf{p}_{\mu^-}) = \bar{u}\,(\mathbf{p}_{e^-})\,\big[\gamma_\mu\,(m_\mu + m_e) - (p_{\mu^-} + p_{e^-})_\mu\big]\,u\,(\mathbf{p}_{\mu^-}) \tag{14}$$

$$\bar{u}\,(\mathbf{p}_{e^{-\prime}})\,\mathrm{i}\,\sigma_{\mu\nu}q'^\nu u\,(\mathbf{p}_{\mu^-}) = \bar{u}\,(\mathbf{p}_{e^{-\prime}})\,\big[\gamma_\mu\,(m_\mu + m_e) - (p_{\mu^-} + p_{e^{-\prime}})_\mu\big]\,u\,(\mathbf{p}_{\mu^-}) \tag{15}$$

There are two types of contributions to $|\mathcal{M}_3|^2$: Those coming from the modulus squared of a diagram

$$\frac{e^4|F_2|^2}{4m_\mu^2}\Big\{ |\bar{u}\,(\mathbf{p}_{e^-})\,\sigma_{\mu\nu}q^\nu u\,(\mathbf{p}_{\mu^-})\,\frac{\mathrm{i}}{q^2}\,\bar{u}'\,(\mathbf{p}_{e^{-\prime}})\,\gamma^\mu v\,(\mathbf{p}_{e^+})|^2$$
$$+\, |\bar{u}\,(\mathbf{p}_{e^{-\prime}})\,\sigma_{\mu\nu}q'^\nu u\,(\mathbf{p}_{\mu^-})\,\frac{\mathrm{i}}{q'^2}\,\bar{u}\,(\mathbf{p}_{e^-})\,\gamma^\mu v\,(\mathbf{p}_{e^+})|^2 \Big\} \tag{16}$$

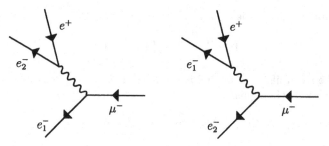

fig. 1

(once integrated over phase space, the two contributions are equivalent) and interference terms

$$\frac{e^4|F_2|^2}{2m_\mu^2} \left(\bar{u}\left(\mathbf{p}_{e^-}\right) \sigma_{\mu\nu} q^\nu u \left(\mathbf{p}_{\mu^-}\right) \frac{i}{q^2} \bar{u}'\left(\mathbf{p}_{e^{-\prime}}\right) \gamma^\mu v \left(\mathbf{p}_{e^+}\right) \right)$$

$$\cdot \left(\bar{u}\left(\mathbf{p}_{e^{-\prime}}\right) \sigma_{\mu\nu} q'^\nu u \left(\mathbf{p}_{\mu^-}\right) \frac{i}{q'^2} \bar{u}\left(\mathbf{p}_{e^-}\right) \gamma^\mu v \left(\mathbf{p}_{e^+}\right) \right) \tag{17}$$

Noting that the three-body phase space element is

$$\mathrm{d}\Phi^{(3)} = \frac{1}{128\pi^3} \frac{dq^2 dq'^2}{m_\mu^2} \tag{18}$$

one expects that the dominant parts of these amplitudes come from small values of q^2 or q'^2. The interference term corresponds physically to a superposition of the wave functions of the two e^-; the farther apart the directions of the two e^-, the more this term will decrease. Now, as q^2 approaches its minimum value, $q^2 \simeq 4m_e^2$, the corresponding e^- travels together with e^+ and in opposite direction (in the center of mass) with respect to the second e^-; therefore, the interference term should be negligible. We verify this fact explicitly, by calculating the full expression in the limit $m_e \to 0$ and setting the lower limit $q^2 = 4m_e^2$ in the final integration. Carrying out the traces of the γ matrices we obtain

$$|\mathcal{M}_3|^2 = \frac{e^4|F_2|^2}{4m_\mu^2} \left\{ \frac{q'^2}{q^2} \left(q^2 + q'^2 - m_\mu^2\right) + \frac{1}{2}\frac{m_\mu^2}{q^2} \left(m_\mu^2 - q'^2\right) \right.$$

$$\left. + \frac{q^2}{q'^2} \left(q^2 + q'^2 - m_\mu^2\right) + \frac{1}{2}\frac{m_\mu^2}{q'^2} \left(m_\mu^2 - q^2\right) + \left(q^2 + q'^2 - m_\mu^2\right) \right\} \tag{19}$$

the last term is the one due to the interference. Clearly, this terms exhibits no singularity for $q^2, q'^2 \to 0$, as expected, so that the most singular contributions come from the other terms. We isolate all singularities and, exploiting the symmetries of phase space, we find

$$\begin{aligned}
\Gamma &\simeq \frac{e^4}{128\pi^3 m_\mu^5} \int_{4m_e^2}^{m_\mu^2} \frac{dq^2}{q^2} \int_0^{m_\mu^2-q^2} dq'^2 \left\{ q'^2 \left(q'^2 - m_\mu^2 \right) + \frac{1}{2}m_\mu^4 \right\} \\
&\simeq \frac{|F_2|^2 \alpha^2}{6\pi} m_\mu \ln\frac{m_\mu^2}{4m_e^2}
\end{aligned} \tag{20}$$

A factor of 1/2, introduced above, takes into account the presence of two identical particles (e^-) in the final state. In conclusion, we find

$$BR\left(\frac{\mu^- \to e^- e^+ e^-}{\mu^- \to e^- \gamma} \right) \simeq \frac{\alpha}{3\pi} \ln\frac{m_\mu^2}{4m_e^2} \tag{21}$$

One can arrive at the same result also in another manner, if the interference term is a priori neglected. In this case, one considers only a single diagram and decompose three-body phase space as follows

$$d\Phi^{(3)}\left(M \to m_1, m_2, m_3 \right) = \frac{d\mu^2}{2\pi} d\Phi^{(2)}\left(M \to m_1, \mu \right) d\Phi^{(2)}\left(\mu \to m_2, m_3 \right) \tag{22}$$

which in our case reduces to

$$d\Phi^{(3)} = \frac{dq^2}{2\pi} \frac{1}{8\pi} \left(1 - \frac{q^2}{m_\mu^2} \right) d\Phi^{(2)}\left(q \to e^+ e'^- \right) \tag{23}$$

We now write the amplitude in the form

$$\mathcal{M}_3 = J_\mu(q) \frac{ie}{q^2} j^\mu\left(e^+ e^- \right) \tag{24}$$

where $J_\mu(q)$ is defined in (2) and $j^\mu(e^+e^-) = \overline{u}(\mathbf{p}_{e-})\gamma^\mu v(\mathbf{p}_{e+})$, obtaining

$$|\mathcal{M}_3|^2 = J_\mu(q) J_\nu^*(q) \frac{e^4}{(q^2)^2} L_{\mu\nu}^{(e^+e^-)} \tag{25}$$

$$L_{\mu\nu}^{(e^+e^-)} = \overline{j_\mu(e^+e^-)j_\nu^+(e^+e^-)}$$

Integrating over two-body phase space (see Appendix F)

$$\int L_{\mu\nu}^{(e^+e^-)} d\Phi^{(2)}\left(q \to e^+ e^- \right) = \frac{1}{6\pi} \left(q_\mu q_\nu - q^2 g_{\mu\nu} \right) \tag{26}$$

leads directly to

$$\frac{\Gamma_3}{\Gamma_2} = \frac{\int \frac{4}{3q^2}|J_\mu(q)|^2 \frac{dq^2}{2\pi} \frac{e^2}{8\pi} \left(1 - \frac{q^2}{m_\mu^2} \right)}{|J_\mu(0)|^2} \tag{27}$$

The most singular part of the above, setting $J_\mu(q) \simeq J_\mu(0)$, results again in

$$\frac{\Gamma_3}{\Gamma_2} \simeq \frac{\alpha}{3\pi} \int_{4m_e^2}^{m_\mu^2} \frac{dq^2}{q^2} \simeq \frac{\alpha}{3\pi} \ln\frac{m_\mu^2}{4m_e^2} \tag{28}$$

Finally,

$$BR\left(\frac{\mu^- \to e^- e^+ e^-}{\mu^- \to \text{anything}}\right) < 3.5 \cdot 10^{-13} \tag{29}$$

Throughout this calculation we have ignored a possible contribution from $F_1(q^2)$ in Eq.(2). For $q^2 \neq 0$ this is in principle present; however, since $F_1(0) = 0$, its effect on the amplitude will be negligible in the region $q^2 \ll m_\mu^2$ which is the one giving the dominant contribution to Γ_3.

Problem 34. $\pi^0 \to e^+ e^-$ in an external electric field

π^0 decays almost excusively to 2γ, with mean life $\tau = .84 \cdot 10^{-16}$ sec; its parity is $P = -1$. In the presence of a static external electric field, the following process may take place

$$\gamma^* + \pi^0 \to e^+ + e^- \tag{1}$$

where γ^* is a virtual photon supplied by the static field.

Calculate the branching ratio induced by the external field

$$BR_{E_0}\left(\frac{\pi^0 \to e^+ e^-}{\pi^0 \to \gamma\gamma}\right) \tag{2}$$

as a function of the field $\mathbf{E}_0$ and the pion momentum $\mathbf{p}$. For which values of these parameters is the branching ratio greater than 10^{-7} ?

Solution

Since the pion has negative intrinsic parity, the amplitude for $\pi^0 \to \gamma\gamma$ must have the form

$$\mathcal{M} = \frac{4C}{m_\pi} \varepsilon_{\mu\nu\rho\sigma} k_1^\mu \varepsilon_1^\nu k_2^\rho \varepsilon_2^\sigma \tag{3}$$

Thus

$$\overline{|\mathcal{M}|^2} = 8|C|^2 m_\pi^2 \tag{4}$$

and the decay width is

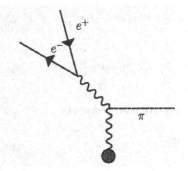

fig. 1

$$\Gamma = \frac{1}{2m_\pi} \overline{|\mathcal{M}|^2} \frac{1}{16\pi} = \frac{|C|^2}{4\pi} m_\pi \qquad (5)$$

From this we obtain

$$\frac{|C|^2}{4\pi} = \frac{\Gamma}{m_\pi} \qquad |C| \simeq .8 \cdot 10^{-3} \qquad (6)$$

Fig. 1 shows the leading diagram for the process $\pi^0 \to e^+e^-$ in an external field; the corresponding amplitude is

$$\mathcal{M} = \frac{2Ce}{m_\pi} \varepsilon_{\mu\nu\rho\sigma} F_{\text{ex}}^{\mu\nu} \frac{iq^\rho}{q^2} \bar{u}\left(e^-\right) \gamma^\sigma v\left(e^+\right) \qquad (7)$$

with $q_\mu = (p_{e^+} + p_{e^-})_\mu$. There follows

$$|\mathcal{M}|^2 = \frac{4|C|^2 e^2}{m_\pi^2} F_{\text{ex}}^{\mu\nu} F_{\text{ex}}^{\mu'\nu'} \varepsilon_{\mu\nu\rho\sigma} \varepsilon_{\mu'\nu'\rho'\sigma'} \frac{q^\rho q^{\rho'}}{\left(q^2\right)^2} \bar{u}\gamma^\sigma v \left(\bar{u}\gamma^{\sigma'} v\right)^* \qquad (8)$$

The leptonic tensor, summed over polarizations,

$$L^{\sigma\sigma'} = \sum_{\text{pol}} \bar{u}\gamma^\sigma v \left(\bar{u}\gamma^{\sigma'} v\right)^* \qquad (9)$$

becomes

$$L^{\sigma\sigma'} = 2\left(q^\sigma q^{\sigma'} - q^2 g^{\sigma\sigma'}\right) - 2\left(p_{e^+} - p_{e^-}\right)^\sigma \left(p_{e^+} - p_{e^-}\right)^{\sigma'} \qquad (10)$$

Since the static field is purely electric, we write

$$F_{\text{ex}}^{0i} = E_{\text{ex}}^i \qquad (11)$$

The amplitude now reads

$$\overline{|\mathcal{M}_{fi}|^2} = \frac{16|C|^2 e^2}{m_\pi^2} E_{\text{ex}}^i E_{\text{ex}}^{i'} \varepsilon_{ijk}\varepsilon_{i'j'k'} \frac{q^j q^{j'}}{\left(q^2\right)^2} L^{kk'} \qquad (12)$$

Integration over the phase space of e^+e^- gives (see Appendix F)

$$\frac{1}{\Phi^{(2)}} \int L^{\sigma\sigma'} \, d\Phi^{(2)} = \frac{4}{3}\left(1 + \frac{2m_e^2}{q^2}\right)\left(q^\sigma q^{\sigma'} - q^2 g^{\sigma\sigma'}\right) \qquad (13)$$

The approximation $q^2 \simeq m_\pi^2$ leads to

$$\begin{aligned}
\overline{|\mathcal{M}_{fi}|^2} &= \frac{16|C|^2 e^2}{m_\pi^2} E_{\text{ex}}^i E_{\text{ex}}^{i'} \varepsilon_{ijk}\varepsilon_{i'j'k'} \frac{1}{m_\pi^2} \frac{4}{3}\left(1 + \frac{2m_e^2}{m_\pi^2}\right) p_\pi^j p_\pi^{j'} \\
&= \frac{64}{3} \frac{|C|^2 e^2}{m_\pi^4}\left(1 + \frac{2m_e^2}{m_\pi^2}\right) |\mathbf{E} \wedge \mathbf{p}_\pi|^2 \qquad (14)
\end{aligned}$$

where $\mathbf{p}_\pi$ is the momentum of the pion entering the electric field. From this we obtain the following decay width

$$\Gamma_{e^+e^-} = \frac{1}{2E_\pi}\overline{|\mathcal{M}|^2}\frac{1}{8\pi}\sqrt{1 - \frac{4m_e^2}{m_\pi^2}} \simeq \frac{16}{3}|C|^2\alpha\frac{|\mathbf{E}\wedge\mathbf{p}_\pi|^2}{m_\pi^4 E_\pi} \tag{15}$$

The ratio between the two decay widths (for the same pion beam) is

$$\frac{\Gamma_{e^+e^-}}{\Gamma_{2\gamma}} \simeq \frac{64\pi\alpha}{3}\frac{|\mathbf{E}\wedge\mathbf{p}_\pi|^2}{m_\pi^6} \tag{16}$$

This ratio will become of the order of 10^{-7} for

$$|\mathbf{E}| \simeq \frac{m_\pi}{|\mathbf{p}_\pi|}\, 4.3\cdot 10^{17}\,\text{Volt/cm} \tag{17}$$

Problem 35.   The decay $p \to e^+ + \pi^0$

There are theoretical reasons to suppose that the decay

$$p \to e^+ + \pi^0 \tag{1}$$

may be possible. This decay has not been observed todate; infact, the following limit is placed by experiments on the width of this decay mode

$$\frac{\Gamma}{\hbar} < \frac{1}{5.5\cdot10^{31}\text{yrs}} \tag{2}$$

a)   Write down the most general matrix element describing process (1), compatible with Lorentz invariance; discuss the bounds imposed by the experimental limit (2) on the parameters of this matrix element. Note that parity may be violated in this process.
b)   Calculate the angular distribution and energy of outgoing positrons in the center-of-mass frame, if the initial protons are nonpolarized. What information can be drawn from measurements of the proton and positron polarizations?
c)   The final π^0 is not observed, since it decays immediately into $\gamma\gamma$. Calculate the distribution in the angles θ_1, θ_2, formed by the momenta of e^+ and of the two photons in the final $e^+\gamma\gamma$ state, knowing that π^0 has spin zero.

Solution

a)   The most general matrix element for the process $p \to e^+ + \pi^0$ is given by the most general Lorentz invariant of the form

$$\mathcal{M} = \bar{u}(\mathbf{p}_{e^+})\, T\, u(\mathbf{p}_p) \tag{3}$$

T is a function of momenta and gamma matrices. Using momentum conservation and the equations of motion, T is reduced to

$$T = A + B\gamma^5 \tag{4}$$

where A and B are constants. Parity invariance would then require either $A = 0$ or $B = 0$.

Let us calculate the decay width as a function of these two parameters.

$$
\begin{aligned}
|\mathcal{M}|^2 &= \frac{1}{2}\,\mathrm{Tr}\left[\left(A + B\gamma^5\right)\left(\not{p}_p + m_p\right)\left(A^* - B^*\gamma^5\right)\left(\not{p}_e + m_e\right)\right] \\
&= \left(|A|^2 + |B|^2\right)\left(m_p^2 - m_{\pi^0}^2 + m_e^2\right) + 2m_e m_p\left(|A|^2 - |B|^2\right)
\end{aligned} \tag{5}
$$

Neglecting m_e, we have

$$\Gamma = \frac{1}{2m_p}\int |\mathcal{M}|^2 \, d\Phi^{(2)} = \frac{\left(|A|^2 + |B|^2\right)}{16\pi} m_p \left(1 - \frac{m_{\pi^0}^2}{m_p^2}\right) \tag{6}$$

Thus, the experimental limit requires

$$\frac{\left(|A|^2 + |B|^2\right)}{16\pi} m_p \left(1 - \frac{m_{\pi^0}^2}{m_p^2}\right) < \frac{1}{5.5 \cdot 10^{31}\mathrm{yrs}} \tag{7}$$

which results in

$$|A|^2 + |B|^2 < 5 \cdot 10^{-62} \tag{8}$$

b) Let us calculate the decay probability in the most general case, in which the initial state is polarized and final state polarization is observed

$$|\mathcal{M}|^2 = \mathrm{Tr}\left[(A + B\gamma^5)(\not{p}_p + m_p)\frac{1 + \gamma^5 \not{w}_p}{2}(A^* - B^*\gamma^5)(\not{p}_e + m_e)\frac{1 + \gamma^5 \not{w}_e}{2}\right] \tag{9}$$

with

$$w_p = (0, \zeta_p) \qquad w_e = \left(\frac{\zeta_e \mathbf{P}_e}{m_e}, \zeta_e + \frac{\mathbf{P}_e(\zeta_e \mathbf{P}_e)}{m_e\left(E_e + m_e\right)}\right) \tag{10}$$

We find

$$
\begin{aligned}
|\mathcal{M}|^2 &= \left(|A|^2 + |B|^2\right)\frac{1}{4}\mathrm{Tr}\left(\not{p}_e \not{p}_p - m_e m_p \not{w}_e \not{w}_p\right) \\
&\quad + \left(|A|^2 - |B|^2\right)\frac{1}{4}\mathrm{Tr}\left(m_e m_p + \not{p}_p \not{w}_p \not{p}_e \not{w}_e\right) \\
&\quad - 2\,\mathrm{Re}\,AB^* \frac{1}{4}\mathrm{Tr}\left(m_e \not{w}_e \not{p}_p - m_p \not{w}_p \not{p}_e\right) \\
&\quad - 2\mathrm{i}\,\mathrm{Im}\,AB^* \frac{1}{4}\mathrm{Tr}\left(\not{p}_p \not{w}_p \not{p}_e \not{w}_e \gamma^5\right)
\end{aligned} \tag{11}
$$

Carrying out the traces

$$
\begin{aligned}
|\mathcal{M}|^2 &= \left(|A|^2 + |B|^2\right)\left[m_p E_e + m_p m_e\left(\boldsymbol{\zeta}_p\boldsymbol{\zeta}_e + \frac{(\boldsymbol{\zeta}_p\mathbf{P_e})\,(\boldsymbol{\zeta}_e\mathbf{P_e})}{m_e\,(E_e + m_e)}\right)\right] \\
&+ \left(|A|^2 - |B|^2\right)\left[m_p m_e + m_p E_e\left(\boldsymbol{\zeta}_p\boldsymbol{\zeta}_e - \frac{(\boldsymbol{\zeta}_p\mathbf{P_e})\,(\boldsymbol{\zeta}_e\mathbf{P_e})}{E_e\,(E_e + m_e)}\right)\right] \\
&- 2\mathrm{Re}\,AB^*\left(m_p\,\boldsymbol{\zeta}_e\mathbf{P_e} + m_p\,\boldsymbol{\zeta}_p\mathbf{P_e}\right) + 2\mathrm{Im}\,AB^* m_p\,\mathbf{P_e}\cdot\left(\boldsymbol{\zeta}_e \wedge \boldsymbol{\zeta}_p\right) \qquad (12)
\end{aligned}
$$

with

$$
E_e = \frac{m_p^2 + m_e^2 - m_{\pi^0}^2}{2m_p} \qquad (13)
$$

If the protons are nonpolarized, the amplitude becomes

$$
|\mathcal{M}|^2 = \left(|A|^2 + |B|^2\right) m_p E_e + \left(|A|^2 - |B|^2\right) m_p m_e - 2\mathrm{Re}\,AB^* m_p \boldsymbol{\zeta}_e\mathbf{P_e} \qquad (14)
$$

The angular distribution reads

$$
\mathrm{d}\Gamma = \frac{1}{2}\left[\left(|A|^2 + |B|^2\right) E_e + \left(|A|^2 - |B|^2\right) m_e - 2\mathrm{Re}\,AB^*\,\boldsymbol{\zeta}_e\mathbf{P_e}\right]\mathrm{d}\Phi^{(2)} \qquad (15)
$$

$$
\mathrm{d}\Phi^{(2)} \simeq \left(1 - \frac{m_\pi^2}{m_p^2}\right)\frac{\mathrm{d}\Omega}{32\pi^2}
$$

In the limit $m_e \to 0$, this is proportional to

$$
1 - \frac{2\mathrm{Re}\,AB^*}{|A|^2 + |B|^2}\,\boldsymbol{\zeta}_e\hat{\mathbf{P_e}} \qquad (16)
$$

corresponding to a positron polarization

$$
\langle\boldsymbol{\zeta}_e\rangle = \frac{-2\mathbf{Re}\,AB^*}{|A|^2 + |B|^2}\hat{\mathbf{P_e}} \qquad (17)
$$

Thus a measurement of the polarization allows us to deduce the value of $\mathrm{Re}\,AB^*$, having already obtained of $|A|^2 + |B|^2$ from the decay width.

If the process is invariant under time reversal T, then A and B are relatively real, and no further measurements are necessary. In general, however, determining A and B completely (up to a physically irrelevant global phase) requires a third measurement, for example that of the following correlation involving polarized protons

$$
\langle\mathbf{P_e}\cdot\left(\boldsymbol{\zeta}_e \wedge \boldsymbol{\zeta}_p\right)\rangle \qquad (18)
$$

If T-invariance holds, then $\mathrm{Im}\,AB^* = 0$ and the above correlation vanishes.

c)   The photon distribution is isotropic in the pion rest frame, and therefore

$$
\frac{\mathrm{d}\Gamma_{2\gamma}}{\mathrm{d}\Omega_0} = \text{constant} \qquad (19)
$$

We go over to the proton rest frame via a Lorentz transformation

$$
\begin{aligned}
\omega_0 \cos\theta_0 &= \gamma\left(\omega\cos\theta - \beta\omega\right) \\
\omega_0 \sin\theta_0 &= \omega\sin\theta
\end{aligned}
\tag{20}
$$

giving

$$
\cos\theta_0 = \frac{\cos\theta - \beta}{1 - \beta\cos\theta}
\tag{21}
$$

The angular distribution in $d\Omega = 2\pi\,d\cos\theta$ becomes

$$
d\Omega_0 = d\Omega\,\frac{1-\beta^2}{(1-\beta\cos\theta)^2} = d\Omega\left(\frac{m_\pi}{E_\pi}\right)^2\frac{1}{\left(1 - \dfrac{p_\pi}{E_\pi}\cos\theta\right)^2}
\tag{22}
$$

θ is the angle between the pion momentum and the momentum of one of the two photons, in the proton rest frame.

Problem 36.   Limits on Z_0 decay into electron plus heavy lepton

Consider the decay

$$
\begin{aligned}
Z_0 \to e^- \;&+\; E^+ \\
&\;\;\llcorner_{e^+ + \gamma}
\end{aligned}
\tag{1}
$$

Z^0 is the neutral vector boson of the standard model, with mass M_Z=91.173 ± 0.020 GeV, and E^+ is a hypothetic excited state of the positron, with spin 1/2 and mass $m_+ > m_e$. The coupling of e^+ and E^+ to Z^0 is taken to be

$$
\frac{g}{4M_Z}\left\{\bar\psi_e i\sigma^{\mu\nu}\psi_E + h.c.\right\}Z^0_{\mu\nu}
\tag{2}
$$
$$
Z^0_{\mu\nu} = \partial_\mu Z^0_\nu - \partial_\nu Z^0_\mu
$$
$$
\sigma_{\mu\nu} = \frac{1}{2i}\left[\gamma_\mu, \gamma_\nu\right]
$$

and similarly for the coupling to the photon

$$
\frac{g'}{2m_+}\left\{\bar\psi_e i\sigma^{\mu\nu}\psi_E + h.c.\right\}F_{\mu\nu}
\tag{3}
$$

a)   Calculate the event distribution on the plane of the invariant masses $m^2_{e^+e^-}$ and $m^2_{e^-\gamma}$.
b)   Calculate the angular distribution of the photon with respect to the positron.
c)   Calculate the decay width for process (1).

Solution

a) The relation (see Appendix A)

$$m_{e^+\gamma}^2 + m_{e^+e^-}^2 + m_{e^-\gamma}^2 = M_Z^2 + 2m_e^2 \tag{4}$$

implies

$$m_{e^+\gamma}^2 = M_Z^2 + 2m_e^2 - m_{e^+e^-}^2 - m_{e^-\gamma}^2 \tag{5}$$

The transition probability for this process is proportional to

$$\delta(m_{e^+\gamma}^2 - m_{E^+}^2) \tag{6}$$

Therefore, all events are concentrated on a straight line segment in the $m_{e^+e}^2 - m_{e^-\gamma}^2$ plane, at 45° with respect to the axes. The event distibution along this segment is determined from the modulus squared of the matrix element.

We can deduce the event distribution through the following line of arguments:

Nonpolarized Z^0's lead to nonpolarized E^+ by invariance under parity of the interaction (2); indeed, if the polarization of e^- is not observed, then the configuration has cylindrical symmetry about the direction of the $E^+ - e^-$ momentum and E^+ can have at most a longitudinal polarization, which in turn is excluded by parity.

Hence, the angular distribution of the decay $E^+ \to e^+ + \gamma$ is isotropic in the rest frame of E^+. Now, the invariant $m_{e^+e^-}^2$ calculated in this frame equals

$$m_{e^+e^-}^2 = A + B\cos\theta \tag{7}$$

where A and B depend only on the particle masses, and θ is the angle between e^+ and e^-. Since $m_{e^+e^-}^2$ is linear in $\cos\theta$, the distribution with respect to $m_{e^+e^-}^2$ is also uniform.

We conclude that the Dalitz plot will show a uniform event density along the segment $m_{e^+\gamma}^2 = M_{E^+}^2$.

b) As discussed above, the angular distribution of γ (and e^+) in the rest frame of E^+ is isotropic; this implies a flat spectrum for γ and e^+ in a frame in which E^+ is in motion. Denoting by E and ε the energies of E^+ and e^+ in such a frame and neglecting the electron mass:

$$m_{e^+\gamma}^2 = M_{E^+}^2 = 2(E - \varepsilon)\varepsilon(1 - \cos\theta) \tag{8}$$

In terms of $\bar{\varepsilon} = \varepsilon/M_{E^+}$ and $\gamma = E/M_{E^+}$

$$1 = 2(\gamma - \bar{\varepsilon})\bar{\varepsilon}(1 - \cos\theta) \tag{9}$$

$$\bar{\varepsilon} = \frac{\gamma}{2} \pm \sqrt{\frac{\gamma^2}{4} - \frac{1}{2(1 - \cos\theta)}} \tag{10}$$

We observe that there exists a minimum angle

$$1 - \cos\theta \geq \frac{2}{\gamma} \tag{11}$$

or

$$\sin^2\frac{\theta}{2} \geq \frac{1}{\gamma} \tag{12}$$

Differentiating Eq.(9), we find

$$\mathrm{d}\cos\theta \, \bar{\varepsilon}(\gamma - \bar{\varepsilon}) = (1 - \cos\theta)(\gamma - 2\bar{\varepsilon}) \, \mathrm{d}\bar{\varepsilon}$$
$$\frac{\mathrm{d}N}{\mathrm{d}\cos\theta} = \frac{\mathrm{d}N}{\mathrm{d}\bar{\varepsilon}} \frac{\mathrm{d}\bar{\varepsilon}}{\mathrm{d}\cos\theta} \tag{13}$$

Since the energy distribution is uniform we obtain, using (9) and (10),

$$\frac{\mathrm{d}N}{\mathrm{d}\cos\theta} \propto \frac{\bar{\varepsilon}(\gamma - \bar{\varepsilon})}{(1 - \cos\theta)(\gamma - 2\bar{\varepsilon})} = \frac{1}{2}\frac{1}{(1 - \cos\theta)^2}\frac{1}{\sqrt{\dfrac{\gamma^2}{4} - \dfrac{1}{2(1 - \cos\theta)}}}$$

$$\simeq \frac{1}{(1 - \cos\theta)^{\frac{3}{2}}\sqrt{\gamma^2(1 - \cos\theta) - 2}} \tag{14}$$

c) Neglecting the electron mass, the decay width is given by

$$\Gamma_{Z \to e^- E^+} = \frac{1}{2M_Z}\left(\frac{g}{2M_Z}\right)^2 \overline{\mathcal{M}^2}\frac{1}{8\pi}\left(1 - \frac{M_{E^+}^2}{M_Z^2}\right) \tag{15}$$

$\overline{\mathcal{M}^2}$ equals

$$\begin{aligned}
\overline{\mathcal{M}^2} &= \frac{1}{3}g_{\mu\mu'}\,\mathrm{Tr}\left\{\slashed{p}_{e^-}\sigma^{\mu\nu}(P_Z)_\nu\,(\slashed{p}_{E^+} - M_{E^+})\,\sigma^{\mu'\nu'}(P_Z)_{\nu'}\right\} \\
&= -\frac{1}{12}\,\mathrm{Tr}\left\{\slashed{p}_{e^-}\,(\gamma^\mu\,\slashed{P}_Z - \slashed{P}_Z\,\gamma^\mu)\,(\slashed{p}_{E^+} - M_{E^+})\,(\gamma_\mu\slashed{P}_Z - \slashed{P}_Z\,\gamma_\mu)\right\} \\
&= \frac{1}{12}\,\mathrm{Tr}\left\{(\slashed{p}_{E^+} - M_{E^+})\left[16(P_Z p_{e^-})\slashed{P}_Z - 4P_Z^2\slashed{p}_{e^-}\right]\right\} \\
&= \frac{2}{3}\left(M_Z^2 - M_{E^+}^2\right)\left(M_Z^2 + 2M_{E^+}^2\right) \tag{16}
\end{aligned}$$

Finally,

$$\Gamma = M_Z\frac{g^2}{96\pi}\left(1 - \frac{M_{E^+}^2}{M_Z^2}\right)^2\left(1 + \frac{2M_{E^+}^2}{M_Z^2}\right) \tag{17}$$

If one were to introduce in Eq.(16) a spin projector

$$\left(1 + \gamma^5\slashed{\psi}_{E^+}\right) \tag{18}$$

the term containing γ^5 would always vanish. This confirms the conclusion of part a): a nonpolarized Z^0 will produce a nonpolarized E^+.

Problem 37.   Possible production of massive neutrinos in triton (H_3) decay

A recent experiment of triton β-decay

$$H^3 \rightarrow e^- + \bar{\nu}_e + He^3 \tag{1}$$

has exhibited an anomaly in the electron spectrum. In order to explain this anomaly one may assume that the antineutrino produced in this reaction is a superposition of the normal $\bar{\nu}_e$ and a heavy antineutrino[1] $\bar{\nu}'$

$$|\bar{\nu}\rangle = \frac{1}{1 + |\alpha|^2} \left\{ |\bar{\nu}_e\rangle + \alpha |\bar{\nu}'\rangle \right\} \tag{2}$$

$$|\alpha| \simeq 0.2, \qquad m_{\bar{\nu}'} \simeq 17 \, \text{KeV}$$

a)   Calculate the spectrum deformation with respect to the case $\alpha = 0$ and show that it depends on $|\alpha|^2$.

b)   Assuming universality of the Fermi interaction, calculate the dependence of the mean life on α. The transition has $\Delta J = 0$, $\Delta T = 0$ and no change in parity (Fermi transition). The two nuclear states have isospin $1/2$ and mass difference $Q + m_e$, with $Q = 18.6$ KeV.

Solution

a)   The distribution of decay products has the general form

$$d\Gamma \sim \sum_f |\langle f | H_I | i \rangle|^2 \, \delta^4(P_i - P_f) \tag{3}$$

In our case the final states are $\bar{\nu}_e$ and $\bar{\nu}'$. If the Fermi interaction is universal, we may write

$$H_I = \frac{G_F}{\sqrt{2}} \left[J_\mu^h \, \bar{\psi}_e \gamma^\mu (1 - \gamma^5) \left(\frac{\psi_{\nu_e}}{\sqrt{1 + |\alpha|^2}} + \frac{\alpha \, \psi_{\nu'}}{\sqrt{1 + |\alpha|^2}} \right) + h.c. \right] \tag{4}$$

where J_μ^h is the hadronic current. The spectrum of the electron is

$$d\Gamma = \frac{1}{1 + |\alpha|^2} \, d\Gamma(\bar{\nu}_e) + \frac{|\alpha|^2}{1 + |\alpha|^2} \, d\Gamma(\bar{\nu}') \tag{5}$$

with

$$d\Gamma(\bar{\nu}_e) = \frac{1}{2M} \frac{G_F^2}{2} \left| \langle He^3 | J_\mu^h | H^3 \rangle \, \bar{u}\,(\mathbf{p}_e) \gamma^\mu (1 - \gamma_5) \, v\,(\mathbf{p}_{\bar{\nu}_e}) \right|^2 d\phi^{(3)}$$

$$d\Gamma(\bar{\nu}') = \frac{1}{2M} \frac{G_F^2}{2} \left| \langle He^3 | J_\mu^h | H^3 \rangle \, \bar{u}\,(\mathbf{p}_e) \gamma^\mu (1 - \gamma_5) \, v\,(\mathbf{p}_{\bar{\nu}'}) \right|^2 d\phi'^{(3)} \tag{6}$$

---

[1] J.J.Simpson, Phys. Rev. Lett. **54** (1985) 1891.

Let us calculate these differential widths for neutrinos of arbitrary mass, treating the nuclei in the nonrelativistic limit. In general, the matrix element of the hadronic current is

$$\langle He^3| \, J_\mu \, |H^3\rangle = \langle V_\mu - A_\mu\rangle \qquad (7)$$

In the nonrelativistic limit, the only contributions allowed by parity come from V_0 and **A**. The CVC hypothesis implies

$$\langle He^3| \, V_0 \, | \, H^3\rangle = 2M \qquad (8)$$

since $\int d^3 x V_0$ is the generator T^+ of isospin

$$T^+ = \begin{pmatrix} 0 & 1 \\ 0 & 0 \end{pmatrix} \qquad (9)$$

The axial current can be written as

$$\langle He^3| \, \mathbf{A} \, | \, H^3\rangle = \boldsymbol{\sigma} \, 2M \, \beta \qquad (10)$$

where β is the ratio of axial to vector decay constants. The full matrix element thus reduces to

$$\mathcal{M} = 2M \frac{G_F}{\sqrt{2}} \left[j^0 - \beta \langle \boldsymbol{\sigma}\rangle \cdot \mathbf{j} \right] \qquad (11)$$

Summing over the final hadronic spin and averaging over the initial one, we obtain

$$|\mathcal{M}|^2 = 2M^2 G_F^2 \left[L_{00} + \beta^2 \sum_i L_{ii} \right] \qquad (12)$$

The leptonic tensor, summed over final polarizations, is

$$\begin{aligned} L^{\mu\nu} &= \operatorname{Tr}\left\{ \gamma^\mu (1 - \gamma_5)(\not{p}_{e^-} + m_e)\, \gamma^\nu (1 - \gamma^5)(\not{p}_{\bar{\nu}_e} - m_\nu)\right\} \\ &= 8 \left[p_{e^-}^\mu \, p_{\bar{\nu}_e}^\nu + p_{e^-}^\nu \, p_{\bar{\nu}_e}^\mu - (p_{\bar{\nu}_e} \cdot p_{e^-}) g^{\mu\nu} + i\,\varepsilon^{\mu\nu}{}_{\rho\sigma}\, p_{e^-}^\rho \, p_{\bar{\nu}_e}^\sigma \right] \end{aligned} \qquad (13)$$

In particular,

$$\begin{aligned} L_{00} &= 8 \left[E_{e^-} E_{\bar{\nu}_e} + \mathbf{p}_{e^-}\mathbf{p}_{\bar{\nu}_e} \right] \\ \sum_i L_{ii} &= 8 \left[\mathbf{p}_{e^-}\mathbf{p}_{\bar{\nu}_e} + 3E_{e^-} E_{\bar{\nu}_e} \right] \end{aligned} \qquad (14)$$

Integrating over the electron and neutrino directions, we find

$$d\Gamma = \frac{G_F^2}{2\pi^3} (1 + 3|\beta|^2)\, E_{e^-} \, E_{\bar{\nu}_e}\, p_{e^-}\, p_{\bar{\nu}_e} \, dE_{e^-} \; dE_{\bar{\nu}_e}\; \delta(\Delta - E_{e^-} - E_{\bar{\nu}_e}) \qquad (15)$$

Δ is the mass difference between the two nuclear states, $\Delta = Q + m_e$. The integral over neutrino energy can now be eliminated against the δ-function, with the result

$$d\Gamma = \frac{G_F^2}{2\pi^3} (1 + 3|\beta|^2)\, E_{e^-} \, (\Delta - E_{e^-})\sqrt{(\Delta - E_{e^-})^2 - m_\nu^2}\, \sqrt{E_{e^-}^2 - m_e^2}\; dE_{e^-} \qquad (16)$$

The kinematic limits are

$$m_e \leq E_{e^-} \leq \Delta - m_\nu \tag{17}$$

Substituting this expression in Eq.(5) we obtain the following electron spectrum

$$d\Gamma = \frac{1}{1 + |\alpha|^2} \frac{G_F^2}{2\pi^3} (1 + 3|\beta|^2) \sqrt{E_{e^-}^2 - m_e^2}\, E_{e^-} (\Delta - E_{e^-}) \cdot$$

$$\cdot \left[(\Delta - E_{e^-})\, \theta(\Delta - E_{e^-}) + |\alpha|^2 \sqrt{(\Delta - E_{e^-})^2 - m_\nu^2}\, \theta(\Delta - m_{\nu'} - E_{e^-}) \right] dE_{e^-} \tag{18}$$

Since all energies are nonrelativistic, it is convenient to use the variables

$$\begin{aligned} T &= E_{e^-} - m_e \\ Q &= \Delta - m_e \end{aligned} \tag{19}$$

Then, Eq.(18) can be rewritten in the form

$$d\Gamma = \frac{1}{1 + |\alpha|^2} \frac{G_F^2}{2\pi^3} (1 + 3|\beta|^2)\, m_e \sqrt{2m_e T}\, (Q - T) \cdot$$

$$\cdot \left[(Q - T)\, \theta(Q - T) + |\alpha|^2 \sqrt{(Q - T)^2 - m_{\nu'}^2}\, \theta(Q - T - m_{\nu'}) \right] dT \tag{20}$$

b) To calculate the mean life we integrate the above expression in T. This is easily done in the approximation $Q - m_{\nu'} \ll Q$, with the result

$$\Gamma = \frac{G_F^2}{2\pi^3} \frac{1 + 3|\beta|^2}{1 + |\alpha|^2}\, m_e \sqrt{2m_e} \left[\frac{16}{105} Q^{\frac{7}{2}} + \frac{\pi}{8} \sqrt{Q + m_{\nu'}}\, (Q - m_{\nu'})^2\, Q |\alpha|^2 \right]$$

$$= \frac{G_F^2}{2\pi^3} (1 + 3|\beta|^2)\, m_e \sqrt{2m_e}\, Q^{\frac{7}{2}} \frac{16}{105}\, \frac{1 + |\alpha|^2 \frac{105\pi}{128} \sqrt{\frac{Q + m_{\nu'}}{Q}} \left(\frac{Q - m_{\nu'}}{Q} \right)^2}{1 + |\alpha|^2}$$

$$= \left[1.4 \cdot 10^{-25}\, \text{eV} \right] \cdot (1 + 3|\beta|^2) \frac{1 + 2.6 \cdot 10^{-2} |\alpha|^2}{1 + |\alpha|^2} \tag{21}$$

For $\alpha \simeq 0.2$, we find

$$\Gamma_\alpha \simeq \Gamma_{\alpha=0} (1 - |\alpha|^2)$$

$$\left| \frac{\Delta\Gamma}{\Gamma} \right| \simeq |\alpha|^2 \simeq 0.04 \tag{22}$$

Problem 38.   On the possibility of a halo in the charge distribution of the proton

Suppose the proton charge distribution has the following form

$$e\rho(r) = \frac{e}{4\pi r} \left[(1 - \varepsilon) \frac{1}{r_0^2} e^{-\frac{r}{r_0}} + \varepsilon \frac{1}{(50\, r_0)^2} e^{-\frac{r}{50\, r_0}} \right] \tag{1}$$

with $\varepsilon \simeq 10^{-2}$ and $r_0 \simeq 2$ fm. This form was proposed some years ago, to explain the discrepancy between theory and experiment in the energy difference (Lamb shift) of the hydrogen levels $2P_{\frac{1}{2}}$ and $2S_{\frac{1}{2}}$.

For $\varepsilon = 0$ one obtains the usual proton charge distribution; the added term is a halo of radius ~ 100 fm, containing approximately 1% of the total charge.

a)   Calculate the effect, if any, of the new term on the level shifts of the hydrogen atom.

b)   Calculate the effect on the angular distribution of $e - p$ scattering, in the Born approximation.

c)   Explain why it is difficult to determine directly the existence of this halo from $e - p$ scattering. (For experimental reasons, we know that this experiment is difficult to be perform at electron energies smaller than 200 MeV.)

Solution

a)   To determine level shifts we must calculate the matrix elements of the perturbation on the states $2S$ and $2P$. This perturbation may be obtained from the charge distribution, starting with the Poisson equation

$$\Delta\phi(\mathbf{r}) = -e\rho(\mathbf{r}) \tag{2}$$

In momentum representation this becomes

$$\tilde{\phi}(\mathbf{k}) = e\,\frac{\tilde{\rho}(\mathbf{k})}{\mathbf{k}^2} \tag{3}$$

where

$$\tilde{\rho}(\mathbf{k}) = \int d^3x\, e^{+i\mathbf{k}\mathbf{x}}\rho(\mathbf{x}) = \frac{1-\varepsilon}{1+(|\mathbf{k}|r_0)^2} + \frac{\varepsilon}{1+(50\,|\mathbf{k}|r_0)^2} \tag{4}$$

We find

$$\phi(\mathbf{r}) = \int \frac{d^3k}{(2\pi)^3}\, e^{-i\mathbf{k}\mathbf{r}}\,\tilde{\phi}(\mathbf{k}) = \frac{e}{4\pi r}\left[1 - (1-\varepsilon)e^{-\frac{r}{r_0}} - \varepsilon\, e^{-\frac{r}{50 r_0}}\right] \equiv \frac{e}{4\pi r} + \phi'(r) \tag{5}$$

Using the explicit form of the radial wave functions

$$
\begin{aligned}
R_{20} &= \frac{1}{\sqrt{2}\,a_0^{\frac{3}{2}}} e^{-\frac{r}{2a_0}}\left(1 - \frac{r}{2a_0}\right) \\
R_{21} &= \frac{1}{2\sqrt{6}\,a_0^{\frac{3}{2}}} e^{-\frac{r}{2a_0}}\,\frac{r}{2a_0}
\end{aligned}
\tag{6}
$$

where a_0 is the Bohr radius, it is straightforward to calculate the matrix elements

$$\Delta E_{n\ell} = -\langle n\ell | e\,\phi'(r) | n\ell \rangle = -\int r^2\, dr\, e\,\phi'(r)\,|R_{n\ell}(r)|^2 \tag{7}$$

with the result

$$\Delta E_{20} = \frac{\alpha}{2a_0}\left[(1-\varepsilon)\frac{r_0^2\left(\frac{r_0^2}{2}+a_0^2\right)}{(r_0+a_0)^4} + \varepsilon(50\,r_0)^2\frac{\left(\frac{50\,r_0}{2}\right)^2+a_0^2}{(50\,r_0+a_0)^4}\right]$$

$$\Delta E_{21} = \frac{\alpha}{2a_0}\left[(1-\varepsilon)\frac{r_0^4}{2(r_0+a_0)^4} + \frac{\varepsilon(50\,r_0)^4}{2(50\,r_0+a_0)^4}\right] \tag{8}$$

To first order in $\left(\frac{r_0}{a_0}\right)^2$ we find

$$\Delta E_{20} \simeq \varepsilon\frac{\alpha}{2a_0}\left(\frac{50\,r_0}{a_0}\right)^2 \tag{9a}$$

$$\Delta E_{21} \simeq 0 \tag{9b}$$

We observe that, since the P wave vanishes at the origin, both the usual proton charge distribution and the halo have a negligible effect; the only relevant correction is a raising of the $2S$ level.

In conclusion,

$$\Delta E_{2S-2P} \simeq \epsilon\frac{\alpha}{2a_0}\left(\frac{50\,r_0}{a_0}\right)^2 \tag{10}$$

Comparing with the fine structure splitting

$$\Delta E_{fs} = \frac{\alpha^3}{8a_0} \tag{11}$$

we have

$$\frac{\Delta E_{2S-2P}}{\Delta E_{fs}} \simeq \frac{4\varepsilon}{\alpha^2}\left(\frac{50\,r_0}{a_0}\right)^2 \simeq 10^{-3} \tag{12}$$

for $\varepsilon \simeq 10^{-2}$; this is of the same order of magnitude as the Lamb shift.

b)   The differential cross section for $e-p$ scattering in the Born approximation is

$$\frac{d\sigma}{d\Omega} = \frac{\alpha^2}{4E^2\sin^4\frac{\theta}{2}}\cos^2\left(\frac{\theta}{2}\right)|\tilde{\rho}(\mathbf{k})|^2 \qquad m_p \gg E \gg m_e \tag{13}$$

where $\tilde{\rho}(\mathbf{k})$ is given in (4).

For $E \simeq 200$ MeV, we write

$$|\mathbf{k}|\,r_0 \simeq 2E\,r_0\sin\frac{\theta}{2} \simeq 4\sin\frac{\theta}{2} \tag{14}$$

and the explicit angular dependence may be read off (13).

c) The term in ε will be significant only if $(50\,|\mathbf{k}|r_0)^2 \leq 1$, which implies

$$200 \sin \frac{\theta}{2} \leq 1 \qquad \text{or} \qquad \theta \leq \frac{1}{100} \tag{15}$$

In this case we will have

$$\tilde{\rho}(\mathbf{k}) = 1 - \varepsilon + \varepsilon \frac{1}{(100\,\theta)^2 + 1} \tag{16}$$

$$|\tilde{\rho}(\mathbf{k})|^2 = 1 - 2\varepsilon \frac{(100\,\theta)^2}{1 + (100\,\theta)^2} \tag{17}$$

The experimental difficulty consists in the fact that one must measure the cross section with a precision of $\sim 1\%$, at angles θ which are a fraction of a degree. The angle at which this effect is visible may increase at lower energies.

Problem 39. Limits on the existence of massless scalar particles coupled to leptons

Consider a massless scalar particle with zero charge, coupling only to electrons and muons by means of the interaction Lagrangian

$$\mathcal{L}_I(x) = g\,\phi(x)\left[\,:\bar{\psi}_e(x)\psi_e(x): + :\bar{\psi}_\mu(x)\psi_\mu(x):\,\right] \tag{1}$$

$\phi(x)$ is the scalar field describing this particle and g is a coupling constant, $g \ll 1$. We wish to study the physical consequences of the possible existence of such a particle.

a) Calculate, to lowest perturbative order, the potential between two electrons at large distances, in the nonrelativistic limit.

b) Calculate, again to lowest order, the contribution of interaction (1) to the total cross section for the process

$$e^+ e^- \to \mu^+ \mu^- \tag{2}$$

at large energies with respect to the muon mass. Knowing that the experimental value for this cross section agrees with the prediction from electrodynamics to within 1%, establish an upper bound for the constant g.

c) What is the effect of interaction (1) on the vertical direction (plumb line) of a body made out of hydrogen and one made out of deuterium? Assume that the earth contains equal amounts of protons and neutrons.

The observed verticals for these two bodies coincide up to an angle of $\Delta\theta \simeq 10^{-14}$; use this fact to set an upper limit on g.

Solution

a) The Born amplitude corresponds to the diagrams of fig.1.

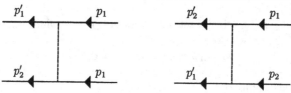

fig. 1

At large distances, the momentum transfer $p'_1 - p_1$ is small; the first diagram dominates, giving

$$- i \mathcal{M} = g^2 \, \bar{u}(\mathbf{p}'_1) \, u(\mathbf{p}_1) \, \frac{-i}{q^2} \, \bar{u}(\mathbf{p}'_2) u(\mathbf{p}_2) \qquad (3)$$

with $q^2 = (p'_1 - p_1)^2$. In the nonrelativistic limit, $u(\mathbf{p}) \simeq \sqrt{2m}\, \mathrm{w}$, where w is the electron spin state. Therefore the amplitude equals

$$- i \mathcal{M} = -i \, (2m)^2 \, \frac{g^2}{q^2} \qquad (4)$$

and is spin-independent.

The relation $E^2 - \mathbf{p}^2 = m^2$ implies

$$\frac{dE}{d|\mathbf{p}|} = v = \frac{|\mathbf{p}|}{E} \qquad (5)$$

and

$$\delta E = \frac{|\mathbf{p}|}{E} \, \delta|\mathbf{p}| \ll \delta|\mathbf{p}| \qquad (6)$$

Thus, for $v \ll 1$, the energy transfer q_0 is negligible compared to the three-momentum transfer and $q^2 \simeq -\mathbf{q}^2$. Using nonrelativistic normalization, Eq.(4) becomes

$$\mathcal{M} = -\frac{g^2}{|\mathbf{q}|^2} \qquad (7)$$

By definition, the potential is the Fourier transform of $\mathcal{M}$ in the Born approximation. Therefore,

$$V(\mathbf{r}) = -\frac{g^2}{4\pi \, |\mathbf{r}|} \qquad (8)$$

This corresponds to an attractive force.

b)   To lowest order, the amplitude for the process $e^+ e^- \to \mu^+ \mu^-$ equals

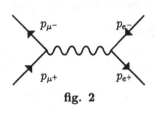

fig. 2

$$g^2\, \bar{v}(\mathbf{p}_{e^+})\, u(\mathbf{p}_{e^-})\, \frac{1}{s}\, \bar{u}(\mathbf{p}_{\mu^-})\, v(\mathbf{p}_{\mu^+}) \tag{9}$$

The corresponding diagram is shown in fig.2. This process goes through a state with angular momentum 0 in the center of mass. Consequently, the expression for the total cross section will contain no interference term with the electromagnetic amplitude, in which the intermediate state has angular momentum 1.

We thus may write

$$\sigma = \sigma_{em} + \sigma_{scalar} \tag{10}$$

$$\sigma_{em} = \frac{4\pi}{3}\,\frac{\alpha^2}{s} \tag{11}$$

$$\sigma_{scalar} = \int \frac{d\Omega}{2s}\,\frac{1}{4}\,\mathrm{Tr}\,\{\not{p}_{e^+}\not{p}_{e^-}\}\,\mathrm{Tr}\,\{\not{p}_{\mu^+}\not{p}_{\mu^-}\}\,\frac{1}{32\pi^2} \tag{12}$$

Carrying out the calculation we obtain

$$\sigma_{scalar} = \frac{\pi\alpha_g^2}{s} \tag{13}$$

where $\alpha_g = g^2/4\pi$.

The limit $\sigma_{scalar}/\sigma_{em} \leq 1\%$ results in

$$\alpha_g^2 \leq \frac{\alpha^2}{100} \tag{14}$$

c)   The potential generated by the electrons in the earth is

$$V = \frac{g\,N_e}{4\pi\,R_E} \tag{15}$$

The number of electrons almost coincides with the number of protons in the earth; it equals

$$N_e = \frac{M_E}{2m_p} \tag{16}$$

since the earth has approximately equal numbers of protons and neutrons. The ratio ρ between the hypothetic force generated by V and the gravitational force is, for hydrogen,

$$\rho_H = \frac{\alpha_g\,M_E}{2m_p}\,\frac{1}{G\,M_E\,m_p} = \frac{\alpha_g}{2G\,m_p^2} \tag{17}$$

For deuterium, the hypothetic force stays the same, since it acts only on the electron, whereas the gravitational force is doubled; thus,

$$\rho_D = \frac{\rho_H}{2} \tag{18}$$

At a given latitude the combination of these forces with the centrifugal force leads to

$$\mathbf{g}_H = \frac{G M_E}{R_E^2} \left(1 + \rho_H\right) \mathbf{n} + \Omega^2 R_E \cos\theta \, \mathbf{n}'$$

$$\mathbf{g}_D = \frac{G M_E}{R_E^2} \left(1 + \rho_D\right) \mathbf{n} + \Omega^2 R_E \cos\theta \, \mathbf{n}' \tag{19}$$

$\mathbf{n}$ is the radial unit vector with respect to the earth's center, and $\mathbf{n}'$ is a radial unit vector on the plane of the parallel.

Assuming that ρ_H, $\rho_D \ll 1$, the angle Δ between the effective vertical and the geometric one $(\mathbf{n})$ obeys

$$\operatorname{tg}\Delta = \frac{\Omega^2 R_E \sin\theta \, \cos\theta}{\dfrac{G M_E}{R_E^2}\left(1 + \rho\right) - \Omega^2 R_E \cos^2\theta} \tag{20}$$

At 45% latitude this becomes

$$\operatorname{tg}\Delta = \frac{1}{\dfrac{2G M_E}{\Omega^2 R_E^3}\left(1 + \rho\right) - 1} \tag{21}$$

The relation

$$\frac{2G M_E}{\Omega^2 R_E^3} \gg 1 \tag{22}$$

simplifies the difference in the deviation of the two bodies, as follows

$$\delta\Delta \simeq \frac{\Omega^2 R_E^3}{2G M_E}\left(\rho_H - \rho_D\right) = \frac{\Omega^2 R_E^3}{2G M_E}\frac{\rho_H}{2} \tag{23}$$

Using also Eq.(17),

$$\alpha_g \simeq 4G m_p^2 \frac{2G M_E}{\Omega^2 R_E^3}\delta\Delta \tag{24}$$

Substituting the numerical values

$$
\begin{aligned}
G &= 6.7 \cdot 10^{-8} \ \mathrm{cm}^3 \ \mathrm{sec}^{-2} \ \mathrm{g}^{-1} \\
\Omega &\simeq 2\pi \left(1.1 \cdot 10^{-5}\right) \mathrm{sec}^{-1} \\
R_E &\simeq 6.3 \cdot 10^8 \ \mathrm{cm} \\
M_E &\simeq 6 \cdot 10^{27} \mathrm{g}
\end{aligned} \tag{25}
$$

we finally obtain

$$\alpha_g \simeq 10^{-34}\, \delta\Delta \tag{26}$$

For $\delta\Delta \le 10^{-14}$ we find

$$\alpha_g \le 10^{-48} \tag{27}$$

This limit is much more stringent than the one obtained through $e^+ e^- \to \mu^+ \mu^-$ in Eq.(14).

Problem 40.   Some consequences of the existence of axions

We want to study some consequences of the existence of a pseudoscalar particle A (axion) whose mass is a few eV. The axion couples to the field ψ of the electron through the interaction Lagrangian

$$\mathcal{L} = i\,g\,\bar{\psi}\gamma_5\psi\,A \tag{1}$$

a)   Write down the amplitude for the process $\gamma + e^- \rightarrow A + e^-$, to lowest order in g and e. Verify gauge invariance and calculate, in the limit of nonrelativistic electrons, the production cross section.

b)   In the same limit, calculate the cross section for $A + e^- \rightarrow A + e^-$.

c)   Let us picture the sun as a sphere, having uniform density $\rho \simeq 1.6\,\mathrm{g/cm^3}$, radius $R = 6.6 \cdot 10^{10}$ cm and temperature $T = 10^7\,\mathrm{K}°$. Estimate the amount of energy radiated by the sun in the form of axions, as a function of g. (Consider only values of g small enough, so that the mean free path is larger than the solar radius.)

d)   Knowing that solar radiation resembles that of a black body at temperature $T_s = 6 \cdot 10^3\,\mathrm{K}°$, how would the life of the sun diminish as a function of g?

Solution

a)   To lowest perturbative order, the process $\gamma + e^- \rightarrow A + e^-$ is described by the diagrams shown in fig. 1.

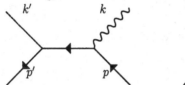

fig. 1

The corresponding amplitude is

$$
\begin{aligned}
\mathcal{M} &= -g\,e\,\bar{u}(\mathbf{p}')\left\{\gamma_5\,\frac{1}{\slashed{p}+\slashed{k}-m_e}\,\gamma_\mu + \gamma_\mu\,\frac{1}{\slashed{p}-\slashed{k}'-m_e}\,\gamma_5\right\}u(\mathbf{p})\,\varepsilon^\mu \\
&= -g\,e\,\bar{u}(\mathbf{p}')\,\gamma_5\left[\frac{\slashed{k}'\slashed{\varepsilon}}{2pk} + \frac{\slashed{\varepsilon}\slashed{k}'}{2p'k}\right]u(\mathbf{p}) \tag{2}
\end{aligned}
$$

Using the equations of motion and the relations $k' = p+k-p'$, $\slashed{\varepsilon}\slashed{k} = \slashed{k}\slashed{\varepsilon}$, Eq.(2) takes the form

$$\mathcal{M} = -g\,e\,\bar{u}(\mathbf{p}')\left[\gamma_5\left(\frac{p\cdot\varepsilon}{p\cdot k} - \frac{p'\cdot\varepsilon}{p'\cdot k}\right) + \gamma_5\frac{\slashed{k}\slashed{\varepsilon}}{2}\left(\frac{1}{p\cdot k} - \frac{1}{p'\cdot k}\right)\right]u(\mathbf{p}) \tag{3}$$

Substituting ε_μ with k_μ in (3) gives immediately zero, thus verifying gauge invariance.

In the nonrelativistic limit we have

$$\frac{p \cdot \varepsilon}{p \cdot k} - \frac{p' \cdot \varepsilon}{p' \cdot k} \simeq \frac{1}{m_e \omega}(p - p') \cdot \varepsilon = -\frac{k'\varepsilon}{m_e \omega}$$

$$\frac{1}{p \cdot k} - \frac{1}{p' \cdot k} = -\frac{1}{m_e^2}(1 - \cos\theta) \tag{4}$$

where

$$\cos\theta = \frac{\mathbf{k}\,\mathbf{k'}}{|\mathbf{k}|\,|\mathbf{k'}|} \tag{5}$$

In terms of the nonrelativistic spinors w we write

$$\bar{u}(\mathbf{p'})\gamma_5 u(\mathbf{p}) \simeq -\mathbf{w'}^* (\mathbf{k} - \mathbf{k'}) \cdot \boldsymbol{\sigma} \mathbf{w}$$

$$\bar{u}(\mathbf{p'})\gamma_5 \slashed{k}\slashed{\varepsilon} u(\mathbf{p}) \simeq -2m_e \omega \mathbf{w'}^* \boldsymbol{\varepsilon} \cdot \boldsymbol{\sigma} \mathbf{w} \tag{6}$$

The amplitude now reads

$$\mathcal{M} \simeq -\frac{\omega}{m_e} g\, e\, \mathbf{w'}^* \{(\mathbf{n} - \mathbf{n'}) \cdot \boldsymbol{\sigma}\, \mathbf{n'}\varepsilon + (\boldsymbol{\varepsilon} \cdot \boldsymbol{\sigma})\,(1 - \cos\theta)\}\,\mathbf{w} \tag{7}$$

Averaging on initial and summing over final spin orientations we obtain

$$\overline{|\mathcal{M}|^2} = \frac{\omega^2}{m_e^2} g^2 e^2\,(1 - \cos\theta)^2 \tag{8}$$

The formula for the differential cross section

$$d\sigma \simeq \frac{1}{4m_e \omega}\overline{|\mathcal{M}|^2}\,d\Phi^{(2)} \tag{9}$$

$$d\Phi^{(2)} \simeq \frac{\omega}{m_e\,16\pi^2}\,d\Omega$$

now becomes

$$\frac{d\sigma}{d\Omega} \simeq \frac{1}{16\pi} g^2\alpha\,\frac{\omega^2}{m_e^4}\,(1 - \cos\theta)^2 \tag{10}$$

The total cross section in the nonrelativistic limit equals

$$\sigma = \frac{1}{3} g^2\alpha\,\frac{\omega^2}{m_e^4} \tag{11}$$

b)   The amplitude for the process $A + e^- \to A + e^-$ is

$$\mathcal{M} = i g^2 \bar{u}(\mathbf{p'})\left\{\gamma_5 \frac{1}{\slashed{p} + \slashed{k} - m_e}\gamma_5 + \gamma_5 \frac{1}{\slashed{p} - \slashed{k'} - m_e}\gamma_5\right\} u(\mathbf{p})$$

$$= -i g^2 \bar{u}(\mathbf{p'})\left\{\frac{1}{m_e + \slashed{p} + \slashed{k}} + \frac{1}{m_e + \slashed{p} - \slashed{k'}}\right\} u(\mathbf{p})$$

$$= -i g^2 \bar{u}(\mathbf{p'})\left\{\frac{\slashed{k}}{2p \cdot k} + \frac{\slashed{k}}{2p' \cdot k}\right\} u(\mathbf{p}) \tag{12}$$

In the nonrelativistic limit, it reduces to

$$\mathcal{M} \simeq 2g^2\, w'^{*}\, w \tag{13}$$

We are thus led to the cross section

$$\frac{d\sigma}{d\Omega} = \left(\frac{g^2}{4\pi}\right)^2 \frac{1}{m_e^2}$$

$$\sigma = \frac{g^4}{4\pi}\frac{1}{m_e^2} \tag{14}$$

c)   The energy distribution of axions produced in the solar interior, per unit volume and time, can be expressed in terms of the equilibrium photon distribution in the sun as follows

$$\frac{dN_A(E)}{dE\, dV\, dt} = \frac{d\Phi_\gamma}{d\omega}\,\rho_e\,\sigma(\omega) \tag{15}$$

Here, $\dfrac{d\Phi_\gamma}{d\omega}$ is the photon flux per unit energy, related to the Planck distribution through the formula

$$\frac{d\Phi_\gamma}{d\omega} = \frac{dN_\gamma}{d\omega\, dV} \tag{16}$$

$$\frac{dN_\gamma}{d\omega\, dV} = \frac{1}{4\pi^2}\frac{\omega^2}{e^{\hbar\omega/kT}-1}$$

We estimate the electron density assuming that the sun is made primarily out of hydrogen

$$\rho_e = \frac{\rho}{m_p} \simeq 10^{24}/\mathrm{cm}^3 \tag{17}$$

Consistently with the approximations made above, we may set $E \simeq \omega$, writing

$$\frac{dN_A}{d\omega\, dV\, dt} = \frac{1}{4\pi^2}\frac{\omega^2}{e^{\hbar\omega/kT}-1}\frac{\rho}{m_p}\frac{g^2\alpha}{3}\frac{\omega^2}{m_e^4} \tag{18}$$

Thus, the energy radiated by the sun per unit time in the form of axions is

$$\frac{dE_A}{dt} = V_0 \int \omega \frac{dN_A}{d\omega\, dV\, dt} = \frac{M_\odot}{m_p}\frac{\alpha\,g^2}{12\pi^2}\frac{(kT)^6}{m_e^4}\int_0^\infty \frac{x^5 dx}{e^x-1} =$$

$$= \frac{M_\odot}{m_p}\frac{\alpha\,g^2}{12\pi^2}\frac{(kT)^6}{m_e^4}\,\Gamma(6)\,\zeta(6) = \frac{M_\odot}{m_p}\frac{2\pi^4}{189}\,\alpha\,g^2\frac{(kT)^6}{m_e^4}$$

$$\simeq 0.75{\cdot}10^{53}\,\mathrm{erg/sec} \times g^2 \tag{19}$$

The solar black body radiation is

$$\frac{dE_\gamma}{dt} = 4\pi R^2 \int \omega \frac{d\Phi_\gamma}{d\omega}\, d\omega = \frac{1}{\pi}R^2\,(kT_s)^4 \int \frac{x^3\, dx}{e^x-1} = R^2 (kT_s)^4\,\Gamma(4)\,\zeta(4)$$

$$= \frac{\pi^3}{15}R^2\,(kT_s)^4 \simeq 3{\cdot}10^{33}\,\mathrm{erg/sec} \tag{20}$$

T_s is the temperature of the solar surface.

The total energy radiated away by the sun per unit time can be written as

$$E = E_\gamma \left(1 + 2.5 \cdot 10^{19}\, g^2\right) \tag{21}$$

Therefore, the mean life of the sun is modified with respect to its standard value τ_0 by a factor of the order of

$$\tau = \frac{\tau_0}{1 + 2.5 \cdot 10^{19}\, g^2} \tag{22}$$

We may express g in terms of the mean free path of the axion

$$d = \frac{1}{\sigma \rho_e} \tag{23}$$

where σ is the cross section of $A + e^- \to \gamma + e^-$; we relate σ to (11) by the principle of detailed balance (setting $\omega \simeq kT$), and obtain

$$g^2 = \frac{3}{2} \frac{m_e^4}{\alpha\,(kT)^2} \frac{1}{N_A \rho} \frac{1}{d} \tag{24}$$

Substituting (24) in (19) we find

$$\frac{dE_A}{dt} = \frac{dE_\gamma}{dt} \frac{20\pi^2}{63} \left(\frac{T}{T_s}\right)^4 \frac{R}{d} \tag{25}$$

Even if $R/d \ll 1$, the energy emitted by axion formation receives a significant enhancement from the factor $(T/T_s)^4$: The temperature T of the solar interior is larger than the surface temperature T_s by an order of magnitude.

Problem 41.   Virtual electron pairs, sign of the e^+ gravitational mass, and the Eötvös experiment

a)   Consider a state corresponding to an electrostatic field with potential $\phi(\mathbf{x})$; calculate, to first perturbative order, the effect of virtual $e^+ e^-$ pair production on this state.

b)   Consider the same effect on a nucleus (Z A); calculate the energy shift (to lowest order) per unit energy of the produced pair. The electrostatic field may be taken to be Coulombic outside a radius of order $r_A = r_N A^{1/3}$ ($r_N = 1.3\,\mathrm{fm}$), and zero inside. Treat the electrons produced as relativistic ($m_e = 0$); justify this approximation.

c)   Compute the percentage shift of the nuclear mass, $\Delta M/M$, taking $Z = A/2$ and assuming that the energy of the pair is cut off at a value of $\Lambda \simeq 1/r_A$.

d)   The percentage shift in the gravitational mass, $\Delta M_G/M_G$, coincides with $\Delta M/M$, provided the gravitational masses of the electron and positron are positive; if instead these are equal and opposite to each other $\Delta M_G = 0$. In the latter case, calculate the angle between the verticals (plumb lines) of a nucleus with $A = 27$ and one with $A = 198$, at a latitude of 45°.

Solution

a) Let us denote by $|0\rangle$ the state of this system. The perturbation caused by interaction with virtual pairs is described by the Hamiltonian

$$H_I = e \int d^3x \, j^0(\mathbf{x}) \, \phi(\mathbf{x}) \tag{1}$$

where j^0 is the time component of the electromagnetic current. To first order in H_I, the perturbed state is

$$|0'\rangle = |0\rangle + \sum_{e^+e^-} |\mathbf{p}_{e^+}\mathbf{p}_{e^-}\rangle \frac{\langle \mathbf{p}_{e^+}\mathbf{p}_{e^-}|H_I|0\rangle}{E_0 - E_{e^+} - E_{e^-}} \tag{2}$$

More explicitly, using standard normalization for the states

$$|0'\rangle = |0\rangle + \int \frac{d^3\mathbf{p}_{e^-}}{2E_-(2\pi)^3} \frac{d^3\mathbf{p}_{e^+}}{2E_+(2\pi)^3} |\mathbf{p}_{e^+}\mathbf{p}_{e^-}\rangle \frac{\langle \mathbf{p}_{e^+}\mathbf{p}_{e^-}|H_I|0\rangle}{E_0 - E_{e^+} - E_{e^-}} \tag{3}$$

The matrix element can be easily calculated, giving

$$\begin{aligned}
\langle \mathbf{p}_{e^+}\mathbf{p}_{e^-}|H_I|0\rangle &= e \int d^3x \, \phi(\mathbf{x}) \, \langle \mathbf{p}_{e^+}\mathbf{p}_{e^-}|j^0(\mathbf{x})|0\rangle \\
&= e \, \tilde{\phi}(\mathbf{q}) \, \bar{u}(\mathbf{p}_{e^-}) \, \gamma^0 v(\mathbf{p}_{e^+})
\end{aligned} \tag{4}$$

where

$$\begin{aligned}
\mathbf{q} &= \mathbf{p}_{e^+} + \mathbf{p}_{e^+} \\
\tilde{\phi}(\mathbf{q}) &= \int d^3\mathbf{x} \, e^{-i\mathbf{q}\mathbf{x}} \, \phi(\mathbf{x})
\end{aligned} \tag{5}$$

b) Given that $\langle 0|H_I|0\rangle = 0$ the energy shift of this state is a second order effect in H_I. Calling M the unperturbed energy (that is, the nuclear mass)

$$\begin{aligned}
\Delta M &= M' - M = \sum \frac{|\langle \mathbf{p}_{e^+}\mathbf{p}_{e^-}|H_I|0\rangle|^2}{E_0 - E_{e^+} - E_{e^-}} \\
&= \sum_{pol} \int \frac{d^3\mathbf{p}_{e^+}}{2E_{e^+}(2\pi)^3} \int \frac{d^3\mathbf{p}_{e^-}}{2E_{e^-}(2\pi)^3} \frac{|\langle \mathbf{p}_+\mathbf{p}_-|H_I|0\rangle|^2}{E_0 - E_{e^+} - E_{e^-}}
\end{aligned} \tag{6}$$

Substituting expression (4) and summing over polarizations

$$\Delta M = \frac{e^2}{(2\pi)^6} \int \frac{d^3\mathbf{p}_{e^+}}{2E_{e^+}} \frac{d^3\mathbf{p}_{e^-}}{2E_{e^-}} \frac{|\tilde{\phi}(\mathbf{q})|^2}{M - E_{e^-} - E_{e^+}} L_{00} \tag{7}$$

$$L_{\mu\nu} = 2\left(q_\mu q_\nu - q^2 g_{\mu\nu} - r_\mu r_\nu\right)$$

with $q = p_{e^+} + p_{e^-}$, $r = p_{e^-} - p_{e^+}$.

We can recast the r.h.s of Eq.(7) in terms of two-body phase space, by inserting the factor

$$1 = \int (2\pi)^4 \, \delta^4 \left(q - p_{e^+} - p_{e^-}\right) \frac{d^4 q}{(2\pi)^4} \tag{8}$$

Using the following phase space average (see Appendix F)

$$\overline{r_\mu r_\nu} = \frac{1}{3} \left(q_\mu q_\nu - q^2 g_{\mu\nu}\right) \left(1 - \frac{4\, m_e^2}{q^2}\right) \tag{9}$$

we find

$$\Delta M = e^2 \int \frac{d^4 q}{(2\pi)^4} \, |\tilde{\phi}(\mathbf{q})|^2 \, \overline{L}_{00} \, \Phi^{(2)}(q^2) \tag{10}$$

with

$$\overline{L}_{\mu\nu} = \frac{4}{3} \left(q_\mu q_\nu - q^2 g_{\mu\nu}\right) \left(1 + \frac{2 m_e^2}{q^2}\right)$$

$$\Phi^{(2)}(q^2) = \frac{1}{8\pi} \left(1 - \frac{4 m_e^2}{q^2}\right) \tag{11}$$

The integral is dominated by large values of q (it diverges as $q \to \infty$); this allows us to set $m_e = 0$. We obtain

$$\begin{aligned}
\Delta M &= \int \frac{d^4 q}{(2\pi)^4} \frac{4}{3} \frac{e^2 \mathbf{q}^2 |\tilde{\phi}(\mathbf{q})|^2}{M - q_0} \, d\Phi^{(2)}(q^2) \\
&= \frac{e^2}{6\pi} \int \frac{d^4 q}{(2\pi)^4} \frac{|\tilde{\phi}(\mathbf{q})|^2 \mathbf{q}^2}{M - q_0} \simeq \frac{e^2}{24\pi^4} \int \frac{d\mu}{M - \mu} \int_0^\mu |\mathbf{q}|^4 \, d|\mathbf{q}| \, |\tilde{\phi}(\mathbf{q})|^2
\end{aligned} \tag{12}$$

In the last step we set $q^0 = \mu$. The integration limit on $|\mathbf{q}|$ stems from the constraint $q^2 \ge 0$ of Eq.(8).

For a Coulomb field truncated at r_A we have

$$\tilde{\phi}(\mathbf{q}) = Z \frac{e \cos q \, r_A}{\mathbf{q}^2}$$

$$\Delta M = \int d\mu \int_0^\mu \frac{Z^2 e^4}{24\pi^4} \, d|\mathbf{q}| \cos^2 |\mathbf{q}| \, r_A \frac{1}{M - \mu}$$

$$\frac{d\Delta M}{d\mu} = \frac{2}{3\pi^2} Z^2 \alpha^2 \frac{1}{M - \mu} \left[\frac{\mu}{2} + \frac{1}{4 r_A} \sin 2\mu \, r_A\right] \tag{13}$$

c) The cutoff $\mu \le \frac{1}{r_A} = \Lambda$ leads to ($\Lambda \ll M$)

$$\begin{aligned}
\Delta M &= \frac{2}{3\pi^2} \frac{Z^2 \alpha^2}{M} \left[\frac{\Lambda^2}{4} + \frac{1}{8 r_A^2} \left(1 - \cos 2\Lambda r_A\right)\right] \\
&\simeq \frac{2}{3\pi^2} \frac{Z^2 \alpha^2}{M} \Lambda^2 \left[\frac{1}{4} + \frac{1}{8} \left(1 - \cos 2\right)\right] \simeq 0.28 \frac{Z^2 \alpha^2}{\pi^2 M} \Lambda^2 \tag{14}
\end{aligned}$$

For $M = A m_N$

$$\frac{\Delta M}{M} \simeq 10^{-8} A^{-2/3} \tag{15}$$

d) The state $|e^+ e^-\rangle$ couples to a (constant) gravitational field through the gravitational mass. If $m_{e^+}^G = -m_{e^-}^G$, the total gravitational mass is zero, hence $\Delta m_G = 0$. The shift of the vertical line for a body with inertial mass M_I and gravitational mass M_G, at a distance r from the earth's axis of rotation (at $45°$, $r = R/\sqrt{2}$), is

$$\operatorname{tg} \alpha = \frac{\omega^2 r}{g} \frac{M_I}{M_G} \tag{16}$$

(R, ω are the earth's radius and angular velocity). If $\Delta M_G = 0$, then, for two bodies with inertial masses M_I^1 and M_I^2, we have

$$\Delta \operatorname{tg} \alpha = \frac{\Delta \alpha}{\cos^2 \alpha} \simeq 2\Delta \alpha = \frac{\omega^2 R}{\sqrt{2}\, g} \left(\frac{\Delta M_I^1}{M_I^1} - \frac{\Delta M_I^2}{M_I^2} \right) \tag{17}$$

For $A = 27$ and $A = 198$, Eqs. (15) and (17) imply

$$\Delta \alpha \simeq 10^{-11} \tag{18}$$

The sensitivity of Eötvös–like experiments is $\Delta M/M < 10^{-11}$; comparing with (15), we conclude that the effects of a hypothetic negative gravitational mass of the positron is detectable.

Problem 42.   The magnetic moment of neutrinos

As is well known, neutrinos are produced in weak decays with negative helicity.
a) Show that, for neutrinos traveling freely, helicity is a constant of the motion.
b) Can a massless neutrino have a magnetic dipole moment?
c) Suppose now that the neutrino has a mass, much smaller than the energy at which it is produced, and a nonzero magnetic dipole moment. We apply a constant magnetic field H_0, at an angle θ with respect to the neutrino momentum. What is the probability of helicity flip as a function of θ, to first perturbative order in the external field?
d) Show that, under the conditions of question c), the neutrino will oscillate between the two helicity states. Calculate the wavelength of this oscillation for a given neutrino energy.

Solution

a) Helicity is defined through the formula

$$h = \frac{\mathbf{J} \cdot \mathbf{p}}{|\mathbf{p}|} \tag{1}$$

Given that **p** and **J** are constants of the motion for a free particle, the same will be true for the helicity.

b)   The magnetic dipole coupling of a charged particle is described by a term having the form

$$f(q^2)\,\bar{u}(\mathbf{p}')\,i\,\sigma_{\mu\nu}q^\nu\,u(\mathbf{p})\,A^\mu(\mathbf{q}) \tag{2}$$

where $q = p' - p$ and A^μ is the external electromagnetic potential. If $q^0 = 0$ (stationary field) and $A^0 = 0$, we write

$$f(q^2)\,\bar{u}(\mathbf{p}')\,\sigma_{ij}q^j\,u(\mathbf{p})\,A^i \tag{3}$$

Since σ_{ij} is antisymmetric and $\mathbf{H} = \boldsymbol{\nabla} \wedge \mathbf{A}$, this term in effect describes the coupling of spin to a magnetic field.

For a massless particle it is convenient to use the Kramers representation; we write for left-handed spinors

$$u(\mathbf{p}) = \begin{pmatrix} \mathbf{w} \\ 0 \end{pmatrix} \tag{4}$$

w is a two-component spinor. In this representation we have

$$\gamma^0 = \begin{pmatrix} 0 & 1 \\ 1 & 0 \end{pmatrix} \qquad \boldsymbol{\gamma} = \begin{pmatrix} 0 & \boldsymbol{\sigma} \\ -\boldsymbol{\sigma} & 0 \end{pmatrix} \tag{5}$$

We see that $\sigma_{\mu\nu}$ is diagonal; hence, $\bar{u}\,\sigma_{ij}q^j\,u\,A^i$ is identically zero and there is no magnetic dipole coupling. Said differently, the coupling $\bar{u}\sigma_{\mu\nu}u$ changes the helicity; if only left (or right) helicity states are available, this coupling must be zero.

c,d)   Massive particles possess both helicity states, and they may therefore have a magnetic dipole moment. For a massive neutrino, the Hamiltonian in a magnetic field is

$$H = \boldsymbol{\alpha}\mathbf{p} + m\beta - \mu\gamma^0\,\boldsymbol{\Sigma}\cdot\mathbf{B} \tag{6}$$

where m is the mass, μ is the magnetic dipole and, in the Pauli representation,

$$\boldsymbol{\alpha} = \begin{pmatrix} 0 & \boldsymbol{\sigma} \\ \boldsymbol{\sigma} & 0 \end{pmatrix} \qquad \beta = \gamma^0 = \begin{pmatrix} 1 & 0 \\ 0 & -1 \end{pmatrix} \qquad \boldsymbol{\Sigma} = \begin{pmatrix} \boldsymbol{\sigma} & 0 \\ 0 & \boldsymbol{\sigma} \end{pmatrix} \tag{7}$$

We choose the z-axis along the magnetic field **B** and denote by θ and ϕ the polar angles of the neutrino momentum **p**. The eigenstates of

$$H_0 = \boldsymbol{\alpha}\mathbf{p} + \beta m \tag{8}$$

are

$$u_\pm(\mathbf{p}) = \begin{pmatrix} \sqrt{E+m}\,\mathbf{w}_\pm \\ \sqrt{E-m}\,\dfrac{\mathbf{p}\boldsymbol{\sigma}}{|\mathbf{p}|}\mathbf{w}_\pm \end{pmatrix} \tag{9}$$

$w_\pm$ are Pauli spinors with helicity ± 1:

$$w_+ = \begin{pmatrix} \cos\frac{\theta}{2} & e^{-i\phi/2} \\ \sin\frac{\theta}{2} & e^{-i\phi/2} \end{pmatrix} \qquad w_- = \begin{pmatrix} -\sin\frac{\theta}{2} & e^{-i\phi/2} \\ \cos\frac{\theta}{2} & e^{-i\phi/2} \end{pmatrix} \tag{10}$$

The matrix elements of $H_I = -\mu\gamma^0\,\boldsymbol{\Sigma}\cdot\mathbf{B}$ are

$$\begin{aligned} \langle u_+|H_I|u_+\rangle &= -\mu B\cos\theta\, 2m \\ \langle u_-|H_I|u_-\rangle &= +\mu B\cos\theta\, 2m \\ \langle u_-|H_I|u_+\rangle &= +\mu B\sin\theta\, 2E \end{aligned} \tag{11}$$

We note that

i) For $m = 0$ the diagonal matrix elements vanish.

ii) If the momentum is parallel to B ($\theta = 0$) there are no transitions between the two helicity states.

Normalizing the helicity eigenstates $|+\rangle$ and $|-\rangle$ to unity, the effective Hamiltonian in this basis becomes

$$\begin{aligned} H &= \begin{pmatrix} E - \delta_1 & \delta_2 \\ \delta_2 & E + \delta_1 \end{pmatrix} \\ \delta_1 &= 2\mu B\cos\theta\,\frac{m}{2E} = \frac{\mu B\cos\theta}{\gamma} \\ \delta_2 &= 2\mu B\sin\theta\,\frac{E}{2E} = \mu B\sin\theta \end{aligned} \tag{12}$$

with $\gamma = E/m$. Defining

$$\Delta = \sqrt{\delta_1^2 + \delta_2^2} = \mu B\sqrt{\sin^2\theta + \frac{1}{\gamma^2}\cos^2\theta} \tag{13}$$

the eigenstates are given by

$$\begin{aligned} |1\rangle &= \cos\frac{\alpha}{2}\,|+\rangle - \sin\frac{\alpha}{2}\,|-\rangle & \text{eigenvalue}: E - \Delta \\ |2\rangle &= \sin\frac{\alpha}{2}\,|+\rangle + \cos\frac{\alpha}{2}\,|-\rangle & \text{eigenvalue}: E + \Delta \end{aligned} \tag{14}$$

with

$$\begin{aligned} \text{tg}\,\frac{\alpha}{2} &= \frac{\Delta - \delta_1}{\delta_2} \\ \cos\frac{\alpha}{2} &= \frac{\delta_2}{\sqrt{2}\sqrt{\Delta^2 - \delta_1\Delta}} \\ \sin\frac{\alpha}{2} &= \frac{\Delta - \delta_1}{\sqrt{2}\sqrt{\Delta^2 - \delta_1\Delta}} \end{aligned} \tag{15}$$

A neutrino state with momentum $\mathbf{p}$ and helicity $|-\rangle$ at time $t = 0$ will evolve in time as follows

$$|\psi(t)\rangle = e^{i\mathbf{p}\mathbf{x} - iEt} \left\{ \cos\frac{\alpha}{2} e^{-i\Delta t}|2\rangle - \sin\frac{\alpha}{2} e^{i\Delta t}|1\rangle \right\} \tag{16}$$

Starting with $|\psi(t = 0)\rangle = |-\rangle$, the amplitude for helicity flip at time t reads

$$A = \langle +|\psi(t)\rangle = -i\sin\alpha\sin\Delta t \tag{17}$$

and the corresponding probability is

$$P = \sin^2\alpha\,\sin^2\Delta t \tag{18}$$

A neutrino traveling through a distance L with velocity $v \simeq 1$ will have

$$P = \sin^2\alpha\,\sin^2\frac{\Delta L}{v} \tag{19}$$

The oscillation has a period

$$T_{osc} = \frac{1}{\Delta} = \frac{1}{\mu B\sqrt{\sin^2\theta + \cos^2\theta/\gamma^2}} \tag{20}$$

Particular cases of interest are

i) $\theta = 0$: We have $\delta_2 = 0$, $\alpha = \pi$ and thus $P = 0$. There is no helicity flip.

ii) $\theta = \pi/2$: We have $\alpha = \pi/2$, $\Delta = \mu B$ and $P = \sin^2(\mu B t)$.

This last result may also be obtained in the following manner: In the rest frame of the neutrino one has

$$H = \begin{pmatrix} m & \mu B' \\ \mu B' & m \end{pmatrix} \tag{21}$$

where $B' = \gamma B$ is the (transverse) magnetic field in this frame. In terms of the proper time τ:

$$\psi(\tau) = e^{-im\tau}\frac{1}{\sqrt{2}}\left(|1\rangle e^{i\mu B'\tau} + |2\rangle e^{-i\mu B'\tau}\right) \tag{22}$$

and $P(\tau) = \sin^2\mu B'\tau$. Using $B' = \gamma B$ and $\tau = t/\gamma$, we recover the previous result

$$P(t) = \sin^2(\mu B t) \tag{23}$$

The period of oscillation is $1/\mu B$.

Problem 43. Neutron-antineutron oscillations

Let ψ be a Dirac field describing a free neutron, and ψ_C the charge conjugate field. Consider an effective Lagrangian $\mathcal{L} = \mathcal{L}_0 + \mathcal{L}_I$ where

$$\mathcal{L}_0 = \bar{\psi}\,(i\slashed{\partial} - m)\,\psi$$
$$\mathcal{L}_I = -\frac{\mu}{2}\,(\bar{\psi}_C\psi + \bar{\psi}\psi_C) \tag{1}$$

The term $\mathcal{L}_I$ violates baryon number (n_B).
a)   State the selection rule imposed by $\mathcal{L}$ on n_B.
b)   Derive and solve the equations of motion stemming from (1). As a suggestion, you may express the Lagrangian in terms of ψ and ψ_C, eliminating $\psi^\dagger$ and $\psi_C^\dagger$. Write down the Hamiltonian of the system, using creation and annihilation operators for neutrons and antineutrons.
c)   Consider, at time $t = 0$, a neutron at rest. How does the corresponding state evolve in time? Show that it oscillates between two stationary states, and write down the expression for these states.
d)   Uranium undergoes fission with a mean time $\tau \sim 10^{10}$ years; its formation time may be taken to be $T = 10^9$ years. Supposing that fission processes are for the most part $n\bar{n}$ annihilations, what bounds can be inferred on the parameter μ?
e)   How is the answer to part c) modified in the presence of a magnetic field H_0, knowing that the magnetic moment of the neutron is $\boldsymbol{\mu} = -1.9\,(e\hbar/2m_p c)\,\boldsymbol{\sigma}$?
Note: The cross section for $n\bar{n}$ annihilation at low energy is $\sigma_{ann} \simeq 1$ barn.

Solution

The charge conjugate field ψ_C is defined by

$$\psi_C = C\,\bar{\psi}^T \quad C = i\gamma^2\gamma^0 \quad C^T = C^{-1} = -C \tag{2}$$

T stands for transpose. Eq.(2) implies

$$\bar{\psi}_C^T = -C\,\psi = \psi^T C \tag{3}$$

a)   Baryon number is associated to the transformation $\psi \to e^{i\alpha}\psi$. Clearly this transformation leaves $\mathcal{L}_0$ invariant, while $\mathcal{L}_I$ becomes

$$\mathcal{L}_I = -\frac{\mu}{2}\left(\psi^T C\psi + \bar{\psi} C\bar{\psi}^T\right) \to -\frac{\mu}{2}\left[e^{2i\alpha}\psi^T C\psi + e^{-2i\alpha}\bar{\psi} C\bar{\psi}^T\right] \tag{4}$$

Thus the selection rule sought is $\Delta n_B = \pm 2$. Under charge conjugation C

$$\psi \to \psi_C \qquad \bar{\psi} \to \bar{\psi}_C \tag{5}$$

the full Lagrangian stays invariant.

b)   The relation $\bar{\psi} = \psi_C^T C$ allows us to express the Lagrangian in terms of ψ and ψ_C as follows

$$\mathcal{L} = \psi_C^T C \left(i \not{\partial} - m \right) \psi - \frac{\mu}{2} \left(\psi^T C \psi + \psi_C^T C \psi_C \right) \tag{6}$$

Varying with respect to ψ and ψ_C we obtain the equations of motion

$$\left(i \not{\partial} - m \right) \psi = \mu \psi_C \tag{7}$$
$$\left(i \not{\partial} - m \right) \psi_C = \mu \psi \tag{8}$$

Using annihilation operators $b_r(\mathbf{p})$ and $d_r(\mathbf{p})$ for neutrons and antineutrons the free Hamiltonian takes the form

$$H_0 = \sum_r \int d\Omega_p \, \varepsilon(\mathbf{p}) \left[b_r^\dagger(\mathbf{p}) b_r(\mathbf{p}) + d_r^\dagger(\mathbf{p}) d_r(\mathbf{p}) \right] \tag{9}$$

As usual,

$$d\Omega_p = \frac{d^3 \mathbf{p}}{\left(2\pi\right)^3 2p^0} \tag{10}$$

For the interaction Hamiltonian $H_I = -\mathcal{L}_I$, Equ. (2) gives

$$(\bar{\psi}_C \psi)^\dagger = \psi^\dagger \gamma^0 \psi_C = \bar{\psi} \, \psi_C \tag{11}$$

$$H_I = \frac{\mu}{2} \left(\psi C \psi + \text{h.c.} \right) \tag{12}$$

Expanding the Dirac fields,

$$H_I = \frac{\mu}{2} \int d^3\mathbf{x} \sum_{r,r'} \int d\Omega_p \int d\Omega_q \left[b_r(\mathbf{p}) u_r(\mathbf{p}) e^{-ipx} + d_r^\dagger(\mathbf{p}) v_{-r}(\mathbf{p}) e^{ipx} \right]$$
$$C \left[b_{r'}(\mathbf{q}) u_{r'}(\mathbf{q}) e^{-iqx} + d_{r'}^\dagger(\mathbf{q}) v_{-r'}(\mathbf{q}) e^{iqx} \right] + \text{h.c.} \tag{13}$$

Making also use of

$$C v_{-r}(\mathbf{p}) = \bar{u}_r^T(\mathbf{p})$$
$$C u_r(\mathbf{p}) = \bar{v}_{-r}^T(\mathbf{p}) \tag{14}$$

and

$$\bar{u}_r(\mathbf{p}) C u_{r'}(-\mathbf{p}) = 0$$
$$\bar{v}_r(\mathbf{p}) C v_{r'}(-\mathbf{p}) = 0 \tag{15}$$

we arrive at

$$H_I = -\mu \sum_r \int d\Omega_p \, \frac{m}{\varepsilon(\mathbf{p})} \left[d_r^\dagger(\mathbf{p}) b_r(\mathbf{p}) + b_r^\dagger(\mathbf{p}) d_r(\mathbf{p}) \right] \tag{16}$$

In conclusion

$$H = H_0 + H_I \;=\; \sum_r \int d\Omega_p \Big\{ \varepsilon(\mathbf{p}) \left[b_r^\dagger(\mathbf{p}) b_r(\mathbf{p}) + d_r^\dagger(\mathbf{p}) d(\mathbf{p}) \right]$$

$$-\mu \frac{m}{\varepsilon(\mathbf{p})} \left[d_r^\dagger(\mathbf{p}) b_r(\mathbf{p}) + b_r^\dagger(\mathbf{p}) d_r(\mathbf{p}) \right] \Big\} \qquad (17)$$

c) When acting on the states of one particle at rest, $|n\rangle = b^\dagger(\mathbf{0})|0\rangle$ and $|\bar{n}\rangle = d^\dagger(\mathbf{0})|0\rangle$, the Hamiltonian (17) gives

$$\begin{aligned} H_0 \,|n\rangle &= m\,|n\rangle & H_I \,|n\rangle &= -\mu\,|\bar{n}\rangle \\ H_0 \,|\bar{n}\rangle &= m\,|\bar{n}\rangle & H_I \,|\bar{n}\rangle &= -\mu\,|n\rangle \end{aligned} \qquad (18)$$

Thus, in this subspace we write

$$H = \begin{pmatrix} m & -\mu \\ -\mu & m \end{pmatrix} \qquad (19)$$

The eigenstates are

$$|S\rangle \;=\; \frac{|n\rangle + |\bar{n}\rangle}{\sqrt{2}} \qquad \text{eigenvalue:} \quad m - \mu \qquad (20)$$

$$|A\rangle \;=\; \frac{|n\rangle - |\bar{n}\rangle}{\sqrt{2}} \qquad \text{eigenvalue:} \quad m + \mu \qquad (21)$$

The state of one neutron at rest is then

$$|n\rangle = \frac{|S\rangle + |A\rangle}{\sqrt{2}} \qquad (22)$$

If the state at time $t = 0$ is $|\psi(0)\rangle = |n\rangle$, at later times t it becomes

$$|\psi(t)\rangle = \frac{1}{\sqrt{2}} \left(|S\rangle \, e^{-i(m-\mu)t} + |A\rangle \, e^{-i(m+\mu)t} \right) \qquad (23)$$

The probability of having an antineutron at time t is

$$P(t) = |\langle \bar{n}|\psi(t)\rangle|^2 = \frac{1}{2} \left| \langle \bar{n}|S\rangle \, e^{i\mu t} + \langle \bar{n}|A\rangle \, e^{-i\mu t} \right|^2 = \sin^2 \mu t \qquad (24)$$

d) $n\bar{n}$ annihilation gives the following contribution to the width Γ

$$\Gamma = v\, \rho_{\bar{n}}\, N_n\, \sigma_{n\bar{n}} \qquad (25)$$

with

$$\begin{aligned} \rho_{\bar{n}} &= \text{antineutron density in the nucleus} \\ v &= \text{typical relative velocity, } v \simeq \hbar/(m \cdot 1\,\text{fm}) \\ m &= \text{neutron mass} \\ N &= \text{number of neutrons in the nucleus (150 for } U_{238}) \end{aligned} \qquad (26)$$

The neutron density is $\rho_n = N/(4\pi R^3/3)$, where $R \simeq 10^{-12}$cm for U_{238}. By (24) we have

$$\rho_{\bar{n}} = \rho_n \sin^2 \mu T \simeq \rho_n (\mu T)^2 \tag{27}$$

The width now reads

$$\Gamma \simeq v \frac{3N_n^2}{4\pi R^3} \sigma_{n\bar{n}} (\mu T)^2 \tag{28}$$

Numerically,

$$\Gamma \simeq \left(3 \cdot 10^{58} \text{sec}\right) \mu^2 \tag{29}$$

The condition $\Gamma \le (10^{10} \text{ yrs})^{-1}$ leads to

$$\mu \le 10^{-32} \sec^{-1} = 6 \cdot 10^{-52} \text{ MeV} \tag{30}$$

e) The interaction of neutrons with a magnetic field is characterized by the parameter

$$\alpha = 1.9 \frac{e\hbar}{2m_n c} B \tag{31}$$

Taking for definiteness $B = 10^{-6}$ Gauss, we find $\alpha = 5.7 \cdot 10^{-24}$ MeV.

The total Hamiltonian is

$$H = H_0 + H_I + H_I' = \begin{pmatrix} m + \alpha & -\mu \\ -\mu & m - \alpha \end{pmatrix} \tag{32}$$

H_I can be treated as a perturbation. The modified eigenstates are

$$|n'\rangle = |n\rangle - \frac{\mu}{2\alpha}|\bar{n}\rangle$$
$$|\bar{n}'\rangle = |\bar{n}\rangle + \frac{\mu}{2\alpha}|n\rangle \tag{33}$$

with

$$\frac{\mu}{2\alpha} \le 2 \cdot 10^{-36} \tag{34}$$

If the initial state is $|\psi(0)\rangle = |n\rangle$, i.e.

$$|\psi(0)\rangle = |n'(0)\rangle + \frac{\mu}{2\alpha}|\bar{n}'(0)\rangle \tag{35}$$

at time t we will have

$$|\psi(t)\rangle = e^{-imt} \left[e^{-i\alpha t}|n'\rangle + \frac{\mu}{2\alpha} e^{i\alpha t}|\bar{n}'\rangle \right] \tag{36}$$

Consequently,

$$P_{\bar{n}}(t) = |\langle \bar{n}|n(t)\rangle|^2 = \frac{\mu^2}{\alpha^2} \sin^2 \alpha t \tag{37}$$

For times $\le \alpha^{-1} = 10^2$ sec, the probability (37) of observing an antineutron coincides with (24). This holds provided the beam length L satisfies

$$L \le \frac{\pi v \alpha^{-1}}{2} \tag{38}$$

Problem 44. Eigenstates of gravitons

a)   Describe the states of a massless particle with spin 2 (graviton).

b)   Show that the wave function of a monochromatic graviton can be written as a wave which propagates a transverse symmetric tensor of rank 2 with zero trace. Show that such a wave corresponds to two independent states.

c)   Suppose the graviton is absorbed by a system whose size is smaller than the wavelength. Discuss the selection rules for angular momentum and parity of the absorbing system.

d)   Compare your answers to those obtained for a spin-1 particle.

Solution

a)   Massless particles of spin s are characterized by helicity: The appropriate irreducible representations of the Poincaré group consist of states with helicity $\pm s$. Parity invariance requires that both helicity states exist, and consequently the Hilbert space is a direct sum of irreducible representations with helicity $+s$ and $-s$.

b)   For a massive particle with integer spin s, the wave function may be written in terms of an irreducible four-tensor with s indices. The tensor will be irreducible if it is symmetric under permutation of its indices, and if it vanishes under the contraction of any pair of indices. In the case of spin 2, the corresponding tensor $G_{\mu\nu}$ obeys

$$G_{\mu\nu} = G_{\nu\mu} \qquad G^\mu_\mu = 0 \tag{1}$$

The transversality condition guarantees the correct angular momentum in the rest frame; for spin 2 it reads

$$\partial^\mu G_{\mu\nu} = 0 \tag{2}$$

This leaves 5 independent components, satisfying the equation

$$\left(\Box + m^2\right) G_{\mu\nu}(x) = 0 \tag{3}$$

The solutions at a given momentum have the form

$$G_{\mu\nu}(x) = u_{\mu\nu}\, e^{-ipx} \tag{4}$$

$u_{\mu\nu}$ must satisfy $p^\mu u_{\mu\nu} = 0$, $u_{\mu\nu} = u_{\nu\mu}$, $u^\mu_\mu = 0$, by virtue of (1) and (2).

In the massless case ($p^2 = 0$), we have the additional freedom to perform transformations on $u_{\mu\nu}$ which respect conditions (1) and (2), namely

$$u_{\mu\nu} \to u'_{\mu\nu} = u_{\mu\nu} + c_\mu p_\nu + c_\nu p_\mu \tag{5}$$

c_μ is a constant four-vector which must obey

$$c_\mu p^\mu = 0 \tag{6}$$

Therefore, $u'_{\mu\nu}$ is defined up to 3 arbitrary constants (the 3 independent components of c_μ). This leaves two independent components in $u_{\mu\nu}$, describing the two helicity states.

c) The intrisic parity of a symmetric tensor is +1. In general, if the total angular momentum of the graviton is $\mathbf{j} = \mathbf{s} + \mathbf{l}$, there always exist, for any choice of j, two states with opposite parity, $(-1)^j$ and $(-1)^{j+1}$, corresponding respectively to

$$\ell = j, j \pm 2 \qquad \ell = j \pm 1 \tag{7}$$

In all cases there is a lowest value for j, $j \geq 2$, since helicity is ± 2.

At very low energies, that is, when the absorbing system is very small compared to the wavelength, the only orbital angular momentum which contributes is $\ell = 0$, implying $\mathbf{j} = \mathbf{s}$. Indicating by J_i, J_f (P_i, P_f) the initial and final values of the total angular momentum (parity), we find the selection rules

$$\begin{aligned}
P_f &= P_i \\
J_f &= J_i, J_i \pm 1, J_i \pm 2 \qquad & J_i \geq 2 \\
J_f &= 1, 2, 3 \qquad & J_i = 1 \\
J_f &= 2 \qquad & J_i = 0
\end{aligned} \tag{8}$$

Furthermore, the component M of the system's angular momentum along the direction of the graviton will obey

$$M_f = M_i \pm 2 \tag{9}$$

according to the helicity state of the graviton.

d) The spin-1 case proceeds in an analogous manner. The photon is described by a tensor $A_\mu = \varepsilon_\mu e^{-ipx}$. Transversality requires $p^\mu \varepsilon_\mu = 0$ and gauge invariance ($\varepsilon_\mu \to \varepsilon_\mu + \alpha p_\mu$) eliminates the longitudinal component.

Again, j has a lowest value, $j \geq 1$. The selection rules are

$$\begin{aligned}
P_f &= -P_i \\
J_f &= J_i, J_i \pm 1 \qquad & J_i \geq 1 \\
J_f &= 1 \qquad & J_i = 0 \\
M_f &= M_i \pm 1
\end{aligned} \tag{10}$$

Problem 45.   The cosmic radiation spectrum

Protons originating in cosmic radiation interact, as they travel through space, with the 3° K background radiation present in the universe.

a) Consider a proton with energy E and a photon with energy $E_\gamma \simeq 3°$ K. What is the corresponding center-of-mass energy?

b) The photoabsorption cross section is roughly constant, $\sigma \simeq 100\,\mu$barn, above the threshold of pion photoproduction. Calculate the mean free path of the proton immersed in background radiation, as a function of its energy.

c)   Supposing that hadrons lose on average 20% of their energy in every collision above threshold, show that the spectrum of cosmic rays is cut off abruptly at energies above the photoproduction threshold. The distance traveled by cosmic rays is taken to be sufficiently long.

Solution

a)   The center-of-mass energy is

$$E_{cm}^2 = s = (p_p + p_\gamma)^2 = m_p^2 + 2(E_p E_\gamma - \mathbf{p}_p \mathbf{p}_\gamma) \tag{1}$$

$$s = m_p^2 + 2E_p E_\gamma (1 - v \cos \theta) \tag{2}$$

At 3° K we have on the average

$$E_\gamma \sim kT = \frac{1}{4 \cdot 10^3} \text{ eV} \tag{3}$$

leading to

$$E_{cm} = m_p \sqrt{1 + 0.5 \cdot 10^{-12} \frac{E_p}{m_p} (1 - v \cos \theta)} \tag{4}$$

b)   The inelastic processes will take place above the threshold

$$s \geq (m_p + m_\pi)^2 \simeq m_p^2 + 2m_\pi m_p \tag{5}$$

Comparing with (2),

$$E_p E_\gamma (1 - v \cos \theta) \geq m_\pi m_p \tag{6}$$

Defining

$$\overline{E} = \frac{m_\pi m_p}{E_\gamma} \simeq 5.2 \cdot 10^{11} \text{ GeV} \tag{7}$$

(6) becomes

$$(1 - v \cos \theta) \geq \frac{\overline{E}}{E_p} \tag{8}$$

Thus the threshold is located at $E_p \simeq \overline{E}/2$.
   The mean free path ℓ of a proton undergoing inelastic collision is

$$\ell = \frac{1}{\sigma n f(E_p)} \tag{9}$$

where $n = 0.244 \, (kT/\hbar c)^3$ is the photon density resulting from the Planck distribution, σ is the photoabsorption cross section and $f(E_p)$ is the fraction of photons whose incidence angle renders the process above threshold. By (8), we have

$$f(E_p) = \frac{1}{2} \int_{-1}^{1 - \frac{\overline{E}}{E_p}} \mathrm{d}x = 1 - \frac{\overline{E}}{2E_p} \tag{10}$$

Substituting numerical values, we find

$$\ell = \frac{2 \cdot 10^7}{\left(1 - \dfrac{\overline{E}}{2E_p}\right)} \text{ light years} \tag{11}$$

The probability of zero collisions in a distance L is

$$e^{-L/\ell(E_p)} \tag{12}$$

Schematically, the energy distribution of protons evolves with distance x as follows

$$\frac{\mathrm{d}N(x, E_p)}{\mathrm{d}x} = -\frac{N(x, E_p)}{\ell(E_p)} + \frac{N(x, \frac{E_p/1 - \eta}{})}{\ell(\frac{E_p/1 - \eta}{})} \tag{13}$$

$\eta = 0.8$ is the inelasticity of the process. Knowing the initial distribution, we can calculate using Eq.(13) the spectrum after a distance L. We have solved (13) numerically, starting from an initial spectrum

$$N(0, E_p) \propto \frac{1}{E^3} \tag{14}$$

Fig.1 shows the logarithm of the spectrum at $L = 100\ell_0$ ($\ell_0 = 1/\sigma n$), as a function of E, expressed in GeV. A steep decrease for $E \gtrsim \overline{E}/2$ is evident.

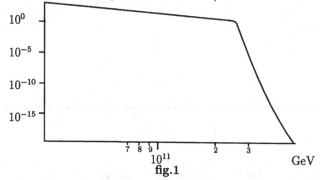

fig.1

Problem 46.   Correlations in sequential atomic decays

a)   An atom undergoes a cascade-type decay, as follows

$$0^+ \;\rightarrow\; 1^- + \gamma$$
$$\hookrightarrow\; 0^+ + \gamma \tag{1}$$

Which polarization states correspond to the two photons being emitted parallel and antiparallel to each other? Write down the helicity amplitudes of the most general configuration.

b) Reconsider the questions in part a) for the following decay sequence

$$0^- \;\rightarrow\; 1^- + \gamma$$
$$\llcorner\!\!\rightarrow 0^+ + \gamma \tag{2}$$

c) How do things change when the intermediate state has no definite parity, being in the form

$$\frac{|1^-\rangle + \varepsilon|1^+\rangle}{\sqrt{1 + |\varepsilon|^2}} \tag{3}$$

d) Discuss angular correlations in the three cases above.

Solution

a) When the two photons are emitted antiparallel to each other, they must have the same helicity: By momentum conservation, the final state atom in the center-of-mass frame moves along the axis of the two photons; angular momentum conservation then implies that the two spins must be opposite.

Let us apply a spatial reflection, followed by a 180-degree rotation about an axis perpendicular to the direction of motion; the momenta of the product particles as well as the amplitude remain invariant, whereas helicities change sign. Since the initial state is invariant under these operations, the spin wave function of the two γ's must be ($|\pm\rangle$ stands for the helicity states)

$$|a\rangle = \frac{|+\rangle|+\rangle + |-\rangle|-\rangle}{\sqrt{2}} \tag{4}$$

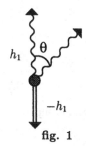 If the γ's come out parallel to each other, the spin wave function is

$$|p\rangle = \frac{|+\rangle|-\rangle + |-\rangle|+\rangle}{\sqrt{2}} \tag{5}$$

In order to write down the helicity amplitudes in a generic configuration, let us call z the axis of emission of one of the two photons, and h_1 the corresponding helicity. The excited atom recoils along $-z$ and has $J_z = -h_1$. The second γ is emitted at an angle θ with respect to the first one; by rotational invariance about z, one can choose z-x to be the decay plane.

fig. 1

The state with $J_z = -h_1$ is a superposition of eigenstates of spin along the direction of the second γ. We denote these states by $|m'\rangle$

$$|m\rangle = d^{(1)}_{m'm}(\theta)|m'\rangle \tag{6}$$

where
$$d^{(1)}_{m'm}(\theta) = \langle m'|R^{\dagger}(\theta)|m\rangle \tag{7}$$

$R(\theta)$ represents a rotation by an angle θ about the y-axis. Thus, the helicity amplitudes are, aside from a multiplicative constant

$$f_{++} = f_{--} = \langle 1|R^{\dagger}(\theta)|-1\rangle = \frac{1}{2}(1 - \cos\theta) \tag{8a}$$

$$f_{+-} = f_{-+} = \langle 1|R^{\dagger}(\theta)| 1\rangle = \frac{1}{2}(1 + \cos\theta) \tag{8b}$$

The state with $m = 0$ does not contribute.

The equalities $f_{++} = f_{--}$ and $f_{+-} = f_{-+}$ are an outcome of the combined invariance under reflections and 180-degree rotations on the decay plane. Calling S the transition operator on the atom for the emission of one photon, and indicating for simplicity only the angular momentum states, we have

$$f_{++} = \sum_{m'}\langle 0|S|m'\rangle d^{(1)}_{m',-1}(\theta)\langle -1|S|0\rangle$$

$$= \langle 0|S|1\rangle\langle -1|S|0\rangle \, d^{(1)}_{1,-1}(\theta) = \frac{1}{2}(1 - \cos\theta)\,\langle 0|S|1\rangle\langle -1|S|0\rangle \tag{9}$$

Only the value $m' = 1$ contributes by conservation of angular momentum. Use has been made above of the rotational invariance of S.

b)   The arguments of part a) apply also in this case for the configurations with parallel and antiparallel γ's, the only difference being a change of sign under the spatial reflection, due to the opposite intrinsic parities of the initial and final state atoms.

Hence in this case we obtain for the spin states $|\bar{a}\rangle$ and $|\bar{p}\rangle$

$$|\bar{a}\rangle = \frac{|+\rangle|+\rangle - |-\rangle|-\rangle}{\sqrt{2}} \tag{10}$$

$$|\bar{p}\rangle = \frac{|+\rangle|-\rangle - |-\rangle|+\rangle}{\sqrt{2}} \tag{11}$$

The argument for the generic configuration leads to

$$f_{++} = -f_{--} = \frac{1}{2}(1 - \cos\theta) \tag{12a}$$

$$f_{+-} = -f_{-+} = \frac{1}{2}(1 + \cos\theta) \tag{12b}$$

c)   The intermediate state of Eq.(3) does not have definite parity; by the superposition principle

$$|\tilde{a}\rangle = \frac{1}{\sqrt{1 + |\varepsilon|^2}}\left[|a\rangle + \varepsilon\,|\bar{a}\rangle\right]$$

$$|\tilde{p}\rangle = \frac{1}{\sqrt{1 + |\varepsilon|^2}}\left[|p\rangle + \varepsilon\,|\bar{p}\rangle\right] \tag{13}$$

$$f_{++} = \frac{1}{2}(1 - \cos\theta) \qquad f_{--} = \frac{1 - \varepsilon}{\sqrt{1 + |\varepsilon|^2}} f_{++}$$

$$f_{+-} = \frac{1}{2}(1 + \cos\theta) \qquad f_{-+} = \frac{1 - \varepsilon}{\sqrt{1 + |\varepsilon|^2}} f_{+-} \qquad (14)$$

We have assumed here that the electromagnetic interaction is parity conserving.

In terms of linear polarizations of the emitted photons, state $|a\rangle$, Eq.(4), corresponds to an amplitude $\varepsilon_1 \varepsilon_2$, and so does state $|p\rangle$, Eq.(5) (parallel polarization). States $|\bar{a}\rangle$ and $|\bar{p}\rangle$ correspond to $(\varepsilon_1 \wedge \varepsilon_2) \cdot \mathbf{n}$, where $\mathbf{n}$ is the direction of one of the two photons (perpendicular linear polarization).

The third case is a superposition of the first two, and corresponds to a relative angle of polarization $\varphi = \text{arctg}\varepsilon$.

As for circular polarization, in processes (1) and (2) left and right photons have equal probabilities of appearing in configuration $|a\rangle$, whereas in process (3) the probabilities of $++$ and $--$ have a ratio of $|(1 + \varepsilon)/(1 - \varepsilon)|^2$: The system distinguishes right from left.

d) To obtain the angular correlation one must sum the modulus squared of the helicity amplitudes over final helicities. We find

$$\frac{1}{\Gamma}\frac{d\Gamma}{d\Omega} = \frac{3}{4}\left(1 + \cos^2\theta\right)\frac{1}{4\pi} \qquad (15)$$

Problem 47. Preparation of polarized electron beams by scattering on laser radiation

In order to obtain longitudinally polarized e-beams in an accelerator, the electrons are made to collide head-on with a laser source of circularly polarized light.

a) Calculate the electron polarization after each passage, assuming that it was zero initially.

At photon energies which are low compared to the electron mass, the Compton amplitude in the rest frame of the initial electron is given by

$$\mathcal{M} = e^2 \mathbf{w}'^\dagger \left[A + \mathbf{B} \cdot \boldsymbol{\sigma}\right] \mathbf{w} \qquad (1)$$

$$A = 2\boldsymbol{\varepsilon}'^* \boldsymbol{\varepsilon}$$

$$\mathbf{B} = \frac{i|\mathbf{k}|}{2m_e} \Big\{ 2(\mathbf{n}' \wedge \boldsymbol{\varepsilon}'^*) \wedge (\mathbf{n} \wedge \boldsymbol{\varepsilon}) + (\mathbf{n} \wedge \boldsymbol{\varepsilon})(\mathbf{n} \wedge \boldsymbol{\varepsilon}'^*)$$

$$+ \quad \mathbf{n}\Big[\boldsymbol{\varepsilon}'^* \cdot (\mathbf{n} \wedge \boldsymbol{\varepsilon})\Big] - (\mathbf{n}' \wedge \boldsymbol{\varepsilon}'^*)(\mathbf{n}'\boldsymbol{\varepsilon}) - \mathbf{n}'\Big[\boldsymbol{\varepsilon} \cdot (\mathbf{n}' \wedge \boldsymbol{\varepsilon}'^*)\Big]\Big\}$$

Here $|\mathbf{k}|$ is the frequency, w and w′ are Pauli spinors, $\mathbf{n}$ and $\mathbf{n}'$ are unit vectors along the momentum of the initial and final photons, $\boldsymbol{\epsilon}$ and $\boldsymbol{\epsilon}'$ are polarization vectors.

What is the range of validity of formula (1)?

Solution

a) Consider a region of size L filled with laser photons, with density ρ. As the beam crosses this region, the probability of scattering is

$$\Pi = \sigma \rho L \qquad (2)$$

Calling $\mathbf{P}_{n-1}$ the polarization before the n-th crossing and $\mathbf{P}'_n$ the polarization of electrons which scatter, $\mathbf{P}_n$ is given by

$$\mathbf{P}_n = (1 - \Pi)\mathbf{P}_{n-1} + \Pi\mathbf{P}'_n \qquad (3)$$

To calculate σ and $\mathbf{P}'_n$ we shall use expression (1) for the scattering amplitude. The approximation in (1) is $\mathcal{O}(|\mathbf{k}|^2/m_e^2)$. The ratio $|\mathbf{k}|/m_e$ ($|\mathbf{k}|$ is the photon momentum in the rest frame of the electron) can be expressed in terms of the photon energy ω in the frame where the accelerator is at rest

$$\frac{|\mathbf{k}|}{m_e} = (1 + v_e)\frac{p_e^0}{m_e}\frac{\omega}{m_e} \simeq \frac{2p_e^0}{m_e}\frac{\omega}{m_e} \qquad (4)$$

For $\omega \sim 5\,\mathrm{eV}$ one has $|\mathbf{k}|/m_e \ll 1$ provided $p_e^0/m_e \ll 5{\cdot}10^4$ ($p_e^0 \ll 25$ GeV).

Let us calculate the invariant cross section, σ, in the rest frame of the incoming electron. Summing over final polarizations we have

$$\begin{aligned}
\sigma &= e^2 \int \frac{d^3p'_e}{(2\pi)^3 2m_e}\frac{d^3k}{(2\pi)^3 2|\mathbf{k}|}(2\pi)^4\delta^{(4)}(p_e + k - p'_e - k') \\
&\quad \sum_{\boldsymbol{\epsilon}'}\mathrm{Tr}\left[(A + \mathbf{B}\cdot\boldsymbol{\sigma})\frac{1}{2}(1 + \mathbf{P}\cdot\boldsymbol{\sigma})(A^* + \mathbf{B}^*\cdot\boldsymbol{\sigma})\right] \\
&= \left(\frac{\alpha}{2m_e}\right)^2 \int d\Omega_\gamma \sum_{\boldsymbol{\epsilon}'}\mathrm{Tr}\left[(A + \mathbf{B}\cdot\boldsymbol{\sigma})\frac{1}{2}(1 + \mathbf{P}\cdot\boldsymbol{\sigma})(A^* + \mathbf{B}^*\cdot\boldsymbol{\sigma})\right]
\end{aligned} \qquad (5)$$

The last integral in (5) is taken over the directions of the outgoing photon. To first order in $|\mathbf{k}|/m$ we write

$$\begin{aligned}
(A + \mathbf{B}\cdot\boldsymbol{\sigma})(1 + \mathbf{P}\cdot\boldsymbol{\sigma})(A^* + \mathbf{B}^*\cdot\boldsymbol{\sigma}) &\simeq AA^* + (A^*\mathbf{B} + A\mathbf{B}^*)\cdot\mathbf{P} + \\
&\quad + [AA^*\mathbf{P} + (A^*\mathbf{B} + A\mathbf{B}^*) + i(A^*\mathbf{B} - A\mathbf{B}^*)\wedge\mathbf{P}]\cdot\boldsymbol{\sigma}
\end{aligned} \qquad (6)$$

The sum over polarizations gives

$$\sum_{pol}A^*A = 4\left(1 - |\mathbf{n}'\boldsymbol{\epsilon}|^2\right) \qquad (7a)$$

$$\sum_{pol} A^*B = i\frac{|\mathbf{k}|}{m_e}\Big\{(\mathbf{n}-\mathbf{n}')\left[(\mathbf{n}-\mathbf{n}')\cdot(\boldsymbol{\varepsilon}\wedge\boldsymbol{\varepsilon}^*)\right]+(\boldsymbol{\varepsilon}^*\mathbf{t})\,\mathbf{t}\wedge\boldsymbol{\varepsilon}-(\mathbf{t}\boldsymbol{\varepsilon})\,\mathbf{t}\wedge\boldsymbol{\varepsilon}^*$$
$$-\boldsymbol{\varepsilon}\left[\boldsymbol{\varepsilon}^*\cdot(\mathbf{n}\wedge\mathbf{n}')\right]-\boldsymbol{\varepsilon}^*\left[\boldsymbol{\varepsilon}\cdot(\mathbf{n}\wedge\mathbf{n}')\right]$$
$$-(\mathbf{n}\mathbf{n}')\left[(\mathbf{n}'\boldsymbol{\varepsilon})\,\mathbf{n}\wedge\boldsymbol{\varepsilon}^*+(\mathbf{n}'\boldsymbol{\varepsilon}^*)\,\mathbf{n}\wedge\boldsymbol{\varepsilon}\right]\Big\} \tag{7b}$$

with $\mathbf{t}=\mathbf{n}'-(\mathbf{n}\mathbf{n}')\,\mathbf{n}$. Performing the angular integration we find

$$\int d\Omega_\gamma \sum_{pol} A^*A = 4\pi\,\frac{8}{3} \tag{8a}$$

$$\int d\Omega_\gamma \sum_{pol} A^*\mathbf{B} = 4\pi\,\frac{i\,|\mathbf{k}|}{m_e}\Big\{\mathbf{n}\left[\mathbf{n}\cdot(\boldsymbol{\varepsilon}\wedge\boldsymbol{\varepsilon}^*)\right]-\frac{1}{3}\boldsymbol{\varepsilon}\wedge\boldsymbol{\varepsilon}^*\Big\} \tag{8b}$$

For circular polarization $\pm$ one has $i\boldsymbol{\varepsilon}\wedge\boldsymbol{\varepsilon}^*=\pm\mathbf{n}$ and thus (8b) becomes

$$\int d\Omega_\gamma \sum_{pol}[A^*\mathbf{B}]_\pm = \pm\,4\pi\,\frac{2}{3}\frac{|\mathbf{k}|}{m_e}\mathbf{n} \tag{9}$$

Using (5) and (6), the cross section for circularly polarized γ's can be written as

$$\sigma_\pm = \frac{8\pi}{3}\left(\frac{\alpha}{m_e}\right)^2\left(1\pm\frac{1}{2}\frac{|\mathbf{k}|}{m_e}\mathbf{Pn}\right) \tag{10}$$

For $\mathbf{P}=0$ one obtains the Thomson cross section. For arbitrary $\mathbf{P}$, as a consequence of parity invariance

$$\sigma_\pm(\mathbf{P})=\sigma_\mp(-\mathbf{P}) \tag{11}$$

The polarization $\mathbf{P}'$ after scattering is determined by the final density matrix in the space of electron spin states

$$\rho_f = \frac{1+\mathbf{P}'\cdot\boldsymbol{\sigma}}{2} = \mathrm{Tr}_f\,S\,\rho_i S^\dagger \tag{12}$$

S is the scattering matrix and Tr_f denotes a summation over all final variables apart from the electron spin, in particular over final particle momenta and the final polarization of γ. By comparison with (5) and (6) there follows

$$\mathbf{P}' = \frac{\displaystyle\int d\Omega_\gamma \sum_{\varepsilon'}\left[AA^*\,\mathbf{P}+(A^*\mathbf{B}+A\mathbf{B}^*)+i(A^*\mathbf{B}-A\mathbf{B}^*)\wedge\mathbf{P}\right]}{\displaystyle\int d\Omega_\gamma \sum_{\varepsilon'}\left[AA^*+(A^*\mathbf{B}+A\mathbf{B}^*)\cdot\mathbf{P}\right]} \tag{13}$$

Using (8) and (10) we obtain

$$\mathbf{P}' = \frac{\mathbf{P}\pm\dfrac{1}{2}\dfrac{|\mathbf{k}|}{m_e}\mathbf{n}}{1\pm\dfrac{1}{2}\dfrac{|\mathbf{k}|}{m_e}\mathbf{Pn}} \simeq \mathbf{P}\pm\frac{1}{2}\frac{|\mathbf{k}|}{m_e}\mathbf{n}\mp\mathbf{P}\frac{1}{2}\frac{|\mathbf{k}|}{m_e}\mathbf{Pn}\equiv\mathbf{P}+\Delta\mathbf{P} \tag{14}$$

In particular, for electron polarization P parallel to the beam and photons circularly polarized along the same direction,

$$P' = P \pm \frac{|\mathbf{k}|}{2m_e} \mp P^2 \frac{|\mathbf{k}|}{2m_e} \tag{15}$$

Substituting in (3) we obtain, to first order in $|\mathbf{k}|/m$,

$$P_{n+1} = P_n + \Pi \Delta P = P_n + \Pi \frac{|\mathbf{k}|}{2m} P_n (1 - P_n) \tag{16}$$

Consistently with approximations made in Eq.(14) and (15), Π can be approximated by $\Pi_0 = \sigma_0 \rho L$, where σ_0 is the Compton cross section. Eq.(16) becomes

$$P_{n+1} = P_n + \frac{\sigma_0 \rho L |\mathbf{k}|}{2m} P_n (1 - P_n) \tag{17}$$

As expected, (16) gives the limiting value $P = 1$, that is, the electrons become completely polarized in the forward direction. Approximating (17) with a differential equation, we find, for large n and $P_n \simeq 1$,

$$P_n \sim 1 - const. e^{-\beta n} \tag{18}$$

with $\beta = \sigma_0 \rho L \frac{|\mathbf{k}|}{2m}$. Thus, typically, it takes

$$\bar{n} = \frac{1}{\beta} = \frac{2m}{|\mathbf{k}|} \frac{1}{\sigma_0 \rho L} \tag{19}$$

beam crossings for the electrons to become completely polarized.

An estimate of the number of photons per unit volume in a laser with mean power $\bar{I} \sim 10^8$ Watt/cm^2 is $\rho \sim 4 \cdot 10^{15}$ cm^{-3}. For 10 GeV electrons interacting with 5 eV laser photons, in an interval of 1 cm, Eq.(19) gives

$$\bar{n} \simeq 1.5 \cdot 10^7 \tag{20}$$

If the electrons follow a circular orbit of length ~ 30 km, they become completely polarized within a time of the order of 1000 sec.

Problem 48.   Decays of the ϕ meson

The ϕ meson ($M = 1019.412 \pm 0.008$ MeV, $\Gamma = 4.41 \pm 0.07$ MeV) decays in the following modes

K^+K^-	49.1±0.8%
$K^0\bar{K}^0$	34.4±0.7%
$\pi^+\pi^-\pi^0$	15.3±1.5%
$\eta\,\gamma$	1.28±0.6%
$e^+\,e^-$	$(3.1\pm0.1)\times10^{-4}$
$\mu^+\,\mu^-$	$(2.5\pm0.31)\times10^{-4}$

a)   Compare the modulus of the decay amplitudes in the first two channels. Take into account corrections coming from phase space and from the final state Coulomb interaction. $m_{K^+} = 493.646 \pm 0.009$ MeV, $m_{K^0} = m_{\bar{K}^0} = 497.671 \pm 0.031$ MeV.
b)   The decay mode $\phi \to \pi^0\pi^0\pi^0$ is not observed. What can we conclude on the isospin and angular momentum of the two-body $\pi^+\pi^-$ system participating in the 3π decay?
c)   Write down and calculate the resonant cross section for the process

$$e^+e^- \to \phi \to \begin{cases} K^+K^- \\ K^0\bar{K}^0 \\ e^+e^- \\ \mu^+\mu^- \end{cases}$$

d) ϕ has a negative G-parity since it decays in 3π; use this fact to conclude that its isospin I is even. From the existence of the K^+K^- decay mode deduce further that $I = 0$.

Solution

a)   Two-body phase space is proportional to the center-of-mass velocity of the decay products, i.e.

$$v = \sqrt{1 - \frac{4m_K^2}{m_\phi^2}} \tag{1}$$

The Coulomb correction C is given by the expression

$$C = \frac{2\pi}{kr_0(1 - e^{-2\pi/kr_0})} \tag{2}$$

where $k = \mu v_R$ is the center-of-mass momentum of the decay products, $\mu = m_K/2$ is the reduced mass and $v_R = 2v$ the relative velocity. r_0 is the Bohr radius for this system, $r_0 = 1/\mu\alpha$.

The ratio between the weak matrix elements is given in terms of the above factors and the branching ratios as follows

$$\frac{|M_{K^+K^-}|^2}{|M_{K^0\bar{K}^0}|^2} = \frac{\Gamma_{K^+K^-}}{\Gamma_{K^0\bar{K}^0}}\sqrt{\frac{m_\phi^2 - 4m_{K^0}^2}{m_\phi^2 - 4m_{K^+}^2}}\cdot\frac{1}{C} \tag{3}$$

Substituting numerical values

$$\sqrt{\frac{m_\phi^2 - 4m_{K^0}^2}{m_\phi^2 - 4m_{K^+}^2}} \simeq 0.87 \tag{4}$$

$$\frac{1}{C} \simeq 1 - \frac{\pi\alpha}{2v} \simeq 1 - \frac{\pi}{137 \cdot 2\sqrt{1 - 4m_{K^+}^2/m_\phi^2}} \simeq 0.95 \tag{5}$$

we finally obtain

$$\frac{|M_{K^+K^-}|^2}{|M_{K^0\bar{K}^0}|^2} \simeq \frac{49.1}{34.4} \cdot 0.95 \cdot 0.87 \cdot (1 \pm 0.05) = 1.18 \pm 0.05 \tag{6}$$

b) The $\pi^+\pi^-$ pair in the decay $\pi^+\pi^-\pi^0$ can have a priori isospin $I = 0, 1, 2$. However, the values $I = 0, 2$ would imply a nonzero amplitude also for the mode $\pi^0\pi^0\pi^0$; therefore, the $\pi^+\pi^-$ pair must have $I = 1$ and hence an antisymmetric isospin wave function. Since pions are bosons, the full wave function must be completely symmetric, and this forces the relative angular momentum of the pair to be odd.

Under charge conjugation, $C|\pi^0\rangle = |\pi^0\rangle$ (π^0 can decay into $\gamma\gamma$) and $C|\pi_\pm\rangle = \eta_\pi|\pi_\mp\rangle$ ($\eta_\pi = \pm 1$); the odd relative angular momentum then leads to $C = -1$ for the final state.

For the K^+K^- system Bose symmetry implies $C = (-)^J$. Having already found that $C = -1$, we deduce that the spin of ϕ is odd.

c) The total cross section for a final state f equals

$$\sigma_f = \frac{\pi}{k^2}(2J + 1)\frac{\Gamma_{e^+e^-}\Gamma_f}{(E - m_\phi)^2 + \frac{1}{4}\Gamma^2} \tag{7}$$

with $J = 1$. Neglecting the electron mass,

$$k^2 = \frac{1}{4}m_\phi^2 \tag{8}$$

At the peak we find

$$\sigma_f \simeq (1.75 \pm .05) \cdot 10^{-29} \left(\frac{\Gamma_f}{\Gamma}\right) \text{cm}^2 \tag{9}$$

We have used the branching ratio for e^+e^- listed in the text.

d) G-parity is given by

$$G = C(-)^I \tag{10}$$

As we have shown, ϕ has $C = -1$, and thus its isospin must be even. The only even value compatible with K^+K^- decay ($K^\pm$ has $I = 1/2$) is $I = 0$.

Problem 49.   The decays of Σ^0

Σ^0 is a neutral particle with spin 1/2 and mass $M = 1192.55 \pm 0.10$ MeV. It decays almost exclusively in

$$\Sigma^0 \to \Lambda^0 + \gamma \qquad (1)$$

with mean life $\tau = (7.4 \pm 0.7) \cdot 10^{-20}$ sec. Λ^0 is also neutral; it has spin 1/2 and mass $M = 1115.63 \pm 0.05$ MeV. The interaction responsible for this decay is electromagnetic

$$\mathcal{L}(x) = j_\mu(x) A^\mu(x) \qquad (2)$$

j_μ is the electromagnetic current operator.

a)   Write down the decay amplitude to first order in $\mathcal{L}$, that is, in terms of the matrix element

$$\langle \Lambda^0 | j_\mu(x) | \Sigma^0 \rangle \qquad (3)$$

b)   Derive the most general form for this matrix element which is compatible with

   i) translational invariance

   ii) Lorentz invariance

   iii) parity invariance

   iv) Hermiticity and current conservation

   v) time reversal invariance

c)   Use the data in the problem to calculate this matrix element.

d)   A small fraction ($\simeq 10^{-3}$) of the decays of Σ^0 are in the channel

$$\Sigma^0 \to \Lambda^0 + e^+ + e^- \qquad (4)$$

Calculate the branching ratio for this channel

$$R = \frac{\Gamma(\Sigma^0 \to \Lambda^0 + e^+ + e^-)}{\Gamma(\Sigma^0 \to \Lambda^0 + \gamma)} \qquad (5)$$

to lowest perturbative order.

The form factors of the $\Sigma^0 \to \Lambda^0$ transition may be considered constant with respect to the invariant mass μ of the $e^+ e^-$ pair. The calculation of the process $\Sigma^0 \to \Lambda^0 + e^+ + e^-$ can be conveniently reduced to a two-body problem (Λ and the $e^+ e^-$ pair), by integrating over the momenta of e^+ and e^- at fixed μ. To do this, we introduce in the phase space integral the following identity

$$\begin{aligned}
1 &= \frac{1}{2\pi} \int d\mu^2 \int \frac{d^4q}{(2\pi)^4} \, 2\pi \, \delta(q^2 - \mu^2) \, (2\pi)^4 \, \delta^4(q - p_{e^+} - p_{e^-}) \\
&= \frac{1}{2\pi} \int d\mu^2 \int \frac{d^3q}{2q_0(2\pi)^3} \, (2\pi)^4 \, \delta^4(q - p_{e^+} - p_{e^-})
\end{aligned} \qquad (6)$$

with $q^0 = \sqrt{\mathbf{q}^2 + \mu^2}$.

Solution

a)   The amplitude can be written as follows

$$\varepsilon_\mu^*(\mathbf{k}) \int d^4x \, e^{ikx} \langle \Lambda^0 | j^\mu(x) | \Sigma^0 \rangle \tag{7}$$

where $\mathbf{k}$ is the photon momentum. Let us now restrict the form of the $\Lambda - \Sigma$ matrix element, consistently with the requirements placed upon it.

   i)

$$\langle \Lambda^0 | j_\mu(x) | \Sigma^0 \rangle = e^{i(p_{\Lambda^0} - p_{\Sigma^0}) \cdot x} \langle \Lambda^0 | j_\mu(0) | \Sigma^0 \rangle \tag{8}$$

   ii)

$$\langle \Lambda^0 | j_\mu(0) | \Sigma^0 \rangle = \overline{u}(\Lambda^0) \Big(AP_\mu + Bq_\mu + C\gamma_\mu + iD\sigma_{\mu\nu}q^\nu + iE\sigma_{\mu\nu}P^\nu +$$
$$+ A'\gamma^5 P_\mu + B'\gamma^5 q_\mu + C'\gamma^5 \gamma_\mu + iD'\gamma^5 \sigma_{\mu\nu}q^\nu + iE'\gamma^5 \sigma_{\mu\nu}P^\nu \Big) u(\Sigma^0) \tag{9}$$

where

$$P_\mu = (p_{\Lambda^0} + p_{\Sigma^0})_\mu \qquad q_\mu = (p_{\Sigma^0} - p_{\Lambda^0})_\mu \tag{10}$$

$A, B \dots E'$ may depend on the invariants made out of the four-vectors present. In process (1) they only depend on particle masses so that they are constant. In process (4) they are functions of $q^2 = \mu^2$.

   iii) Parity invariance eliminates terms with γ^5. The identities

$$\overline{u}(\Lambda^0)\{i\sigma_{\mu\nu}q^\nu - (M_{\Sigma^0} + M_{\Lambda^0})\gamma_\mu + P_\mu\}u(\Sigma^0) = 0 \tag{11a}$$
$$\overline{u}(\Lambda^0)\{i\sigma_{\mu\nu}P^\nu - (M_{\Sigma^0} - M_{\Lambda^0})\gamma_\mu + q_\mu\}u(\Sigma^0) = 0 \tag{11b}$$

allow us to write

$$J_\mu(q) \equiv \langle \Lambda^0 | j_\mu(x) | \Sigma^0 \rangle = \overline{u}(\Lambda^0) \left(AP_\mu + Bq_\mu + iD\sigma_{\mu\nu}q^\nu \right) u(\Sigma^0) \tag{12}$$

   iv) Current conservation, $q^\mu J_\mu = 0$, implies

$$A(M_{\Sigma^0}^2 - M_{\Lambda^0}^2) + Bq^2 = 0 \tag{13}$$

In particular, $A = \mathcal{O}(q^2)$ at $q^2 \to 0$. Eliminating B, Eq.(12) becomes

$$J_\mu(q) = \overline{u}(\Lambda^0) \left[A\left(P_\mu - (M_{\Sigma^0}^2 - M_{\Lambda^0}^2)\frac{q_\mu}{q^2} \right) + iD\sigma_{\mu\nu}q^\nu \right] u(\Sigma^0) \tag{14}$$

Let us now consider the matrix element of the inverse transition

$$\tilde{J}_\mu(q) = \langle \Sigma^0 | j_\mu(0) | \Lambda^0 \rangle \tag{15}$$

Retracing the steps leading to (14), we find

$$\tilde{J}_\mu(q) = \overline{u}(\Sigma^0) \left[\tilde{A}\left(P_\mu - (M_{\Sigma^0}^2 - M_{\Lambda^0}^2)\frac{q_\mu}{q^2} \right) - i\tilde{D}\sigma_{\mu\nu}q^\nu \right] u(\Lambda^0) \tag{16}$$

Hermiticity of the current, $j_\mu^\dagger = j_\mu$, requires that $\tilde{J}_\mu^*(q) = J_\mu(q)$. Therefore,

$$\tilde{A} = A^* \qquad \tilde{D} = D^* \tag{17}$$

v) Under time reversal

$$j_\mu(0) \rightarrow g_{\mu\mu} j_\mu(0) \tag{18}$$

Under TP

$$(TP)^\dagger j_\mu(0)(TP) = j_\mu(0) \tag{19}$$

Since the current is Hermitian, a TP transformation gives

$$\begin{aligned}
\langle \Lambda^0 |(TP)^\dagger j_\mu(0)(TP)|\Sigma^0\rangle &= \langle j_\mu(0)(TP)\Sigma^0|(TP)\Lambda^0\rangle \\
&= \langle (TP)\Sigma^0|j_\mu(0)|(TP)\Lambda^0\rangle
\end{aligned} \tag{20}$$

We combine (19) and (20) with the result

$$\langle \Lambda^0 | j_\mu(0)|\Sigma^0\rangle = \langle (TP)\Sigma^0|j_\mu(0)|(TP)\Lambda^0\rangle \tag{21}$$

The right hand side of (14) equals

$$\langle (TP)\Sigma^0|j_\mu(0)|(TP)\Lambda^0\rangle = $$
$$\overline{u}_{TP}(\Sigma^0)\Big(\tilde{A}\Big[P_\mu - (M_{\Sigma^0}^2 - M_{\Lambda^0}^2)\frac{q_\mu}{q^2}\Big] - i\tilde{D}\sigma_{\mu\nu}q^\nu\Big)u_{TP}(\Lambda^0) \tag{22}$$

Following our conventions for the Dirac matrices (see Appendix E),

$$\begin{aligned}
u_{TP}(\Lambda_0) &= \eta_{\Lambda^0}\gamma^2\gamma^5 u^*(\Lambda_0) \\
u_{TP}(\Sigma_0) &= \eta_{\Sigma^0}\gamma^2\gamma^5 u^*(\Sigma_0)
\end{aligned} \tag{23}$$

η_{Λ^0} and η_{Σ^0} are phase factors. To render PT compatible with $SU(3)$ flavour symmetry, we must have $\eta_{\Lambda^0} = \eta_{\Sigma^0}$. Equ. (22) now reads

$$\langle (TP)\Sigma^0|j_\mu(0)|(TP)\Lambda^0\rangle = $$
$$-\big[\gamma^2\gamma^5 u(\Sigma^0)\big]^T \gamma^0\Big(\tilde{A}\Big[P_\mu - (M_{\Sigma^0}^2 - M_{\Lambda^0}^2)\frac{q_\mu}{q^2}\Big] - i\tilde{D}\sigma_{\mu\nu}q^\nu\Big)\gamma^2\gamma^5 u^*(\Lambda^0) \tag{24}$$

This equation can be simplified further using the relations: $\gamma^2 = (\gamma^2)^T$, $\gamma^5 = (\gamma^5)^T$, $\sigma_{\mu\nu}^T = -\gamma^2\gamma^0\sigma_{\mu\nu}\gamma^2\gamma^0$. Substituting in Eq.(21) we obtain

$$\overline{u}(\Lambda^0)\Big(A\Big[P_\mu - (M_{\Sigma^0}^2 - M_{\Lambda^0}^2)\frac{q_\mu}{q^2}\Big] + iD\sigma_{\mu\nu}q^\nu\Big)u(\Sigma^0) = $$
$$\overline{u}(\Lambda^0)\Big(\tilde{A}\Big[P_\mu - (M_{\Sigma^0}^2 - M_{\Lambda^0}^2)\frac{q_\mu}{q^2}\Big] + i\tilde{D}\sigma_{\mu\nu}q^\nu\Big)u(\Sigma^0) \tag{25}$$

We conclude that $A = \tilde{A}$, $D = \tilde{D}$ and, comparing with (17),

$$A = A^* \qquad D = D^* \tag{26}$$

The term proportional to A does not contribute to process (1), because $q_\mu \varepsilon^\mu = 0$ and $q^2 = 0$. Defining $D = ge/(M_{\Lambda^0} + M_{\Sigma^0})$ we have

$$\langle \Lambda^0 | j_\mu(x) | \Sigma^0 \rangle = \frac{ge}{M_{\Lambda^0} + M_{\Sigma^0}} \bar{u}(p_{\Lambda^0}) \left(i\sigma_{\mu\nu} q^\nu \right) u(p_{\Sigma^0}) \tag{27}$$

It is reasonable to expect that A varies with q^2 at the scale of 1 GeV^2, so that it stays practically zero also for the process $\Sigma^0 \to \Lambda^0 + e^+ + e^-$, similarly D should stay unchanged.

c)   The width of the decay $\Sigma^0 \to \Lambda^0 + \gamma$ is

$$\Gamma = \frac{1}{2M_{\Sigma^0}} \sum_i \varepsilon_\mu^{(i)*} \varepsilon_\nu^{(i)} \frac{g^2 e^2}{(M_{\Lambda^0} + M_{\Sigma^0})^2} \int \frac{d\Omega}{8(2\pi)^2} \left(1 - \frac{M_{\Lambda^0}^2}{M_{\Sigma^0}^2} \right)$$
$$\frac{1}{2}\mathrm{Tr}\left\{ (\not{p}_{\Lambda^0} + M_{\Lambda^0})\sigma^{\mu\rho} q_\rho (\not{p}_{\Sigma^0} + M_{\Sigma^0})\sigma^{\nu\tau} q_\tau \right\} \tag{28}$$

Setting $\sum_i \varepsilon_\mu^{(i)*} \varepsilon_\nu^{(i)} = -g_{\mu\nu}$ (the term with $q_\mu q_\nu$ vanishes upon contraction), we are led to

$$\Gamma = \frac{1}{2M_{\Sigma^0}} \frac{2g^2 e^2}{(M_{\Lambda^0} + M_{\Sigma^0})^2} (M_{\Lambda^0}^2 - M_{\Sigma^0}^2)^2 \frac{1}{8\pi} \left(1 - \frac{M_{\Lambda^0}^2}{M_{\Sigma^0}^2} \right)$$
$$\simeq \frac{\alpha}{8} g^2 M_{\Sigma^0} \left(1 - \frac{M_{\Lambda^0}^2}{M_{\Sigma^0}^2} \right)^3 \qquad \left(\alpha = \frac{e^2}{4\pi} \right) \tag{29}$$

More precisely, we have

$$g^2 = \frac{\Gamma}{M_{\Sigma^0}} \frac{2}{\alpha} \left(1 - \frac{M_{\Lambda^0}^2}{M_{\Sigma^0}^2} \right)^{-3} \left(1 + \frac{M_\Lambda}{M_\Sigma} \right)^2 \tag{30}$$

The numerical value of g is thus

$$g = 1.979 \pm 0.10 \tag{31}$$

The corresponding transition magnetic moment is

$$\mu = \frac{g}{M_\Lambda + M_\Sigma} = g\mu_N \frac{2m_p}{M_\Lambda + M_\Sigma} = (1.60 \pm .08)\,\mu_N \tag{32}$$

d)   The width of the decay $\Sigma^0 \to \Lambda^0 e^+ e^-$ is, making use of the matrix element (27),

$$\Gamma_{\Sigma^0 \to \Lambda^0 e^+ e^-} =$$
$$\frac{1}{2M_{\Sigma^0}} \frac{g^2 e^2}{(M_{\Lambda^0} + M_{\Sigma^0})^2} \int \frac{1}{2}\mathrm{Tr}\left\{ (\not{p}_{\Lambda^0} + M_{\Lambda^0})\sigma^{\mu\rho} q_\rho (\not{p}_{\Sigma^0} + M_{\Sigma^0})\sigma^{\nu\tau} q_\tau \right\}$$
$$\frac{e^2}{q^4}\mathrm{Tr}\left\{ (\not{p}_{e^-} + m_e)\gamma_\mu (\not{p}_{e^+} - m_e)\gamma_\nu \right\}$$
$$(2\pi)^4 \delta^4(p_{\Sigma^0} - p_{\Lambda^0} - p_{e^-} - p_{e^+}) \frac{d^3 p_{\Lambda^0}}{2E_{\Lambda^0}(2\pi)^3} \frac{d^3 p_{e^+}}{2E_{e^+}(2\pi)^3} \frac{d^3 p_{e^-}}{2E_{e^-}(2\pi)^3} \tag{33}$$

As suggested, we define

$$T_{\mu\nu}(q) = \int (2\pi)^4 \delta^4(q - p_{e+} - p_{e-}) \frac{d^3 p_{e+}}{2E_{e+}(2\pi)^3} \frac{d^3 p_{e-}}{2E_{e-}(2\pi)^3} \cdot$$
$$\cdot \mathrm{Tr}\left\{ (\not{p}_{e-} + m_e)\gamma_\mu (\not{p}_{e+} - m_e)\gamma_\nu \right\} \tag{34}$$

$T_{\mu\nu}$ can be easily calculated in the rest frame of the e^+e^- pair, with the result

$$T_{\mu\nu} = \frac{4}{3}(q^2 + 2m_e^2)\frac{1}{8\pi}\sqrt{1 - \frac{4m_e^2}{q^2}}\left(g_{\mu\nu} - \frac{q_\mu q_\nu}{q^2}\right) \tag{35}$$

Using the identity provided in the text, we write

$$\Gamma_{\Sigma^0 \to \Lambda^0 e^+ e^-} = \frac{1}{2M_{\Sigma^0}}\frac{g^2 e^2}{(M_{\Lambda^0} + M_{\Sigma^0})^2}\int \frac{dq^2}{2\pi} H^{\mu\nu}(q)\, T_{\mu\nu}\frac{e^2}{q^4}$$
$$\times \frac{1}{8\pi}\sqrt{1 - \frac{(q + M_{\Lambda^0})^2}{M_{\Sigma^0}^2}}\sqrt{1 - \frac{(q - M_{\Lambda^0})^2}{M_{\Sigma^0}^2}} \tag{36}$$

The last factor is the phase space volume for two bodies with masses $q = \sqrt{q^2}$ and M_{Λ^0}. We have also defined

$$H^{\mu\nu}(q) = \frac{1}{2}\mathrm{Tr}\left\{ (\not{p}_{\Lambda^0} + M_{\Lambda^0})\sigma^{\mu\rho}q_\rho(\not{p}_{\Sigma^0} + M_{\Sigma^0})\sigma^{\nu\tau}q_\tau \right\} \tag{37}$$

The term proportional to $q_\mu q_\nu$ in $T_{\mu\nu}$ gives zero when contracted with $H_{\mu\nu}$; there remains only the term with $g_{\mu\nu}$ which is easier to calculate. We find

$$g_{\mu\nu}H^{\mu\nu} = 2\left[(M_{\Sigma^0} + M_{\Lambda^0})^2 + \frac{q^2}{2}\right]\left[(M_{\Sigma^0} - M_{\Lambda^0})^2 - q^2\right] \tag{38}$$

The branching ratio is

$$R = \frac{e^2}{2\left(1 - \frac{M_{\Lambda^0}^2}{M_{\Sigma^0}^2}\right)^3}\int_{4m_e^2}^{(M_{\Sigma^0} - M_{\Lambda^0})^2}\frac{dq^2}{2\pi q^4}\frac{4}{3}(q^2 + 2m_e^2)\sqrt{1 - \frac{4m_e^2}{q^2}}$$
$$\times 2\left[\left(1 + \frac{M_{\Lambda^0}}{M_{\Sigma^0}}\right)^2 + \frac{q^2}{2M_{\Sigma^0}^2}\right]\left[\left(1 - \frac{M_{\Lambda^0}}{M_{\Sigma^0}}\right)^2 - \frac{q^2}{M_{\Sigma^0}^2}\right]$$
$$\times \frac{1}{8\pi}\sqrt{1 - \frac{(q + M_{\Lambda^0})^2}{M_{\Sigma^0}^2}}\sqrt{1 - \frac{(q - M_{\Lambda^0})^2}{M_{\Sigma^0}^2}} \tag{39}$$

In terms of the variables $y = q^2/M_{\Sigma^0}^2$, $\varepsilon = M_{\Lambda^0}/M_{\Sigma^0}$ this becomes

$$R \simeq \frac{8\alpha}{3}\frac{1}{(1 - \varepsilon^2)^3}\int_{4m_e^2/M_{\Sigma^0}^2}^{(1-\varepsilon)^2}\frac{dy}{y}(1 + \varepsilon)^3\{(1 - \varepsilon)^2 - y\}^{\frac{3}{2}} \times \frac{1}{8\pi} \tag{40}$$

To arrive at (40) we dropped $q^2/2M_{\Sigma^0}^2$ compared to $(1 + M_{\Lambda^0}/M_{\Sigma^0})^2$, and used the following approximations

$$(q^2 + 2m_e^2)\sqrt{1 - \frac{4m_e^2}{q^2}} \simeq q^2 \tag{41}$$

$$\sqrt{1 - \frac{(q + M_{\Lambda^0})^2}{M_{\Sigma^0}^2}}\sqrt{1 - \frac{(q - M_{\Lambda^0})^2}{M_{\Sigma^0}^2}} \equiv$$

$$\equiv \sqrt{\left(1 + \frac{M_{\Lambda^0}}{M_{\Sigma^0}}\right)^2 - \frac{q^2}{M_{\Sigma^0}^2}}\sqrt{\left(1 - \frac{M_{\Lambda^0}}{M_{\Sigma^0}}\right)^2 - \frac{q^2}{M_{\Sigma^0}^2}} \simeq (1 + \varepsilon)\sqrt{(1 - \varepsilon)^2 - y} \tag{42}$$

Carrying out the integral, we finally obtain

$$
\begin{aligned}
R &= \frac{\alpha}{3\pi}\frac{1}{(1 - \varepsilon^2)^3}\int_{\frac{4m_e^2}{M_{\Sigma^0}^2}}^{(1-\varepsilon)^2}\frac{dy}{y}(1 + \varepsilon)^3\{(1 - \varepsilon)^2 - y\}^{\frac{3}{2}} = \\
&= \frac{\alpha}{3\pi}\int_{\frac{4m_e^2}{M_{\Sigma^0}^2(1-\varepsilon)^2}}^{1}\frac{dy}{y}[1 - y]^{3/2} \\
&\simeq \frac{\alpha}{3\pi}\left[\ln\left(\frac{M_{\Sigma^0}^2(1 - \varepsilon)^2}{4m_e^2}\right) - \frac{3}{2}\left(1 - \frac{4m_e^2}{M_{\Sigma^0}^2(1 - \varepsilon)^2}\right)\right] \simeq 5.5 \cdot 10^{-3} \tag{43}
\end{aligned}
$$

One may also integrate expression (39) numerically, finding

$$R = 5.45 \cdot 10^{-3} \tag{44}$$

This value is in agreement with experiment[1].

Problem 50. l_2 decays of vector and pseudoscalar mesons

A meson A^- of mass m has the following decay modes

$$A^- \to \mu^- \bar{\nu}_\mu \tag{1}$$

$$A^- \to e^- \bar{\nu}_e \tag{2}$$

a)   Assume that the weak interaction has the form

$$H_w = \frac{G}{\sqrt{2}}\left\{J_\mu^{(h)\dagger}J_{(l)}^\mu + h.c.\right\} \tag{3}$$

---

[1]Particle Data Group, *Phys. Lett.* **B239** 1990

with

$$J_\rho^{(l)} = \bar{\nu}_\mu \gamma_\rho (1 - \gamma^5)\mu + \bar{\nu}_e \gamma_\rho (1 - \gamma^5)e \qquad (4)$$

Using symmetry arguments, find the most general expression for the matrix element

$$\langle 0|J_\mu^h(x)|A^-\rangle \qquad (5)$$

for pseudoscalar and vector mesons.
b)   Calculate in the two cases the longitudinal polarization of the emitted lepton.
c)   Discuss the possibility of determining the spin of the meson from the branching ratio of the decay modes (1) and (2).

Solution

a)   The decay amplitude is

$$\mathcal{M} = \frac{G}{\sqrt{2}} \bar{u}(\mathbf{p}_{\mu^-}) \gamma^\mu (1 - \gamma^5) v(\mathbf{p}_{\bar{\nu}_\mu}) \langle 0|J_\mu^{(h)}(0)|A^-\rangle \qquad (6)$$

We write for the hadronic matrix element

$$\langle 0|J_\mu^{(h)}(x)|A^-\rangle = e^{-ip_A \cdot x} \langle 0|J_\mu^{(h)}(0)|A^-\rangle \qquad (7)$$

by translational invariance. $\langle 0|J_\mu^{(h)}(0)|A^-\rangle$ must behave as a four-vector, and hence it has the form

$$
\begin{array}{lll}
\text{(i)} & \langle 0|J_{(h)}^\mu(0)|A^-\rangle = 2A p_{A^-}^\mu & \text{(pseudoscalar)} \\
\text{(ii)} & \langle 0|J_{(h)}^\mu(0)|A^-\rangle = 2B \varepsilon^\mu & \text{(vector)}
\end{array} \qquad (8)
$$

In the pseudoscalar case, $p^\mu{}_{A^-}$ is the only available four-vector. In the vector case, the matrix element must be linear in the meson polarization ε^μ by the superposition principle; since $p \cdot \varepsilon = 0$, the only possible dependence is the one in (8).
b)   A pseudoscalar meson has zero angular momentum in its rest frame. Given that the antineutrino has positive helicity, conservation of angular momentum along the decay axis requires that the charged lepton has also positive helicity. The decay amplitude in this case becomes

$$
\begin{aligned}
\mathcal{M} &= 2A \frac{G}{\sqrt{2}} \bar{u}(\mathbf{p}_{\mu^-}) \left[\not{p}_{\mu^-} + \not{p}_{\bar{\nu}_\mu} \right] (1 - \gamma^5) v(\mathbf{p}_{\bar{\nu}_\mu}) \\
&= 2A \frac{G}{\sqrt{2}} m_\mu \bar{u}(\mathbf{p}_{\mu^-})(1 - \gamma^5) v(\mathbf{p}_{\bar{\nu}_\mu})
\end{aligned} \qquad (9)
$$

We form the modulus squared and project onto muon states of helicity h, finding

$$|\mathcal{M}(h)|^2 = 2|A|^2 G^2 m_\mu^2 \quad \text{Tr}\left\{ (\not{p}_{\mu^-} + m_\mu)(1 + \gamma^5 \not{s})(1 - \gamma^5)\not{p}_{\bar{\nu}_\mu} \right\} \qquad (10)$$

$$w = h \left\{ \frac{|\mathbf{p}_{\mu^-}|}{m_\mu}, \frac{E}{m_\mu} \frac{\mathbf{p}_{\mu^-}}{|\mathbf{p}_{\mu^-}|} \right\} \qquad (h = \pm 1)$$

This leads to

$$|\mathcal{M}(h)|^2 = 4A^2 G^2 m_\mu^2 (1+h)(m^2 - m_\mu^2) \tag{11}$$

As expected, the average helicity is

$$\langle h \rangle = \frac{|\mathcal{M}_+|^2 - |\mathcal{M}_-|^2}{|\mathcal{M}_+|^2 + |\mathcal{M}_-|^2} = 1 \tag{12}$$

In the vector case we have

$$\mathcal{M} = 2B \frac{G}{\sqrt{2}} \bar{u}(\mathbf{p}_{\mu^-}) \displaystyle{\not{\epsilon}}(1 - \gamma^5) v(\mathbf{p}_{\bar{\nu}_\mu}) \tag{13}$$

Proceeding as above,

$$|\mathcal{M}(h)|^2 = 4|B|^2 G^2 \frac{1}{2} \text{Tr} \left\{ (\displaystyle{\not{p}}_{\mu^-} + m_\mu)(1 + \gamma^5 \displaystyle{\not{w}}) \displaystyle{\not{\epsilon}}(1 - \gamma^5) \displaystyle{\not{p}}_{\bar{\nu}_\mu} \displaystyle{\not{\epsilon}}^* \right\} \tag{14}$$

The sum over meson polarizations gives

$$\sum_i \varepsilon_i^\mu \varepsilon_i^{\nu *} = - \left(g^{\mu\nu} - \frac{p_{A^-}^\mu p_{A^-}^\nu}{m^2} \right) \tag{15}$$

Finally, the amplitude becomes

$$|\mathcal{M}(h)|^2 = \frac{4}{3} m^2 B^2 G^2 \left\{ 2(1-h) + \frac{m_\mu^2}{m^2}(1+h) \right\} (m^2 - m_\mu^2) \tag{16}$$

and the average polarization is

$$\langle h \rangle = -1 + \frac{m_\mu^2}{m^2} \frac{1}{\left(1 + \frac{m_\mu^2}{2m^2} \right)} \tag{17}$$

c)   For spin zero, the branching ratio of the two decay modes can be found from Eq.(11). Summing over h, and assuming m_μ, $m_e \ll m$, we obtain

$$\frac{B(2)}{B(1)} \simeq \frac{m_e^2}{m_\mu^2} \tag{18}$$

For spin one, we use Eq.(16) and find

$$\frac{B(2)}{B(1)} \simeq 1 \tag{19}$$

Indeed, the spin of the meson can be deduced from the branching ratio.

Problem 51. K_{l3}^0 decay

The decay $K^0 \rightarrow \pi^- \ell^+ \nu_\ell$, where ℓ is a lepton (μ^+ or e^+), is well described by the Fermi interaction

$$H = \frac{G}{\sqrt{2}} J_\mu^h \overline{\psi}_{\nu_\ell} \gamma^\mu (1 - \gamma^5) \psi_\ell \tag{1}$$

J_μ^h is the hadronic current, $J_\mu^h = V_\mu - A_\mu$.

a)   Find the most general expression for the matrix element of the hadronic current, compatible with Lorentz and time reversal invariance. Neglect final state interaction.

b)   Show that J_μ^h contributes to this process only through its vector part V_μ.

c)   Assuming the form factors are constant, calculate the energy spectrum of π^- and ℓ. Calculate the width of the decay into e^+; you may neglect the lepton mass in this case.

d)   Show that the lepton polarization perpendicular to the decay plane vanishes, provided time reversal invariance holds and final state interaction can be neglected.

Solution

a)   The matrix element of the hadronic current can be written as follows

$$\langle \pi | J_\mu^h(0) | K \rangle = f_1 P_\mu + f_2 q_\mu \tag{2}$$

where

$$P_\mu = p_{K\mu} + p_{\pi\mu} \qquad q_\mu = p_{K\mu} - p_{\pi\mu} = p_{\ell\mu} + p_{\nu_\ell\mu} \tag{3}$$

The form factors are functions of q^2.

Time reversal invariance requires the two form factors to be mutually real. We render f_1 real by an appropriate choice of phase of the matrix element.

b) π and K have the same parity; consequently, only the vector current will contribute.

The charge associated to the vector current is a generator of the symmetry group $SU(3)$ of strong interactions, by virtue of CVC. In the limit in which the symmetry is exact this current is conserved, $q^\mu V_\mu = 0$. This implies that

$$\langle \pi | J_\mu^h | K \rangle = f_1 P_\mu \tag{4}$$

since $P_\mu q^\mu = m_K^2 - m_\pi^2 \simeq 0$ in the symmetric limit.

More generally, the amplitude

$$\mathcal{M} = \frac{G}{\sqrt{2}} (f_1 P_\mu + f_2 q_\mu) \, \overline{u}(\mathbf{p}_{\nu_\ell}) \gamma^\mu (1 - \gamma^5) v(\mathbf{p}_\ell) \tag{5}$$

becomes, using the equations of motion,

$$\mathcal{M} = \frac{G}{\sqrt{2}} \left\{ f_1 P_\mu \overline{u}(\mathbf{p}_{\nu_\ell}) \gamma^\mu (1 - \gamma^5) v(\mathbf{p}_\ell) - f_2 m_\ell \overline{u}(\mathbf{p}_{\nu_\ell})(1 + \gamma^5) v(\mathbf{p}_\ell) \right\} \tag{6}$$

Thus, in any case, the contribution of the second term is only proportional to the lepton mass.

c)   Summing over final polarizations we obtain

$$
\begin{aligned}
|\mathcal{M}|^2 \;=\; 4G^2 \Big\{ & f_1^2 [2(P\cdot p_{\nu_\ell})(P\cdot p_\ell) - P^2(p_{\nu_\ell}\cdot p_\ell)] \\
& + 2\mathrm{Re}(f_1^* f_2) m_\ell^2 (P\cdot p_{\nu_\ell}) + |f_2|^2 m_\ell^2 (p_{\nu_\ell}\cdot p_\ell) \Big\}
\end{aligned}
\tag{7}
$$

We parameterize the invariants in terms of the center-of-mass energies of the lepton and the pion, as follows

$$
\begin{aligned}
(P\cdot p_{\nu_\ell}) &= \frac{3m_K^2 - m_\pi^2 + m_\ell^2}{2} - m_K(2E_\ell + E_\pi) \\
(P\cdot p_\ell) &= -\frac{m_K^2 + m_\pi^2 + m_\ell^2}{2} + m_K(2E_\ell + E_\pi) \\
(p_{\nu_\ell}\cdot p_\ell) &= \frac{m_K^2 + m_\pi^2 - m_\ell^2}{2} - m_K E_\pi \\
P^2 &= m_K^2 + m_\pi^2 + 2m_K E_\pi
\end{aligned}
\tag{8}
$$

The general form of the decay width is

$$
d\Gamma = \frac{1}{2m_K} |\mathcal{M}|^2 \, d\Phi^{(3)}
\tag{9}
$$

The differential element of three-body phase space is, after angular integrations,

$$
d\Phi^{(3)} = \frac{1}{32\pi^3} \, dE_\ell \, dE_\pi
\tag{10}
$$

In our case, the decay width reads

$$
\begin{aligned}
d\Gamma \;=\; \frac{G^2}{32\pi^3 m_K} \Big\{ & f_1^2 \Big\{ \big[3m_K^2 - m_\pi^2 + m_\ell^2 - 2m_K(2E_\ell + E_\pi) \big] \\
& \times \big[2m_K(2E_\ell + E_\pi) - m_K^2 - m_\pi^2 - m_\ell^2 \big] \\
& - \big(m_K^2 + m_\pi^2 + 2m_K E_\pi \big)\big(m_K^2 + m_\pi^2 - m_\ell^2 - 2m_K E_\pi \big) \Big\} \\
& + 2\mathrm{Re}(f_1^* f_2) m_\ell^2 \big[3m_K^2 - m_\pi^2 + m_\ell^2 - 2m_K(2E_\ell + E_\pi) \big] \\
& + |f_2|^2 m_\ell^2 \big[m_K^2 + m_\pi^2 - m_\ell^2 - 2m_K E_\pi \big] \Big\} \, dE_\ell \, dE_\pi
\end{aligned}
\tag{11}
$$

The last two terms are proportional to m_ℓ^2; if the lepton is a positron they can be left out, leading to

$$
\begin{aligned}
d\Gamma &= \frac{G^2 f_1^2 m_K}{8\pi^3} \Big\{ 4E_\ell m_K + 2m_K E_\pi - 4E_\ell^2 - 4E_\ell E_\pi - m_K^2 - m_\pi^2 \Big\} \, dE_\ell \, dE_\pi \\
&= \frac{G^2 f_1^2 m_K}{8\pi^3} \Big\{ (m_K - 2E_\ell)(2E_\pi + 2E_\ell - m_K) - m_\pi^2 \Big\} \, dE_\ell \, dE_\pi
\end{aligned}
\tag{12}
$$

Equivalently,

$$\mathrm{d}\Gamma = \frac{G^2 f_1^2 m_K}{8\pi^3} \left\{ \mathbf{p}_\pi^2 - (m_K - E_\pi - 2E_\ell)^2 \right\} \mathrm{d}E_\ell \, \mathrm{d}E_\pi \qquad (13)$$

The kinematic limits on the domain of integration are

$$\frac{1}{2}\left(m_K - E_\pi - |\mathbf{p}_\pi|\right) \le E_\ell \le \frac{1}{2}\left(m_K - E_\pi + |\mathbf{p}_\pi|\right)$$
$$m_\pi \le E_\pi \le \frac{m_K^2 + m_\pi^2}{2m_K} \qquad (14)$$

The integral over $\mathrm{d}E_\ell$ gives

$$\Gamma = \frac{G^2 f_1^2 m_K}{12\pi^3} \left(E_\pi^2 - m_\pi^2\right)^{3/2} \mathrm{d}E_\pi \qquad (15)$$

Integrating also over $\mathrm{d}E_\pi$ we find

$$\mathrm{d}\Gamma = \frac{G^2 f_1^2 m_K^5}{768\pi^3} \left\{ \left(1 - \frac{m_\pi^2}{m_K^2}\right)^3 \left(1 + \frac{m_\pi^2}{m_K^2}\right) - 6\frac{m_\pi^2}{m_K^2} \left(1 - \frac{m_\pi^2}{m_K^2}\right) \left(1 + \frac{m_\pi^2}{m_K^2}\right) \right.$$
$$\left. - 12 \left(\frac{m_\pi^2}{m_K^2}\right)^2 \ln \frac{m_\pi^2}{m_K^2} \right\} \qquad (16)$$

We may simplify this result by keeping only leading orders in $\left(\frac{m_\pi^2}{m_K^2}\right)$

$$\Gamma \simeq \frac{G^2 f_1^2 m_K^5}{768\pi^3} \left(1 - \frac{8m_\pi^2}{m_K^2}\right) \qquad (17)$$

d) Let us denote by $\boldsymbol{\zeta}$ the polarization of the lepton in its rest frame. The corresponding four-vector polarization a^ℓ in an arbitrary frame is

$$a^\ell = \left(\frac{\boldsymbol{\zeta} \mathbf{p}_\ell}{m_\ell}, \ \boldsymbol{\zeta} + \frac{\mathbf{p}_\ell \left(\mathbf{p}_\ell \boldsymbol{\zeta}\right)}{m_\ell \left(E_\ell + m_\ell\right)} \right) \qquad (18)$$

The decay amplitude is given by

$$|\mathcal{M}|^2 = \frac{G^2}{4} \mathrm{Tr} \left\{ \not{p}_{\nu_\ell} \gamma^\mu \left(1 - \gamma^5\right) \left(\not{p}_\ell - m_\ell\right) \left(1 + \gamma^5 \not{a}\right) \gamma^\nu \left(1 - \gamma^5\right) \right\}$$
$$\cdot \left(f_1 P_\mu + f_2 q_\mu\right) \left(f_1^* P_\nu + f_2^* q_\nu\right)$$
$$= \frac{G^2}{4} \mathrm{Tr} \left\{ \not{p}_{\nu_\ell} \gamma^\mu \left(1 - \gamma^5\right) \left(\not{p}_\ell + m_\ell \not{a}\right) \gamma^\nu \left(1 - \gamma^5\right) \right\}$$
$$\cdot \left(f_1 P_\mu + f_2 q_\mu\right) \left(f_1^* P_\nu + f_2^* q_\nu\right) \qquad (19)$$

We carry out the traces, with the result

$$
\begin{aligned}
|\mathcal{M}|^2 \;=\; & \tfrac{1}{2}\overline{|\mathcal{M}|^2} + 2G^2 m_\ell \Big\{ |f_1|^2 [2(P\cdot p_{\nu_\ell})(P\cdot a) - P^2(p_{\nu_\ell}\cdot a)] \\
& + |f_2|^2 [2(q\cdot p_{\nu_\ell})(q\cdot a) - q^2(p_{\nu_\ell}\cdot a)] \\
& + 2\mathrm{Re} f_1 f_2^* [(p_{\nu_\ell}\cdot P)(q\cdot a) + (p_{\nu_\ell}\cdot q)(P\cdot a) - (P\cdot a)(p_{\nu_\ell}\cdot a)] \\
& + 2\mathrm{Im} f_1 f_2^* \, \varepsilon_{\mu\nu\rho\sigma} P^\mu q^\nu p_{\nu_\ell}{}^\rho a^\sigma \Big\}
\end{aligned}
\tag{20}
$$

where $\overline{|\mathcal{M}|^2}$ is the modulus squared of the matrix element, summed over final polarizations. We note the following relations

$$
(q\cdot a) = (p_{\nu_\ell}\cdot a) \qquad\quad (P\cdot a) = 2(k\cdot a) - (p_{\nu_\ell}\cdot a)
$$

$$
(k\cdot a) = m_K \frac{\boldsymbol{\zeta}\mathbf{p}_\ell}{m_\ell} \qquad p_{\nu_\ell}\cdot a = E_\nu \frac{\boldsymbol{\zeta}\mathbf{p}_\ell}{m_\ell} - \boldsymbol{\zeta}\mathbf{p}_{\nu_\ell} \frac{(\mathbf{p}_{\nu_\ell}\mathbf{p}_\ell)(\boldsymbol{\zeta}\mathbf{p}_\ell)}{m_\ell(E_\ell + m_\ell)}
\tag{21}
$$

All but the last term in (20) involve components of the lepton polarization on the decay plane ($\mathbf{p}_{\nu_\ell}\boldsymbol{\zeta}$ and $\mathbf{p}_\ell\boldsymbol{\zeta}$). Only the last term,

$$
\varepsilon_{\mu\nu\rho\sigma} P^\mu q^\nu p_{\nu_\ell}{}^\rho a^\sigma = 2m_K (\mathbf{p}_{\nu_\ell}\wedge \mathbf{p}_\ell)\cdot\boldsymbol{\zeta}
\tag{22}
$$

would bring about a polarization perpendicular to that plane. However, this term changes sign under time reversal and its presence would violate T. Indeed, as discussed above, T invariance requires f_1 and f_2 to be mutually real, so that $\mathrm{Im} f_1 f_2^*$ vanishes.

Problem 52. π^0 decay into $\gamma\gamma$ and Dalitz pairs

a) Write down the amplitude for $\pi^0 \to \gamma\gamma$ decay in terms of the decay width Γ, assuming that π^0 is scalar or pseudoscalar.

b) Calculate the probability that one of the two photons forms an e^+e^- pair (Dalitz pair). What is the distribution of its invariant mass? Study the distribution of the angle formed between the linear polarization of the second photon and the direction of the e^+e^- plane, for each parity assignment assumed for the pion.

c) The invariant mass distribution found in part b) is strongly peaked at small μ ($\mu \simeq 2m_e$). Using this fact to simplify your calculations, find the probability for the process $\pi^0 \to e^+e^-e^+e^-$. Calculate the correlation between the planes of the two pairs, for each of the two parity assignments.

Solution

a) The decay $\pi^0 \to \gamma\gamma$ can be described by an effective Lagrangian which has the form, in the scalar and pseudoscalar case,

$$
\mathcal{L}_S \;=\; \frac{C_S}{m_\pi} \pi^0 F_{\mu\nu} F^{\mu\nu}
\tag{1a}
$$

$$\mathcal{L}_{PS} = \frac{C_{PS}}{m_\pi} \pi^0 \frac{1}{2} \varepsilon_{\mu\nu\rho\sigma} F^{\mu\nu} F^{\rho\sigma} \tag{1b}$$

where π^0 is the pion field and $F^{\mu\nu}$ is the electromagnetic field tensor. The amplitudes derived from (1) are[2]

$$\mathcal{M}_S = \frac{4iC_S}{m_\pi} \{(k_1{\cdot}k_2)(\varepsilon_1{\cdot}\varepsilon_2) - (k_1{\cdot}\varepsilon_2)(k_2{\cdot}\varepsilon_1)\} \tag{2a}$$

$$\mathcal{M}_{PS} = \frac{4iC_{PS}}{m_\pi} \varepsilon_{\mu\nu\rho\sigma} k_1^\mu k_2^\nu \varepsilon_1^\rho \varepsilon_2^\sigma \tag{2b}$$

Summing over photon polarizations in the final state, and using the relation

$$2(k_1{\cdot}k_2) = m_\pi^2 \tag{3}$$

we obtain

$$|\mathcal{M}|^2_{S,(PS)} = 8m_\pi^2 |C_{S,(PS)}|^2 \tag{4}$$

The line width follows immediately

$$\Gamma_{S,(PS)} = \frac{m_\pi}{4\pi} |C_{S,(PS)}|^2 \tag{5}$$

Eq.(5) allows us to deduce the modulus of the coupling constants C_S and C_{PS}, given the total decay width.[3]

b) The transition amplitude with production of one Dalitz pair is

$$\mathcal{M}_S = \frac{4C_S}{m_\pi}(k^\mu \varepsilon^\nu - k^\nu \varepsilon^\mu) \frac{eq_\mu}{q^2} \bar{u}(p_{e-})\gamma_\nu v(p_{e+}) \tag{6a}$$

$$\mathcal{M}_{PS} = \frac{4C_{PS}}{m_\pi} \varepsilon^{\mu\nu\rho\sigma} k_\mu \varepsilon_\nu q_\rho \bar{u}(p_{e-})\gamma_\sigma v(p_{e+}) \frac{e}{q^2} \tag{6b}$$

with

$$\begin{array}{ll} p_\pi^\mu = k^\mu + q^\mu & q^\mu = p_{e+}{}^\mu + p_{e-}{}^\mu \\ (\varepsilon{\cdot}k) = 0 & q_\mu \bar{u}(p_{e-})\gamma^\mu v(p_{e+}) = 0 \end{array} \tag{7}$$

The last two relations are a consequence of gauge invariance. The modulus squared of the amplitude is

$$|\mathcal{M}_S|^2 = \frac{16|C_S|^2 e^2}{m_\pi^2(q^2)^2} \{(q{\cdot}k)\varepsilon^\mu - (q{\cdot}\varepsilon)k^\mu\} \{(q{\cdot}k)\varepsilon^{\nu*} - (q{\cdot}\varepsilon^*)k^\nu\} L_{\mu\nu} \tag{8a}$$

$$|\mathcal{M}_{PS}|^2 = \frac{16|C_{PS}|^2 e^2}{m_\pi^2(q^2)^2} \varepsilon_{\mu\nu\rho\sigma} k^\mu \varepsilon^\nu q^\rho \varepsilon_{\mu'\nu'\rho'\sigma'} k^{\mu'} \varepsilon^{\nu'} q^{\rho'} L^{\sigma\sigma'} \tag{8b}$$

---

[2] A factor of 2 in Eq.(2) corresponds to the possibility of creating a photon from each of the two powers of $F_{\mu\nu}$ appearing in (1).

[3] Eq.(5) is gotten from (4) by means of the formula

$$\Gamma = \frac{1}{2!} \int \frac{|\mathcal{M}|^2}{2m_\pi} d\Phi^{(2)}$$

The factor of 1/2! is due to the presence of two identical particles in the final state.

We have set

$$L_{\mu\nu} = \text{Tr}\left\{\gamma^\mu(m_e - \not{p}_{e^+})\gamma^\nu(\not{p}_{e^-} + m_e)\right\} = 2\left(g^{\mu\nu}q^2 - q^\mu q^\nu\right) + 2r^\mu r^\nu \tag{9}$$

$$r^\mu = p_{e^+}{}^\mu - p_{e^-}{}^\mu \qquad (q\cdot r) = 0 \tag{10}$$

We can exploit gauge freedom to replace ε_μ with ε'_μ, defined as follows

$$\varepsilon'_\mu = \varepsilon_\mu - \frac{(q\cdot\varepsilon)}{(q\cdot k)}k_\mu \tag{11}$$

ε'_μ obeys

$$\varepsilon'_\mu\cdot q^\mu = 0 \tag{12}$$

It is easy to verify that the above condition is simply the Coulomb gauge in the pion rest frame. Eq.(8) now simplifies to

$$|\mathcal{M}_S|^2 = \frac{32|C_S|^2 e^2}{m_\pi^2(q^2)^2}(q\cdot k)^2\left\{q^2(\varepsilon'\cdot\varepsilon'^*) + (r\cdot\varepsilon')(r\cdot\varepsilon'^*)\right\} \tag{13a}$$

$$|\mathcal{M}_{PS}|^2 = \frac{16|C_{PS}|^2 e^2}{m_\pi^2(q^2)^2}\Big[(q\cdot k)^2\left\{q^2(\varepsilon'\cdot\varepsilon'^*) - (r\cdot\varepsilon')(r\cdot\varepsilon'^*)\right\} \tag{13b}$$

$$+ (\varepsilon'\cdot\varepsilon'^*)\left\{q^2(r\cdot k)^2 + r^2(q\cdot k)^2\right\}\Big]$$

The photon polarization satisfies $\varepsilon'\cdot\varepsilon'^* = -1$. In order to simplify further Eq.(13), we go over to the pion rest frame. We write

$$(q\cdot k) = q^0 k^0 - \mathbf{qk} = q^0 k^0 + \mathbf{k}^2 = k^0\left(q^0 + k^0\right) = m_\pi k^0 = m_\pi|\mathbf{q}| \tag{14}$$

hence $(q\cdot k)^2/m_\pi^2 = \mathbf{q}^2$. The decay plane of the lepton pair is identified by the two orthogonal vectors

$$\mathbf{q}, \qquad \mathbf{r}_\perp = \mathbf{r} - \frac{(\mathbf{rq})}{\mathbf{q}^2}\mathbf{q} \tag{15}$$

The Coulomb gauge condition (11) implies $\boldsymbol{\varepsilon}\mathbf{q} = -\boldsymbol{\varepsilon}\mathbf{k} = 0$ and

$$(\varepsilon\cdot r)^2 = (\boldsymbol{\varepsilon}\mathbf{r})^2 = (\boldsymbol{\varepsilon}\mathbf{r}_\perp)^2 = |\mathbf{r}_\perp|^2\sin^2\varphi \tag{16}$$

where φ is the angle between $\boldsymbol{\varepsilon}$ and the normal to the decay plane

$$\cos\phi = \boldsymbol{\varepsilon}'\cdot\frac{(\mathbf{q}\wedge\mathbf{r})}{|\mathbf{q}\wedge\mathbf{r}|} \tag{17}$$

Finally, we make use of the relation

$$\mathbf{r}_\perp^2 = \mathbf{r}^2 - \frac{(\mathbf{rq})^2}{\mathbf{q}^2} = \frac{|\mathbf{q}\wedge\mathbf{r}|^2}{\mathbf{q}^2} \tag{18}$$

The amplitudes can now be written in the simple form

$$|\mathcal{M}_S|^2 = 32 C_S e^2 \frac{q^2}{q^2} \left[\frac{(\mathbf{q} \wedge \mathbf{r})^2}{q^2 \mathbf{q}^2} \sin^2 \phi - 1 \right] \tag{19}$$

$$|\mathcal{M}_{PS}|^2 = 32 C_{PS} e^2 \frac{q^2}{q^2} \left[\frac{(\mathbf{q} \wedge \mathbf{r})^2}{q^2 \mathbf{q}^2} \cos^2 \phi - 1 \right] \tag{20}$$

To determine the invariant mass distribution of the lepton pair, it is convenient to decompose three-body phase space $(\gamma e^+ e^-)$ by the formula (see Appendix A)

$$d\Phi^{(3)} = \frac{dq^2}{2\pi} \frac{1}{8\pi} \left(1 - \frac{q^2}{m_\pi^2} \right) d\Phi^{(2)} \left(q \to e^+ e^- \right) \tag{21}$$

leading to

$$d\Gamma = \frac{1}{2\pi} \sum \overline{|\mathcal{M}|^2} \frac{1}{8\pi} \left(1 - \frac{q^2}{m_\pi^2} \right) \sqrt{1 - \frac{4m_e^2}{q^2}} \frac{dq^2}{2\pi} \tag{22}$$

In the above, $\overline{|\mathcal{M}|^2}$ is obtained from (8), replacing $L_{\mu\nu}$ by its phase space average $\bar{L}_{\mu\nu}$ (see Appendix F)

$$\bar{L}_{\mu\nu} = \frac{\int d\Phi^{(2)}(e^+ e^-) L_{\mu\nu}}{\int d\Phi^{(2)}(e^+ e^-)} = \frac{4}{3} \left(1 + \frac{2m_e^2}{q^2} \right) \left(q^2 g_{\mu\nu} - q_\mu q_\nu \right) \tag{23}$$

A simple calculation gives

$$\overline{|\mathcal{M}|^2} = -\frac{16}{3} \frac{C^2 e^2}{m_\pi^2 q^2} \left(1 + \frac{2m_e^2}{q^2} \right) \left(m_\pi^2 - q^2 \right)^2 (\varepsilon \cdot \varepsilon^*) \tag{24}$$

and

$$\Gamma = \frac{C^2}{4\pi} m_\pi \frac{2\alpha}{3\pi} \int_{4m_e^2}^{m_\pi^2} \frac{dq^2}{q^2} \left(1 - \frac{q^2}{m_\pi^2} \right)^3 \left(1 + \frac{2m_e^2}{q^2} \right) \left(1 - \frac{4m_e^2}{q^2} \right)^{1/2} \tag{25}$$

In the limit $m_e \to 0$ the distribution of the invariant mass is

$$\frac{d\Gamma(\gamma e^+ e^-)}{dq^2} = \Gamma \frac{2\alpha}{3\pi} \frac{1}{q^2} \left(1 - \frac{q^2}{m_\pi^2} \right)^3 \tag{26}$$

We integrate over q^2 in the kinematically allowed region, with the result

$$B(\gamma e^+ e^-) = \frac{\Gamma(\gamma e^+ e^-)}{\Gamma(\gamma\gamma)} = \frac{2\alpha}{3\pi} \left[\ln \frac{m_\pi^2}{4m_e^2} - \frac{11}{6} \right] \tag{27}$$

Substituting numerical values[4]

$$B(\gamma e^+ e^-) = 1.23 \cdot 10^{-2} \tag{28}$$

---

[4] A numerical integration of the exact expression (25) gives $B(\gamma e^+ e^-) = 1.19 \cdot 10^{-2}$.

in agreement[5] with the experimental result $(1.198 \pm 0.032)10^{-2}$.

c)   The fact that the invariant mass is concentrated at low values allows us to neglect interference terms, corresponding to an exchange of e^- between the two pairs. In terms of the 2γ decay amplitude

$$\mathcal{M} = \varepsilon_\mu^* \varepsilon_\nu'^* Q^{\mu\nu}(q, q') \tag{29}$$

the amplitude for decay into two pairs is approximately given by

$$\mathcal{M} = e^2 j_\mu j_\nu' Q_{\mu\nu}(q, q') \frac{1}{q^2} \frac{1}{q'^2} \tag{30}$$

with

$$
\begin{aligned}
j_\mu &= \bar{u}\gamma_\mu v & q \cdot j &= 0 \\
j_\mu' &= \bar{u}'\gamma_\mu v' & q' \cdot j' &= 0
\end{aligned} \tag{31}
$$

Equations (2) give, for the scalar and pseudoscalar cases,

$$Q_{\mu\nu}^{(S)}(q, q') = \frac{4iC_S}{m_\pi} \left[(q \cdot q')g_{\mu\nu} - q_\nu q_\mu' \right] \tag{32a}$$

$$Q_{\mu\nu}^{(PS)}(q, q') = \frac{4iC_{PS}}{m_\pi} \varepsilon_{\alpha\beta\mu\nu} q^\alpha q'^\beta \tag{32b}$$

We decompose four body phase space as follows

$$
\begin{aligned}
&d\Phi^{(4)}\left(\pi \to e^+e^-e^+e^-\right) = \\
&\frac{dq^2}{2\pi} \frac{dq'^2}{2\pi} d\Phi^{(2)}\left(\pi \to q + q'\right) d\Phi^{(2)}\left(q \to e^+e^-\right) d\Phi^{(2)}\left(q' \to e^+e^-\right)
\end{aligned} \tag{33}
$$

The differential width is

$$
\begin{aligned}
d\Gamma &= \frac{e^4}{4m_\pi} Q_{\mu\nu}(q, q') Q_{\rho\sigma}^*(q, q') L^{\mu\rho} L^{\nu\sigma} \frac{1}{q^4} \frac{1}{q'^4} \frac{dq^2}{2\pi} \frac{dq'^2}{2\pi} \\
&\quad d\Phi^{(2)}\left(\pi \to q + q'\right) d\Phi^{(2)}\left(q \to e^+e^-\right) d\Phi^{(2)}\left(q' \to e^+e^-\right)
\end{aligned} \tag{34}
$$

The transversality of $Q_{\mu\nu}$ with respect to q_μ and q_ν' allows us to write

$$
\begin{aligned}
d\Gamma &= \frac{e^4}{4m_\pi} \left(1 + \frac{2m_e^2}{q^2}\right) \left(1 + \frac{2m_e^2}{q'^2}\right) \sqrt{1 - \frac{4m_e^2}{q^2}} \sqrt{1 - \frac{4m_e^2}{q'^2}} \\
&\quad Q_{\mu\nu} Q^{*\mu\nu} \frac{4}{3q^2} \frac{4}{3q'^2} \frac{dq^2}{2\pi} \frac{dq'^2}{2\pi} \frac{1}{(8\pi)^2} d\Phi^{(2)}\left(\pi \to q + q'\right)
\end{aligned} \tag{35}
$$

---

[5]Particle Data Group, *Phys. Lett.* **B239** 1990

This expression receives its dominant contribution from the phase space region $4m_e^2 \leq q^2 \ll m_\pi^2$. In this region we may treat the invariant $Q_{\mu\nu}^* Q^{\mu\nu}$ as constant and approximate it with its value at $q^2 = q'^2 = 0$ (on-shell photons). Comparing (35) with the width of $\pi^0 \rightarrow \gamma\gamma$

$$\Gamma = \frac{1}{32\pi m_\pi} Q_{\mu\nu}^* Q^{\mu\nu} \qquad (36)$$

we find

$$\frac{\Gamma\left(e^+ e^- e^+ e^-\right)}{\Gamma(2\gamma)} = \frac{8\alpha^2}{9\pi} \int \frac{dq^2}{q^2} \frac{dq'^2}{q'^2}$$

$$d\Phi^{(2)}\left(\pi \rightarrow q + q'\right)\left(1 + \frac{2m_e^2}{q^2}\right)\left(1 + \frac{2m_e^2}{q'^2}\right)\sqrt{1 - \frac{4m_e^2}{q^2}}\sqrt{1 - \frac{4m_e^2}{q'^2}} \qquad (37)$$

We estimate this integral keeping only dominant terms in m_π/m_e and find

$$\frac{\Gamma\left(e^+ e^- e^+ e^-\right)}{\Gamma(2\gamma)} \simeq \left(\frac{\alpha}{3\pi}\right)^2 \int_{4m_e^2}^{m_\pi^2} \frac{dq^2}{q^2} \int_{4m_e^2}^{m_\pi^2} \frac{dq'^2}{q'^2} = \left(\frac{\alpha}{3\pi}\right)^2 \ln^2 \frac{m_\pi^2}{4m_e^2} = 5.7 \cdot 10^{-5} \qquad (38)$$

In this approximation we have set $\Phi^{(2)} = \frac{1}{8\pi}$ and integrated over the range $4m_e^2 \leq q^2 \leq m_\pi^2$, taking q and q' to be independent. A numerical integration of expression (37) over the kinematic region $2m_e \leq q + q' \leq m_\pi$ gives instead

$$\frac{\Gamma\left(e^+ e^- e^+ e^-\right)}{\Gamma(2\gamma)} \simeq 3.49 \cdot 10^{-5} \qquad (39)$$

The experimental value for this ratio is[6]

$$\frac{\Gamma\left(e^+ e^- e^+ e^-\right)}{\Gamma(2\gamma)} = (3.24 \pm 0.3)10^{-5} \qquad (40)$$

We recall that interference terms have been neglected in (37).

To study the angular correlation between the decay planes, we consider again the expression (30) for the amplitude. For definiteness, let us concentrate on the scalar case. The amplitude is proportional to

$$\frac{1}{q^2 q'^2}\left\{(q \cdot q')j^\mu j'_\mu - (q'_\mu j^\mu)(q^\nu j'_\nu)\right\} \qquad (41)$$

Since low values of q^2 and q'^2 dominate, we may set in the numerator $q^2 = q'^2 = 0$. Let $\mathbf{q}$ point along direction 3. Current conservation in this approximation implies

$$q^\mu j_\mu = q^0 j^0 - q^3 j^3 = 0 \quad \Rightarrow \quad j^3 = j^0$$
$$q'^\mu j'_\mu = q^0 j'^0 - q'^3 j'^3 = 0 \quad \Rightarrow \quad j'^3 = -j'^0 \qquad (42)$$

---

[6]Particle Data Group, *Phys. Lett.* **B239** 1990

This results in

$$Q_{\mu\nu}(q,q')j^{\mu}j'^{\nu} = 2q_0q'_0(j_0j'_0 - j_3j'_3 - \mathbf{j}_\perp\mathbf{j}'_\perp) - 4q_0q'_0j_0j'_0 = -2q_0q'_0\mathbf{j}_\perp\mathbf{j}'_\perp \quad (43)$$

where $\mathbf{j}_\perp$ denotes the component of the current perpendicular to $\mathbf{q}$. Thus the amplitude squared is proportional to

$$\sum_{i,j=1}^{2} L_{ij}(q,r)L^{ij}(q',r') \quad (44)$$

Using Eq.(9), at $q^2 = 0$,

$$\sum_{i,j=1}^{2} L_{ij}(q,r)L^{ij}(q',r') = (\mathbf{r}_\perp\cdot\mathbf{r}'_\perp)^2 \quad (45)$$

The orthogonal vectors $\mathbf{r}_\perp$ and $\mathbf{q}$ uniquely determine the decay plane of a pair; we have

$$(\mathbf{r}_\perp\cdot\mathbf{r}'_\perp)^2 \propto \cos^2\theta \quad (46)$$

where θ is the angle between the normals to the decay planes. Thus, in the scalar case, the correlation between the decay planes is given by

$$\frac{1}{\Gamma}\frac{d\Gamma}{d\theta} = 2\cos^2\theta \quad (47)$$

In the pseudoscalar case it equals

$$\frac{1}{\Gamma}\frac{d\Gamma}{d\theta} = 2\sin^2\theta \quad (48)$$

(We used the fact that the phase space is proportional to $d\theta$.)

Problem 53.   Narrow band ν beams

A beam of scalar particles with positive charge ($90\%\pi^+, 10\%K^+$) and momentum $\mathbf{p}$ ($|\mathbf{p}| = 400$ GeV/c) travels down a vacuum tube of length $L = 100$ m; at the end of the tube the beam is deviated away (see fig.1).

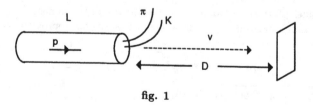

fig. 1

A bubble chamber, centered on the beam axis at a distance $D = 2000$ m from the tube, is exposed to neutrinos produced by π's and K's inside the tube.

a) Neutrinos are produced by the following decays

$$\begin{aligned}
\pi^+ &\to \mu^+ \nu_\mu \qquad BR \simeq 100\% \\
K^+ &\to \mu^+ \nu_\mu \qquad BR = 63.5\%
\end{aligned} \tag{1}$$

($m_{\pi^+} = 139.6\,\mathrm{MeV}$, $\tau_{\pi^+} = 2.6 \cdot 10^{-8}\,\mathrm{sec}$; $m_{K^+} = 493.7\,\mathrm{MeV}$, $\tau_{K^+} = 1.24 \cdot 10^{-8}\,\mathrm{sec}$) Calculate the energy spectrum of ν in the laboratory frame. (You may neglect ν's produced by other decay modes of K^+.)

b) Determine the neutrino energy as a function of the transverse distance r from the center of the bubble chamber and plot this function.

c) Suppose the beam has a momentum uncertainty $\Delta p/p = 1\%$, and an angular spread $\Delta\theta = 1.2 \cdot 10^{-3}$. What is the resulting uncertainty on the energy distribution?

Solution

a) The neutrino energy spectrum is a weighted sum of the spectra in the two decays, with weights

$$W_\pi = 0.9 \qquad W_K = 0.1 \tag{2}$$

Given that π and K have spin zero, their decay is isotropic in the rest frame; consequently, the energy spectrum of neutrinos in the laboratory frame will be flat, in the range

$$\frac{E-p}{2}\left(1 - \frac{\mu^2}{m^2}\right) \le E_\nu \le \frac{E+p}{2}\left(1 - \frac{\mu^2}{m^2}\right) \tag{3}$$

where m, E and p are the mass, energy and momentum of the meson; μ is the mass of the charged lepton. In terms of the mean life τ, the decay probability per unit time Γ, in the laboratory frame, is

$$\Gamma = \frac{m}{E}\frac{1}{\tau} \tag{4}$$

The probability of decaying into the particular mode considered is given by

$$\Gamma_\nu = \Gamma\, B \tag{5}$$

where B is the corresponding branching ratio. Now, the time of flight of mesons inside the tube is

$$t = \frac{E}{p}L \tag{6}$$

Thus, the total number of decays per particle of the incoming beam is

$$N = \frac{L}{\tau}\frac{m}{p}B \tag{7}$$

In terms of the normalized energy distribution

$$
\frac{1}{p\left(1-\frac{\mu^2}{m^2}\right)}\,\theta\left[E_\nu - \frac{E-p}{2}\left(1-\frac{\mu^2}{m^2}\right)\right]\theta\left[\frac{E+p}{2}\left(1-\frac{\mu^2}{m^2}\right) - E_\nu\right] \tag{8}
$$

we obtain, using $E \simeq p$,

$$
\frac{\mathrm{d}N}{\mathrm{d}E_\nu} \simeq \frac{L}{p}\left\{\frac{m_\pi}{\tau_\pi}\frac{B_\pi W_\pi}{E_\pi\left(1-\frac{\mu^2}{m_\pi^2}\right)}\,\theta\left[E_\pi\left(1-\frac{\mu^2}{m_\pi^2}\right) - E_\nu\right]\right.
$$
$$
\left. + \frac{m_K}{\tau_K}\frac{B_K W_K}{E_K\left(1-\frac{\mu^2}{m_K^2}\right)}\,\theta\left[E_K\left(1-\frac{\mu^2}{m_K^2}\right) - E_\nu\right]\right\} \tag{9}
$$

Numerically we find

$$
p\frac{\mathrm{d}N}{\mathrm{d}E_\nu} \simeq 9.4{\cdot}10^{-3}\,\theta(\bar{E}_{\nu_\pi} - E_\nu) + 2.2{\cdot}10^{-3}\,\theta(\bar{E}_{\nu_K} - E_\nu) \tag{10}
$$

with

$$
\bar{E}_{\nu_\pi} = E_\pi\left(1-\frac{m_\mu^2}{m_\pi^2}\right) \simeq 172\,\mathrm{GeV} \qquad \bar{E}_{\nu_K} = E_K\left(1-\frac{m_\mu^2}{m_K^2}\right) \simeq 380\,\mathrm{GeV} \tag{11}
$$

b)   The neutrino energy in the laboratory frame is related to the center-of-mass energy E_ν^{cm} as follows

$$
E_\nu^{\mathrm{cm}} = \gamma\,(1 - \beta\cos\theta)\,E_\nu \tag{12}
$$

with

$$
\gamma = \frac{E}{m} \quad \beta = \frac{p}{E} \quad E_\nu^{\mathrm{cm}} = \frac{m^2 - \mu^2}{2m} \tag{13}
$$

Eq.(12) gives the neutrino energy as a function of the direction θ

$$
E_\nu(\theta) = \frac{1}{2}\left(1-\frac{\mu^2}{m^2}\right)\frac{m^2}{E - p\cos\theta} = \frac{1}{2}\left(1-\frac{\mu^2}{m^2}\right)\frac{E + p\cos\theta}{1 + \frac{p^2}{m^2}\sin^2\theta} \tag{14}
$$

If $r \ll D$ we write approximately

$$
\cos\theta \simeq 1 \qquad \sin\theta \simeq r/D \tag{15}
$$

The r-dependence of the energy then reads

$$
E_\nu(r) = \frac{1}{2}\left(1-\frac{\mu^2}{m^2}\right)\frac{E + p}{1 + \frac{p^2 r^2}{m^2 D^2}} \simeq \frac{\bar{E}_\nu}{1 + \frac{p^2 r^2}{m^2 D^2}} \tag{16}
$$

A plot of this function is shown in fig.2.

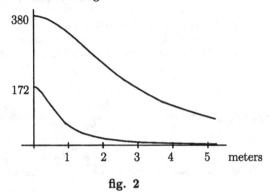

fig. 2

The two curves correspond to $\bar{E}_{\nu_\pi}$ and $\bar{E}_{\nu_K}$. The Lorentzian half-widths are of the order of

$$\Delta r \simeq \frac{mD}{p} \tag{17}$$

At $p = 400$ GeV/c, we have $\Delta r \simeq 0.7$ m for pions and $\Delta r \simeq 2.6$ m for kaons.

c)   In order to find the error induced on $E_\nu(r)$ by the momentum uncertainty, we differentiate Eq.(16)

$$\frac{\Delta E_\nu(r)}{E_\nu(r)} \simeq \frac{\Delta p}{p} \left\{ 1 - \frac{2\dfrac{p^2 r^2}{m^2 D^2}}{1 + \dfrac{p^2 r^2}{m^2 D^2}} \right\} \tag{18}$$

The numerical values in this problem lead to

$$\frac{\Delta E_\nu(r)}{E_\nu(r)} \simeq \frac{\Delta p}{p} \tag{19}$$

The beam spread introduces a further uncertainty Δr of the form

$$\frac{\Delta r}{D} \simeq \Delta\theta \simeq 1.2 \cdot 10^{-3} \tag{20}$$

Making use of (16) once again, we finally obtain for the total error in energy

$$\frac{\Delta E_\nu(r)}{E_\nu(r)} \simeq \sqrt{\left(\frac{\Delta p}{p} \frac{1 - \dfrac{p^2 r^2}{m^2 D^2}}{1 + \dfrac{p^2 r^2}{m^2 D^2}} \right)^2 + \left(\frac{2\dfrac{p^2}{m^2}\dfrac{r}{D}}{1 + \dfrac{p^2 r^2}{m^2 D^2}} \Delta\theta \right)^2} \tag{21}$$

Problem 54.   Positronium decay into $\gamma\gamma$ and $\nu_e\bar{\nu}_e$

Positronium pairs (e^+e^-) are produced in the ground state (S-wave) as a uniform statistical mixture of singlet and triplet. They can decay into $\gamma\gamma$, $\nu_e\bar{\nu}_e$ and $\gamma\gamma\gamma$. The effective interaction leading to two-photon decay is the nonrelativistic limit of the diagrams shown in fig. 1.

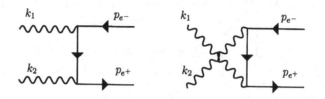

fig. 1

The second decay mode ($\nu_e\bar{\nu}_e$) is described by the effective interaction

$$\mathcal{L} = -\frac{G_F}{\sqrt{2}}\bar{\nu}_e\gamma^\mu(1-\gamma^5)\nu\,\bar{e}\gamma_\mu(a+b\gamma^5)e$$

$$a = \frac{1}{2}+\sin^2\theta_W \qquad b = -\frac{1}{2} \tag{1}$$

with $G_F\,m_p^2 \simeq 10^{-5}$. This effective lagrangian includes the contribution of charged currents, corresponding to Fermi interaction, and of neutral currents (see also Prob.102).

a) Calculate the width for the decay into $\gamma\gamma$, in sec^{-1}. Show that only the singlet state contributes to this mode, and only the triplet contributes to the three-photon decay.

b) Taking into account neutrino helicities, show that only the triplet state can undergo a weak decay.

c) Calculate the branching ratio

$$R\left(\frac{e^+e^-\to\nu_e\bar{\nu}_e}{e^+e^-\to\gamma\gamma}\right) \tag{2}$$

for the given statistical mixture.

Solution

a) The quantum numbers of e^+e^- states are determined by the formulae

$$\begin{aligned}
P &= (-1)^{L+1}\\
C &= (-1)^{L+S}\\
CP &= (-1)^{S+1}
\end{aligned} \tag{3}$$

There exist two S-wave states, a singlet $^1S_0$ and a triplet $^3S_1$, having charge conjugation eigenvalues $C = 1$ and $C = -1$, respectively. An N-photon state has $C = (-1)^N$; therefore the singlet can decay only into an even number of photons, while the triplet can decay only into an odd number.

We may express the bound state e^+e^- as a superposition of momentum eigenstates

$$|\psi\rangle = \int \frac{d^3q}{(2\pi)^3}\, \tilde{\psi}(\mathbf{q})\, |\frac{\mathbf{P}}{2} + \mathbf{q}\rangle|\frac{\mathbf{P}}{2} - \mathbf{q}\rangle \tag{4}$$

where $\mathbf{P}$ and $\mathbf{q}$ denote the total and relative momentum of the leptons, and $\tilde{\psi}(\mathbf{q})$ is the Fourier transform of the Schrödinger wave function. In the nonrelativistic limit this state is normalized as follows

$$\langle\psi|\psi'\rangle = 4E_{e^-}E_{e^+}(2\pi)^3\delta^3(\mathbf{P}-\mathbf{P}')\int \frac{d^3q}{(2\pi)^3}\left|\tilde{\psi}(\mathbf{q})\right|^2 \simeq 4m_e^2(2\pi)^3\delta^3(\mathbf{P}-\mathbf{P}') \tag{5}$$

We have left out spin variables for the time being.

Let us now consider the decay of the singlet state into a pair of photons with momenta $\mathbf{k}$ and $\mathbf{k}' = -\mathbf{k}$ in the center-of-mass frame. To lowest perturbative order, the corresponding decay width is

$$\Gamma_{2\gamma} = \frac{1}{4m_e^2}\sum_{pol}\int \frac{1}{2!}\,d\Phi^{(2\gamma)}\left|\int \frac{d^3q}{(2\pi)^3}\mathcal{M}_S(\mathbf{k},\mathbf{q})\tilde{\psi}(\mathbf{q})\right|^2 \tag{6}$$

$\mathcal{M}_S(\mathbf{k},\mathbf{q})$ is the scattering amplitude for the process[7] $e^+e^- \to \gamma\gamma$. Eq.(6) is written in the nonrelativistic limit $E_e \simeq m_e$ with the normalization (5).

$\tilde{\psi}(\mathbf{q})$ falls off at momenta of the order of $m_e\alpha$ (the Bohr momentum of the system); these are small compared to characteristic energies of the problem, $E_\gamma \sim m_e$, and therefore we may neglect the $\mathbf{q}$-dependence in the amplitude $\mathcal{M}(\mathbf{k},\mathbf{q})$. In other words, the decay takes place at distances of the order of $1/m_e$, much smaller than the Bohr radius $1/m_e\alpha$. This allows us to simplify (6)

$$\Gamma_{2\gamma} = |\psi(0)|^2 \frac{1}{4m_e^2}\sum_{pol}\int \frac{1}{2!}\,d\Phi^{(2\gamma)}\left|\mathcal{M}_S(\mathbf{k},0)\right|^2 \tag{7}$$

We made use of

$$\int \frac{d^3q}{(2\pi)^3}\tilde{\psi}(\mathbf{q}) = \psi(0) \tag{8}$$

We want to express the above in terms of the cross section for $e^+e^- \to \gamma\gamma$ scattering. Averaging over initial spins, we write

$$\bar{\sigma} = \frac{1}{v_{rel}}\frac{1}{4E_{e^-}E_{e^+}}\sum_{pol}\frac{1}{4}\sum_{spin}\int \frac{1}{2!}\,d\Phi^{(2\gamma)}\left|\mathcal{M}\right|^2 \tag{9}$$

---

[7]Eq.(6) is the Born approximation for the decay amplitude. The factor of 2! is due to the presence of two identical particles in the final state.

Denoting by $\mathcal{M}_S$ and $\mathcal{M}_T$ the spin singlet and triplet amplitudes, we have

$$\bar{\sigma} \propto \frac{1}{4}\left(3|\mathcal{M}_T|^2 + |\mathcal{M}_S|^2\right) \tag{10}$$

Only the singlet amplitude contributes to the process considered; comparing (9) and (10) to Eq.(7) we find

$$\Gamma_{2\gamma} = 4|\psi(0)|^2 \left(v_{\text{rel}}\sigma_{2\gamma}\right)_{v_{\text{rel}}\to 0} \tag{11}$$

To evaluate this expression, we note that the wave function of the $1S$ state of positronium is

$$\psi(x) = \frac{1}{\sqrt{\pi a^3}}e^{-r/a} \qquad a = \frac{2}{\alpha m_e} \tag{12}$$

while the amplitude for $e^+e^- \to 2\gamma$ is given by

$$\mathcal{M} = e^2 \varepsilon_\mu^*(\mathbf{k})\varepsilon_\nu'^*(\mathbf{k}')\,\bar{v}(\mathbf{p}_{e+})Q_{\mu\nu}u(\mathbf{p}_{e-}) \tag{13}$$

$$Q_{\mu\nu} = \gamma_\mu \frac{1}{\not{p}_{e-} - \not{k} - m_e}\gamma_\nu + \gamma_\nu \frac{1}{\not{p}_{e-} - \not{k}' - m_e}\gamma_\mu \tag{14}$$

$$k^\mu + k'^\mu = p_{e-}{}^\mu + p_{e+}{}^\mu \tag{15}$$

k and k' are the four-momenta of the photons. We now perform the sum over final polarizations and average over initial spins, with the result

$$\overline{|\mathcal{M}|^2} = \frac{e^4}{4}\text{Tr}\left\{(\not{p}_{e-} + m_e)\,Q_{\mu\nu}\,(\not{p}_{e+} - m_e)\,\overline{Q}^{\mu\nu}\right\} \tag{16}$$

The total cross section becomes

$$d\sigma_{2\gamma} = \frac{1}{v_{\text{rel}}}\frac{1}{4E_{e-}E_{e+}}\overline{|\mathcal{M}|^2}\,d\Phi^{(2)} = \frac{1}{16\pi}\overline{|\mathcal{M}|^2}\frac{dt}{s\left(s - 4m_e^2\right)} \tag{17}$$

$$s = (p_{e+} + p_{e-})^2 \qquad t = (p_{e-} - k)^2$$

To evaluate $\Gamma_{2\gamma}$ in Eq.(11)) we need the nonrelativistic limit of (17); setting $p_{e-} \simeq (m_e, \mathbf{0}) \simeq p_{e+}$ we obtain

$$\sigma_{2\gamma} \simeq \frac{\pi\alpha^2}{m_e^2}\frac{1}{v_{\text{rel}}} \tag{18}$$

A direct expansion of the amplitude (13) leads to the same result: The nonrelativistic limit corresponds to the kinematic configuration

$$p_{e-} \simeq (m_e, \mathbf{0}) \simeq p_{e+} \qquad k = (\omega, \mathbf{k}) \quad k' = (\omega, -\mathbf{k}) \tag{19}$$

where $\omega = m_e/2$ is the photon energy in the center of mass. In the Coulomb gauge only the spatial components of Q_{ij} are present; using (19) and the equations of motion it is easy to find

$$\mathcal{M} = e^2 \frac{1}{2m_e\omega}\bar{v}(\mathbf{p}_{e+})\left(\not{\varepsilon}\not{k}\not{\varepsilon}' + \not{\varepsilon}'\not{k}'\not{\varepsilon}\right)u(\mathbf{p}_{e-}) \tag{20}$$

The above is simplified further, using the nonrelativistic expression for u and v in terms of the Pauli spinors $w_\pm$

$$
\begin{aligned}
\mathcal{M} &= e^2 \frac{1}{2m_e\omega} \bar{v} \left[2k^0\gamma^0 \boldsymbol{\varepsilon}\boldsymbol{\varepsilon}' + \mathbf{k} \cdot \boldsymbol{\gamma}(\not{q}\not{q}' - \not{q}'\not{q}) \right] u \\
&\simeq 2m_e e^2 \frac{i}{m_e\omega} w_+^\dagger \left[(\mathbf{k} \cdot \boldsymbol{\sigma})(\boldsymbol{\varepsilon} \wedge \boldsymbol{\varepsilon}' \cdot \boldsymbol{\sigma}) \right] w_- \\
&= 2m_e e^2 \frac{i}{m_e\omega} w_+^\dagger \, \mathbf{k} \cdot \boldsymbol{\varepsilon} \wedge \boldsymbol{\varepsilon}' \, w_-
\end{aligned} \tag{21}
$$

Averaging over initial and summing over final spin orientations we obtain

$$
\overline{|\mathcal{M}|^2} = 32\pi^2\alpha^2 \qquad e^2 = 4\pi\alpha \tag{22}
$$

There follows

$$
\sigma_{2\gamma} = \frac{1}{v_{\rm rel}} \int \frac{1}{4E_{e-}E_{e+}} \overline{|\mathcal{M}|^2} \, d\Phi^{(2)} = \frac{\pi\alpha^2}{m_e^2} \frac{1}{v_{\rm rel}} \tag{23}
$$

in agreement with (18).

Eqs. (18), (11) and (12) yield

$$
\Gamma = \frac{1}{2}\alpha^5 m_e = 0.8 \cdot 10^{10} \text{sec}^{-1} \tag{24}
$$

b) In the center-of-mass frame the two final state neutrinos have opposite momenta. Their helicities are

$$
h_\nu = -\frac{1}{2} \qquad h_{\bar{\nu}} = \frac{1}{2} \tag{25}
$$

Thus, the total spin of the system has a projection $J_z = -1$ along the direction of the neutrino. This is turn implies that $J \geq 1$ so that the spin state of positronium is a triplet. The same conclusion may be reached considering discrete quantum numbers: Indeed, the final state has $CP = 1$ (because $S = 1$); since the Fermi interaction conserves CP, the singlet state of positronium cannot decay in $\nu\bar{\nu}$.

c) Proceeding as in part a), we write for the width of the decay into $\nu\bar{\nu}$

$$
\Gamma_{\nu\bar{\nu}} = \frac{4}{3}|\psi(0)|^2 \left(v_{\rm rel}\sigma_{\nu\bar{\nu}} \right)_{v_{\rm rel}\to 0} \tag{26}
$$

Let us calculate $\sigma_{\nu\bar{\nu}}$. The amplitude for $e^+e^- \to \nu_e\bar{\nu}_e$ is

$$
\mathcal{M} = \frac{G_F}{\sqrt{2}} \bar{u}(\mathbf{p}_\nu)\gamma^\mu(1-\gamma^5)v(\mathbf{p}_{\bar{\nu}})\bar{v}(\mathbf{p}_{e+})\gamma_\mu(a+b\gamma^5)u(\mathbf{p}_{e-}) \tag{27}
$$

$$
\begin{aligned}
\overline{|\mathcal{M}|^2} &= \frac{G_F^2}{4} \text{Tr}\left\{ \not{p}_\nu \gamma^\mu \not{p}_{\bar{\nu}} \gamma^\nu (1-\gamma^5) \right\} \text{Tr}\left\{ (\not{p}_{e+} - m)\gamma_\mu(a+b\gamma^5)(\not{p}_{e-}+m)(a+b\gamma^5) \right\} \\
&= 8G_F^2 \left[(a-b)^2(p_\nu p_{e+})(p_{\bar{\nu}}p_{e-}) + (a+b)^2(p_\nu p_{e-})(p_{\bar{\nu}}p_{e+}) - m^2(a^2-b^2)(p_\nu p_{\bar{\nu}}) \right] \\
&= 8G_F^2 E_e^4 \left[(a-b)^2(1+v\cos\theta)^2 + (a+b)^2(1-v\cos\theta)^2 - 2\frac{m^2}{E_e^2}(a^2-b^2) \right] \tag{28}
\end{aligned}
$$

with

$$\mathbf{v} = \frac{\mathbf{p}_e}{E_e} \qquad \cos\theta = \hat{\mathbf{p}}_{\mathbf{e}} \cdot \hat{\mathbf{p}}_{\nu_{\mathbf{e}}} \tag{29}$$

E_e is the electron energy in the center of mass frame. We may neglect v_e ($v_e \ll 1$) in positronium, writing

$$\overline{|\mathcal{M}|^2} \simeq 8G_F^2 m^4 \tag{30}$$

$$d\sigma_{\nu\bar{\nu}} = \frac{1}{4m_e^2 v_{\text{rel}}} \overline{|\mathcal{M}|^2} \frac{d\Omega}{32\pi^2} \tag{31}$$

and

$$\sigma_{\nu\bar{\nu}} = \frac{G_F^2 m_e^2}{4\pi v_{\text{rel}}} \tag{32}$$

$$\Gamma_{\nu\bar{\nu}} = \frac{1}{3\pi a^3} \frac{G_F^2 m_e^2}{\pi} = G_F^2 m_e^5 \frac{\alpha^3}{24\pi^2} \simeq 0.2 \cdot 10^{-3} \text{sec}^{-1} \tag{33}$$

Finally, given that the statistical mixture contains 3 triplet and one singlet states, all with the same probability, we find

$$R = \frac{3\Gamma_{\nu\bar{\nu}}}{\Gamma_{\gamma\gamma}} = \frac{G_F^2 m_e^2}{4\pi^2 \alpha^2} \simeq 0.4 \cdot 10^{-13} \tag{34}$$

Problem 55.   Nuclear β-decay between 1^+ and 0^+ states

A spin-1 nucleus with positive parity and isospin assignment $(T=1, T_3)$ decays into a spin-0 nucleus with positive parity and isospin $(T=1, T_3'=T_3+1)$. The decay proceeds through the Fermi interaction Hamiltonian

$$\mathcal{L}_I = \frac{G_F}{\sqrt{2}} \left(V_\mu(x) - A_\mu(x) \right) \bar{\psi}_e(x) \gamma_\mu (1 - \gamma_5) \psi_{\nu_e}(x) \tag{1}$$

a)   What is the most general form for the matrix element of V_μ and A_μ between the initial and final state, in the nonrelativistic limit? Point out explicitly the dependence on T_3.

b)   Calculate the angular distribution and the spectrum of outgoing $\bar{\nu}_e$'s in the decay of a nucleus, completely polarized along the z-axis.

c)   Consider a specimen containing radioactive nuclei, polarized as above; the specimen is assumed thick enough to absorb all electrons produced. Find the torque along the z-axis induced on the specimen, knowing the total quantity of radioactive nuclei present in the specimen and their mean life.

Solution

a) The matrix element of the hadronic current is a four-vector; as such, it is expressible in terms of the four-vectors describing the initial and final states, namely p_μ, p'_μ (nuclear momenta) and ε_μ (polarization of spin-1 state). Given the positive parity of the spin-1 state, ε_μ is an axial vector.

For the vector current, the most general matrix element is

$$\langle A'|V_\mu|A\rangle = F\varepsilon_{\mu\nu\rho\sigma}\varepsilon^\nu p^\rho p'^\sigma \tag{2}$$

For the axial current,

$$\langle A'|A_\mu|A\rangle = G_1\varepsilon_\mu + G_2(\varepsilon\cdot p')p_\mu + G_3(\varepsilon\cdot p')p'_\mu \tag{3}$$

We have required linearity of the wave function in ε_μ and made use of $\varepsilon\cdot p = 0$.

In the center of mass we find

$$
\begin{aligned}
\langle A'|V_0|A\rangle &= 0 \\
\langle A'|\mathbf{V}|A\rangle &= FM_A\,(\boldsymbol{\varepsilon}\wedge\mathbf{p}') \\
\langle A'|A_0|A\rangle &= -G_2\,(\boldsymbol{\varepsilon}\mathbf{p}')\,M_A - G_3\,(\boldsymbol{\varepsilon}\mathbf{p}')\,P'_0 \\
\langle A'|\mathbf{A}|A\rangle &= G_1\boldsymbol{\varepsilon} - G_3\,(\boldsymbol{\varepsilon}\mathbf{p}')\,\mathbf{p}'
\end{aligned}
\tag{4}
$$

The only term surviving the nonrelativistic limit ($\mathbf{p}'\to 0$) is

$$\langle A'|\mathbf{A}|A\rangle = G_1\boldsymbol{\varepsilon} \tag{5}$$

Terms neglected in (5) are of the order of v/c; they are typically suppressed by a factor 10^{-3}-10^{-4}, for a mass difference of a few MeV. To isolate the isospin dependence, we note that the current transforms as the T_+ component of an isospin vector; by the Wigner-Eckart theorem, its matrix elements are proportional to a Clebsch-Gordan coefficient

$$\langle T'=1, T'_3 = T_3 \pm 1|T_\pm|T=1, T_3\rangle = \sqrt{(T\pm T_3 + 1)\,(T\mp T_3)} \tag{6}$$

Hence,

$$G_1 = g\,\sqrt{2M_A\,2M_{A'}}\sqrt{(2+T_3)\,(1-T_3)} \tag{7}$$

We have written out explicitly the normalization factors ($2M$) of the nuclear states.

b) In the nonrelativistic limit, the decay matrix element is

$$\mathcal{M} = \frac{G_F}{\sqrt{2}}\,\sqrt{(2+T_3)\,(1-T_3)}\,g\varepsilon_\mu\bar{u}(\mathbf{p}_{e^-})\gamma^\mu\left(1-\gamma^5\right)v(\mathbf{p}_{\bar{\nu}_e})\,\sqrt{2M_A\,2M_{A'}} \tag{8}$$

Summing over final polarizations,

$$
\begin{aligned}
|\mathcal{M}|^2 &= \frac{G_F^2}{2} \left(2 + T_3\right) \left(1 - T_3\right) 2M_A\, 2M_{A'} \\
&\quad \cdot |g|^2 \varepsilon_\mu \varepsilon_\nu^* \operatorname{Tr}\left\{\gamma^\mu \left(1 - \gamma^5\right) \slashed{p}_{\bar{\nu}_e} \gamma^\nu \left(1 - \gamma^5\right) \left(\slashed{p}_{e^-} + m_e\right)\right\} \\
&= 4G_F^2 |g|^2 \left(2 + T_3\right) \left(1 - T_3\right) 2M_A\, 2M_{A'} \\
&\quad \cdot \varepsilon_\mu \varepsilon_\nu^* \left[p_{\bar{\nu}_e}{}^\mu p_{e^-}{}^\nu + p_{\bar{\nu}_e}{}^\nu p_{e^-}{}^\mu - g^{\mu\nu}\left(p_{\bar{\nu}_e} \cdot p_{e^-}\right) + i\varepsilon^{\mu\nu}{}_{\rho\sigma} p_{\bar{\nu}_e}{}^\rho p_{e^-}{}^\sigma\right] \\
&= 4G_F^2 |g|^2 \left(2 + T_3\right) \left(1 - T_3\right) 2M_A\, 2M_{A'} \\
&\quad \cdot \Big[\left(\boldsymbol{\varepsilon} \mathbf{p}_{\bar{\nu}_e}\right)\left(\boldsymbol{\varepsilon}^* \mathbf{p}_{e^-}\right) + \left(\boldsymbol{\varepsilon} \mathbf{p}_{e^-}\right)\left(\boldsymbol{\varepsilon}^* \mathbf{p}_{\bar{\nu}_e}\right) + \\
&\qquad + \left(E_{e^-} E_{\bar{\nu}_e}\right) - \left(\mathbf{p}_{e^-} \mathbf{p}_{\bar{\nu}_e}\right) - i\left(\boldsymbol{\varepsilon} \wedge \boldsymbol{\varepsilon}^*\right) \cdot \left(E_{\bar{\nu}_e} \mathbf{p}_{e^-} - E_{e^-} \mathbf{p}_{\bar{\nu}_e}\right)\Big]
\end{aligned} \tag{9}
$$

For nonrelativistic nucleons, the phase space reduces to

$$
d\Phi^{(3)} = \frac{d^3\mathbf{p}_{e^-}}{2E_{e^-}(2\pi)^3} \frac{d^3\mathbf{p}_{\bar{\nu}}}{2E_{\bar{\nu}_e}(2\pi)^3} \frac{1}{2M_{A'}} 2\pi\delta\left(\Delta - E_{e^-} - E_{\bar{\nu}_e}\right) \tag{10}
$$

with $\Delta = M_A - M_{A'}$. Terms linear in $\mathbf{p}_{e^-}$ vanish upon integration over the electron momentum, leaving us with

$$
\begin{aligned}
d\Gamma &= \frac{1}{2M_A} |\mathcal{M}|^2\, d\Phi^{(3)} \\
&= 4G_F^2 |g|^2 \left(2 + T_3\right) \left(1 - T_3\right) \left[E_{\bar{\nu}_e} + i\mathbf{p}_{\bar{\nu}_e} \cdot \left(\boldsymbol{\varepsilon} \wedge \boldsymbol{\varepsilon}^*\right)\right] \\
&\quad \cdot \frac{\left(\Delta - E_{\bar{\nu}_e}\right)\sqrt{\left(\Delta - E_{\bar{\nu}_e}\right)^2 - m_e^2}}{2\pi} \frac{d^3\mathbf{p}_{\bar{\nu}}}{2E_{\bar{\nu}_e}(2\pi)^3}
\end{aligned} \tag{11}
$$

Using the relation

$$
i\left(\boldsymbol{\varepsilon} \wedge \boldsymbol{\varepsilon}^*\right) = \hat{\mathbf{n}}_\varepsilon \tag{12}
$$

where $\hat{\mathbf{n}}_\varepsilon$ is a unit vector along the direction of polarization, we obtain

$$
\begin{aligned}
d\Gamma &= \frac{G_F^2 |g|^2 \left(2 + T_3\right) \left(1 - T_3\right)}{8\pi^4} \\
&\quad \left(\Delta - E_{\bar{\nu}_e}\right) E_{\bar{\nu}_e}^2 \sqrt{\left(\Delta - E_{\bar{\nu}_e}\right)^2 - m_e^2}\, dE_{\bar{\nu}_e}\left(1 + \hat{\mathbf{n}} \cdot \hat{\mathbf{n}}_\varepsilon\right) d\Omega
\end{aligned} \tag{13}
$$

We can read off Eq.(13) the angular distribution

$$
d\Gamma = \Gamma\left(1 + \cos\theta\right)\frac{d\Omega}{4\pi} \tag{14}
$$

and the spectrum

$$
\frac{d\Gamma}{dE_{\bar{\nu}_e}} = \frac{1}{2\pi^3}\, G_F^2 |g|^2 \left(2 + T_3\right)\left(1 - T_3\right)\left(\Delta - E_{\bar{\nu}_e}\right) E_{\bar{\nu}_e}^2 \sqrt{\left(\Delta - E_{\bar{\nu}_e}\right)^2 - m_e^2} \tag{15}
$$

If the electron mass is negligible compared to Δ, the total width Γ becomes

$$\Gamma = \frac{1}{60\pi^3} G_F^2 |g|^2 \left(2 + T_3\right) \left(1 - T_3\right) \Delta^5 \tag{16}$$

c)   The torque induced is the angular momentum transferred on the specimen per unit time; this is opposite to the one carried away by neutrinos

$$\mathbf{M} = \frac{d\mathbf{J}}{dt} = -\frac{d\mathbf{J}_{\bar{\nu}_e}}{dt} \tag{17}$$

We note that the neutrino distribution is axially symmetric with respect to the z-axis; therefore, the z-component of the neutrino angular momentum is due exclusively to neutrino helicity. Since the $\bar{\nu}_e$ spin is always parallel to its momentum, we find

$$(\mathbf{J}_{\bar{\nu}_e})_3 = \frac{1}{2}\cos\theta \qquad (\mathbf{J}_{\bar{\nu}_e})_{1,2} = 0 \tag{18}$$

Averaging over the angular distribution

$$M_3 = \frac{dJ_3}{dt} = -\int \frac{\cos\theta}{2} \, d\Gamma = -\Gamma \int_{-1}^{1} d\cos\theta \, \frac{1+\cos\theta}{2} \frac{\cos\theta}{2} = -\frac{\Gamma}{6} \tag{19}$$

In conclusion, denoting by N_0 the number of radioactive nuclei in the specimen at $t = 0$, we have at time t (in units $\hbar = 1$)

$$M_3(t) = -\frac{\Gamma}{6} N_0 e^{-\Gamma t} \tag{20}$$

Problem 56.   *W* decays

The $W^{\pm}$ boson has spin 1 and mass $M_W = 80.22 \pm .26$ GeV. Some of its leptonic decay modes are

$$W^+ \to \mu^+ \nu_\mu \tag{1}$$
$$W^+ \to e^+ \nu_e \tag{2}$$
$$W^- \to \mu^- \bar{\nu}_\mu \tag{3}$$
$$W^- \to e^- \bar{\nu}_e \tag{4}$$

The interaction Lagrangian causing these decays is

$$\mathcal{L}_I = g \sum_\ell \left\{ W_\mu \bar{\psi}_{\nu_\ell} \gamma^\mu \left(1 - \gamma^5\right) \psi_\ell + \text{h.c.} \right\} \tag{5}$$

with $g^2 = 0.08$

a)   Derive the angular distribution in the decays (1) - (4), for W completely polarized.

b)   Calculate the decay probability per unit time, expressed in $\sec^{-1}$.

c)   Calculate the branching ratios

$$R_+ = \frac{W^+ \to \mu^+ + \nu_\mu}{W^+ \to e^+ + \nu_e} \qquad R_- = \frac{W^- \to \mu^- + \bar{\nu}_\mu}{W^- \to e^- + \bar{\nu}_e} \tag{6}$$

d)   Show that the interaction (5) is CP-invariant. Discuss the results of parts a) and c) in view of this invariance.

Solution

a)   The decay matrix elements can be read off the interaction Lagrangian (5)

$$\mathcal{M}^+ = g\varepsilon_\mu^{(+)}\bar{u}(\mathbf{p}_{\nu_\ell})\gamma^\mu\left(1-\gamma^5\right)v(\mathbf{p}_{\bar{\ell}}) \tag{7a}$$

$$\mathcal{M}^- = g\varepsilon_\mu^{(-)}\bar{u}(\mathbf{p}_\ell)\gamma^\mu\left(1-\gamma^5\right)v(\mathbf{p}_{\bar{\nu}_\ell}) \tag{7b}$$

Summing over final polarizations, we find

$$|\mathcal{M}^+|^2 = g^2\varepsilon_\mu^{(+)}\varepsilon_\nu^{(+)*}\,\text{Tr}\left\{\gamma^\mu\left(1-\gamma^5\right)(\not{p}_\ell - m_\ell)\gamma^\nu\left(1-\gamma^5\right)\not{p}_{\nu_\ell}\right\} \tag{8a}$$

$$|\mathcal{M}^-|^2 = g^2\varepsilon_\mu^{(-)}\varepsilon_\nu^{(-)*}\,\text{Tr}\left\{\gamma^\mu\left(1-\gamma^5\right)\not{p}_{\nu_\ell}\gamma^\nu\left(1-\gamma^5\right)(\not{p}_\ell + m_\ell)\right\} \tag{8b}$$

which simplifies to

$$|\mathcal{M}^\pm|^2 = 8g^2\varepsilon_\mu^{(\pm)}\varepsilon_\nu^{(\pm)*}\left[p_\ell^\mu p_{\nu_\ell}^\nu + p_\ell^\nu p_{\nu_\ell}^\mu - (p_\ell\cdot p_{\nu_\ell})g^{\mu\nu} \pm i\,\varepsilon^{\mu\nu}{}_{\rho\sigma}p_\ell^\rho p_{\nu_\ell}^\sigma\right] \tag{9}$$

In the center-of-mass frame this expression becomes

$$|\mathcal{M}^\pm|^2 = 4g^2\left[M_W^2 - m_\ell^2 - 4\left|(\mathbf{p}_\ell\varepsilon^{(\pm)})\right|^2 \mp 2i\,M_W\mathbf{p}_\ell\cdot(\boldsymbol{\varepsilon}^{(\pm)*}\wedge\boldsymbol{\varepsilon}^{(\pm)})\right] \tag{10}$$

Polarization along the z-axis is described by

$$\boldsymbol{\varepsilon} = -\frac{i}{\sqrt{2}}\,(1,i,0) \tag{11}$$

satisfying

$$i\,\boldsymbol{\varepsilon}\wedge\boldsymbol{\varepsilon}^* = \hat{\mathbf{z}} \tag{12}$$

Substituting the above in Eq.(10), we obtain the following angular distribution of leptons

$$|\mathcal{M}^\pm|^2 = 2\,\frac{\left(M_W^2 - m_\ell^2\right)}{M_W^2}\,g^2\left[M_W^2 + m_\ell^2 \pm \cos\theta\left(M_W^2 - m_\ell^2\right)\right](1\pm\cos\theta) \tag{13}$$

We observe that a W^- polarized along z gives the same angular distribution as a W^+ polarized along $-z$.

b)   The general expression for the decay width

$$d\Gamma = \frac{1}{2M_W}\,|\mathcal{M}|^2\,d\Phi^{(2)} \tag{14}$$

leads to

$$\Gamma_+ = \Gamma_- \equiv \Gamma$$

$$\Gamma = \frac{(M_W^2 - m_\ell^2)}{M_W^2}\,g^2\left[M_W^2 + m_\ell^2 + \frac{1}{3}(M_W^2 - m_\ell^2)\right]\frac{(M_W^2 - m_\ell^2)}{8\pi M_W^3} \tag{15}$$

Expanding in the ratio $(m_\ell/M_W)^2$

$$\Gamma \simeq \frac{g^2}{6\pi} M_W \left(1 - \frac{3}{2}\frac{m_\ell^2}{M_W^2}\right) \tag{16}$$

The term in $(m_\ell/M_W)^2$ is negligible to all purposes. Substituting numerical values, we obtain

$$\Gamma \simeq 178 \text{ MeV} \qquad \frac{1}{\Gamma} \simeq 4 \cdot 10^{-24}\text{sec} \tag{17}$$

Γ is the partial width for the decay into one lepton pair.

c)   Eq.(16) yields directly the branching ratios

$$R_+ = R_- \simeq 1 - \frac{3}{2}\frac{m_\mu^2 - m_e^2}{M_W^2} \simeq 1 + \mathcal{O}(10^{-5}) \tag{18}$$

d)   Vector and axial vector currents, defined by

$$V_{ab}^\mu =: \bar\psi_a \gamma^\mu \psi_b : \qquad A_{ab}^\mu =: \bar\psi_a \gamma^\mu \gamma^5 \psi_b : \tag{19}$$

undergo the following transformations under C and P[8]

$$
\begin{array}{llll}
C & : & V_{ab}^\mu \to -V_{ba}^\mu & A_{ab}^\mu \to A_{ba}^\mu \\
P & : & V_{ab}^\mu \to g_{\mu\mu}V_{ab}^\mu & A_{ab}^\mu \to -g_{\mu\mu}A_{ab}^\mu \\
CP & : & V_{ab}^\mu \to -g_{\mu\mu}V_{ba}^\mu & A_{ab}^\mu \to -g_{\mu\mu}A_{ba}^\mu
\end{array}
\tag{20}
$$

Thus, any linear combination J_{ab}^μ of vector and axial vector currents transforms as

$$CP: \qquad J_{ab}^\mu \to -g_{\mu\mu}J_{ba}^\mu = -g_{\mu\mu}\left(J_{ab}^\mu\right)^\dagger \tag{21}$$

We now define the CP transformation for the W_μ field as

$$CP: \qquad W_\mu \to -g_{\mu\mu}W_\mu{}^\dagger \tag{22}$$

Then the interaction Lagrangian

$$\mathcal{L}_I = W^\mu J_\mu + W^\dagger{}_\mu J^{\mu\dagger} \tag{23}$$

remains invariant under CP. An immediate consequence of this invariance is

$$\langle \mathbf{p}_{\ell^-}, \mathbf{p}_{\bar\nu_\ell}|\mathcal{L}_I|W^-, M\rangle = \langle -\mathbf{p}_{\ell^+}, -\mathbf{p}_{\nu_\ell}|\mathcal{L}_I|W^+, M\rangle \tag{24}$$

where M is the projection of the W spin along z. Applying a rotation by π on the $(\hat{\mathbf{z}}\text{-}\mathbf{p}_\ell)$ plane, rotational invariance gives

$$\langle \mathbf{p}_{\ell^-}, \mathbf{p}_{\bar\nu_\ell}|\mathcal{L}_I|W^-, M\rangle = \langle +\mathbf{p}_{\ell^+}, +\mathbf{p}_{\nu_\ell}|\mathcal{L}_I|W^+, -M\rangle \tag{25}$$

This relation verifies the results found previously

$$\frac{d\Gamma_{W^+}^{(\pm)}}{d\Omega} = \frac{d\Gamma_{W^-}^{(\mp)}}{d\Omega} \qquad R_+ = R_- \tag{26}$$

The superscript $\pm$ denotes the sign of the polarization along z. Spin indices of the final state were left implicit in Eq.(24); these are left invariant by CP, and are summed over in the observables of Eq.(26).

---

[8]In this relation, repeated indices are not summed over; $g_{\mu\mu}$ is the metric $(+, -, -, -)$.

Problem 57.   Muon capture by p

A μ-mesic hydrogen atom can decay in the following mode

$$\mu^- + p \to n + \nu_\mu \tag{1}$$

This process is known as muon capture.

a)   Calculate the neutron energy in the final state.

b)   To an excellent approximation, the interaction responsible for process (1) is the Fermi interaction

$$H_I = \frac{G_F}{\sqrt{2}}\left(V_\mu - A_\mu\right) J^\mu \tag{2}$$

Derive the corresponding matrix element. You may assume that the hadronic currents have the form

$$\begin{aligned}
\langle n|V_\mu(0)|p\rangle &= F_V\,\bar{u}(\mathbf{p}_n)\gamma_\mu u(\mathbf{p}_p) \\
\langle n|A_\mu(0)|p\rangle &= F_A\,\bar{u}(\mathbf{p}_n)\gamma_\mu\gamma^5 u(\mathbf{p}_p)
\end{aligned} \tag{3}$$

with $F_V \simeq 1$ and $F_A \simeq 1.2$.

c)   Calculate the mean life of the nS-state of this atom. Find the angular distribution, averaged over initial and summed over final spins orientations.

Solution

a)   In the center-of-mass frame, energy and momentum conservation read

$$\begin{aligned}
E_0 \equiv m_\mu + m_p - \Delta E &= E_n + E_\nu \\
\mathbf{p}_n + \mathbf{p}_\nu &= 0
\end{aligned} \tag{4}$$

$\Delta E \simeq \alpha^2 m_\mu \ll m_\mu$ is the binding energy of the mesic atom. Furthermore,

$$E_n^2 - |\mathbf{p}_n|^2 = m_n^2 \qquad E_\nu = |\mathbf{p}_\nu| = |\mathbf{p}_n| \tag{5}$$

There follows

$$\begin{aligned}
E_n + p_n &= E_0 \\
E_n - p_n &= \frac{E_n^2 - p_n^2}{E_n + p_n} = \frac{m_n^2}{E_0}
\end{aligned} \tag{6}$$

Thus the energy and momentum of the neutron are

$$E_n = \frac{E_0^2 + m_n^2}{2E_0} \qquad p_n = \frac{E_0^2 - m_n^2}{2E_0} \tag{7}$$

We may write approximately $m_n \simeq m_p$, $m_\mu/m_n \ll 1$ and $E_0 \simeq m_p + m_\mu \simeq m_n + m_\mu$ with the result

$$E_n \simeq m_n + \frac{m_\mu^2}{2m_n}$$

$$|\mathbf{p}_n| \simeq m_\mu \tag{8}$$

The neutron velocity is then

$$\frac{v}{c} \simeq \frac{m_\mu}{m_n} \tag{9}$$

b) The matrix element for this process equals

$$\mathcal{M} = \frac{G_F}{\sqrt{2}}\, \bar{u}(\mathbf{p}_n)\gamma_\mu \left(F_V - \gamma^5 F_A\right) u(\mathbf{p}_p)\, \bar{u}(\mathbf{p}_\nu)\gamma^\mu \left(1 - \gamma^5\right) u(\mathbf{p}_\mu) \tag{10}$$

Averaging over initial and summing over final spin orientations we obtain

$$\begin{aligned}
\overline{|\mathcal{M}|^2} &= \frac{G_F^2}{2} \operatorname{Tr}\left\{\gamma^\mu \left(F_V - \gamma^5 F_A\right) \frac{\not{p}_p + m_p}{2}\gamma^\nu \left(F_V^* - \gamma^5 F_A^*\right)(\not{p}_n + m_n)\right\} \\
&\quad \cdot \operatorname{Tr}\left\{\gamma_\mu \left(1 - \gamma^5\right) \frac{\not{p}_{\mu^-} + m_\mu}{2}\gamma_\nu \left(1 - \gamma^5\right)\not{p}_{\nu_\mu}\right\} \\
&= 4G_F^2\Big[\left(|F_V|^2 + |F_A|^2\right)\left(p_p^\nu p_n^\mu + p_p^\mu p_n^\nu - (p_p \cdot p_n)g^{\mu\nu}\right) \\
&\quad + \left(|F_V|^2 - |F_A|^2\right) m_p m_n g^{\mu\nu} + 2\mathrm{i}\mathrm{Re}\, F_A F_V^* \varepsilon^{\mu\nu}{}_{\alpha\beta}p_p^\alpha p_n^\beta\Big] \\
&\quad \Big[(p_{\mu^-})_\mu(p_{\nu_\mu})_\nu + (p_{\mu^-})_\nu(p_{\nu_\mu})_\mu - (p_{\mu^-} \cdot p_{\nu_\mu})g_{\mu\nu} + \mathrm{i}\varepsilon_{\mu\nu\alpha\beta}(p_{\mu^-})^\alpha(p_{\nu_\mu})^\beta\Big] \\
&= 8G_F^2\Big[\left(|F_V|^2 + |F_A|^2\right)\left\{(p_p \cdot p_{\mu^-})(p_n \cdot p_{\nu_\mu}) + (p_p \cdot p_{\nu_\mu})(p_n \cdot p_{\mu^-})\right\} \\
&\quad + 2\mathrm{Re}\, F_V F_A^* \left\{(p_p \cdot p_{\mu^-})(p_n \cdot p_{\nu_\mu}) - (p_p \cdot p_{\nu_\mu})(p_n \cdot p_{\mu^-})\right\} \\
&\quad - \left(|F_V|^2 - |F_A|^2\right) m_p m_n(p_{\mu^-} \cdot p_{\nu_\mu})\Big] \tag{11}
\end{aligned}$$

In the nonrelativistic approximation we have

$$\begin{aligned}
p_{\mu^-} &= (m_\mu, 0) & p_p &= (m_n, 0) \\
p_{\nu_\mu} &= (m_\mu, m_\mu\hat{\mathbf{p}}) & p_n &= (m_n, -m_\mu\hat{\mathbf{p}})
\end{aligned} \tag{12}$$

and the scalar products become

$$\begin{aligned}
p_p \cdot p_{\mu^-} &= m_n m_\mu & p_p \cdot p_{\nu_\mu} &= m_\mu m_n \\
p_n \cdot p_{\mu^-} &= m_n m_\mu & p_n \cdot p_{\nu_\mu} &= m_\mu(m_n + m_\mu) & p_{\mu^-} \cdot p_{\nu_\mu} &= m_\mu^2
\end{aligned} \tag{13}$$

Hence the amplitude simplifies to

$$\begin{aligned}
\overline{|\mathcal{M}|^2} &= 8G_F^2 m_n m_\mu^2 \Big[|F_V|^2 (m_n + m_\mu) + |F_A|^2 (3m_n + m_\mu) + 2\mathrm{Re}\, F_V F_A^* m_\mu\Big] \\
&\simeq 8G_F^2 m_n^2 m_\mu^2 \Big[|F_V|^2 + 3|F_A|^2 + v|F_V + F_A|^2\Big] \tag{14}
\end{aligned}$$

with $v = m_\mu / m_n$.

c) The mean life of the nS-state of the mesic atom can be expressed in terms of the cross section σ for the process

$$\mu^- + p \to n + \nu_\mu \qquad (15)$$

by means of the formula (see Prob.54)

$$\Gamma = \lim_{v_r \to 0} v_r \sigma \, |\psi_{ns}(0)|^2 \qquad (16)$$

Since the angular distribution is spherically symmetric, we have

$$d\Gamma = \Gamma \frac{d\Omega}{4\pi} \qquad (17)$$

where v_r is the relative velocity between μ^- and p. The cross section is given by

$$d\sigma = \frac{|\mathcal{M}|^2}{4 m_p m_\mu v_r} \, d\Phi^{(2)}$$

$$d\Phi^{(2)} \simeq \frac{d\Omega}{32\pi^2} \frac{2m_\mu}{m_n} \qquad (18)$$

leading to

$$\sigma = \frac{1}{16\pi} \frac{|\mathcal{M}|^2}{m_p^2 v_r} \qquad (19)$$

Using also the expression for the normalized wave function of the nS-state

$$|\psi_{ns}(0)|^2 = \frac{1}{\pi a^3 n^3} \qquad (20)$$

($a = 1/\alpha m_\mu$ is the Bohr radius of the atom), we finally obtain

$$\Gamma = \frac{m_\mu^3 \alpha^3}{2\pi^2 n^3} G_F^2 m_\mu^2 \left[|F_V|^2 + 3|F_A|^2 \right] \simeq \frac{2 \cdot 10^{-19} \, \text{MeV}}{n^3}$$

$$\tau \simeq 3.36 \cdot 10^{-3} \text{sec} \cdot n^3 \qquad (21)$$

Problem 58.   Rare decay modes of K_S^0

Some of the decay modes of the K_S meson (the short-lived component of K_0) are:

$$\begin{aligned}
K_S &\to \pi^+ \pi^- & BR(1) &= (68.61 \pm .28)10^{-2} & (1) \\
K_S &\to \pi^0 \pi^0 & BR(2) &= (31.39 \pm .28)10^{-2} & (2) \\
K_S &\to \gamma\gamma & BR(3) &= (2.4 \pm 1.2)10^{-6} & (3) \\
K_S &\to \mu^+ \mu^- & BR(4) &< 3.2 \cdot 10^{-7} & (4) \\
K_S &\to e^+ e^- & BR(5) &< 1.0 \cdot 10^{-5} & (5)
\end{aligned}$$

The mean life of K_S is $\tau_S = (89.22 \pm 0.20) \cdot 10^{-12}$ sec and its mass is $M_K = 497.67 \pm 0.03$ MeV.
K^+ has the following partial decay mode:

$$K^+ \to \pi^+ \pi^0 \qquad\qquad BR(1') = (21.17 \pm 0.16)10^{-2} \qquad\qquad (1')$$

$$\tau_{K^+} = 1.24 \cdot 10^{-8} \text{sec} \qquad M_{K^+} = 493.65 \pm 0.1 \,\text{MeV}$$

There are reasons to expect the amplitude for modes (3) (4) and (5) to be strictly zero, neglecting electromagnetic final state interactions.

a) What conclusions can be drawn from the measured values of $BR(1)$, $BR(2)$ and $BR(1')$, regarding the scattering phases of the S-wave of $\pi^+ \pi^-$ with isospin $I = 0$ and $I = 2$?

Assume that the weak interaction Hamiltonian H_W consists of an isospin doublet $H_W^{(1/2)}$ and an isospin quadruplet $H_W^{(3/2)}$.

b) How do the above conclusions get modified by the Coulomb interaction in the $\pi^+ \pi^-$ final state?

c) Calculate the branching ratio for decay mode (3) coming from final state electromagnetic interaction, and estimate it for modes (4) and (5).

Solution

a,b) K has spin zero; therefore, the π's are emitted in an S-wave. Since these are bosons, the isospin state must be even under particle interchange, leading to total isospin $I = 0$ or $I = 2$. In particular for $K^+ \to \pi^+ \pi^0$ we have $I = 2$ since $I_3 = 1$.

We write

$$A\left(K^+ \to \pi^+ \pi^0\right) = \langle I = 2, I_3 = 1 | R | K^+ \rangle \qquad\qquad (6)$$

where R is the decay operator. By time reversal invariance (T):

$$\langle I = 2, I_3 = 1 | R | K^+ \rangle = \sum_n \langle T\left(K^+\right) | H_W | n_- \rangle \langle n_- | T\left(I = 2, I_3 = 1\right)_+ \rangle \qquad (7)$$

$|n_-\rangle$ is a complete set of final state interaction eigenstates whose boundary condition at $t \to +\infty$ is the free particle state $|n\rangle$.

$\langle n_- | T\left(I = 2, I_3 = 1\right)_+ \rangle = \langle n | S_{FS} | T\left(I = 2, I_3 = 1\right) \rangle$ is the S-matrix element of final state interaction between the observed 2π state and n. If one can neglect final state interaction, then $S_{FS} = 1$ and the only state n which contributes is the 2π state with $I = 2, I_3 = 1$; given that a spin zero state, such as $|K\rangle$, coincides at rest with $|TK\rangle$ aside from a phase, Eq.(7) implies that in this case the phase can be chosen in a way which makes the decay matrix element real.

The strong interaction entering the final state commutes with I, I_3, energy and angular momentum; thus, the only contribution to the r.h.s. of (7) comes from the state $|n_-\rangle = |I = 2, I_3 = 1, l = 0\rangle$ with

$$\langle n_- | T\left(I = 2, I_3 = 1\right)_+ \rangle = e^{2i\delta_2(M_K)} \delta_{n,(I=2,I_3=1)} \qquad\qquad (8)$$

$\delta_2\left(M_K\right)$ is the S-wave scattering phase of $\pi\pi$ at energy equal to the mass of K and $I = 2$. Only up to 2 pions are allowed in the final state; 3π states are forbidden by G-parity invariance, while energy conservation excludes channels with more pions.

Thus, with the choice of phase: $|TK^+\rangle = |K^+\rangle$, we have

$$A\left(K^+ \to \pi^+\pi^0\right) = A^*\left(K^+ \to \pi^+\pi^0\right) e^{2i\delta_2} \tag{9}$$

We set

$$A = \sqrt{\frac{3}{2}} a_2 e^{i\delta_2} \tag{10}$$

where a_2 is a real number.

Similarly, it can be shown that T-invariance implies

$$A\left(K^0 \to (2\pi)_I\right) = a_I e^{i\delta_I(M_K)} \tag{11}$$

where the a_I's are real numbers.

By CTP-invariance we obtain

$$A\left(\bar{K}^0 \to (2\pi)_I\right) = a_I e^{i\delta_I(M_K)} \tag{11'}$$

The factor $\sqrt{3/2}$ in Eq.(10) is a matter of convention, which is consistent with the standard notation of (11) and (11') for neutral K decay. Indeed, using the Wigner-Eckart theorem and the isospin content of H_W, we obtain for the K^+ decay amplitude

$$\frac{\langle\left(\pi^+\pi^0\right)_{I=2}|H_W^{3/2}|K^+\rangle}{\langle(\pi\pi)_{I=2}|H_W^{3/2}|K^0\rangle} = \frac{C_{\frac{1}{2}\frac{1}{2}1}^{\frac{1}{2}\frac{3}{2}2}}{C_{-\frac{1}{2}\frac{1}{2}0}^{\frac{1}{2}\frac{3}{2}2}} = \sqrt{\frac{3}{2}} \tag{12}$$

The definition

$$|K_S\rangle = \frac{|K^0\rangle + |\bar{K}^0\rangle}{\sqrt{2}} \tag{13}$$

together with (11) and (11') lead to

$$A\left(K_S \to (2\pi)_I\right) = \sqrt{2} a_I e^{i\delta_I(M_K)} \tag{14}$$

Writing for the final states

$$\begin{aligned}
|\pi^0\,\pi^0\rangle &= \sqrt{\frac{2}{3}}|T=2\rangle - \sqrt{\frac{1}{3}}|T=0\rangle \\
\frac{|\pi^+\,\pi^-\rangle + |\pi^-\,\pi^+\rangle}{\sqrt{2}} &= \sqrt{\frac{1}{3}}|T=2\rangle + \sqrt{\frac{2}{3}}|T=0\rangle \\
\frac{|\pi^0\,\pi^+\rangle + |\pi^+\,\pi^0\rangle}{\sqrt{2}} &= |T=2\rangle
\end{aligned} \tag{15}$$

we obtain

$$A\left(K_S \to \pi^+\,\pi^-\right) = \sqrt{2}\left(\sqrt{\frac{2}{3}}a_0 e^{i\delta_0} + \sqrt{\frac{1}{3}}a_2 e^{i\delta_2}\right) \tag{16a}$$

$$A\left(K_S \to \pi^0\,\pi^0\right) = \sqrt{2}\left(-\sqrt{\frac{1}{3}}a_0 e^{i\delta_0} + \sqrt{\frac{2}{3}}a_2 e^{i\delta_2}\right) \tag{16b}$$

Since the S-wave angular distribution is isotropic, there follows

$$\Gamma = \frac{1}{2M_K}|A|^2\, C\, \Phi^{(2)} \tag{17}$$

where C is the final state Coulomb correction and $\Phi^{(2)}$ is the volume of two-body phase space; for K_S

$$\Phi^{(2)} = \frac{1}{8\pi}\sqrt{1 - \frac{4m_\pi^2}{M_{K_0}^2}} \tag{18}$$

and for K^+

$$\Phi^{(2)} = \frac{1}{8\pi}\sqrt{1 - \frac{(m_{\pi^0}+m_{\pi^+})^2}{M_{K^+}^2}}\sqrt{1 - \frac{(m_{\pi^0}-m_{\pi^+})^2}{M_{K^+}^2}} \tag{19}$$

Numerically for the three processes (1) (2) ($1'$)

$$8\pi\,\Phi^{(2)}_{(1)} = 0.828$$
$$8\pi\,\Phi^{(2)}_{(2)} = 0.840$$
$$8\pi\,\Phi^{(2)}_{(1')} = 0.831 \tag{20}$$

The Coulomb correction is present only in process (1) and equals

$$C = \frac{2\pi}{x}\,\frac{1}{1-e^{-\frac{2\pi}{x}}} \qquad \left(x = \frac{2p_\pi}{\alpha m_\pi}\right) \tag{21}$$

p_π is the π momentum in the rest frame of K. Numerically $p_\pi = 207\,\text{MeV}/c$ and

$$C(1) = 1.008$$
$$C(2) = 1.$$
$$C(1') = 1. \tag{22}$$

Eq.(17), together with the known width of process $(1')$, gives us $|a_2|$. Neglecting, as a first approximation, the phase space differences (20) and the correction (22), we obtain from (17) (16) and (10)

$$\frac{\Gamma\,(1')}{\Gamma\,(1)+\Gamma\,(2)} = \frac{3}{4}\frac{a_2^2}{a_0^2+a_2^2} \tag{23}$$

The experimental value for this ratio is

$$BR(1')\frac{\tau(K_S)}{\tau(K^+)} = 1.5{\cdot}10^{-3} \qquad (24)$$

whence $|a_2/a_0| \simeq 4.5{\cdot}10^{-2}$.

With the same approximations the ratio $R = \dfrac{\Gamma(1) - 2\Gamma(2)}{\Gamma(1)}$ equals

$$R = 3\sqrt{2}\frac{a_2}{/a_0}\,\cos\left(\delta_2 - \delta_0\right) \qquad (25)$$

neglecting terms of order $(a_2/a_0)^2$. The experimental value of R is $0.085 \pm .008$ and thus

$$\cos\left(\delta_2 - \delta_0\right) = 0.44 \pm .04 \qquad (26)$$

The phase space and Coulomb corrections of eqs.(20), (22) lead to a fractional change of $\simeq 4{\cdot}10^{-3}$ in a_2/a_0, which is negligible within errors.

c)   The amplitude for process (3) is approximately

$$A(3) \simeq A(1)\cdot\alpha \qquad (27)$$

that is, one expects $\Gamma(3)/\Gamma(2) \simeq O(\alpha^2)$.

The diagrams responsible for processes (4) and (5), shown in fig.1, lead to

$$\begin{aligned} A(4) &\simeq A(5) \simeq \alpha^2 A(1) \\ \frac{\Gamma(5)}{\Gamma(1)} &\simeq \frac{\Gamma(4)}{\Gamma(1)} \simeq \alpha^4 \end{aligned} \qquad (28)$$

in agreement with observation.

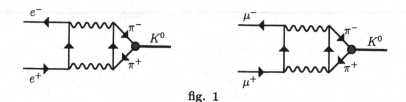

fig. 1

Let us compute $BR(3)$. The decay probability into 2γ is given by

$$\Gamma_{K\to 2\gamma} = \frac{1}{2M_K}\frac{1}{2}\sum_{pol}\int d\Phi^{(2)}\,|\langle 2\gamma_-|H_W|K\rangle|^2 =$$

$$\frac{1}{4M_K}\sum_{pol}\int \frac{d^3\mathbf{q}}{(2\pi)^3}\frac{d^3\mathbf{q'}}{(2\pi)^3}\frac{1}{2q_0}\frac{1}{2q_0'}(2\pi)^4\delta^{(4)}\left(P_K - q - q'\right)|\langle \mathbf{q}, \mathbf{q'}|H_W|K\rangle|^2 \qquad (29)$$

A factor of $1/2$ in (29) is due to the presence of two identical final state particles.

Denoting the photon polarization vectors by ε, ε' and their center-of-mass momentum by $\mathbf{q}$, we express the matrix element $\langle 2\gamma_- | H_W | K \rangle$ in the general form

$$\langle 2\gamma_- | H_W | K \rangle = F\varepsilon\varepsilon' + G\mathbf{q} \cdot (\varepsilon \wedge \varepsilon') \tag{30}$$

The two form factors, F and G, correspond to a final state with positive and negative parity, respectively. Since the decay is CP-conserving, and since K_S has $CP = +1$ and the 2γ's have $C = +1$, there follows that the final state must have $P = +1$ and therefore $G = 0$.

In terms of F, Eq.(29) is written

$$\Gamma_{K \to 2\gamma} = \frac{1}{2M_K} \frac{1}{16\pi} \sum_{pol} |\varepsilon\varepsilon'|^2 |F|^2 = \frac{1}{16\pi M_K} |F|^2 \tag{31}$$

We now evaluate the imaginary part of F; in this way, we set a lower limit to $\Gamma_{K \to 2\gamma}$.

Let us introduce a complete set of "out" states $|n_+\rangle$:

$$\langle 2\gamma_- | H_W | K \rangle = \sum_{n_+} \langle 2\gamma_- | n_+ \rangle \langle n_+ | H_W | K \rangle \tag{32}$$

$\langle 2\gamma_- | n_+ \rangle$ is an S-matrix element

$$\langle 2\gamma_- | n_+ \rangle = \delta_{n,2\gamma} - \mathrm{i}(2\pi)^4 \delta(P_n - k_{\gamma_1} - k_{\gamma_2}) \mathcal{M}_{2\gamma,n} \tag{33}$$

To lowest order in the electromagnetic interaction, the only contributions to the sum (32) come from 2γ- and 2π-states, generated by the weak K decay. In the first case, the only S-matrix term contributing to (33) is the identity. We thus have

$$\langle 2\gamma_- | H_W | K \rangle - \langle 2\gamma_+ | H_W | K \rangle = \sum_{(2\pi)_+} \langle 2\gamma_- | (2\pi)_+ \rangle \langle (2\pi)_+ | H_W | K \rangle \tag{34}$$

By T-invariance

$$\langle 2\gamma_+ | H_W | K \rangle = \langle K | H_W | 2\gamma_- \rangle = (\langle 2\gamma_- | H_W | K \rangle)^* \tag{35a}$$

$$\langle (2\pi)_+ | H_W | K \rangle = \langle K | H_W | (2\pi)_- \rangle \equiv A^*(K, (2\pi)) \tag{35b}$$

In these decays the 2π-states are in an S-wave and $A(K, (2\pi))$ is a momentum independent number. Using (35) we can rewrite (34) as

$$2\mathrm{Im}\,\mathcal{M}_{K \to 2\gamma} =$$
$$= \int \frac{\mathrm{d}^3\mathbf{p}}{(2\pi)^3} \frac{\mathrm{d}^3\mathbf{p}'}{(2\pi)^3} \frac{\mathcal{M}(2\pi \to 2\gamma)}{2p_0\, 2p'_0} A^*(K, 2\pi)(2\pi)^4 \delta^{(4)}(q + q' - p - p')$$
$$= A(K, 2\pi) \int \mathcal{M}(2\pi \to 2\gamma)\, \mathrm{d}\Phi^{(2\pi)} \tag{36}$$

p and **p**′ are the pion momenta in the intermediate state.

The phase space measure $d\Phi^{(2\pi)}$ factorizes as follows

$$d\Phi^{(2\pi)} = \frac{1}{8\pi}\sqrt{1 - \frac{4m_\pi^2}{M_K^2}}\frac{d\Omega_p}{4\pi} = \Phi^{(2\pi)}\frac{d\Omega_p}{4\pi} \tag{37}$$

To evaluate $\text{Im}\,\mathcal{M}_{K\to 2\gamma}$ we must calculate the amplitude $M(2\pi \to 2\gamma)$; the corresponding diagrams, to lowest perturbative order, are shown in fig. 2, together with momentum definitions.

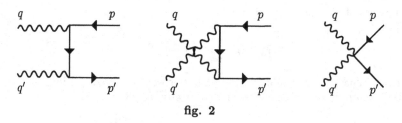

fig. 2

We write

$$\mathcal{M}(2\pi \to 2\gamma) = 2e^2\varepsilon_\mu^*\varepsilon_\nu'^* \left[\frac{p^\mu p'^\nu}{(p\cdot q)} + \frac{p^\nu p'^\mu}{(p'\cdot q)} - g^{\mu\nu}\right] \tag{38}$$

and, in the Coulomb gauge,

$$\varepsilon_\mu = (0, \boldsymbol{\varepsilon}) \qquad \boldsymbol{\varepsilon}\mathbf{q} = 0 \tag{39}$$

$$\mathcal{M}(2\pi \to 2\gamma) = 2e^2\varepsilon_i^*\varepsilon_j'^* \left[\frac{p_i p_j'}{p^0 q^0 - \mathbf{pq}} + \frac{p_j p_i'}{p^0 q^0 + \mathbf{pq}} + \delta_{ij}\right] = M_{ij}\varepsilon_i^*\varepsilon_j'^* \tag{40}$$

By symmetry

$$\int \frac{d\Omega_p}{4\pi} M_{ij} = B_1\delta_{ij} + B_2 q_i q_j \tag{41}$$

The coefficient B_2 does not contribute by virtue of Eq.(39), whereas B_1 can be expressed through (41) via a projection

$$B_1 = \frac{1}{2}\left(\delta_{ij} - \frac{q_i q_j}{\mathbf{q}^2}\right)\int \frac{d\Omega_p}{4\pi} M_{ij} \tag{42}$$

We thus obtain

$$B_1 = -\int \frac{d\Omega_p}{4\pi}e^2 \left\{\left[\mathbf{p}^2 - \frac{(\mathbf{pq})^2}{\mathbf{q}^2}\right]\frac{2p_0 q_0}{(p_0 q_0)^2 - (\mathbf{pq})^2} - 2\right\} \tag{43}$$

and using the relationship

$$|\mathbf{q}| = |q^0| = p_0 = \frac{M_K}{2}$$

$$B_1 = -\int \frac{d\Omega_p}{4\pi} 2e^2 \frac{\left(\mathbf{p}^2 - p_0^2\right) p_0^2}{p_0^4 - (\mathbf{pq})^2} \tag{44}$$

Finally

$$\hat{\mathbf{p}}\hat{\mathbf{q}} = \cos\theta \qquad \frac{|\mathbf{p}|}{p_0} = \sqrt{1 - \frac{4m_\pi^2}{M_K^2}} = v = 0.828$$

$$B_1 = -\int \frac{d\phi \, d\cos\theta}{4\pi} 2e^2 \frac{v^2 - 1}{1 - v^2 \cos^2\theta} = e^2 \frac{\left(1 - v^2\right)}{v} \ln \frac{1+v}{1-v} \tag{45}$$

Eq.(36) can now be rewritten, using (30)

$$
\begin{aligned}
\operatorname{Im}\mathcal{M}_{K\to 2\gamma} &= \frac{1}{2}\Phi^{(2\pi)} A(K, 2\pi) e^2 \frac{\left(1 - v^2\right)}{v} \ln \frac{1+v}{1-v} \varepsilon\varepsilon' \\
&= \varepsilon\varepsilon' \operatorname{Im} F
\end{aligned}
\tag{46}
$$

Inserting this expression into (31) we have

$$\Gamma_{K\to 2\gamma} \geq \frac{1}{16\pi M_K} |\operatorname{Im} F|^2 = \Gamma(K \to 2\pi)\frac{\Phi^{(2\pi)}}{32\pi}\left[e^2\frac{\left(1 - v^2\right)}{v}\ln\frac{1+v}{1-v}\right]^2 \tag{47}$$

that is,

$$\frac{\Gamma_{K\to 2\gamma}}{\Gamma} \geq BR(1)\frac{v}{256\pi^2}\left[e^2\frac{\left(1 - v^2\right)}{v}\ln\frac{1+v}{1-v}\right]^2 = (1.52 \pm 0.01)\cdot 10^{-6} \tag{48}$$

We can also evaluate the real part of F, using a dispersion relation. Calling s the square of the center-of-mass energy:

$$\operatorname{Re} F(s) = \frac{1}{\pi}\int_{4m_\pi^2}^{\infty} \frac{\operatorname{Im} F(s')}{s' - s}\, ds'$$

In terms of s

$$v = \sqrt{1 - \frac{4m_\pi^2}{s}}$$

$$\operatorname{Im} F(s) = A\frac{\alpha}{4}\left(1 - v^2\right)\ln\frac{1+v}{1-v}$$

The amplitude A of the decay $K \to 2\pi$ is taken to be constant, independent of s. Changing variables

$$\frac{s'}{m_\pi^2} = \frac{(1+z)^2}{z}$$

$$\operatorname{Re} F(s) = A\frac{\alpha}{\pi}\int_0^1 \frac{1-z}{1+z}\ln\frac{1}{z}\frac{1}{(1+z)^2 - \dfrac{s}{m_\pi^2}z}\, dz$$

This integral can be performed numerically for $s = m_K^2$, giving

$$\mathrm{Re}\, F(m_K^2) = 0.39 \cdot 10^{-3}\, A \tag{49}$$

With this numerical value we have

$$\frac{\Gamma_{K \to 2\gamma}}{\Gamma} = (1.65 \pm 0.01) \cdot 10^{-6} \tag{50}$$

in agreement[9] with the experimental result $(2.4 \pm 1.2) \cdot 10^{-6}$.

Problem 59.   Leptonic decays of the ρ meson

The leptonic decays of ρ, $\rho^+ \to e^+ \nu_e$, $\rho^+ \to \mu^+ \nu_\mu$, have never been observed due to insufficient statistics.

a)   How large must the ρ sample be in order to observe these decays with reasonable accuracy? You may use the CVC hypothesis, and the following data on ρ: $M_\rho = 768.1 \pm 0.5\,\mathrm{MeV}$, $\Gamma(\rho \to \pi\pi) \simeq 151.5 \pm 1.2\,\mathrm{MeV}$, $\sigma_{e^+e^- \to e^+e^-} \simeq (0.5 \pm .1)10^{-34}\,\mathrm{cm}^2$ (at the ρ peak).
b)   Assuming universality of the weak interaction, what is the expected width of the decay $\tau^+ \to \rho^+ \bar{\nu}_\tau$? ($M_\tau = 1784 \pm 3\,\mathrm{MeV}$).

Solution

a)   The decay amplitude of $\rho \to \ell^+ \nu_\ell$ is

$$A = \frac{G_F}{\sqrt{2}}\, \langle 0 | J_\mu^\dagger(0) | \rho^+ \rangle\, \bar{u}(\mathbf{p}_{\nu_\ell}) \gamma^\mu \left(1 - \gamma^5\right) v(\mathbf{p}_{\ell^+}) \tag{1}$$

The parity of ρ is negative. By rotational invariance, only spatial components of the current contribute to the matrix element in the center of mass. Thus, only the vector current V_μ is present; the hadronic part of the matrix element takes the form

$$\langle 0 | J_\mu^\dagger(0) | \rho^+ \rangle = \langle 0 | V_\mu^\dagger(0) | \rho^+ \rangle = F_\rho \varepsilon_\mu \tag{2}$$

Here F_ρ is a constant with dimensions of mass squared, and ε_μ is the polarization vector of ρ.
    The decay width equals

$$\Gamma = \frac{1}{2M_\rho}\, |\mathcal{M}|^2\, \Phi^{(2)} \tag{3}$$

where the amplitude $\mathcal{M}$ is

$$
\begin{aligned}
|\mathcal{M}|^2 &= \frac{G_F^2}{2} |F_\rho|^2 \frac{1}{3} \sum_i |\bar{u}(\mathbf{p}_{\nu_\ell}) \gamma^\mu \left(1 - \gamma^5\right) v(\mathbf{p}_{\ell^+}) \varepsilon_\mu^{(i)}|^2 \\
&= \frac{4 G_F^2 |F_\rho|^2}{3} \left(M_\rho^2 - m_\ell^2\right) \left(1 + \frac{m_\ell^2}{2 M_\rho^2}\right)
\end{aligned}
\tag{4}
$$

---

[9]Particle Data Group, *Phys. Lett.* **B239** 1990.

Eqs. (3) and (4) lead readily to

$$\Gamma = \frac{G_F^2 |F_\rho|^2}{12\pi} M_\rho \left(1 - \frac{m_\ell^2}{M_\rho^2}\right)^2 \left(1 + \frac{m_\ell^2}{2M_\rho^2}\right) \tag{5}$$

We have made use of the formulae

$$\Phi^{(2)} = \frac{1}{8\pi}\left(1 - \frac{m_\ell^2}{m_\rho^2}\right) \qquad \sum_i \varepsilon_\mu^{(i)}(k)\varepsilon_\nu^{(i)*}(k) = -\left(g_{\mu\nu} - \frac{k_\mu k_\nu}{k^2}\right) \tag{6}$$

The only unknown quantity above is $|F_\rho|^2$. To determine it, consider the expression for the cross section $\sigma_{e^+e^- \to e^+e^-}$ at the ρ peak

$$\sigma = \frac{12\pi}{M_\rho^2} \frac{\left(\Gamma_{\rho \to e^+e^-}^2/4\right)}{\left(\Gamma_\rho^2/4\right)} \tag{7}$$

The factor of 4 in the denominator comes directly from the Breit-Wigner formula, while the one in the numerator is due to the average over initial spins. At the same time, the CVC hypothesis implies

$$\langle 0|V_3^\mu|\rho^0\rangle = \frac{1}{\sqrt{2}} \langle 0|V_-^\mu|\rho^+\rangle \tag{8}$$

Therefore, we find

$$\langle 0|V_3^\mu|\rho^0\rangle = \frac{F_\rho}{\sqrt{2}}\varepsilon^\mu \tag{9}$$

and

$$\Gamma_{\rho \to e^+e^-} = \frac{2\pi}{3}\alpha^2 |F_\rho|^2 M_\rho^{-3} \tag{10}$$

Eqs. (7) and (10) allow us to express $|F_\rho|^2$ in terms of σ and Γ_ρ.

We now substitute F_ρ in (5) with the result

$$\frac{\Gamma_{\rho \to \nu \ell}}{\Gamma_\rho} = \frac{G_F^2 M_\rho^5}{8\pi^2\alpha^2} \sqrt{\frac{\sigma_{e^+e^- \to e^+e^-}}{12\pi}} \simeq 10^{-13} \tag{11}$$

We conclude that the number of ρ's at our disposal must be greater than 10^{13}, in order for the leptonic decay to be observable.

b) The amplitude for the decay $\tau^+ \to \rho^+ \bar{\nu}_\tau$ is

$$A = \frac{G_F}{\sqrt{2}} \langle \rho^+|V_\mu|0\rangle \, \bar{v}(\mathbf{p}_\tau)\gamma_\mu\left(1 - \gamma^5\right) v(\mathbf{p}_{\bar{\nu}_\tau}) \tag{12}$$

Using our result on the current matrix element, we obtain

$$\Gamma_{\tau \to \rho^+ \bar{\nu}_\tau} = \frac{G_F^2 |F_\rho|^2}{8\pi} m_\tau \left(1 - \frac{M_\rho^2}{m_\tau^2}\right)^2 \left(1 + \frac{m_\tau^2}{2M_\rho^2}\right) \tag{13}$$

The corresponding numerical value is

$$\Gamma_{\tau \to \rho^+ \bar{\nu}_\tau} \simeq (0.6 \pm .06) \cdot 10^{-3} \, \text{eV} \tag{14}$$

From Eq.(7) and from the relative error on σ given in the text (20 %), the relative error on $\Gamma_{e^+e^-}$ is $\sim$ 10%. Eq.(13) fixes then to 10% the relative error of $\Gamma_{\tau \to \rho^+ \bar{\nu}_\tau}$. Errors in the masses which appear in the equations are much smaller and irrelevant.

Problem 60.   The decay of positronium into two photons

Positronium in the $1S$ orbital can decay into two or three photons, according to its spin state.
a)   Which state decays into two photons? Explain.
b)   Calculate the correlation between the linear polarizations of the two photons.

Solution

a)   A fermion-antifermion system with orbital angular momentum L and spin S has charge conjugation eigenvalue

$$C = (-1)^{L+S} \tag{1}$$

and parity

$$P = (-1)^{L+1} \tag{2}$$

Bound states of e^+ and e^- must be eigenstates of P and C, since electrodynamics is invariant under these two operations. Therefore, the ground state has a definite S.

By charge conjugation invariance, the state with $L = 0$, $S = 0$ (parapositronium) can decay into an even number of photons ($C = +1$); to lowest order in α it decays into two photons. Conversely, the state with $L = 0$, $S = 1$ (orthopositronium) decays into an odd number of photons ($C = -1$). Since decay into a single photon is forbidden by energy and momentum conservation, orthopositronium decays into three photons to lowest order. Amplitudes with higher numbers of photons are suppressed in both cases by a factor of $\alpha \simeq 10^{-2}$ per photon.

b)   The two photon state has $J = 0$ and $P = -1$. The most general gauge- and Lorentz-invariant decay amplitude is given by

$$A = C_1 \, F^{(1)}_{\mu\nu} F^{(2)\mu\nu} + C_2 \, \varepsilon^{\mu\nu\rho\sigma} \, F^{(1)}_{\mu\nu} F^{(2)}_{\rho\sigma} \tag{3}$$

Parity invariance requires the first term to vanish; the amplitude becomes

$$A = C \, \varepsilon^{\mu\nu\rho\sigma} \, F^{(1)}_{\mu\nu} F^{(2)}_{\rho\sigma} = C \left(\mathbf{E}^{(1)} \, \mathbf{H}^{(2)} + \mathbf{H}^{(1)} \, \mathbf{E}^{(2)} \right) \tag{4}$$

where $\mathbf{E}$ and $\mathbf{H}$ are the electric and magnetic field operators. Denoting by $\mathbf{k}_1$ the center-of-mass momentum of one of the two photons, the amplitude takes the form

$$A = C \left|\mathbf{k}_1\right| \mathbf{k}_1 \cdot (\varepsilon_1 \wedge \varepsilon_2) \tag{5}$$

We see that the two photons are polarized perpendicular to each other; consequently, we have

$$\langle \varepsilon_1 \, \varepsilon_2 \rangle = 0 \tag{6}$$

Problem 61. Neutron β-decay

The β-decay of neutrons

$$n \to p + e^- + \bar{\nu}_e \tag{1}$$

is described by the Fermi interaction

$$H = \frac{G_F}{\sqrt{2}} J_\mu j^\mu \tag{2}$$

where

$$j_\mu = \overline{\psi}_{\nu_e} \gamma_\mu \left(1 - \gamma^5\right) \psi_e \qquad J_\mu = J_\mu^V - J_\mu^A \tag{3}$$

G_F is the Fermi constant, $G_F = 1.027 \cdot 10^{-5}\, m_p^{-2}$. For the range of momentum transfer in this process we may assume

$$J_\mu^V = \overline{\psi}_p \gamma_\mu \psi_n \qquad J_\mu^A = C_A \overline{\psi}_p \gamma_\mu \gamma^5 \psi_n \tag{4}$$

a) What is the maximum kinetic energy imparted on the proton?
b) Write down the matrix elements of the currents V_μ and A_μ in the nonrelativistic limit.
c) In the same limit, calculate the spectrum of the emitted electrons and the decay width.
d) Show that an imaginary part in C_A, $C_A = |C_A| e^{i\phi}$ implies violation of time reversal invariance T. If T remains intact, show that

$$\langle \mathbf{s}_n \cdot (\mathbf{p}_\nu \wedge \mathbf{p}_e) \rangle = 0 \tag{5}$$

Calculate the above quantity for $\phi \neq 0$.

Solution

a)   The proton kinetic energy is maximized when the invariant mass of the $e\bar{\nu}$ pair takes its minimum value, m_e, corresponding to $E_\nu = 0$. In this configuration, energy and momentum conservation imply

$$m_n = \sqrt{p^2 + m_e^2} + \sqrt{p^2 + m_p^2} \tag{6}$$

which gives

$$E_p = m_p + \frac{(m_n - m_p)^2 - m_e^2}{2m_n} \tag{7}$$

Thus the maximum kinetic energy equals

$$\frac{(m_n - m_p)^2 - m_e^2}{2m_n} = 0.75 \,\text{KeV} \tag{8}$$

b) In the nonrelativistic limit we have

$$\bar{u}_p \gamma_\mu u_n \simeq \left(2m_n\, u_p^\dagger u_n, \mathbf{0}\right)$$
$$\bar{u}_p \gamma_\mu \gamma^5 u_n \simeq \left(0, 2m_n\, u_p^\dagger \boldsymbol{\sigma} u_n\right) \tag{9}$$

c) The differential width is given by

$$d\Gamma = \frac{1}{2m_n} \frac{G_F^2}{2} H_{\mu\nu} L^{\mu\nu} \, d\Phi^{(3)} \tag{10}$$

The leptonic tensor is defined by

$$L_{\mu\nu} = \sum_{pol} j_\mu j_\nu^* \tag{11}$$

and similarly for the hadronic tensor $H_{\mu\nu}$. We find

$$
\begin{aligned}
L_{\mu\nu} &= \text{Tr}\left\{ (\not{p}_e + m_e)\gamma_\mu \left(1 - \gamma^5\right) \not{p}_{\bar{\nu}} \gamma_\nu \left(1 - \gamma^5\right) \right\} \\
&= 8\left\{ p_{e\mu} p_{\bar{\nu}_e \nu} + p_{e\nu} p_{\bar{\nu}_e \mu} - (p_e \cdot p_{\bar{\nu}_e})\, g_{\mu\nu} \right\} - 8i\varepsilon_{\mu\nu\rho\sigma} p_e^\rho p_{\bar{\nu}_e}^\sigma
\end{aligned} \tag{12}
$$

In the nonrelativistic limit, denoting by $\mathbf{P}$ the neutron polarization, we obtain

$$H_{\mu\nu} L^{\mu\nu} = 4m_n^2 \text{Tr}\left\{ \frac{1 + \mathbf{P} \cdot \boldsymbol{\sigma}}{2} \left[j_0 + C_A \boldsymbol{\sigma} \cdot \mathbf{j}\right]\left[j_0^* + C_A^* \boldsymbol{\sigma} \cdot \mathbf{j}^*\right] \right\} \tag{13}$$

The trace acts on the nucleon spin degrees of freedom. Eq.(13) leads to the following expression for the width

$$
\begin{aligned}
d\Gamma &= \frac{G_F^2}{4m_n} 4m_n^2 \left\{ j_0 j_0^* + |C_A|^2 \left(\mathbf{j}\,\mathbf{j}^* + i\mathbf{P} \cdot (\mathbf{j} \wedge \mathbf{j}^*)\right) + \mathbf{P} \cdot 2\text{Re}\left(C_A j_0^* \mathbf{j}\right) \right\} d\Phi^{(3)} \\
&= 8G_F^2 \Big\{ E_e E_{\bar{\nu}} \left(1 + 3|C_A|^2\right) + \mathbf{p_e}\, \mathbf{p}_{\bar{\nu}_e} \left(1 - |C_A|^2\right) \\
&\quad + 2\mathbf{P} \cdot \left[E_{\bar{\nu}} \mathbf{p_e} \left(\text{Re}\, C_A + |C_A|^2\right) + E_e \mathbf{p}_{\bar{\nu}_e} \left(\text{Re}\, C_A - |C_A|^2\right) \right] \\
&\quad - 2\text{Im}\, C_A \mathbf{P} \cdot \left(\mathbf{p_e} \wedge \mathbf{p}_{\bar{\nu}_e}\right) \Big\} \frac{p_{\bar{\nu}_e} p_e}{128\pi^5} \, dE_e \, d\Omega_e \, d\Omega_{\bar{\nu}}
\end{aligned} \tag{14}
$$

Only the first term survives angular integration, with the result

$$\frac{d\Gamma}{dE_e} = \frac{G_F^2}{\pi^3} \left(1 + 3|C_A|^2\right) \sqrt{E_e^2 - m_e^2} \, E_e \, (Q - E_e)^2 \qquad (15)$$

$$\Gamma = \frac{\left(1 + 3|C_A|^2\right) G_F^2 Q^5}{\pi^3} \int_0^1 (1 - x)^2 \, x \sqrt{x^2 - \left(\frac{m_e}{Q}\right)^2} \, dx \qquad (16)$$

where $Q = m_n - m_p$. The integral in (16) equals $1/30 = 0.033$ for $m_e = 0$ and 0.0157 for the physical electron mass.

d) Under time reversal

$$T\!:\!\bar{\psi}\gamma_\mu\psi\!:\!T^\dagger = g_{\mu\mu}\!:\!\bar{\psi}\gamma_\mu\psi\!:$$
$$T\!:\!\bar{\psi}\gamma^5\gamma_\mu\psi\!:\!T^\dagger = g_{\mu\mu}\!:\!\bar{\psi}\gamma^5\gamma_\mu\psi\!: \qquad (17)$$

Given that T is antilinear, invariance under T requires $C_A = C_A^*$.

For $\mathrm{Im}\, C_A = 0$, there follows immediately from (14) that

$$\langle \mathbf{s}_n \cdot (\mathbf{p}_\nu \wedge \mathbf{p}_e) \rangle = 0 \qquad (18)$$

Eq.(18) can be proved from first principles, invoking time reversal invariance followed by a 180° rotation on the decay plane (neglecting final state interaction).

For $\mathrm{Im}\, C_A = |C_A| \sin \phi \neq 0$, we find

$$\langle \mathbf{s}_n \cdot (\hat{\mathbf{p}}_\nu \wedge \hat{\mathbf{p}}_e) \rangle = \frac{2|C_A| \sin \phi}{1 + 3|C_A|^2} \frac{I_N}{I_D} \qquad (19)$$

with

$$I_N = \int p_e^2 p_{\bar{\nu}_e}^2 \, dE_e \qquad I_D = \int E_e p_e p_{\bar{\nu}_e}^2 \, dE_e \qquad (20)$$

A numerical integration gives

$$\frac{I_N}{I_D} = 0.731$$

Since $|C_A| \simeq 1.23$, Eq.(19) gives

$$\langle \mathbf{s}_n \cdot (\hat{\mathbf{p}}_\nu \wedge \hat{\mathbf{p}}_e) \rangle \simeq 0.32 \sin \phi \qquad (21)$$

Problem 62.   Some characteristics of $\Sigma \to N - \pi$ decays

Σ^+ and Σ^- decay predominantly into $N - \pi$. The corresponding branching ratios are:

$$\Sigma^- \;\to\; n\pi^- \qquad \text{BR } 100\% \qquad (1)$$
$$\Sigma^+ \;\to\; p\pi^0 \qquad \text{BR } 51.6\% \qquad (2)$$
$$\Sigma^+ \;\to\; n\pi^+ \qquad \text{BR } 48.4\% \qquad (3)$$

Remaining decay modes have BR $\sim 10^{-3}$. The quantum numbers of the decay operator R are $T = 1/2$, $T_3 = -1/2$, $S = -1$.

a)   The above decays violate parity. Knowing that Σ has spin 1/2, what are the allowed values of orbital angular momentum in the final state?

b)   Express the amplitude for each decay in terms of transition amplitudes between states of given isospin.

c)   Assuming that time reversal invariance T holds, relate the phases of the decay amplitude to those of $\pi - N$ scattering. The decay amplitude may be parametrized as follows

$$A = a + b \frac{\mathbf{p} \cdot \boldsymbol{\sigma}}{|\mathbf{p}|} \tag{4}$$

where the operator $\boldsymbol{\sigma}$ acts between spin states of the initial particle and the final nucleon.

d)   Determine the decay parameters in terms of the $\pi - N$ scattering phases, the given branching ratios, and the asymmetry α of the angular distribution in the decay of polarized Σ,

$$\frac{d\sigma}{d\cos\theta} \simeq 1 + \alpha \cos\theta \tag{5}$$

Numerically, α equals -0.07 ± 0.01, $-0.98 \pm .015$, 0.07 ± 0.01 for decay modes (1), (2), (3), respectively.

Solution

a)   By angular momentum conservation, the allowed final states are superpositions of $S_{\frac{1}{2}}$ and $P_{\frac{1}{2}}$.

b)   The isospin content of the final states is, in standard notation,

$$|n\pi^-\rangle = \left|\frac{3}{2}, -\frac{3}{2}\right\rangle \tag{6a}$$

$$|p\pi^0\rangle = \sqrt{\frac{2}{3}} \left|\frac{3}{2}, \frac{1}{2}\right\rangle + \sqrt{\frac{1}{3}} \left|\frac{1}{2}, \frac{1}{2}\right\rangle \tag{6b}$$

$$|n\pi^+\rangle = \sqrt{\frac{1}{3}} \left|\frac{3}{2}, \frac{1}{2}\right\rangle - \sqrt{\frac{2}{3}} \left|\frac{1}{2}, \frac{1}{2}\right\rangle \tag{6c}$$

The Wigner-Eckart theorem implies, for $\Delta T = 1/2$,

$$A\left(\Sigma^- \to n\pi^-\right) = A_{\frac{3}{2}}$$
$$A\left(\Sigma^+ \to \frac{3}{2}, \frac{1}{2}\right) = \sqrt{\frac{1}{3}} A_{\frac{3}{2}} \tag{7}$$

Similarly,

$$A\left(\Sigma^+ \to \frac{1}{2}, \frac{1}{2}\right) = \sqrt{\frac{2}{3}} A_{\frac{1}{2}} \tag{8}$$

There follows

$$A\left(\Sigma^- \to n\pi^-\right) = A_{\frac{3}{2}} \tag{9a}$$

$$A\left(\Sigma^+ \to p\pi^0\right) = \frac{\sqrt{2}}{3}\left(A_{\frac{3}{2}} + A_{\frac{1}{2}}\right) \tag{9b}$$

$$A\left(\Sigma^+ \to n\pi^+\right) = \frac{1}{3}\left(A_{\frac{3}{2}} - 2A_{\frac{1}{2}}\right) \tag{9c}$$

We see that all three amplitudes are expressible in terms of two unknowns; this leads to the condition

$$A\left(\Sigma^- \to n\pi^-\right) = \sqrt{2}A\left(\Sigma^+ \to p\pi^0\right) + A\left(\Sigma^+ \to n\pi^+\right) \tag{10}$$

c) The spin dependence of the amplitudes has the general form

$$A_{\frac{3}{2}} = a_{\frac{3}{2}} + b_{\frac{3}{2}}\frac{\boldsymbol{\sigma} \cdot \mathbf{p}}{|\mathbf{p}|} \tag{11a}$$

$$A_{\frac{1}{2}} = a_{\frac{1}{2}} + b_{\frac{1}{2}}\frac{\boldsymbol{\sigma} \cdot \mathbf{p}}{|\mathbf{p}|} \tag{11b}$$

If time reversal invariance holds, we may choose the phases of the T-transformed states in such a way that

$$\arg a_{\frac{3}{2}} = \delta_{1,3}^-$$
$$\arg b_{\frac{3}{2}} = \delta_{1,3}^+$$
$$\arg a_{\frac{1}{2}} = \delta_{1,1}^-$$
$$\arg b_{\frac{1}{2}} = \delta_{1,1}^+$$

where $\delta_{2J,2T}^{(\pm)}$ is the $\pi - N$ scattering phase in a state with parity $\pm$, angular momentum J and isospin T. Experimental data for $\pi - N$ at energies equal to the Σ mass lead to negligible values for all of these phases (of the order of a degree). We may thus take the amplitudes to be real.

d) Polarized Σ's lead to the following distribution

$$\frac{1}{|a|^2 + |b|^2}\text{Tr}\left\{\frac{1 + \mathbf{n} \cdot \boldsymbol{\sigma}}{2}\left(a + b\frac{\mathbf{p} \cdot \boldsymbol{\sigma}}{|\mathbf{p}|}\right)\left(a^* + b^*\frac{\mathbf{p} \cdot \boldsymbol{\sigma}}{|\mathbf{p}|}\right)\right\} = 1 + \frac{2\,\text{Re}\,ab^*}{|a|^2 + |b|^2}\cos\theta \tag{12}$$

For the three decay modes considered we find

$$\alpha_1 = \frac{2a_{\frac{3}{2}}b_{\frac{3}{2}}}{|a_{\frac{3}{2}}|^2 + |b_{\frac{3}{2}}|^2} \simeq 0.07 \pm 0.01 \tag{13a}$$

$$\alpha_2 = \frac{2\left(a_{\frac{3}{2}} + a_{\frac{1}{2}}\right)\left(b_{\frac{3}{2}} + b_{\frac{1}{2}}\right)}{|a_{\frac{3}{2}} + a_{\frac{1}{2}}|^2 + |b_{\frac{3}{2}} + b_{\frac{1}{2}}|^2} \simeq -0.98 \pm 0.015 \tag{13b}$$

$$\alpha_3 = \frac{2\left(a_{\frac{3}{2}} - 2a_{\frac{1}{2}}\right)\left(b_{\frac{3}{2}} - 2b_{\frac{1}{2}}\right)}{|a_{\frac{3}{2}} - 2a_{\frac{1}{2}}|^2 + |b_{\frac{3}{2}} - 2b_{\frac{1}{2}}|^2} \simeq -0.07 \pm 0.01 \tag{13c}$$

Further conditions on a and b come from the branching ratios; in particular,

$$R_1 \equiv \varepsilon_{PS} \frac{\Gamma\left(\Sigma^+ \to p\pi^0\right)}{\Gamma\left(\Sigma^+ \to n\pi^0\right)} = 2\,\frac{|a_{\frac{3}{2}} + a_{\frac{1}{2}}|^2 + |b_{\frac{3}{2}} + b_{\frac{1}{2}}|^2}{|a_{\frac{3}{2}} - 2a_{\frac{1}{2}}|^2 + |b_{\frac{3}{2}} - 2b_{\frac{1}{2}}|^2} \tag{14}$$

The phase space correction ε_{PS} equals

$$\varepsilon_{PS} = \sqrt{\frac{\left(M_{\Sigma^+}^2 + m_n^2 - m_{\pi^+}^2\right)^2 - 4\,m_n^2 M_{\Sigma^+}^2}{\left(M_{\Sigma^+}^2 + m_p^2 - m_{\pi^0}^2\right)^2 - 4\,m_p^2\,M_{\Sigma^+}^2}} \simeq 0.987 \tag{15}$$

and thus it is negligible. The ratio between Σ^+ and Σ_- decay widths gives the condition

$$R_2 = \frac{\Gamma\left(\Sigma^+ \to N\pi\right)}{\Gamma\left(\Sigma^- \to N\pi\right)} = \frac{\frac{1}{3}\left(|a_{\frac{3}{2}}|^2 + |b_{\frac{3}{2}}|^2\right) + \frac{2}{3}\left(|a_{\frac{1}{2}}|^2 + |b_{\frac{1}{2}}|^2\right)}{|a_{\frac{3}{2}}|^2 + |b_{\frac{3}{2}}|^2} \tag{16}$$

Let us solve Eqs.(13) for $a_{\frac{3}{2}}$, $a_{\frac{1}{2}}$, $b_{\frac{3}{2}}$ and $b_{\frac{1}{2}}$. For an approximate solution we take

$$\alpha_1 = \alpha_3 = 0 \qquad \alpha_2 = -1 \tag{17}$$

Eqs. (13) lead to

$$\left(a_{\frac{3}{2}} + a_{\frac{1}{2}}\right) = -\left(b_{\frac{3}{2}} + b_{\frac{1}{2}}\right) \tag{18}$$

and to

$$\begin{cases} b_{\frac{3}{2}} = 0 \\ a_{\frac{3}{2}} = 2a_{\frac{1}{2}} \\ b_{\frac{1}{2}} = -3a_{\frac{1}{2}} \end{cases} \tag{19}$$

or alternatively

$$\begin{cases} a_{\frac{3}{2}} = 0 \\ b_{\frac{3}{2}} = 2b_{\frac{1}{2}} \\ a_{\frac{1}{2}} = -3b_{\frac{1}{2}} \end{cases} \tag{20}$$

The resulting values for $R_{1,2}$ are

$$R_1 = 1 \qquad R_2 = 2 \tag{21}$$

in reasonable agreement with data (after the phase space correction)

$$R_1 = 1.04 \pm 0.01 \qquad R_2 = 1.85 \pm 0.01 \tag{22}$$

The available data do not make it possible to distinguish between the two solutions (19), (20): correlations between initial and final spins would allow it.

Problem 63.   The decays $W^\pm \to \pi^\pm + \pi^0$

Let us consider the following decay mode

$$W^\pm \to \pi^\pm + \pi^0 \tag{1}$$

for the intermediate vector bosons of weak interactions.

a)   Show that the final state has isospin $I = 1$; show that only the vector term of the standard weak Lagrangian

$$\mathcal{L}_I = g\, W_\mu \left(V^\mu - A^\mu\right) \tag{2}$$

contributes to the decay.

b)   Assuming the validity of the conserved vector current hypothesis, calculate the decay width for this process, making use of the following:

(i) $\sigma_{e^+e^- \to \pi^+\pi^-}$ is dominated by a Breit-Wigner resonance centered on the ρ mass; $M_\rho = 768.1 \pm 0.5$ MeV, $\Gamma_\rho = 130$ MeV, $\Gamma_{\rho \to e^+e^-} = (4.44 \pm 0.21) \cdot 10^{-5} \Gamma_\rho$

(ii) $M_W = 80.22 \pm 0.26$ GeV.

c)   Calculate the branching ratio $\left(W^\pm \to \pi^\pm \pi^0\right) / \left(W^\pm \to \text{hadrons}\right)$ knowing that

$$\sigma^{(\gamma)}_{e^+e^- \to \text{hadrons}\ (I=1)} \simeq \frac{2\pi\alpha^2}{s} \tag{3}$$

The cross section in (3) refers to the process $e^+e^- \to \gamma \to \text{hadrons}$.

Solution

a)   Since the final state must have angular momentum 1, the orbital wavefunction will be antisymmetric under exchange of the two pions. To respect Bose symmetry, the isospin wave function must also be antisymmetric, leading to $I = 1$.

The amplitude is proportional to

$$\langle \pi^+ \pi^0 | \left(V_\mu - A_\mu\right) |0\rangle \, \varepsilon^\mu \tag{4}$$

In the center of mass only the spatial part contributes, since the time component of ε_μ vanishes. Also the parity of the state will be negative, because the pions are in a P-wave. Consequently, only V_μ will contribute.

b)   Let us denote the matrix element by V_μ^+

$$\langle \pi^+ \pi^0 | V_\mu^+ |0\rangle = V_\mu^+ \tag{5}$$

V_μ^+ is the component $V_\mu^1 + iV_\mu^2$ of an isospin triplet whose third component is the isovector electromagnetic current; by the Wigner-Eckart theorem, we have

$$\langle (\pi^+\pi^0)_{I=1} | V_\mu^+ |0\rangle = \sqrt{2} \langle (\pi^+\pi^-)_{I=1} | V_\mu^3 |0\rangle \tag{6}$$

Under the assumption of ρ-dominance, we write

$$\sigma_{e^+e^- \to \pi^+\pi^-} = \frac{4\pi}{s} \frac{(2J+1)}{(2S_e+1)^2} \frac{\Gamma_{ee}\Gamma_{\pi\pi} 4M_\rho^2}{\left(s-M_\rho^2\right)^2 + \Gamma_{\pi\pi}^2 M_\rho^2} \tag{7}$$

where Γ_{ee} and $\Gamma_{\pi\pi}$ are the partial widths for ρ decay into e^+e^- and $\pi^+\pi^-$. We can also calculate the differential cross section for this process, mediated by a single photon, finding

$$d\sigma_{e^+e^- \to \pi^+\pi^-} = \frac{e^4}{2s} \left| \frac{l_\mu V^{3\mu}}{s} \right|^2 \frac{d\Omega}{32\pi^2} \sqrt{1 - \frac{4m_\pi^2}{s}} \tag{8}$$

with

$$l_\mu = \bar{u}(\mathbf{p}_{e^-})\gamma_\mu v(\mathbf{p}_{e^+}) \tag{9}$$

Averaging over initial spins we obtain

$$\langle l^\mu l^\nu \rangle \simeq \frac{1}{4}\mathrm{Tr}\left(\not{p}_{e^-}\gamma^\mu \not{p}_{e^+}\gamma^\nu\right) = p_{e^-}{}^\mu p_{e^+}{}^\nu + p_{e^-}{}^\nu p_{e^+}{}^\mu - g^{\mu\nu} p_{e^+}\cdot p_{e^-} \tag{10}$$

Neglecting the electron mass, only spatial components in (10) survive, giving

$$\langle l_i l_j \rangle \simeq \frac{s}{2}(\delta_{ij} - n_i n_j) \tag{11}$$

$\mathbf{n} = \mathbf{p}_{e^-}/|\mathbf{p}_{e^-}|$. If we also neglect m_π^2/s, we obtain

$$\sigma_{e^+e^- \to \pi^+\pi^-} = \frac{\alpha^2}{8s^2}\int d\Omega\left[|\mathbf{V}^3|^2 - |\mathbf{V}^3\cdot\mathbf{n}|^2\right] = \frac{\pi\alpha^2}{3s^2}|\mathbf{V}^3|^2 \tag{12}$$

Eqs. (6), (7) and (12) now yield

$$|\mathbf{V}_+|^2 = 2|\mathbf{V}^3|^2 = \frac{6s^2}{\pi\alpha^2}\sigma_{e^+e^- \to \pi^+\pi^-} \underset{s\gg M_\rho^2}{\simeq} \frac{72\Gamma_{ee}\Gamma_{\pi\pi} M_\rho^2}{\alpha^2 s} \tag{13}$$

The width for $W^\pm \to \pi^\pm \pi^0$ equals

$$\Gamma_{W^\pm \to \pi^\pm \pi^0} = \frac{1}{2M_W} g^2 \int \sum_i |\varepsilon_\mu^{(i)} V_+^\mu|^2 \frac{d\Omega}{32\pi^2} = \frac{1}{6M_W} g^2 \frac{|\mathbf{V}_+|^2}{8\pi} \tag{14}$$

Use of (13) leads to

$$\Gamma_{W^\pm \to \pi^\pm \pi^0} = \frac{3}{2\pi} \frac{g^2}{\alpha^2} \frac{M_\rho^2}{M_W^3} \Gamma_{ee}\Gamma_{\pi\pi} \tag{15}$$

In the standard model, the coupling constant g is given by

$$g^2 = \frac{e^2}{2\sin^2\theta_W} = \frac{2\pi\alpha}{\sin^2\theta_W} \tag{16}$$

with $\sin^2 \theta_W = 0.2325(8)$. In conclusion,

$$\Gamma_{W^{\pm} \to \pi^{\pm} \pi^0} \simeq 2.06 \cdot 10^{-6} \text{MeV} \tag{17}$$

c) Under the CVC hypothesis, we write

$$\frac{\Gamma_{W \to \text{hadrons}}}{\Gamma_{W \to \pi^+ \pi^0}} = \frac{\sigma^{(\gamma)}_{e^+ e^- \to \text{hadrons}(I=1)}}{\sigma^{(\gamma)}_{e^+ e^- \to \pi^+ \pi^-}} \Big|_{s = M_W^2} \tag{18}$$

There follows

$$\frac{\Gamma_{W \to \pi^+ \pi^0}}{\Gamma_{W \to \text{hadrons}}} \simeq 6 \frac{\Gamma_{ee} \Gamma_{\pi\pi} M_\rho^2}{\alpha^2 M_W^4} \simeq 1.64 \cdot 10^{-9} \tag{19}$$

Problem 64. $\Lambda \to N + \pi$ and the $\Delta T = 1/2$ rule

The Λ baryon (spin 1/2, $M = 1115$ MeV, $T = 0$) decays almost exclusively into $\pi -$ nucleon

$$\begin{aligned} BR(\Lambda \to p\pi^-) &= (64.2 \pm .5)\% \\ BR(\Lambda \to n\pi^0) &= (35.8 \pm .5)\% \end{aligned} \tag{1}$$

a) Assuming that this decay is induced by an isospin 1/2 operator, calculate the ratio

$$\frac{BR(\Lambda \to p\pi^-)}{BR(\Lambda \to n\pi^0)} \tag{2}$$

Take into account phase space corrections and final state Coulomb effects.

b) Give an estimate of the $\Delta T = 3/2$ amplitude, and compare to the data of (1).

c) The pion distribution with respect to the direction of polarization of Λ has the form

$$P(\theta) = 1 + \alpha \cos\theta \tag{3}$$

For the decay $p - \pi^-$ $\alpha == 0.642 \pm 0.013$. If the $\Delta T = 1/2$ assumption holds, show that α is the same for the two decay modes. Calculate the fraction of P-wave in the final state.

Solution

a) Λ has zero isospin; if $\Delta T = 1/2$, the state produced in this decay will have isospin $T = 1/2$, $T_3 = -1/2$. We write

$$|\frac{1}{2}, -\frac{1}{2}\rangle = \sqrt{\frac{2}{3}} \ |\pi^- p\rangle - \sqrt{\frac{1}{3}} \ |\pi^0 n\rangle \tag{4}$$

Ignoring final state interactions, we have

$$\frac{BR(\Lambda \to p\pi^-)}{BR(\Lambda \to n\pi^0)} = 2 \tag{5}$$

The phase space correction to this ratio is

$$\varepsilon_{PS} = \sqrt{\frac{\left(M_\Lambda^2 + m_p^2 - m_{\pi^-}^2\right)^2 - 4m_p^2 M_\Lambda^2}{\left(M_\Lambda^2 + m_n^2 - m_{\pi^0}^2\right)^2 - 4m_n^2 M_\Lambda^2}} = 0.967 \tag{6}$$

The Coulomb correction equals

$$\varepsilon_{\text{coul}} = 1 + \frac{\alpha\pi}{v_{rel}} = 1.033 \tag{7}$$

Thus, if $\Delta T = 1/2$, the expected branching ratio is

$$\frac{BR(\Lambda \to p\pi^-)}{BR(\Lambda \to n\pi^0)} = 1.999 \tag{8}$$

The two corrections cancel against each other in this case.

b)   Suppose there exists also a $\Delta T = 3/2$ component

$$|\frac{3}{2}, -\frac{1}{2}\rangle = \sqrt{\frac{1}{3}} \, |\pi^- p\rangle + \sqrt{\frac{2}{3}} \, |\pi^0 n\rangle \tag{9}$$

Let us consider a final state of the form

$$|f\rangle = \alpha|\frac{1}{2}, -\frac{1}{2}\rangle + \beta|\frac{3}{2}, -\frac{1}{2}\rangle \qquad |\alpha|^2 + |\beta|^2 = 1$$

$$|f\rangle = \left(\alpha\sqrt{\frac{2}{3}} + \beta\sqrt{\frac{1}{3}}\right) |\pi^- p\rangle + \left(\beta\sqrt{\frac{2}{3}} - \alpha\sqrt{\frac{1}{3}}\right) |\pi^0 n\rangle$$

A comparison with the observed branching ratio gives (assuming that the errors on the widths are independent)

$$\left|\frac{\alpha - \beta\sqrt{2}}{\sqrt{2}\alpha + \beta}\right|^2 \simeq \frac{35.8}{64.2} \left(1 \pm \sqrt{\left(\frac{0.5}{35.8}\right)^2 + \left(\frac{0.5}{64.2}\right)^2}\right) \tag{10}$$

To first order in β/α, we find

$$1 - 3\sqrt{2} \, \text{Re}\left[\frac{\beta}{\alpha}\right] \simeq 1.11 \pm 0.03 \tag{11}$$

that is,

$$\text{Re}\left[\frac{\beta}{\alpha}\right] \simeq -0.026 \pm 0.010 \tag{12}$$

c) For $\Delta T = 1/2$ the final state has definite isospin, and its projection on the two possible channels is a mere Clebsch-Gordan coefficient; therefore α is the same for both channels. If Λ is polarized $(+1/2)$, the angular probability amplitude is

$$\alpha_S \langle \theta \varphi | S, \frac{1}{2} \rangle + \alpha_P \langle \theta \varphi | P, \frac{1}{2} \rangle =$$
$$= \alpha_S Y_0^0(\hat{\mathbf{p}}) | \uparrow \rangle + \alpha_P \left[-\sqrt{\frac{1}{3}} Y_1^0(\hat{\mathbf{p}}) | \uparrow \rangle + \sqrt{\frac{2}{3}} Y_1^1(\hat{\mathbf{p}}) | \downarrow \rangle \right] \qquad (13)$$

where $| \uparrow \rangle$ and $| \downarrow \rangle$ are proton spin states, and $|\alpha_S|^2 + |\alpha_P|^2 = 1$.

The angular distribution, summed over final spins, is

$$
\begin{aligned}
\frac{4\pi}{\Gamma} \frac{d\Gamma}{d\Omega} &= \left| \alpha_S Y_0^0 - \alpha_P \sqrt{\frac{1}{3}} Y_1^0 \right|^2 + |\alpha_P|^2 \frac{2}{3} \left| Y_1^1 \right|^2 \\
&= |\alpha_S|^2 + |\alpha_P|^2 + 2\,\mathrm{Re}\,[\alpha_S^* \alpha_P] \cos\theta \\
&= 1 + 2\,\mathrm{Re}\,[\alpha_S^* \alpha_P] \cos\theta
\end{aligned}
\qquad (14)
$$

The known value of the coefficient α leads to

$$2\,\mathrm{Re}\,[\alpha_S^* \alpha_P] = 0.642 \pm 0.013 \qquad (15)$$

Neglecting $\pi - N$ scattering phases at the Λ mass, the coefficients α_S and α_P are relatively real, and we obtain

$$\alpha_{S,P} \sqrt{1 - \alpha_{S,P}^2} = 0.321 \pm 0.006 \qquad (16)$$

leading to $\alpha_P \simeq 0.34$ or $\alpha_S \simeq 0.34$.

An additional measurement (e.g., the proton polarization) is needed in order to discriminate between these two solutions.

Problem 65. Electron helicity in $K_{\ell 3}$

The decay

$$K^0 \to \pi^- + e^+ + \nu_e \qquad (1)$$

is described by the effective interaction

$$H = \frac{G}{\sqrt{2}} \left[J_\mu^h \, J_\mu^{e\,\dagger} + J_\mu^e \, J_\mu^{h\,\dagger} \right] \qquad (2)$$

where $J_\mu^e = \bar{\psi}_e \gamma_\mu (1 - \gamma_5) \psi_\nu$ is the weak leptonic current and $J_\mu^h = V_\mu - A_\mu$ is the hadronic current.

a) Show that the hadronic current contributes to decay (1) only through its vector part V_μ and not through the axial vector A_μ.

b)   Write down the most general form for the matrix element

$$\langle \pi^- | \, V_\mu(x) \, | K^0 \rangle \tag{3}$$

compatible with vector current conservation.

c)   Assuming constant form factors, derive the event distribution on the plane of the invariant masses

$$m^2_{\pi^- e}, \quad m^2_{\pi^- \nu} \tag{4}$$

The electron mass can be neglected.

d)   Calculate the positron helicity in the final state.

Solution

a)   The matrix element of a (vector or axial) current J_μ between K^0 and π^- is, by Lorentz invariance,

$$\langle \pi^- | \, J_\mu \, | K^0 \rangle = A P_\mu + B q_\mu \tag{5}$$

where A and B are invariants, and

$$P_\mu = (p_{K^0} + p_{\pi^-})_\mu \qquad q_\mu = (p_{K^0} - p_{\pi^-})_\mu \tag{6}$$

Under parity, we have

$$\langle P\pi^- | \, J_\mu \, | P K_0 \rangle = \langle \pi^- | \, P J_\mu P^{-1} \, | K^0 \rangle \tag{7}$$

The left hand side of (7), using the parameterization (5), becomes

$$(A P_\mu + B q_\mu) \, g_{\mu\mu} \tag{8}$$

since A and B are parity invariant, being functions of $(K \cdot \pi)$, and π and K have equal intrinsic parities. The right hand side equals $\pm (A P_\mu + B q_\mu) \, g_{\mu\mu}$ for a vector (axial) current. There follows that only the vector part contributes to the matrix element.

b)   Current conservation requires

$$q^\mu \, \langle \pi^- | \, J_\mu \, | K^0 \rangle = 0 \tag{9}$$

which implies

$$B = -\frac{A}{q^2} \left(m_K^2 - m_\pi^2 \right) \tag{10}$$

Thus the matrix element reads

$$\langle \pi^- | \, V^\mu \, | K^0 \rangle = A \left[P_\mu - \frac{(P \cdot q)}{q^2} q_\mu \right] \tag{11}$$

c)   The event density on the $m_{\pi\nu}^2 - m_{\pi e}^2$ plane (the Dalitz plot) is equal, up to a normalization, to the matrix element squared; this is so because the area differential on this plane is proportional to the phase space element. More precisely, we write

$$\frac{\mathrm{d}\Gamma}{\mathrm{d}m_{\pi\nu}^2\,\mathrm{d}m_{\pi e}^2} = \frac{1}{256\pi^3}\,\frac{1}{m_K^3}\,|\mathcal{M}|^2 \tag{12}$$

In the above, use was made of the expression for three-body phase space

$$\mathrm{d}\Phi^{(3)}(\pi,\nu,e) = \frac{\mathrm{d}m_{\pi\nu}^2\,\mathrm{d}m_{\pi e}^2}{16 m_K^2\,(2\pi)^3} \tag{13}$$

Eq.(11) now leads to

$$\mathcal{M} = \frac{G_F}{\sqrt{2}}\,A\left[P_\mu - \frac{(P\cdot q)}{q^2} q_\mu\right]\bar{u}(\nu)\gamma^\mu(1-\gamma_5)v(e^+) \tag{14}$$

with the result

$$\frac{\mathrm{d}\Gamma}{\mathrm{d}m_{\pi\nu}^2\,\mathrm{d}m_{\pi e}^2} = \frac{1}{32\pi^3}\frac{G_F^2}{m_K^3}\left[m_{\pi e}^2\,m_{\pi\nu}^2 - m_K^2\,m_\pi^2\right] \tag{15}$$

d)   The transition probability for a given kinematic configuration and a given positron polarization is proportional to

$$P_W \propto H^{\mu\nu} L_{\mu\nu} \tag{16}$$

with (cf. (11))

$$H^{\mu\nu} = \left[P^\mu - \frac{(P\cdot q)}{q^2}\,q^\mu\right]\left[P^\nu - \frac{(P\cdot q)}{q^2}\,q^\nu\right] \tag{17}$$

and

$$L_{\mu\nu} = \mathrm{Tr}\left\{\gamma_\mu(1-\gamma_5)\slashed{p}_{\nu_e}\,\gamma_\nu(1-\gamma_5)(\slashed{p}_e - m_e)\frac{(1+\gamma_5\slashed{W})}{2}\right\} \tag{18}$$

We now shift γ_5 across other γ matrices, and use $\gamma_5^2 = 1$ to find

$$L_{\mu\nu} = \mathrm{Tr}\left\{\gamma_\mu\slashed{p}_{\nu_e}\,\gamma_\nu(1-\gamma_5)(\slashed{p}_e + m_e\slashed{W})\right\} \tag{19}$$

The term proportional to γ_5 drops out in Eq.(16) because $H^{\mu\nu}$ is symmetric; in any case, such a term would only contribute to a polarization perpendicular to the decay plane, thus violating time reversal invariance (in the absence of a final state interaction). We obtain for P_W

$$P_W \propto 2\left[(P\cdot p_{\nu_e}) - \frac{(P\cdot q)(q\cdot p_{\nu_e})}{q^2}\right]\left[(P\cdot K) - \frac{(P\cdot q)}{q^2}(q\cdot K)\right] - $$
$$-(p_{\nu_e}\cdot K)\left(P^2 - \frac{(P\cdot q)^2}{q^2}\right) \tag{20}$$

where

$$K^\mu = p_e{}^\mu + m_e W^\mu \tag{21}$$

For final states with helicity λ, W_μ is given by

$$mW_\mu = \lambda(|\mathbf{p}|, \, p_0\,\hat{\mathbf{p}}) \tag{22}$$

$\mathbf{p}$ is the electron momentum, $\hat{\mathbf{p}}$ its direction and p_0 is the electron energy. K_μ correspondingly becomes

$$K_\mu = (p_0 + \lambda|\mathbf{p}|)\,(1, \, \lambda\hat{\mathbf{p}}) \tag{23}$$

Inserting Eq.(23) in (20) yields

$$P_W \propto (1 + \lambda\cos\theta)E_\nu$$

$$\cdot \left[4m_K^2(p_0 + \lambda|\mathbf{p}|) + m_e^2(p_0 - \lambda|\mathbf{p}|)\left(1 + \frac{P\cdot q}{q^2}\right)^2 - 4m_K\, m_e^2\left(1 + \frac{P\cdot q}{q^2}\right)\right] \tag{24}$$

θ is the angle formed by the positron and neutrino momenta. The positron helicity is

$$h = \frac{P_+ - P_-}{P_+ + P_-} \tag{25}$$

Setting $v = |\mathbf{p}|/p_0$ in (24) we find

$$h = 1 -$$

$$(1 - \cos\theta)\,\frac{4(1 - v) + \dfrac{m_e^2}{m_K^2}\,R^2(1 + v) - \dfrac{4m_e^2}{m_K p_0}\,R}{4(1 + v\cos\theta) + \dfrac{m_e^2}{m_K^2}\,R^2(1 - v\cos\theta) - \dfrac{4m_e^2}{m_K p_0}\,R} \tag{26}$$

Here, R equals

$$R = 1 + \frac{P\cdot q}{q^2} = 1 + \frac{m_K^2 - m_\pi^2}{m_e^2 - 2p_0 E_\nu + 2|\mathbf{p}|E_\nu\cos\theta}\,\cos\theta \tag{27}$$

Problem 66.   The decay $\tau^- \to \pi^- + \nu_\tau$

The τ lepton (spin $1/2$, $m = (1784.1 \pm 3)$ MeV) can decay into a pion plus a neutrino

$$\tau^- \to \pi^- + \nu_\tau \tag{1}$$

a)   Write down the transition amplitude resulting from the Fermi interaction.
b)   Knowing that π undergoes the decay

$$\pi^- \to \mu^- + \bar\nu_\mu \tag{2}$$

with a mean life of $2.6\cdot10^{-8}$ sec, calculate the decay width for process (1).
c)   Calculate the angular distribution of pions produced in the decay of polarized τ's.
   $\left[m_{\pi^-} = 139.568 \text{ MeV}, \; m_\mu = 105.658 \text{ MeV}, \; G_F = 1.027\cdot10^{-5}m_p^{-2}\right]$

Solution

a)   The Fermi interaction gives rise to the following amplitude

$$\mathcal{M}_\tau = \frac{G_F}{\sqrt{2}} \langle \nu_\tau | J_\mu^{lept}(0) | \tau^- \rangle \langle \pi^- | J_\mu^{h\dagger}(0) | 0 \rangle \tag{3}$$

The leptonic matrix element equals

$$\langle \nu_\tau | J_\mu^{lept}(0) | \tau^- \rangle = \bar{u}(\mathbf{p}_{\nu_\tau}) \gamma_\mu (1 - \gamma_5) u(\mathbf{p}_{\tau^-}) \tag{4}$$

By Lorentz invariance, the hadronic matrix element is given by

$$\langle \pi^- | J_\mu^{h\dagger}(0) | 0 \rangle = F_\pi (p_{\pi^-})_\mu \tag{5}$$

F_π is a constant with the dimensions of a mass. Four-momentum conservation

$$p_\tau^\mu = p_{\pi^-}^\mu + p_\nu^\mu \tag{6}$$

and use of the Dirac equation allow us to rewrite (3) in the form

$$\mathcal{M}_\tau = \frac{G_F}{\sqrt{2}} F_\pi\, m_\tau\, \bar{u}(\mathbf{p}_{\nu_\tau})(1 + \gamma_5) u(\mathbf{p}_{\tau_-}) \tag{7}$$

b)   The amplitude for π decay can be written in a similar way

$$\mathcal{M}_\pi = \frac{G_F}{\sqrt{2}} \langle \bar{\nu}_\mu \mu^- | J_\mu^{lept\dagger}(0) | 0 \rangle \langle 0 | J_\mu^h(0) | \pi^- \rangle \tag{8}$$

Using Eq.(5) we obtain

$$\mathcal{M}_\pi = \frac{G_F}{\sqrt{2}} F_\pi\, m_\mu\, \bar{u}(\mathbf{p}_{\mu^-})(1 - \gamma_5) v(\mathbf{p}_{\bar{\nu}_\mu}) \tag{9}$$

The total decay width of π^- is

$$\Gamma_\pi = \frac{1}{2m_\pi} \int \overline{|\mathcal{M}_\pi|^2}\, d\Phi^{(2)} \tag{10}$$

Writing

$$d\Phi^{(2)} = \frac{d\Omega}{32\pi^2} \left(1 - \frac{m_\mu^2}{m_\pi^2} \right) \tag{11}$$

and

$$\begin{aligned}
\overline{|\mathcal{M}_\pi|^2} &= \frac{G_F^2}{2} |F_\pi|^2\, m_\mu^2\, \mathrm{Tr}\left\{ (\slashed{p}_{\mu^-} + m_\mu)(1 - \gamma_5)\slashed{p}_{\bar{\nu}_\mu}(1 + \gamma_5) \right\} \\
&= 2 G_F^2 |F_\pi^2|\, m_\mu^2 \left(m_\pi^2 - m_\mu^2 \right)
\end{aligned} \tag{12}$$

we find

$$\Gamma_\pi = G_F^2 \frac{|F_\pi|^2}{8\pi} m_\mu^2 m_\pi \left(1 - \frac{m_\mu^2}{m_\pi^2}\right)^2 \tag{13}$$

We note that F_π is related to the usual pion decay constant, f_π, as follows

$$F_\pi = \sqrt{2} f_\pi \cos\theta \tag{14}$$

where θ is the Cabibbo angle and $f_\pi = 93$ MeV.

The amplitude of Eq.(7) leads to

$$\Gamma_{\tau\to\pi\nu} = \frac{G_F^2}{16\pi} |F_\pi|^2 m_\tau^3 \left(1 - \frac{m_\pi^2}{m_\tau^2}\right)^2 \tag{15}$$

An additional factor of $1/2$ as compared to Eq.(13) is present here, coming from the average over the τ spin. The ratio of the two widths equals

$$\frac{\Gamma_{\tau\to\pi\nu}}{\Gamma_{\pi\to\mu\nu}} = \frac{1}{2} \frac{m_\tau^3}{m_\mu^2 m_\pi} \frac{\left(1 - \frac{m_\pi^2}{m_\tau^2}\right)^2}{\left(1 - \frac{m_\mu^2}{m_\pi^2}\right)^2} = 0.98 \cdot 10^4 \tag{16}$$

From the known value of Γ_π, $\Gamma_\pi = 2.53 \cdot 10^{-8}$ eV, we conclude

$$\Gamma_{\tau\to\pi\nu} = 2.5 \cdot 10^{-4}\, \text{eV} \tag{17}$$

The experimental value is[10].

$$\Gamma_{\tau\to\pi\nu} = (2.5 \pm 0.17) \cdot 10^{-4}\, \text{eV} \tag{18}$$

c)   The decay probability of polarized τ's is proportional to

$$|\mathcal{M}|^2 \propto \text{Tr} \left\{ \not{p}_{\nu_\tau} (1 + \gamma_5)(\not{p}_\tau + m_\tau)(1 + \gamma_5 \not{W})(1 - \gamma_5) \right\} \tag{19}$$

Carrying out the traces

$$|\mathcal{M}|^2 \propto (p_\tau \cdot p_{\nu_\tau}) + m_\tau (W \cdot p_{\nu_\tau}) \tag{20}$$

We see that the angular distribution of the pion, relative to the direction of τ polarization, has the form

$$\frac{d\Gamma}{d\Omega} = \frac{\Gamma}{4\pi} (1 + \cos\theta) \tag{21}$$

This distribution vanishes at $\theta = \pi$, that is, when the neutrino direction coincides with the polarization of τ; this is a consequence of the fact that the neutrino helicity is -1. In fact, knowing that the angular distribution is of the form $1 + \alpha\cos\theta$, the helicity of ν fixes α to be -1.

---

[10]Particle Data Group, *Phys. Lett.* **B239** 1990

Problem 67. Lepton-antilepton decays of Z_0

The neutral vector boson Z_0 couples to leptons through the interaction Lagrangian

$$\mathcal{L}_I = e\, Z_\mu \left(aV^\mu + bA^\mu\right) \tag{1}$$

with

$$V_\mu = \bar{\psi}\gamma_\mu\psi \qquad A_\mu = \bar{\psi}\gamma_\mu\gamma_5\psi \tag{2}$$

$$a = \left(\sin^2\theta_W - \frac{1}{4}\right)\frac{1}{\sin\theta_W\cos\theta_W} \qquad b = \frac{1}{4\sin\theta_W\cos\theta_W} \qquad \sin^2\theta_W = 0.2325 \pm 0.0008$$

The mass of Z_0 is 91 GeV; e is the electron charge.

a) Calculate the partial width in MeV of the following decays

$$Z_0 \to e^+ e^- \qquad Z_0 \to \mu^+\mu^- \qquad Z_0 \to \tau^+\tau^- \tag{3}$$

b) Calculate the angular distribution of leptons in decays of polarized Z_0's.
c) Discuss the transformation properties of (1) under P, C, CP, and the consequences of these properties on observable quantities.
d) Discuss Z decays into pion pairs, $Z \to \pi\pi$. Which decay modes of this type are possible? How can they be parameterized?

Solution

The width for Z_0 decay into an electron-positron pair can be written in the form

$$\Gamma\left(Z_0 \to e^+ e^-\right) = \frac{1}{2M_Z}\sqrt{1 - \frac{4m^2}{M_Z^2}}\int \overline{|\mathcal{M}|^2}\,\frac{\mathrm{d}\Omega}{32\pi^2} \tag{4}$$

By rotational invariance, the above does not depend on the spin state of Z_0. In view of question b), let us calculate the width for Z_0 completely polarized along the z axis. The amplitude reads

$$\mathcal{M} = e\,\varepsilon_\mu\,\bar{u}(\mathbf{p}_\ell)\left(a\,\gamma^\mu + b\,\gamma^\mu\gamma^5\right)v(\mathbf{p}_{\bar{\ell}}) \tag{5}$$

Summing over final lepton polarizations

$$\overline{|\mathcal{M}|^2} = e^2\,\varepsilon_\mu\varepsilon_\nu^*\,\mathrm{Tr}\left\{(\not{p}_\ell + m)(a\,\gamma^\mu + b\,\gamma^\mu\gamma^5)(\not{p}_{\bar{\ell}} - m)(a^*\gamma^\nu + b^*\gamma^\nu\gamma^5)\right\} \tag{6}$$

Polarization along the z axis implies $\varepsilon_\mu = (0,\ 1/\sqrt{2},\ \mathrm{i}/\sqrt{2},\ 0)$ and

$$\varepsilon_\mu\varepsilon_\nu^* = \frac{1}{2}\begin{pmatrix} 0 & 0 & 0 & 0 \\ 0 & 1 & -\mathrm{i} & 0 \\ 0 & \mathrm{i} & 1 & 0 \\ 0 & 0 & 0 & 0 \end{pmatrix} \tag{7}$$

Let us denote by $L_{\mu\nu}$ the trace in Eq.(6); in the center of mass frame we have

$$
\begin{aligned}
\frac{1}{4}L^{\mu\nu} &= g^{\mu\nu}\left(2m^2|b|^2 - \frac{M^2}{2}(|a|^2 + |b|^2)\right) + \frac{1}{2}q^\mu q^\nu(|a|^2 + |b|^2) \\
&\quad -2P^\mu P^\nu(|a|^2 + |b|^2) + 2i\,\mathrm{Re}(a^*b)\,M_Z\,\varepsilon^{0\mu\sigma\nu}P_\sigma
\end{aligned}
\tag{8}
$$

with $P^\mu = (p_\ell - p_{\bar\ell})^\mu/2$. In terms of the angle θ, formed by the lepton momentum $\mathbf{p}_\ell$ and the polarization axis, we find

$$
\begin{aligned}
\overline{|\mathcal{M}|^2} &= 2e^2 M_Z^2 \left\{ (|a|^2 + |b|^2)\left[1 - \left(1 - \frac{4m^2}{M_Z^2}\right)\frac{\sin^2\theta}{2}\right] \right. \\
&\quad \left. -\frac{4m^2}{M_Z^2}|b|^2 + \sqrt{1 - \frac{4m^2}{M_Z^2}}\,2\mathrm{Re}\,a^*b\,\cos\theta \right\}
\end{aligned}
\tag{9}
$$

Using Eq.(4) we obtain for the total width

$$
\Gamma = \frac{\alpha M_Z}{2}\sqrt{1 - \frac{4m^2}{M_Z^2}}\left\{(|a|^2 + |b|^2)\left[1 - \frac{1}{3}\left(1 - \frac{4m^2}{M_Z^2}\right)\right] - \frac{4m^2}{M_Z^2}|b|^2\right\}
\tag{10}
$$

a)   We substitute numerical values and neglect $4m^2/M_Z^2 \ll 1$ in all cases; the result is $\Gamma/M_Z = 0.88\cdot10^{-3}$ for all three decays, corresponding to $\Gamma = 79$ MeV.

b)   The angular distribution is

$$
\frac{1}{\Gamma}\frac{d\Gamma}{d\Omega} = \frac{1}{4\pi}\frac{1 - \left(1 - \frac{4m^2}{M_Z^2}\right)\sin^2\frac{\theta}{2} - \frac{4m^2}{M_Z^2}\frac{|b|^2}{|a|^2 + |b|^2} + \frac{2\mathbf{Re}(a^*b)}{|a|^2 + |b|^2}\sqrt{1 - \frac{4m^2}{M_Z^2}}\cos\theta}{1 - \frac{1}{3}\left(1 - \frac{4m^2}{M_Z^2}\right) - \frac{4m^2}{M_Z^2}\frac{|b|^2}{|a|^2 + |b|^2}}
\tag{11}
$$

c)   The effective Lagrangian (1) is not invariant under P and C separately; indeed, the vector and axial current have opposite behaviour under these transformations, so that no single eigenvalue assigned to Z_0 can render the Lagrangian invariant. The product CP will be conserved provided the coefficients a and b are relatively real.

d)   In the decay $Z \to \pi\pi$ pions are produced in a P-wave. Bose statistics rules out the mode $\pi^0\pi^0$; only the mode $\pi^+\pi^-$ is allowed, in a state with $C = -1$, $P = -1$. The most general parameterization for the decay amplitude is

$$
M_{Z\to\pi\pi} = C\,\varepsilon_\mu\,P^\mu
\tag{12}
$$

where $P^\mu = (p_{\pi^+} - p_{\pi^-})^\mu/2$ and C is a constant.

Problem 68.   Verifications of the CVC hypothesis

We would like to test the conserved vector current (CVC) hypothesis, by comparing the following decays

$$\pi^+ \rightarrow \quad \pi^0 + e^+ + \nu_e \qquad Q = 4.09 \text{ MeV} \qquad (1)$$

$$O^{14} \rightarrow \quad N^{14*} + e^+ + \nu_e \qquad Q = 1.81 \text{ MeV} \qquad (2)$$

$$Cl^{34} \rightarrow \quad S^{34*} + e^+ + \nu_e \qquad Q = 4.46 \text{ MeV} \qquad (3)$$

$$K^0 \rightarrow \quad K^+ + e^- + \bar{\nu}_e \qquad Q = 3.70 \text{ MeV} \qquad (4)$$

All of these decays correspond to transitions between spinless hadrons of the same parity, within the same isospin multiplet ($T = 1$ for the first three transitions and $T = 1/2$ for the last one). The energy released by these decays in the form of kinetic energy (Q) is much smaller than the hadronic scale.

a)   Show that only the vector part of the hadronic current contributes to the above processes.

b)   The hadronic matrix element may be taken proportional to the isospin charge. What is the approximation involved?

c)   Compare the electron spectrum in these processes; calculate the ratios between decay widths.

Solution

a)   Let us construct the most general matrix element of a current between two spinless hadronic states, $|i\rangle$ and $|f\rangle$. There are two vectors at our disposal

$$
\begin{aligned}
P^\mu &= p_i^\mu + p_f^\mu \\
q^\mu &= p_f^\mu - p_i^\mu
\end{aligned}
\qquad (5)
$$

and one Lorentz scalar variable, q^2, on which form factors may depend. The matrix element will read

$$\langle f, \mathbf{p}_f | J_\mu | i, \mathbf{p}_i \rangle = f_1(q^2) P_\mu + f_2(q^2) q_\mu \qquad (6)$$

Among the quantum numbers describing hadronic states, the momentum has been explicitly indicated, because of its nontrivial properties under reflection.

In all cases considered, initial and final states have the same parity; thus, under parity we have

$$
\begin{aligned}
\langle f, \mathbf{p}_f | V_\mu | i, \mathbf{p}_i \rangle &= \langle f, -\mathbf{p}_f | \tilde{V}_\mu | i, -\mathbf{p}_i \rangle = \\
&= \left(f_1^{(V)}(q^2) P_\mu + f_2^{(V)}(q^2) q_\mu \right) \\
\langle f, \mathbf{p}_f | A_\mu | i, \mathbf{p}_i \rangle &= \langle f, -\mathbf{p}_f | - \tilde{A}_\mu | i, -\mathbf{p}_i \rangle = \qquad (7) \\
&= - \left(f_1^{(A)}(q^2) P_\mu + f_2^{(A)}(q^2) q_\mu \right) \qquad (8)
\end{aligned}
$$

We have used the notation

$$\tilde{J}_\mu = (J_0, -\mathbf{J}) \tag{9}$$

Comparing Eqs. (8) and (7) to (6), we conclude that only the vector current can induce transitions without parity flip.

b)   For a current transforming as an isospin triplet, the Wigner-Eckart theorem implies

$$\langle A' | V_\mu^\pm | A \rangle = \langle A' | j_\mu | A \rangle \sqrt{(T + 1 \pm T_3)\,(T \mp T_3)} \tag{10}$$

$\langle A' | j_\mu | A \rangle$ represents a reduced matrix element, independent of T_3. Clearly, for Eq.(10) to be valid, $|A\rangle$ and $|A'\rangle$ must have a definite isospin.

Identifying the weak interaction vector current with the conserved isospin Noether current leads to

$$q^\mu \langle A' | V_\mu | A \rangle = 0 \tag{11}$$

There follows, from Eq.(6), that

$$f_2 = 0 \tag{11'}$$

Furthermore, the normalization of the current is fixed from the requirement that the corresponding charge be equal to isospin

$$\langle A' | \int \mathrm{d}^3 x V_0^\pm(\mathbf{x}) | A \rangle = \sqrt{(T + 1 \pm T_3)\,(T \mp T_3)}\, 2E_A (2\pi)^3 \delta^{(3)}(\mathbf{p}_A - \mathbf{p}'_{A'}) =$$

$$= \int d^3 x e^{-i\mathbf{q}\mathbf{x}} f_1^\pm(q^2) P_0 (2\pi)^3 \delta^{(3)}(\mathbf{q}) = f_1^\pm(0) 2E_A (2\pi)^3 \delta^{(3)}(\mathbf{p}_A - \mathbf{p}'_{A'}) \tag{12}$$

that is,

$$f_1^\pm(0) = \sqrt{(T + 1 \pm T_3)\,(T \mp T_3)} \tag{13}$$

The states connected by the conserved charge in Eq.(12) have the same energy, since the charge commutes with the Hamiltonian.

c)   The transition amplitude for the first three processes has the form

$$\mathcal{M} = \frac{G}{\sqrt{2}} J_\mu^{(h)} \bar{u}(\mathbf{p}_\nu) \gamma^\mu (1 - \gamma^5) v(\mathbf{p}_e) \tag{14}$$

and, for the last process,

$$\mathcal{M} = \frac{G}{\sqrt{2}} J_\mu^{(h)} \bar{u}(\mathbf{p}_e) \gamma^\mu (1 - \gamma^5) v(\mathbf{p}_\nu) \tag{15}$$

The energy released in all cases is small compared to the hadronic mass scale. We may thus neglect the q^2-dependence of f_1; Eqs. (6), (11') and (13) allow us to set

$$J_\mu^{(h)} = C_{T,T_3} P_\mu \tag{16}$$

with

$$C_{T,T_3} = \sqrt{(T + 1 \pm T_3)\,(T \mp T_3)} \tag{17}$$

The transition probability is given by

$$d\Gamma = \sum_{\text{pol}} \frac{1}{2M_A} |\mathcal{M}|^2 \, d\Phi^{(3)} \tag{18}$$

Substituting for $\mathcal{M}$ we obtain

$$
\begin{aligned}
d\Gamma &= C_{T,T_3}^2 \frac{1}{2M_A} 8G^2 \left[2(P \cdot p_\nu)(P \cdot p_e) - P^2(p_e \cdot p_\nu) \right] d\Phi^{(3)} = \\
&= C_{T,T_3}^2 16G^2 M_A \left[\varepsilon_e \varepsilon_\nu + \mathbf{p}_e \mathbf{p}_\nu \right] d\Phi^{(3)}
\end{aligned} \tag{19}
$$

In the limit of nonrelativistic hadrons, the phase space element becomes

$$d\Phi^{(3)} = \frac{1}{(2\pi)^3} \frac{1}{2M_A} p_e \, d\varepsilon_e \, \varepsilon_\nu \, d\varepsilon_\nu \, \delta(Q + m_e - \varepsilon_e - \varepsilon_\nu) \tag{20}$$

with the result

$$d\Gamma = C_{T,T_3}^2 \frac{G^2}{\pi^3} \varepsilon_e \sqrt{\varepsilon_e^2 - m_e^2} \, (m_e + Q - \varepsilon_e)^2 \, d\varepsilon_e \tag{21}$$

Eq.(21) gives the electron spectrum in each of the 4 processes. Integrating on ε_e gives, after change of variables to $y = \varepsilon_e/Q$

$$\Gamma = C_{T,T_3}^2 \frac{G^2}{\pi^3} Q^5 \, \Phi\left(\frac{m_e}{Q}\right)$$

$$\Phi\left(\frac{m_e}{Q}\right) = \int_{\frac{m_e}{Q}}^1 dy \left(1 - y + \frac{m_e}{Q}\right)^2 y \sqrt{y^2 - \left(\frac{m_e}{Q}\right)^2}$$

For small values of m_e/Q

$$\Phi\left(\frac{m_e}{Q}\right) \simeq \frac{1}{30}\left[1 + \mathcal{O}\left(\frac{m_e}{Q}\right)^2\right]$$

A confirmation of the CVC hypothesis now consists in verifying that, for all processes considered,

$$\frac{\Gamma}{\frac{G^2}{\pi^3} Q^5 \Phi\left(\frac{m_e}{Q}\right)} = C_{T,T_3}^2 \tag{22}$$

Problem 69.   Some consequences of Vector Meson Dominance

A good phenomenological description of the photon-hadron coupling consists in assuming that it is always mediated by vector mesons (Vector Meson Dominance, VMD). The Lagrangian which couples photons to vector mesons is taken to be

$$\mathcal{L}_{\text{VMD}} = \frac{e\, m_\rho^2}{g} \left(\rho_\mu^0 + \frac{1}{3}\, \omega_\mu \right) A^\mu \tag{1}$$

with

$$M_\rho = (768.1 \pm 0.5)\ \text{MeV} \qquad M_\omega = (781.95 \pm 0.14)\ \text{MeV} \tag{2}$$

g is the strong coupling constant appearing in the Hamiltonian describing the $\rho \to \pi\pi$ decay

$$\mathcal{L}_{\rho\pi\pi} = g\rho \cdot (\pi \wedge \partial_\mu \pi) \tag{3}$$

(The vector product refers to isospin). Finally, the effective interaction which couples ρ, π and ω is

$$\mathcal{L}_{\omega\rho\pi} = g_{\omega\rho\pi}\, \varepsilon^{\mu\nu\rho\sigma} \left(\pi^a\, \partial_\mu \rho_\nu^a \right) \partial_\rho \omega_\sigma \tag{4}$$

a)   Knowing the width $\Gamma(\rho \to \pi\pi) = (151.5 \pm 1.2)$ MeV, determine the absolute value of g.

b)   Compare VDM with the Lagrangian which describes the decay $\pi^0 \to \gamma\gamma$

$$
\begin{aligned}
\mathcal{L}_{\pi^0\gamma\gamma} &= \frac{1}{f_\pi} \frac{e^2}{32\pi^2}\, \varepsilon_{\mu\nu\rho\sigma}\, F^{\mu\nu} F^{\rho\sigma}\, \pi^0 \\
f_\pi &= 93\ \text{MeV} \\
m_{\pi^0} &= (134.9743 \pm 0.0008)\ \text{MeV}
\end{aligned}
\tag{5}
$$

Determine $g_{\omega\rho\pi}$ in terms of g and f_π.

c)   The process $\omega \to 3\pi$ is dominated by $\omega \to \rho\pi$, followed by $\rho \to \pi\pi$. Estimate $\Gamma(\omega \to 3\pi)$ as a function of g, $g_{\omega\rho\pi}$ and of the masses; estimate the branching ratio

$$\frac{\Gamma(\omega \to \gamma\pi^0)}{\Gamma(\omega \to 3\pi)} \tag{6}$$

N.B. In answering part c), it is legitimate to treat phase space nonrelativistically.

Solution

a)   The coupling of Eq.(3) is isospin invariant. All ρ's have therefore equal widths, and it suffices to calculate one of them, for example the width of ρ^0. The definition of the charged pion fields

$$\pi^\pm = \frac{\pi^1 \pm i\, \pi^2}{\sqrt{2}} \tag{7}$$

leads to the following interaction term for ρ^0 decay

$$\mathcal{L}_{\rho^0 \to \pi\pi} = i\,g\,\rho_\mu^0\,(\pi^+ \partial_\mu\,\pi^- - \pi^- \partial_\mu\,\pi^+) \tag{8}$$

The corresponding matrix element is

$$M = g\,\varepsilon_\mu\,(p_+ - p_-)^\mu \tag{9}$$

We may now calculate the total width, with the result

$$
\begin{aligned}
d\Gamma(\rho^0 \to \pi^+\pi^-) &= \frac{1}{2m_\rho}|\mathcal{M}|^2\,d\Phi^{(2)} \\
&= -\frac{1}{2m_\rho}\frac{g^2}{3}\left(g_{\mu\nu} - \frac{k_\mu k_\nu}{M_\rho^2}\right)(p_+ - p_-)^\mu (p_+ - p_-)^\nu\,d\Phi^{(2)} \quad (10)
\end{aligned}
$$

$$\Gamma = \frac{g^2}{48\pi}M_\rho\left(1 - \frac{4m_\pi^2}{M_\rho^2}\right)^{\frac{3}{2}} \tag{11}$$

k is the four-momentum of ρ. Substituting numerical values we find $g^2/4\pi \simeq 3$ (strong coupling).

b) VMD asserts that the decay of π^0 proceeds through Eq.(4) and the diagram of fig.1.

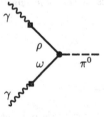

fig. 1

The corresponding amplitude is

$$2g_{\omega\rho\pi}\,\varepsilon_{\mu\nu\rho\sigma}\,k_1^\mu\,\varepsilon_1^\nu\,k_2^\rho\,\varepsilon_2^\sigma\,\frac{e^2}{3g^2} \tag{12}$$

A factor of 2 arises from possibility of exchanging the two photons. Comparing this amplitude to the one gotten from $\mathcal{L}_{\pi^0\gamma\gamma}$

$$\frac{1}{f_\pi}\frac{e^2}{4\pi^2}\,\varepsilon_{\mu\nu\rho\sigma}\,k_1^\mu\,\varepsilon_1^\nu\,k_2^\rho\,\varepsilon_2^\sigma \tag{13}$$

we obtain

$$\frac{2}{3g^2}\,g_{\omega\rho\pi}\,e^2 = \frac{1}{f_\pi}\frac{e^2}{4\pi^2} \tag{14}$$

and hence,

$$g_{\omega\rho\pi} = \frac{3g^2}{8\pi^2 f_\pi} \tag{15}$$

c)   Let us calculate $\Gamma(\omega \to \pi^0\gamma)$, as shown in fig. 2. The decay amplitude is

$$\mathcal{M} = \frac{e}{g}\, g_{\omega\rho\pi}\, \varepsilon_{\mu\nu\rho\sigma}\, k^{\mu} \varepsilon^{*\nu}_{(\gamma)}\, p^{\rho}_{\omega}\, \varepsilon^{\sigma}_{(\omega)} \qquad (16)$$

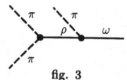

fig. 2

leading to

$$|\mathcal{M}|^2 = \left(g_{\omega\rho\pi}\,\frac{e}{g}\right)^2 \frac{1}{6}\,(M^2_\omega - m^2_\pi) \qquad (17)$$

and

$$\Gamma(\omega \to \gamma\pi^0) = \frac{\alpha}{24}\, M^3_\omega \left(\frac{g_{\omega\rho\pi}}{g}\right)^2 \left(1 - \frac{m^2_\pi}{M^2_\omega}\right)^3 \qquad (18)$$

As for the process $\omega \to 3\pi$, its width receives contributions from the diagram of fig. 3, summed over all possible charge assignments of π and ρ. The resulting amplitude is

$$\mathcal{M} = g_{\omega\rho\pi}\, \varepsilon_{\mu\nu\rho\sigma}\, p^{\mu}_0\, p^{\nu}_+\, p^{\rho}_-\, \varepsilon^{\sigma}_{(\omega)} \sum_{i\neq j} \frac{1}{(p_i + p_j)^2 - M^2_\rho} \qquad (19)$$

The indices i and j run over the three possible values of the pion charge. In the nonrelativistic approximation we may set

$$(p_i + p_j)^2 \simeq 4m^2_\pi \qquad (20)$$

whence

$$\mathcal{M} = 6g\, \frac{g_{\omega\rho\pi}}{M^2_\rho - 4m^2_\pi}\, \varepsilon_{\mu\nu\rho\sigma}\, p^{\mu}_0\, p^{\nu}_+\, p^{\rho}_-\, \varepsilon^{\sigma}_{(\omega)} \qquad (21)$$

In the center of mass frame, the modulus squared of the amplitude reduces to

$$|\mathcal{M}|^2 = \left(\frac{6g\, g_{\omega\rho\pi}}{M^2_\rho - 4m^2_\pi}\right)^2 \frac{M^2_\omega}{3} \left[(\mathbf{p}_+)^2(\mathbf{p}_-)^2 - (\mathbf{p}_+\mathbf{p}_-)^2\right] \qquad (22)$$

A further simplification in the nonrelativistic limit

$$\frac{M_\omega - 3m_\pi}{3} \ll m_\pi \qquad (23)$$

yields

$$\Gamma = \left(\frac{6g\, g_{\omega\rho\pi}}{M^2_\rho - 4m^2_\pi}\right)^2 \frac{M_\omega}{6}\, \frac{(M_\omega - 3m_\pi)^4 m^2_\pi}{3^{\frac{3}{2}}\, 24\,(4\pi)^2} \qquad (24)$$

In the above we made use of the integral

$$\int dE_1 \, dE_2 \left\{ 4E_1 E_2 - [2(E_1 + E_2) - \Delta]^2 \right\} = \frac{\pi \Delta^4}{3^{\frac{3}{2}} \, 12} \tag{25}$$

An exact treatment of the phase space integration[11] introduces a correction factor of ~ 3.5.

Finally, the ratio between the two decay widths assumes the form

$$\frac{\Gamma(\omega \to \pi^0 \gamma)}{\Gamma(\omega \to 3\pi)} = \frac{3^{\frac{3}{2}} \alpha (4\pi)^2}{6 g^4} \frac{\left(1 - \frac{m_\pi^2}{M_\omega^2}\right)^3 \left(1 - \frac{4m_\pi^2}{M_\rho^2}\right)^2}{\left(1 - \frac{3m_\pi}{M_\omega}\right)^4} \frac{M_\rho^4}{m_\pi^2 M_\omega^2} \tag{26}$$

Problem 70. Possible β-decays of $\rho^{\pm}$ and $D^{*\pm}$

a)   Calculate the partial width and branching ratio of the decay

$$\rho^+ \to \rho^0 + e^+ + \nu_e \tag{1}$$

$m_\rho = (768.1 \pm 0.5)$ MeV, $\Gamma_\rho = (151.5 \pm 1.)$ MeV, $m_{\rho^+} - m_{\rho^0} = (0.3 \pm 2)$ MeV.

Assume at first that the ρ's are infinitely narrow and $m_{\rho^+} - m_{\rho^0} = 2$ MeV. Subsequently, estimate the corrections due to the finite widths.

b)   Calculate the width and branching ratio of the decays

$$\rho^+ \to e^+ + \nu_e \tag{2}$$
$$\rho^+ \to \mu^+ + \nu_\mu \tag{3}$$

knowing that $\mathrm{BR}(\rho_0 \to e^+ e^-) = (4.6 \pm 0.2) \, 10^{-5}$.

The branching ratios for processes (2) and (3) are almost equal, while for $\pi^+ \to e^+ + \nu_e$ and $\pi^+ \to \mu^+ + \nu_\mu$ they are quite different

$$\frac{\Gamma(\pi^+ \to e^+ + \nu_e)}{\Gamma(\pi^+ \to \mu^+ + \nu_\mu)} \simeq 1.2 \cdot 10^{-4} \tag{4}$$

Explain.

c)   The particles D_+^* and D_0^* have spin 1 and quark content $(c\bar{d})$ and $(c\bar{u})$ respectively.

$$m_{D_+^*} = 2010.1 \pm 0.6 \text{ MeV}; \quad \Gamma_{D_+^*} < 1.1 \text{ MeV} \tag{5}$$

$$m_{D_0^*} = 2007.1 \pm 1.4 \text{ MeV}; \quad \Gamma_{D_0^*} < 2.1 \text{ MeV} \tag{6}$$

Derive the partial width for the decay

$$D_+^* \to D_0^* + e^+ + \nu_e \tag{7}$$

---

[11]M.Gell-Mann, D.Sharp, W.Wagner, *Phys. Rev. Lett.* **8**,261, (1962)

d) Calculate the ratios

$$\frac{\Gamma(D_+^* \to e^+ + \nu_e)}{\Gamma(D_+^* \to \mu^+ + \nu_\mu)} \tag{8}$$

and

$$\frac{\Gamma(D_+ \to e^+ + \nu_e)}{\Gamma(D_+ \to \mu^+ + \nu_\mu)} \tag{9}$$

D_+ is the pseudoscalar partner of D_+^*, in a manner analogous to π^+ and ρ^+.

$$m_{D+} = 1869.3 \pm 0.5 \, \text{MeV} \tag{10}$$

Solution

a) To an excellent approximation, the decay amplitude is given by the Fermi Lagrangian

$$M = \frac{G_F}{\sqrt{2}} (V - A)_\mu \, \bar{u}(\mathbf{p}_{\nu_e}) \gamma_\mu (1 - \gamma^5) v(\mathbf{p}_{e^+}) \tag{11}$$

where

$$(V - A)_\mu = \langle \rho^0 | \, J_\mu \, | \rho^+ \rangle \tag{12}$$

By the CVC hypothesis, the vector part of the current is proportional to the isovector component of the electromagnetic current

$$\langle \rho^0 | \, V_\mu \, | \rho^+ \rangle = \sqrt{2} \, \langle \rho^+ | \, j_\mu^{em} \, | \rho^+ \rangle \tag{13}$$

Consequently,

$$\langle \rho^0 | \, V_\mu \, | \rho^+ \rangle = -\sqrt{2} \, (\varepsilon_0^* \varepsilon_+) \, (p_0 + p_+)_\mu \tag{14}$$

Given the smallness of the mass difference $\rho^+ - \rho^0$, it is reasonable to treat the matrix elements of the hadronic currents in the nonrelativistic limit. Then, by parity considerations, the only contributions to these matrix elements come from the time component of the vector current

$$\langle \rho^0 | \, V_0 \, | \rho^+ \rangle = \sqrt{2} \, (\varepsilon_0^* \varepsilon_+) \, 2m_\rho \tag{15}$$

and the spatial part of the axial current, which can be defined as follows

$$\langle \rho^0 | \, \mathbf{A} \, | \rho^+ \rangle = \beta \sqrt{2} \, (\varepsilon_0^* \wedge \varepsilon_+) \, 2m_\rho \tag{16}$$

β is the ratio between the axial and vector form factors at zero momentum transfer. We are thus led to the amplitude

$$\mathcal{M} = 2G_F \, m_\rho \left[\varepsilon_0^* \varepsilon_+ \, j_e^0 + \beta \, (\varepsilon_0^* \wedge \varepsilon_+) \, \mathbf{j}_e \right] \tag{17}$$

with

$$j_e^\mu = \bar{u}(\mathbf{p}_{\nu_e})\gamma^\mu(1 - \gamma^5)v(\mathbf{p}_{e^+}) \tag{18}$$

The modulus squared of the amplitude, summed over final polarizations and averaged over initial ones, is

$$|\mathcal{M}|^2 = \frac{4}{3}G_F^2 m_\rho^2 \left[3L_{00} + 2|\beta|^2 \sum_{i=1}^3 L_{ii}\right] \tag{19}$$

where

$$L_{\mu\nu} = \text{Tr}\left[\gamma_\mu(1 - \gamma_5)(\not{p}_{e^+} - m_e)\gamma_\nu(1 - \gamma_5)\not{p}_{\nu_e}\right] \tag{20}$$

In particular, we have

$$L_{00} = 8\left[E_{e^+}E_\nu + \mathbf{p}_{e^+}\mathbf{p}_{\nu_e}\right]$$
$$\sum_{i=1}^3 L_{ii} = 8\left[3E_{e^+}E_\nu - \mathbf{p}_{e^+}\mathbf{p}_{\nu_e}\right]$$

Eq.(19) now reads

$$|\mathcal{M}|^2 = \frac{32}{3}G_F^2 m_\rho^2 \left[3E_{e^+}E_\nu(1 + 2|\beta|^2) + \mathbf{p}_{e^+}\mathbf{p}_{\nu_e}(3 - 2|\beta|^2)\right] \tag{21}$$

This expression can now be integrated over three body phase space; in the nonrelativistic limit (for ρ) the latter becomes

$$d\Phi^{(3)} = \frac{1}{16\pi^3 m_\rho}p_e E_\nu \, dE_{e^+} \, dE_\nu \, \delta(\Delta - E_{e^+} - E_\nu) \tag{22}$$

$$m_e \leq E_{e^+} \leq \Delta = m_{\rho^+} - m_{\rho^0}$$

The result, calculated in the limit $m_e \to 0$, is

$$\Gamma_0 = \frac{G_F^2 \Delta^5}{30\pi^3}(1 + 2|\beta|^2) = 7 \cdot 10^{-24} \text{ MeV } (1 + 2|\beta|^2) \tag{23}$$

To take into account the dependence on the electron mass, one may write

$$\Gamma = c(\frac{m_e}{\Delta})\Gamma_0 \tag{24}$$

where

$$c(x) = (1 - x^2)^{\frac{1}{2}}\left(1 - \frac{9}{2}x^2 - 4x^4\right) + \frac{15}{2}x^4 \ln\frac{1 + (1 - x^2)^{\frac{1}{2}}}{x} \tag{25}$$

In the case at hand, $c = 0.74$.

The effect of a finite width can be estimated by introducing a weight $P(E)$ on the ρ^+ and ρ^0 masses, à la Breit-Wigner

$$\Gamma(\rho^+ \to \rho^0 + e^+ + \nu_e) =$$

$$= \int P(E_+)\, dE_+ \int P(E_0)\, dE_0\, \theta(E_+ - E_0 - m_e)\, \Gamma(\rho^+(E_+) \to \rho^0(E_0) + e^+ + \nu_e)$$

$$= \frac{|G|^2(1 + 2|\beta|^2)}{30\pi^3} \cdot$$

$$\int \frac{\Gamma_+}{(E_+ - m_{\rho^+})^2 + \Gamma_+^2/4} \frac{dE_+}{2\pi} \frac{\Gamma_0}{(E_0 - m_{\rho^0})^2 + \Gamma_0^2/4} \frac{dE_0}{2\pi} \theta(E_+ - E_0)\, (E_+ - E_0)^5 \quad (26)$$

The last equality is again valid in the limit $m_e \to 0$. We now set, consistently with data,

$$\Gamma_+ \simeq \Gamma_0 \qquad m_{\rho^+} \simeq m_{\rho^0} \tag{27}$$

and change variables to

$$x = E_+ - E_0 \qquad y = \frac{E_+ + E_0}{2} \tag{28}$$

It is easy to show that integration in dy leads to

$$\Gamma = \frac{G_F^2 \left(1 + 2|\beta|^2\right)}{30\pi^3} \left(\frac{\Gamma_0}{\pi} \int \frac{dx}{x^2 + \Gamma_0^2} x^5 \right) \tag{29}$$

Evidently, the integral needs to be cut off, at a value of the order of the half-width of the distribution (in this case, $2\Gamma_0$). Thus the order of magnitude of the integral is Γ_0^5 and we conclude that

$$\Gamma \simeq \frac{G_F^2 \left(1 + 2|\beta|^2\right)}{30\pi^3} \Gamma_0^5 \tag{30}$$

Thus, for $\Gamma_+, \Gamma_0 \gg \Delta$, the correction induced by Eq.(30) on the width (23) is given by

$$\frac{\Gamma(\text{wide})}{\Gamma(\text{narrow})} \sim \mathcal{O}\left(\frac{\Gamma_+}{\Delta}\right)^5 \tag{31}$$

b)   The decay amplitude for processes (2) and (3) is

$$\mathcal{M} = \frac{G_F}{\sqrt{2}} \langle 0|\, J_\mu^\dagger\, |\rho_+\rangle\, \bar{u}(\mathbf{p}_{\nu_\ell})\gamma^\mu(1 - \gamma_5)v(\mathbf{p}_{\ell^+}) \tag{32}$$

The corresponding line width is

$$\Gamma_\ell = \frac{1}{2M} \frac{G_F^2}{2} \int \langle 0|\, J_\mu^\dagger\, |\rho_+\rangle\langle\rho_+|\, J_\nu\, |0\rangle\, L^{\mu\nu}\, d\Phi^{(2)} \tag{33}$$

Integrating over phase space we find

$$\int L^{\mu\nu}\, d\Phi^{(2)} = \frac{1}{3}\left(1 - \frac{m_\ell^2}{p^2}\right)\left[p_\mu\, p_\nu - p^2\, g_{\mu\nu}\right]\left(1 + 2\frac{m_\ell^2}{p^2}\right)\frac{1}{8\pi}\left(1 - \frac{m_\ell^2}{p^2}\right) \tag{34}$$

where $p = p_\nu + p_{\ell^+} = p_\rho$. Thus the lepton mass may be neglected, with an error of $\mathcal{O}(m_\ell^2/m_\rho^2)$.

The matrix element of the hadronic current is purely vector-like and it is proportional to the electromagnetic current

$$\langle 0 | \, J_\mu^\dagger \, | \rho_+ \rangle = \sqrt{2} \langle 0 | \, J_\mu^{em} \, | \rho_0 \rangle = \sqrt{2} \, A \, \varepsilon_\mu \tag{35}$$

A is the coupling of ρ to the electromagnetic current.

We also calculate the decay probability for $\rho^0 \to e^+ e^-$, as follows

$$
\begin{aligned}
\Gamma_{\rho^0 \to e^+ e^-} &= \frac{e^2}{2 m_\rho} \langle 0 | \, J_\mu^{em} \, | \rho_0 \rangle \langle \rho_0 | \, J_\nu^{em} \, | 0 \rangle \frac{1}{m_\rho^4} \int L_{(e^+ e^-)}^{\mu\nu} \, d\Phi^{(2)} \\
&= \frac{e^2}{2 m_\rho} \langle 0 | \, J_\mu^{em} \, | \rho_0 \rangle \langle \rho_0 | \, J_\nu^{em} \, | 0 \rangle \frac{1}{m_\rho^4} \frac{4}{3} (p^\mu p^\nu - p^2 g^{\mu\nu}) \frac{1}{8\pi} \\
&= -\frac{e^2}{12 \pi \, m_\rho^3} \langle 0 | \, J_\mu^{em} \, | \rho_0 \rangle \langle \rho_0 | \, J^{em\,\mu} \, | 0 \rangle = \frac{e^2 A^2}{12 \pi \, m_\rho^3}
\end{aligned} \tag{36}
$$

We make use of this result in Eq.(33) and obtain

$$\Gamma_\ell = \frac{G_F^2}{12\pi} m_\rho A^2 = \frac{G_F^2 m_\rho^4}{e^2} \Gamma_{\rho^0 \to e^+ e^-} = \frac{G_F^2 m_\rho^4}{e^2} \Gamma_\rho \, \mathrm{BR} \left(\frac{\rho_0 \to e^+ e^-}{\rho_0 \to \text{all}} \right) \tag{37}$$

Substituting numerical values we find

$$\Gamma_\ell = 3.8 \cdot 10^{-6} \, \text{eV} \tag{38}$$

Let us now turn to the branching ratio

$$\frac{\Gamma(\pi^+ \to e^+ \nu_e)}{\Gamma(\pi^+ \to \mu^+ \nu_\mu)} \tag{39}$$

A main physical difference from the previous case comes from the fact that the pion is a pseudoscalar; hence, the total angular momentum, as well as its projection on the direction of the decay products, is zero. Given that e^+ is ultrarelativistic, its helicity is opposite to the neutrino helicity due to $V - A$ coupling, so that the angular momentum along the direction of flight would be 1, and the amplitude zero, up to terms of $\mathcal{O}(m_e/m_\rho)$. This mechanism is less effective in the case of the muon, which is much slower. In terms of formulae, we set

$$\langle 0 | \, J^\mu \, | \pi^+ \rangle = f \, p^\mu = f \, (p_{e^+}^\mu + p_{\nu_e}^\mu) \tag{40}$$

and write for the decay matrix element

$$\mathcal{M} = \frac{G}{\sqrt{2}} f \, m_e \, \bar{u}(\mathbf{p}_{\nu_e})(1 + \gamma_5) v(\mathbf{p}_{e^+}) \tag{41}$$

Eq.(41) leads to the following ratio of decay probabilities

$$\left(\frac{m_e}{m_\mu}\right)^2 \frac{\left(1 - \frac{m_e^2}{m_\pi^2}\right)^2}{\left(1 - \frac{m_\mu^2}{m_\pi^2}\right)^2} \tag{42}$$

In the above, the first factor comes from the mass in front of the amplitude (41), while one power of the second factor comes from phase space.

c) The process

$$D_+^* \to D_0^* + e^+ + \nu_e \tag{43}$$

has the same quantum numbers as the one in part a), hence the line width can be read off directly from the formula

$$\Gamma = \frac{G^2}{30\pi^3} \left(1 + 2|\beta|^2\right) \left(m_{D_+^*} - m_{D_0^*}\right)^5 \tag{44}$$

Its numerical value is

$$\Gamma = \left(1 + 2|\beta|^2\right) 3.8 \cdot 10^{-23} \, \text{MeV} \tag{45}$$

d) Neglecting corrections of order $(m_e/m_D)^2$, by virtue of the arguments presented above, we find

$$\frac{\Gamma(D_+^* \to e^+ \nu_e)}{\Gamma(D_+^* \to \mu^+ \nu_\mu)} \simeq 1 \tag{46}$$

Similarly,

$$\frac{\Gamma(D_+ \to e^+ \nu_e)}{\Gamma(D_+ \to \mu^+ \nu_\mu)} = \left(\frac{m_e}{m_\mu}\right)^2 \frac{\left(1 - \frac{m_e^2}{m_D^2}\right)^2}{\left(1 - \frac{m_\mu^2}{m_D^2}\right)^2} \simeq \left(\frac{m_e}{m_\mu}\right)^2 \simeq 2 \cdot 10^{-5} \tag{47}$$

Problem 71.   The $\theta - \tau$ puzzle

It has been thought for years that the particles θ^+ and τ^+, having equal mass, charge and mean life, were actually distinct. We know by now that these are simply different decay modes of one and the same spinless particle, K^+; parity is violated in the decay.

$$\theta^+ \to \pi^+ \pi^0 \qquad \tau^+ \to \pi^+ \pi^+ \pi^- \tag{1}$$

a) Suppose θ^+ and τ^+ are the same particle (K^+) and that the decays conserve parity. Knowing that π is a pseudoscalar, derive the parity assignments and the lowest spin value compatible with the assumptions made.

b)   Using the reply to part a), calculate the angular distribution of π^- with respect to the difference of the π^+ momenta, in the τ mode.

c)   Experimentally the above distribution is isotropic. Deduce from this fact that either θ^+ and τ^+ are not identical, or parity is violated.

d)   Discuss why low values of angular momentum are preferred in three-body decay, given that the radius of K^+ is about 1 fm.

Solution

a)   Let us denote the spin of K^+ by S.

In the decay of θ^+, angular momentum conservation requires that $S = l$, where l is the relative angular momentum of π^+ and π^0. The decay of τ^+ imposes $|l_+ - L| \leq S \leq L + l_+$, where l_+ is the relative angular momentum of the $\pi^+ - \pi^+$ pair and L is the angular momentum of π^- about that pair.

Assume now that parity is conserved. Denoting by P its eigenvalue on the state K^+, we have

$$P = (-)^{l+2} = (-)^S \tag{2}$$

as well as

$$P = (-)^{l_+ + L + 3} \tag{3}$$

since pions are pseudoscalars. Now l_+ must be even because the π^+'s are identical bosons; consequently,

$$P = (-)^S = (-)^{L+1} \tag{4}$$

Eq.(4) excludes the value $S = 0$: Indeed, $S = 0$ implies that L is odd and also that $l_+ = L$; this is impossible, since l_+ must be even. For $S = 1$, there exists a family of allowed solutions, with $L = l_+$. For $S = 2$, the solutions obey $L = l_+ \pm 1$.

In conclusion, the lowest admissible values are

$$\begin{array}{cccc} S = 1 & L = 2 & l_+ = 2 & P = -1 \\ S = 2 & L = 1 & l_+ = 2 & P = +1 \end{array} \tag{5}$$

b)   The angular part of the final state wave function can be expressed in the basis $|l_+ m_+ LM\rangle$, noting that

$$|SS_z\rangle = \sum_{\substack{m_+ M \\ m_+ + M = S_z}} \langle l_+ m_+ LM|SS_z\rangle |l_+ m_+ LM\rangle \tag{6}$$

$\langle l_+ m_+ LM|SS_z\rangle$ are Clebsch-Gordan coefficients.

The quantity which we wish to observe is the representation of this wave function in polar coordinates about the direction of relative motion of the two π^+. In these coordinates, the only component of the intrinsic angular momentum of the $\pi^+ \pi^+$

system is necessarily $m_+ = 0$. (The projection of orbital angular momentum along the direction of relative motion is zero.) Therefore,

$$\langle \theta_{++} = 0 | \langle \theta\phi | \psi \rangle = \sum_{M=S_z} \langle l_+ 0\, LM | SS_z \rangle \langle \theta\phi | LM \rangle \langle \theta_{++} = 0 | l_+ 0 \rangle \qquad (7)$$

Making use of

$$\langle \theta\phi | LM \rangle = Y_L^M(\theta, \phi) \qquad (8)$$

the modulus squared of the wave function, averaged over initial spin, becomes

$$|\psi(\theta)|^2 \propto \frac{1}{2S+1} \sum_{S_z} |Y_L^{S_z}(\theta, \phi)|^2 \, |\langle l_+ 0\, LS_z | SS_z \rangle|^2 \qquad (9)$$

In particular, for $S = 1$ ($l_+ = L = 2$), we have

$$|\psi(\theta)|^2 \; \propto \; \frac{1}{3} \sum_{M=-1}^{1} \left| Y_2^M(\theta, \phi) \right|^2 |\langle 20\, 2M | 1M \rangle|^2$$

$$= \frac{1}{10} \left\{ |Y_2^1(\theta, \phi)|^2 + |Y_2^{-1}(\theta, \phi)|^2 \right\} = \frac{3}{8\pi} \sin^2\theta \, \cos^2\theta \qquad (10)$$

and for $S = 2$ ($l_+ = 2$, $L = 1$),

$$|\psi(\theta)|^2 = \frac{1}{5} |Y_1^1|^2 = \frac{3}{40\pi} \sin^2\theta \qquad (11)$$

c) The request for isotropic distribution corresponds to a constant $|\psi(\theta)|^2$ in the previous calculation. This is only possible for $L = 0$, in which case $l_+ = S$. As a consequence, the parity of the two-pion final state will be $(-)^S$, while the parity of the three-pion state will be $(-)^{S+1}$. There follows that either parity is violated or θ^+ and τ^+ are two different particles.

d) The radial part of the wave function in a state of two free particles behaves at small r as $R_l(r) \simeq (kr)^l$. The interaction responsible for the decay is short range ($a \simeq 1$ fm). The decay matrix element is proportional to $(ka)^{l_+} (ka)^L$ where k is the relative momentum.

In the three-body decay $k_{max} = 125$ MeV. For $k \simeq 50$ MeV/c, we have $ka \simeq 0.25$ and hence large values of angular momentum are suppressed.

Problem 72.   Some aspects of muon decay

Muon decay is well described by the Fermi effective interaction

$$H = \frac{G_F}{\sqrt{2}} \left[J_\lambda^{(\mu)} J_\lambda^{(e)\dagger} + \text{hc} \right] \qquad (1)$$

$$J_\lambda^{(\ell)} = \bar{\psi}_\ell \gamma_\lambda (1 - \gamma_5) \psi_{\nu_\ell} \quad (\ell = \mu, e) \quad G_F = (1.02684 \pm 0.00002) \cdot 10^{-5} \, m_p^{-2}$$

a) Calculate the energy spectrum of the electron and the mean life of μ.
b) Derive the angular distribution of electrons emitted by polarized muons.
c) Calculate the electron helicity as a function of its energy.
d) How do the answers to parts a), b) and c) change in the case of μ^+ decay?
e) How does this analysis apply to τ^- leptonic decays?

Solution

a) The muon decay probability is given by the formula

$$d\Gamma = \frac{1}{2m_\mu} |\mathcal{M}|^2 \, d\Phi^{(3)} \tag{2}$$

with

$$\mathcal{M} = \frac{G_F}{\sqrt{2}} \bar{u}(\mathbf{p}_{\nu_\mu})\gamma^\mu(1-\gamma_5)u(\mathbf{p}_\mu)\,\bar{u}(\mathbf{p}_e)\gamma^\mu(1-\gamma_5)v(\mathbf{p}_{\bar{\nu}_e}) \tag{3}$$

There follows

$$
\begin{aligned}
|\mathcal{M}|^2 &= \frac{G_F^2}{2} \operatorname{Tr}\left\{ (\slashed{p}_{e^-} + m_e) \frac{1+\gamma_5\slashed{w}_e}{2} \gamma^\mu(1-\gamma_5)\slashed{p}_{\bar{\nu}_e}\gamma^\nu(1-\gamma_5) \right\} \\
&\quad \cdot \operatorname{Tr}\left\{ (\slashed{p}_{\mu^-} + m_\mu) \frac{1+\gamma_5\slashed{w}_\mu}{2} \gamma^\nu(1-\gamma_5)\slashed{p}_{\nu_\mu}\gamma^\mu(1-\gamma_5) \right\} \\
&= \frac{G_F^2}{2} L^{(e)}_{\mu\nu} L^{(\mu)\nu\mu}
\end{aligned} \tag{4}
$$

$$
\begin{aligned}
L^{(\ell)\mu\nu} &= 4\left[L^\mu p^\nu_{\bar{\nu}_\ell} + L^\nu p^\mu_{\bar{\nu}_\ell} - (L\cdot p_{\bar{\nu}_\ell})g^{\mu\nu} + i\,\varepsilon^{\mu\nu}{}_{\rho\sigma} L^\rho p^\sigma_{\bar{\nu}_\ell} \right] \\
L^\mu &= p^\mu_\ell - m_\ell w^\mu_\ell
\end{aligned} \tag{5}
$$

Carrying out all contractions we obtain

$$|\mathcal{M}|^2 = 32 G_F^2 \left[(p_{e^-}\cdot p_{\nu_\mu}) - m_{e^-}(w_e\cdot p_{\nu_\mu}) \right] \left[(p_{\mu^-}\cdot p_{\bar{\nu}_e}) - m_\mu(w_\mu\cdot p_{\bar{\nu}_e}) \right] \tag{6}$$

We may now integrate over the neutrino's phase space, using the relation

$$\int p^\alpha_{\bar{\nu}_e} p^\beta_{\nu_\mu} \, d\Phi^{(3)} = \frac{1}{(2\pi)^5} \frac{d^3\mathbf{p}_{e^-}}{8E_e} \int \frac{p^\alpha_{\bar{\nu}_e} p^\beta_{\nu_\mu}}{E_{\bar{\nu}_e} E_{\nu_\mu}} \delta^4(p_{\bar{\nu}_e} + p_{\nu_\mu} - q) \, d^3\mathbf{p}_{\bar{\nu}_e} \, d^3\mathbf{p}_{\nu_\mu} \tag{7}$$

with $q = p_{\mu^-} - p_{e^-}$. Since the integral can only depend on q, we write

$$\int p^\alpha_{\bar{\nu}_e} p^\beta_{\nu_\mu} \, d\Phi^{(3)} = \frac{1}{(2\pi)^5} \frac{d^3\mathbf{p}_{e^-}}{8E_e} I^{\alpha\beta} \tag{8}$$

where

$$I^{\alpha\beta} = A\,q^2\,g^{\alpha\beta} + B\,q^\alpha q^\beta \tag{9}$$

The relationships

$$(p_{\bar{\nu}_e} \cdot p_{\nu_\mu}) = q^2/2 \qquad (p_{\bar{\nu}_e} \cdot q) = (p_{\nu_\mu} \cdot q) = q^2/2 \tag{10}$$

now lead to

$$I_\alpha{}^\alpha = (4A + B)q^2 = \frac{1}{2} I \, q^2$$

$$I_{\alpha\beta} \, q^\alpha q^\beta = (A + B) \, (q^2)^2 = \frac{1}{4} I \, (q^2)^2 \tag{11}$$

I is defined by

$$I = \int \frac{d^3 p_{\bar{\nu}_e} \, d^3 p_{\nu_\mu}}{E_{\bar{\nu}_e} \, E_{\nu_\mu}} \, \delta^4 \left(p_{\bar{\nu}_e} + p_{\nu_\mu} - q \right) = \int \frac{d^3 p_{\bar{\nu}_e}}{E_{\bar{\nu}_e} \, E_{\nu_\mu}} \, \delta \left(E_{\bar{\nu}_e} + E_{\nu_\mu} - q_0 \right) \tag{12}$$

We evaluate I in the center of mass frame of the two neutrinos, where $\mathbf{p}_{\bar{\nu}_e} + \mathbf{p}_{\nu_\mu} = 0$, $E_{\bar{\nu}_e} = E_{\nu_\mu}$, with the result

$$I = 4\pi \int \frac{E_{\bar{\nu}_e}^2 \, dp_{\bar{\nu}_e}}{E_{\bar{\nu}_e}^2} \, \delta \left(2E_{\bar{\nu}_e} - q_0 \right) = 2\pi \tag{13}$$

Finally,

$$I^{\alpha\beta} = \frac{\pi}{6} \left[q^2 \, g^{\alpha\beta} + 2q^\alpha q^\beta \right] \tag{14}$$

Substituting into Eq.(2) yields

$$d\Gamma = \frac{G^2}{6(2\pi)^4 \, m_\mu} \, (p_{e^-} - m_e \, w_{e^-})_\alpha \, (p_{\mu^-} - m_\mu \, w_{\mu^-})_\beta \left[q^2 \, g^{\alpha\beta} + 2q^\alpha q^\beta \right] \frac{d^3 p_{e^-}}{E_{e^-}} \tag{15}$$

The above may be simplified by neglecting the mass of the electron in comparison to its energy

$$
\begin{aligned}
q^2 &\simeq m_\mu^2 - 2m_\mu \, E_{e^-} \\
(p_{e^-} \cdot q) &= (p_{e^-} \cdot p_{\mu^-}) = m_\mu \, E_{e^-} \\
(p_{\mu^-} \cdot q) &= m_\mu^2 - m_\mu \, E_{e^-}
\end{aligned}
$$

In the muon center-of-mass frame we have

$$w_\mu = (0, \boldsymbol{\zeta}_\mu) \qquad w_e = \left(\frac{\mathbf{p}_{e^-} \boldsymbol{\zeta}_e}{m_e}, \, \boldsymbol{\zeta}_e + \frac{(\mathbf{p}_{e^-} \boldsymbol{\zeta}_e) \cdot \mathbf{p}_{e^-}}{m_e(m_e + E_{e^-})} \right) \tag{16}$$

Eq.(15) now becomes

$$
d\Gamma = \frac{G_F^2 \, m_\mu^2}{6(2\pi)^4} \left(1 - \frac{\mathbf{p}_{e^-} \boldsymbol{\zeta}_e}{E_{e^-}} \right)
$$
$$
\left[\left(3 - \frac{4E_{e^-}}{m_\mu} \right) + \frac{\boldsymbol{\zeta}_\mu \mathbf{p}_{e^-}}{E_{e^-}} \left(1 - \frac{4E_{e^-}}{m_\mu} \right) \right] E_{e^-}^2 \, dE_{e^-} \, d\Omega \tag{17}
$$

This is the energy spectrum of the polarized electron. We also sum over electron polarizations and perform the angular integrations

$$d\Gamma = G_F^2 m_\mu^2 \left(3 - \frac{4E_{e^-}}{m_\mu} \right) \frac{E_{e^-}^2 \, dE_{e^-}}{12\pi^3} \tag{18}$$

To obtain the mean life, we must integrate further over the electron energy, in the kinematically allowed range

$$0 \leq E_{e^-} \leq \frac{m_\mu}{2} \tag{19}$$

The result is

$$\Gamma = \frac{G_F^2 m_\mu^5}{192\pi^3} \tag{20}$$

b) By Eq.(17), the angular distributions in the decay of polarized and unpolarized muons differ by a factor of

$$1 + \frac{\dfrac{\boldsymbol{\zeta}_\mu \mathbf{p}_{e^-}}{E_{e^-}} \left(1 - 4\dfrac{E_{e^-}}{m_\mu} \right)}{3 - 4\dfrac{E_{e^-}}{m_\mu}} \tag{21}$$

Integration over energies leads to

$$\frac{d\Gamma}{d\Omega} = \left(\frac{d\Gamma}{d\Omega} \right)_{\text{unpol}} \left(1 - \frac{1}{3} \boldsymbol{\zeta}_\mu \hat{\mathbf{p}}_{e^-} \right) \tag{22}$$

We have neglected $(m_e/m_\mu)^2$.

c) The expectation value of the electron polarization, stemming from the factor

$$1 - \frac{\mathbf{p}_{e^-}}{E_{e^-}} \boldsymbol{\zeta}_e \tag{23}$$

of Eq.(17), is given by

$$\langle \boldsymbol{\zeta}_e \rangle = -\frac{\mathbf{p}_{e^-}}{E_{e^-}} \tag{24}$$

In the ultrarelativistic limit the helicity equals -1.

d) Invariance under CP implies that all observables in μ^+ decay are equal to the corresponding observables in the decay of μ^-, with the same values of spin and opposite values of momenta. Therefore,

$$\left. \frac{d\Gamma}{d\Omega} \right|_{\mu^+} = \left(\frac{d\Gamma}{d\Omega} \right)_{\text{unpol}} \left(1 + \frac{1}{3} \boldsymbol{\zeta}_\mu \hat{\mathbf{p}}_{e^+} \right) \tag{25}$$

and

$$\langle \boldsymbol{\zeta}_e \rangle = +\frac{\mathbf{p}_{e^+}}{E_{e^+}} \tag{26}$$

e)   The analysis of the leptonic decays of τ

$$\tau^- \;\to\; \mu^- \;+ \bar{\nu}_\mu \;+ \nu_\tau$$
$$\tau^- \;\to\; e^- \;+ \bar{\nu}_e \;+ \nu_\tau \tag{27}$$

is essentially identical to the case of μ, apart from the following substitution of masses

$$m_\mu \;\to\; m_\tau$$
$$m_e \;\to\; m_\mu \,, m_e \tag{28}$$

Given that both masses m_μ, m_e, are negligible compared to m_τ, the previous results change only by an overall mass scale. In particular, we deduce the ratio

$$\frac{\Gamma_\tau}{\Gamma_\mu} \simeq \left(\frac{m_\tau}{m_\mu}\right)^5 \tag{29}$$

in the partial widths of decay into electrons.

Problem 73. $\gamma - p$ scattering through a Δ resonance

The cross section for the process

$$\gamma + p \to \Delta^+ \to \pi^+ + n \tag{1}$$

at the peak of the Δ^+ resonance ($M_\Delta = 1232$ MeV; spin $J = 3/2$; isospin $T = 3/2$, $T_3 = 1/2$; parity $+$ relative to the proton) is

$$\sigma_{\text{peak}}(\gamma + p \to \pi^+ + n) \simeq 170 \ \mu\text{b} \tag{2}$$

within a 10% precision. The elastic $\pi^+ p$ cross section at the peak is

$$\sigma_{\text{peak}}(\pi^+ + p \to \pi^+ + p) \simeq 200 \ \text{mb} \tag{3}$$

a)   Calculate the branching ratio

$$\frac{\Gamma(\Delta^+ \to p + \gamma)}{\Gamma(\Delta^+ \to p + \pi^0)} \tag{4}$$

b)   Find the cross section for the process

$$\pi^+ + n \to \pi^0 + p \tag{5}$$

on the resonance.

c)   Calculate the branching ratio

$$\frac{\Gamma(\Delta^+ \to n + e^+ + \nu_e)}{\Gamma(\Delta^+ \to p + \gamma)} \tag{6}$$

Assume that the electromagnetic and weak currents, appearing in the decays $\Delta^+ \to p + \gamma$ and $\Delta^+ \to n + e^+ + \nu_e$, respectively, are different components of the same isospin vector.

Solution

a)   Near resonance we have

$$\sigma(\gamma + p \rightarrow \pi^+ + n) \simeq \frac{\pi}{k_\gamma^2} \frac{(2J_\Delta + 1)}{2 \cdot 2} \frac{\Gamma_{p\gamma} \Gamma_{\pi^+ n}}{(E - M_\Delta)^2 + \Gamma_\Delta^2/4} \tag{7}$$

The factors of 2 in the denominator come from averaging over the polarizations of the initial particles; k_γ is the center-of-mass momentum. Similarly, for the elastic channel $\pi^+ p$,

$$\sigma(\pi^+ + p \rightarrow \pi^+ + p) = \frac{\pi}{k_\pi^2} \frac{(2J_\Delta + 1)}{2} \frac{(\Gamma_{\pi^+ p})^2}{(E - M_\Delta)^2 + \Gamma_\Delta^2/4} \tag{8}$$

On the peak, the ratio of the two cross sections is

$$\frac{k_\pi^2}{k_\gamma^2} \frac{1}{2} \frac{\Gamma_{p\gamma}}{\Gamma_{\pi^+ p}} \frac{\Gamma_{\pi^+ n}}{\Gamma_{\pi^+ p}} = \frac{170 \ \mu b}{200 \ mb} \tag{9}$$

The ratio k_π^2/k_γ^2 is given by

$$\frac{k_\pi^2}{k_\gamma^2} = \frac{\left[M_\Delta^2 - (m_p + m_\pi)^2\right]\left[M_\Delta^2 - (m_p - m_\pi)^2\right]}{\left[M_\Delta^2 - m_p^2\right]^2} \simeq 0.78 \tag{10}$$

leading to

$$\frac{\Gamma_{\pi^+ n}}{\Gamma_{\pi^+ p}} \frac{\Gamma_{p\gamma}}{\Gamma_{\pi^+ p}} \simeq 2.2 \cdot 10^{-3} \tag{11}$$

Isospin invariance requires that

$$\frac{\Gamma(\Delta^+ \rightarrow p + \pi^0)}{\Gamma(\Delta^{++} \rightarrow p + \pi^+)} = \left|C_{\frac{1}{2}0\frac{1}{2}}^{\frac{1}{2}1\frac{3}{2}}\right|^2 = \frac{2}{3} \qquad \frac{\Gamma_{\pi^+ n}}{\Gamma_{\pi^+ p}} = \frac{1}{3} \tag{12}$$

Eqs. (11) and (12) yield

$$\frac{\Gamma_{p\gamma}}{\Gamma_{\pi^0 p}} \simeq \left(2.2 \cdot 10^{-3}\right) \frac{\Gamma_{\pi^+ p}}{\Gamma_{\pi^+ n}} \frac{\Gamma_{\pi^+ p}}{\Gamma_{\pi^0 p}} = 3\left(\frac{3}{2}\right) 2.2 \cdot 10^{-3} = 1.0 \cdot 10^{-2} \tag{13}$$

b)   The cross section for $\pi^+ + n \rightarrow \pi^0 + p$ can be obtained from that of the process $\pi^+ + p \rightarrow \pi^+ + p$, multiplying by the factor

$$\left|C_{-\frac{1}{2}1\frac{1}{2}}^{\frac{1}{2}1\frac{3}{2}}\right|^2 \left|C_{\frac{1}{2}0\frac{1}{2}}^{\frac{1}{2}1\frac{3}{2}}\right|^2 = \frac{2}{9} \tag{14}$$

c)   Under the assumption made in the problem, the currents appearing in the two decays are

$$\langle \Delta^+ | j_0^\mu(0) | p \rangle = J_\mu^3 \tag{15}$$

and

$$\langle \Delta^+ | j_+^\mu(0) | n \rangle = J_\mu^+ \tag{16}$$

The two differ by a Clebsch-Gordan coefficient

$$J_\mu^+ = J_\mu^3 \, \frac{C_{-\frac{1}{2}\,1\,\frac{1}{2}}^{\frac{1}{2}\,1\,\frac{3}{2}}}{C_{-\frac{1}{2}\,0\,-\frac{1}{2}}^{\frac{1}{2}\,1\,\frac{3}{2}}} = \frac{1}{\sqrt{2}} J_\mu^3 \tag{17}$$

The width of the weak decay is

$$\Gamma_{\text{weak}} = \frac{1}{2M_\Delta} \left(\frac{G_F}{\sqrt{2}} \right)^2 \frac{1}{2} \, J_\mu^3 J_\nu^3 \, \text{Tr}\{ \not{p}_{e^+} \gamma^\mu (1 - \gamma_5) \not{p}_{\nu_e} \gamma^\nu (1 - \gamma_5) \} \, d\Phi^{(3)} \tag{18}$$

Let us reduce this expression to the form of a two-body decay: we write

$$
\begin{aligned}
d\Phi^{(3)} &= (2\pi)^4 \, \delta^4(p_\Delta - p_n - p_e - p_{\nu_e}) \frac{d^3\mathbf{p}_n}{2p_n^0 (2\pi)^3} \frac{d^3\mathbf{p}_{e^+}}{2p_{e^+}^0 (2\pi)^3} \frac{d^3\mathbf{p}_\nu}{2p_{\nu_e}^0 (2\pi)^3} \\
&\equiv \int \frac{d\mu^2}{2\pi} \int \frac{d^4q}{(2\pi)^4} \, (2\pi)^4 \delta^4(p_\Delta - p_n - q) \, 2\pi \, \delta(q^2 - \mu^2) \\
&\quad \cdot (2\pi)^4 \delta^4(q - p_{e^+} - p_{\nu_e}) \frac{d^3\mathbf{p}_{e^+}}{2p_{e^+}^0 (2\pi)^3} \frac{d^3\mathbf{p}_\nu}{2p_\nu^0 (2\pi)^3} \frac{d^3\mathbf{p}_n}{2p_n^0 (2\pi)^3}
\end{aligned} \tag{19}
$$

We now define the tensor

$$
\begin{aligned}
L^{\mu\nu}(q) &= \int (2\pi)^4 \, \delta^4(q - p_{e^+} - p_{\nu_e}) \frac{d^3\mathbf{p}_{e^+}}{2p_{e^+}^0 (2\pi)^3} \frac{d^3\mathbf{p}_\nu}{2p_\nu^0 (2\pi)^3} \\
&\quad \cdot \text{Tr}\{ \not{p}_{e^+} \gamma^\mu (1 - \gamma_5) \not{p}_{\nu_e} \gamma^\nu (1 - \gamma_5) \}
\end{aligned} \tag{20}
$$

and integrate over lepton momenta; this is most easily done in the center of mass of the lepton pair, with the result

$$L^{\mu\nu}(q) = \frac{1}{3\pi} (q^\mu q^\nu - q^2 \, g^{\mu\nu}) \tag{21}$$

Conservation of J_μ implies $q^\mu J_\mu = 0$. Eq.(18) becomes

$$
\begin{aligned}
\Gamma_{\text{weak}} &= \frac{1}{2M_\Delta} \left(\frac{G_F}{\sqrt{2}} \right)^2 \frac{1}{2} \int \frac{dq^2}{2\pi} \int \frac{d^3\mathbf{q}}{2q^0 (2\pi)^3} \int \frac{d^3\mathbf{p}_n}{2p_n^0 (2\pi)^3} \frac{1}{3\pi} \\
&\quad J_\mu J^\mu \, q^2 \, (2\pi)^4 \delta^4(p_\Delta - p_n - q) \\
&= \frac{1}{2M_\Delta} \left(\frac{G_F}{\sqrt{2}} \right)^2 \frac{1}{2} \frac{J^\mu J_\mu}{3\pi} \int \frac{dq^2}{2\pi} \frac{q^2}{8\pi} \frac{\sqrt{(M_\Delta + m_n)^2 - q^2} \sqrt{(M_\Delta - M_n)^2 - q^2}}{M_\Delta^2}
\end{aligned} \tag{22}
$$

The width of the decay into $p\gamma$ is

$$\Gamma_{p\gamma} = \frac{1}{2M_\Delta} e^2 J^\mu J_\mu \frac{1}{8\pi} \left(1 - \frac{M_n^2}{M_\Delta^2} \right) \qquad (23)$$

The branching ratio in question is thus given by

$$B = \left(\frac{G_F}{\sqrt{2}} \right)^2 \frac{M_\Delta^4}{3\pi\alpha} \frac{1}{1-\varepsilon^2} \frac{1}{(4\pi)^2} \int_0^{(1-\varepsilon)^2} dy\, y\, \sqrt{(1+\varepsilon)^2 - y}\, \sqrt{(1-\varepsilon)^2 - y} \qquad (24)$$

where we have set $\varepsilon = \frac{M_n}{M_\Delta}$ and $y = \frac{q^2}{M_\Delta^2}$. Using the approximation $\sqrt{(1+\varepsilon)^2 - y} \simeq 1 + \varepsilon$, we find

$$B = \left(\frac{GM_\Delta^2}{\sqrt{2}} \right)^2 \frac{1}{3\pi\alpha} \frac{(1-\varepsilon)^4}{(4\pi)^2} \int_0^1 dy\, y\, \sqrt{1-y} \qquad (25)$$

Substituting $\int_0^1 dy\, y\, \sqrt{1-y} = \frac{4}{15}$ and $G = 10^{-5} M_p^{-2}$ in the above, we arrive at $B \simeq 1.2 \cdot 10^{-14}$.

Problem 74. The process $e^+ e^- \to \mu^+ \mu^-$

a) Construct, to lowest order in α, the amplitude for the process $e^+ + e^- \to \mu^+ + \mu^-$, due to the electromagnetic interaction

$$\mathcal{L}_I = -eA_\mu j^\mu$$

$$j^\alpha =: \bar{\psi}_e \gamma^\alpha \psi_e : + : \bar{\psi}_\mu \gamma^\alpha \psi_\mu : \qquad (1)$$

b) Calculate the cross section, averaged over initial spins and summed over final spins, as a function of the center-of-mass energy. You may ignore lepton masses.

Solution

a,b) Let us denote by q, $\bar{q}$, p and $\bar{p}$ the momenta of μ^-, μ^+, e^- and e^+, respectively. The amplitude sought equals

$$\mathcal{M} = \frac{e^2}{k^2} \bar{v}(\bar{\mathbf{p}}) \gamma^\mu u(\mathbf{p})\, \bar{u}(\mathbf{q}) \gamma_\mu v(\bar{\mathbf{q}})$$

$$\overline{|\mathcal{M}|^2} = \frac{1}{4} \sum_{s_i, s_f} |M|^2 = \frac{e^4}{4k^4} \operatorname{Tr}\{\slashed{p}\gamma^\mu \slashed{\bar{p}}\gamma^\rho\} \operatorname{Tr}\{\slashed{\bar{q}}\gamma_\mu \slashed{q}\gamma_\rho\} \qquad (2)$$

All particle masses have been set to zero ($m_e \ll E$, $m_\mu \ll E$). Carrying out traces, we obtain

$$\overline{|\mathcal{M}|^2} = \frac{4e^4}{k^4} \left(\bar{p}^\mu p^\rho + \bar{p}^\rho p^\mu - \frac{k^2}{2} g^{\mu\rho} \right) \left(q_\mu \bar{q}_\rho + \bar{q}_\mu q_\rho - \frac{k^2}{2} g_{\mu\rho} \right) = \frac{e^4}{k^4} 2(t^2 + u^2) \qquad (3)$$

$$k = p + \bar{p} = q + \bar{q} \qquad t = (p - q)^2 \qquad u = (p - \bar{q})^2$$

In the center-of-mass frame, the dependence on the scattering angle becomes

$$\overline{|\mathcal{M}|^2} = e^4 (1 + \cos^2 \theta) \tag{4}$$

Using the expression for density times flux $\rho_1 \rho_2 \sqrt{(\mathbf{v}_{e^-} - \mathbf{v}_{e^+})^2} = 2\,(2p_0)^2 = 2k^2$, we find for the differential cross section

$$d\sigma = \frac{1}{2k^2} e^4 \left(1 + \cos^2 \theta\right) \frac{d\Omega}{32\pi^2} \tag{5}$$

The two-body phase space element $d\Phi = v_{\mathrm{CM}} \dfrac{d\Omega}{32\pi^2}$ (v_{CM} is the center-of-mass velocity of the outgoing particles) has been approximated by setting $v_{\mathrm{CM}} = 1$. Finally, writing $e^2 = 4\pi\,\alpha$, we have

$$\frac{d\sigma}{d\Omega} = \frac{\alpha^2}{4k^2}(1 + \cos^2 \theta) \qquad \sigma = \int d\Omega \, \frac{d\sigma}{d\Omega} = \frac{4\pi}{3} \frac{\alpha^2}{k^2} \tag{6}$$

k^2 is the square of the center of mass energy, and is usually called s.

Problem 75. Measuring the π electromagnetic form factor by electron scattering

We would like to measure the electromagnetic form factor of the π^+ meson.

a)   Consider at first a pointlike pion. Show that its interaction with the photon field A_μ, to first order in e, is given by

$$H_I(t) = e \int j_\mu(\mathbf{x}, t)\, A_\mu(\mathbf{x}, t)\, d^3 x \tag{1}$$

where

$$j_\mu(x) = \mathrm{i}\left(\phi^\dagger(x)\partial_\mu\phi(x) - \phi(x)\partial_\mu\phi^\dagger(x)\right) \tag{2}$$

ϕ is the free scalar field describing π^+.

   Calculate the Born amplitude for the transition between two pion states with momenta $\mathbf{p}$ and $\mathbf{p}'$, induced by an external field $A_\mu(x)$.

b)   For a non-pointlike pion the interaction (1) remains the same, whereas the current is no longer that of a free particle (Eq.(2)). Write down the most general form for the matrix element of the current between states $\mathbf{p}$ and $\mathbf{p}'$, using its properties under symmetry transformations. (Definition of form factor.)

c)   Show that the forward amplitude ($\mathbf{p} = \mathbf{p}'$) always coincides with the pointlike case. What is the physical reason for this?

d)   In order to measure the form factor, a charged pion beam is made to collide on an atom. One then selects those events in which a recoil electron is present; that is, the scattering process is $e^- + \pi^+ \rightarrow e^- + \pi^+$ and the nucleus does not participate. Assuming that atomic electrons are at rest, calculate the energy in the $e^-\pi^+$ center of mass, for 200 GeV pions. Up to which value of momentum transfer can the form factor be measured, at the given center-of-mass energy? How does one extract the form factor from the angular distribution?

Solution

a)   The Lagrangian for a charged scalar field is

$$\mathcal{L} = \partial_\mu \phi^\dagger \partial_\mu \phi - m^2 \phi^\dagger \phi \tag{3}$$

The minimal substitution $\partial_\mu \rightarrow \partial_\mu - i e A_\mu$ leads to the coupling

$$\mathcal{L} \rightarrow \partial_\mu \phi^\dagger \partial_\mu \phi + e j_\mu A^\mu - m^2 \phi^\dagger \phi + \mathcal{O}(e^2) \tag{4}$$

We can read from this Lagrangian the matrix element of the current

$$\langle \mathbf{p}'|j_\mu(x)|\mathbf{p}\rangle = e^{-i(p'-p)x}(p'+p)_\mu \tag{5}$$

and the scattering amplitude in an external field $A_\mu(x)$

$$\begin{aligned} \mathcal{M} &= e\,(p+p')_\mu \, \tilde{A}^\mu(q) \qquad q = p' - p \\ \tilde{A}_\mu(q) &= \int e^{-iqx} A_\mu(x)\, \mathrm{d}^4 x \end{aligned} \tag{6}$$

b)   Translational and Lorentz invariance reduce the form of the matrix element to

$$\begin{aligned} \langle \mathbf{p}'|j_\mu(x)|\mathbf{p}\rangle &= e^{i(p-p')x}\langle p'|j_\mu(0)|p\rangle \\ \langle \mathbf{p}'|j_\mu(0)|\mathbf{p}\rangle &= A\,P_\mu + B\,q_\mu \\ P_\mu &= (p+p')_\mu \end{aligned} \tag{7}$$

where A and B are functions of q^2. Gauge invariance further requires

$$q^\mu(A\,P_\mu + B\,q_\mu) = 0 \tag{8}$$

Given that $P \cdot q = 0$, the above becomes simply $B\,q^2 = 0$ or $B = 0$. In conclusion,

$$\langle \mathbf{p}'|j_\mu(0)|\mathbf{p}\rangle = A P_\mu \tag{9}$$

Again, A is a function of q^2.

c)   The electric charge operator is defined by

$$Q = \int \mathrm{d}^3 x \, j_0(0, \mathbf{x}) \tag{10}$$

The pion has charge 1 and therefore

$$\langle \mathbf{p'}|Q|\mathbf{p}\rangle = 2p^0(2\pi)^3\delta^3(\mathbf{p}-\mathbf{p'}) \tag{11}$$

On the other hand, we also obtain

$$\begin{aligned}
\langle \mathbf{p'}|Q|\mathbf{p}\rangle &= \langle \mathbf{p'}|\int d^3x\, j_0(0,\mathbf{x})|\mathbf{p}\rangle \\
&= \int d^3x\, e^{i(\mathbf{p'}-\mathbf{p})\mathbf{x}}\langle \mathbf{p'}|j_0(0,0)|\mathbf{p}\rangle \\
&= 2p^0(2\pi)^3\delta^3(\mathbf{p}-\mathbf{p'})A(0)
\end{aligned} \tag{12}$$

Comparing now Eqs. (9) and (11) we conclude

$$A(0) = 1 \tag{13}$$

The physical reason behind this is that in processes with small momentum transfer (i.e., with a length scale which is large compared to the pion radius) the photon interacts only with the global charge of the particle and does not distinguish its distribution.

d)   The Lorentz invariant quantity $s = (p_\pi + p_e)^2$ is equal to the square of the center-of-mass energy, while in the laboratory frame it equals $2m_e E_\pi$. Thus, if $E_\pi \simeq 200\,\text{GeV}$, we find

$$s \simeq 10^{-3} \cdot 200\,\text{GeV}^2 \simeq 0.2\,\text{GeV}^2 \tag{14}$$

and hence $E_{c.m.} \simeq 0.45\,\text{GeV}$. The momentum transfer in the center of mass is, neglecting the electron mass

$$q^2 = -\frac{(s-m_\pi^2)^2}{2s}(1-\cos\theta) \tag{15}$$

The maximum value for q^2, corresponding to backward scattering, is given by

$$-q_{max}^2 \simeq s \simeq 0.2\,\text{GeV}^2 \tag{16}$$

The cross section is related to the pointlike case through

$$\frac{d\sigma}{d\Omega} = |A(q^2)|^2 \left(\frac{d\sigma}{d\Omega}\right)_0 \tag{17}$$

Using the standard expression for electron-photon coupling, and neglecting the electron mass, we are led to

$$\begin{aligned}
\left(\frac{d\sigma}{d\Omega}\right)_0 &= \frac{e^4}{2s}\frac{P_\mu P_\nu}{q^4}\frac{1}{2}\operatorname{Tr}\{\not{p}_{e'}\gamma^\mu\not{p}_e\gamma^\nu\}\frac{1}{32\pi^2} \\
&= \frac{\alpha^2 s}{2q^4}(1+\cos\theta) = \frac{2\alpha^2}{s}\frac{1+\cos\theta}{(1-\cos\theta)^2}
\end{aligned} \tag{18}$$

The form factor can be extracted from data by means of

$$|A(q^2)|^2 = \frac{\dfrac{d\sigma}{d\Omega}}{\left(\dfrac{d\sigma}{d\Omega}\right)_0} \qquad (19)$$

Problem 76.   Threshold behaviour of the process $e^+e^- \to Q^+Q^-$ and the spin of Q

Photons couple to pointlike particles of different spins as follows:
Spin 1/2

$$\mathcal{L}_I = e\, A_\mu\, \bar{\psi}\gamma^\mu\psi \qquad (1)$$

Spin 0 (to lowest order in e)

$$\mathcal{L}_I = e\, A_\mu\, i\left(\phi^\dagger\, \partial^\mu\phi - \phi\, \partial^\mu\phi^\dagger\right) \qquad (2)$$

Spin 1

$$\mathcal{L}_I = e\, A_\mu\, i\left(W_\rho^\dagger\, \partial^\mu W^\rho - W^\rho\, \partial^\mu W_\rho^\dagger\right) \qquad (3)$$

a)   Consider the process $e^+e^- \to Q\bar{Q}$. Calculate the behaviour of the cross section ($\sigma_{Q\bar{Q}}$) just above the threshold for pair production, for each of the three possible values of the spin of Q. The mass M_Q is assumed to be known.
b)   Calculate the ratio

$$R(E) = \frac{\sigma_{Q\bar{Q}}(E)}{\sigma_{\mu^+\mu^-}(E)} \qquad (4)$$

as a function of energy.
c)   Discuss the possibility of determining the spin of Q from measured values of $R(E)$ in the vicinity of the threshold.

Solution

In all three cases the modulus squared of the matrix element can be written in the form

$$|\mathcal{M}|^2 = \frac{e^4}{(Q^2)^2}\, L_{\mu\nu}\, T^{\mu\nu} \qquad (5)$$

where the tensor

$$L_{\mu\nu} = \frac{1}{4}\, \mathrm{Tr}\,\{\slashed{p}_{e^-}\gamma_\mu\slashed{p}_{e^+}\gamma_\nu\} = p_{e^-_\mu}\, p_{e^+_\nu} + p_{e^-_\nu}\, p_{e^+_\mu} - (p_{e^+} \cdot p_{e^-})g_{\mu\nu} \qquad (6)$$

originates from the electron-positron vertex. We denote the electron and positron four-momenta by $p_{e^-_\mu}$ and $p_{e^+_\mu}$, and $Q_\mu = (p_{e^+} + p_{e^-})_\mu$. In the above, the electron mass has been neglected, and a factor of 1/4 comes from averaging over initial spins.

$T^{\mu\nu}$ is defined by

$$T^{\mu\nu} = \sum_{spin} \langle 0|j^\nu(0)|Q\,\bar{Q}\rangle \langle Q\,\bar{Q}|j^\mu(0)|0\rangle \tag{7}$$

Let us evaluate $T_{\mu\nu}$ for each value of the spin of Q, in terms of p_Q^μ and $p_{\bar{Q}}^\mu$ (the four-momenta of Q and $\bar{Q}$ in the final state). For spin $1/2$, 0 and 1, respectively, we obtain

$$
\begin{aligned}
T^{\mu\nu} &= \mathrm{Tr}\left\{(\not{p}_Q + M_Q)\gamma^\mu(\not{p}_{\bar{Q}} - M_Q)\gamma^\nu\right\} \\
&= 4\left\{p_Q^\mu p_{\bar{Q}}^\nu + p_Q^\nu p_{\bar{Q}}^\mu - (M_Q^2 + p_Q\cdot p_{\bar{Q}})g^{\mu\nu}\right\}
\end{aligned}
\tag{8a}
$$

$$T^{\mu\nu} = (p_Q - p_{\bar{Q}})^\mu (p_Q - p_{\bar{Q}})^\nu \tag{8b}$$

$$
\begin{aligned}
T^{\mu\nu} &= \sum_{i,j} \bar{\varepsilon}_{(i)}^\rho\, \bar{\varepsilon}_{(i)}^{\sigma*}\, \varepsilon_{(j)\rho}^*\, \varepsilon_{(j)\sigma}\, (p_Q - p_{\bar{Q}})^\mu (p_Q - p_{\bar{Q}})^\nu \\
&= \left(g_{\rho\sigma} - \frac{p_{Q\rho}p_{Q\sigma}}{M_Q^2}\right)\left(g^{\rho\sigma} - \frac{p_{\bar{Q}}^\rho p_{\bar{Q}}^\sigma}{M_Q^2}\right)(p_Q - p_{\bar{Q}})^\mu (p_Q - p_{\bar{Q}})^\nu \\
&= \left(2 + \frac{(p_Q\cdot p_{\bar{Q}})^2}{M_Q^4}\right)(p_Q - p_{\bar{Q}})^\mu (p_Q - p_{\bar{Q}})^\nu
\end{aligned}
\tag{8c}
$$

The above expressions can be simplified, setting

$$p_{e^-}{}^\mu = \frac{(Q+P)^\mu}{2} \qquad p_{e^+}{}^\mu = \frac{(Q-P)^\mu}{2} \tag{9}$$

We find

$$Q\cdot P = 0 \tag{10}$$

$$L^{\mu\nu} = \frac{1}{2}\left(Q^\mu Q^\nu - Q^2 g^{\mu\nu}\right) - \frac{1}{2}P^\mu P^\nu \tag{11}$$

Similarly, setting

$$p_Q^\mu = \frac{(Q+S)^\mu}{2} \qquad p_{\bar{Q}}^\mu = \frac{(Q-S)^\mu}{2} \qquad Q^\mu = (p_{e^+} + p_{e^-})^\mu = \left(p_Q + p_{\bar{Q}}\right)^\mu \tag{12}$$

we have

$$
\begin{aligned}
T^{\mu\nu} &= 2\left(Q^\mu Q^\nu - Q^2 g^{\mu\nu}\right) - 2S^\mu S^\nu &&(\text{spin } 1/2) &\tag{13a}\\
T^{\mu\nu} &= S^\mu S^\nu &&(\text{spin } 0) &\tag{13b}\\
T^{\mu\nu} &= \left\{3 + \frac{Q^2}{M_Q^2}\left(\frac{Q^2}{4\,M_Q^2} - 1\right)\right\}S^\mu S^\nu &&(\text{spin } 1) &\tag{13c}
\end{aligned}
$$

Let us now write the modulus squared of the amplitude explicitly: For spin $1/2$, it equals

$$
\begin{aligned}
|\mathcal{M}|^2 &= \frac{e^4}{(Q^2)^2}\left\{3(Q^2)^2 + Q^2 P^2 + Q^2 S^2 + (P\cdot S)^2\right\} \\
&= e^4\left\{1 + \cos^2\theta + \frac{4M_Q^2}{Q^2}(1 - \cos^2\theta)\right\}
\end{aligned}
\tag{14}
$$

We have approximated

$$
\begin{aligned}
P^2 &= -Q^2 + 4m_e^2 \simeq -Q^2 & S^2 &= -Q^2 + 4M_Q^2 \\
(P \cdot S)^2 &= Q^2(Q^2 - 4M_Q^2)\cos^2\theta
\end{aligned}
$$

θ is the scattering angle.

For spin 0

$$
|\mathcal{M}|^2 = \frac{e^4}{(Q^2)^2}\left\{-\frac{1}{2}Q^2 S^2 - \frac{1}{2}(P \cdot S)^2\right\} = \frac{e^4}{2}(1 - \cos^2\theta)\left(1 - \frac{4M_Q^2}{Q^2}\right) \tag{15}
$$

For spin 1

$$
|\mathcal{M}|^2 = \frac{e^4}{2}(1 - \cos^2\theta)\left(1 - \frac{4M_Q^2}{Q^2}\right)\left\{3 + \frac{Q^2}{M_Q^2}\left(\frac{Q^2}{4M_Q^2} - 1\right)\right\} \tag{16}
$$

The total cross sections,

$$
\sigma = \frac{1}{2Q^2}\int |M|^2 \frac{d\Omega}{32\pi^2}\sqrt{1 - \frac{4M_Q^2}{Q^2}} \tag{17}
$$

become, respectively:

$$
\sigma = \frac{e^4}{12\pi\,Q^2}\left(1 - \frac{4M_Q^2}{Q^2}\right)^{\frac{1}{2}}\left(1 + \frac{2M_Q^2}{Q^2}\right) \qquad \text{(spin 1/2)} \tag{18a}
$$

$$
\sigma = \frac{e^4}{48\pi\,Q^2}\left(1 - \frac{4M_Q^2}{Q^2}\right)^{\frac{3}{2}} \qquad \text{(spin 0)} \tag{18b}
$$

$$
\sigma = \frac{e^4}{48\pi\,Q^2}\left(1 - \frac{4M_Q^2}{Q^2}\right)^{\frac{3}{2}}\left\{3 + \frac{Q^2}{M_Q^2}\left(\frac{Q^2}{4M_Q^2} - 1\right)\right\} \quad \text{(spin 1)} \tag{18c}
$$

Near threshold,

$$
\sigma \simeq \frac{2\pi\alpha^2}{Q^2}\left(1 - \frac{4\,M_Q^2}{Q^2}\right)^{\frac{1}{2}} \qquad \text{(spin 1/2)} \tag{19a}
$$

$$
\sigma = \frac{\pi\alpha^2}{3Q^2}\left(1 - \frac{4M_Q^2}{Q^2}\right)^{\frac{3}{2}} \qquad \text{(spin 0)} \tag{19b}
$$

$$
\sigma \simeq \frac{\pi\alpha^2}{Q^2}\left(1 - \frac{4M_Q^2}{Q^2}\right)^{\frac{3}{2}} \qquad \text{(spin 1)} \tag{19c}
$$

b,c)   The ratio $R(E)$ near threshold becomes

$$R(E) \;=\; \frac{3}{2}\left(1 - \frac{4M_Q^2}{Q^2}\right)^{\frac{1}{2}} \qquad (\text{spin}\,1/2) \qquad\qquad (20a)$$

$$R(E) \;=\; \frac{1}{4}\left(1 - \frac{4M_Q^2}{Q^2}\right)^{\frac{3}{2}} \qquad (\text{spin}\,0) \qquad\qquad (20b)$$

$$R(E) \;=\; \frac{3}{4}\left(1 - \frac{4M_Q^2}{Q^2}\right)^{\frac{3}{2}} \qquad (\text{spin}\,1) \qquad\qquad (20c)$$

Indeed, the spin of Q can be determined from $R(E)$.

Problem 77.   Low energy limits of the Compton effect

a)   Write the amplitude for the Compton effect on electrons, to lowest perturbative order; show its invariance under gauge transformations.

b)   Write the same amplitude in terms of two-component spinors in the Foldy representation. Show that it has the form

$$u'^{\dagger}(A + \mathbf{B}\cdot\boldsymbol{\sigma})u \qquad\qquad (1)$$

where A and $\mathbf{B}$ are bilinears in the photon polarization. What is the explicit expression A and $\mathbf{B}$ in the laboratory frame?

c)   Consider the limit of low frequency photons; calculate A and $\mathbf{B}$ to first order in the frequency.

d)   Construct the matrix element for this process, squared and summed over final polarizations. Discuss the possibility of measuring the circular polarization of γ's through scattering on polarized electrons.

Solution

a)   The lowest perturbative order corresponds to the diagrams

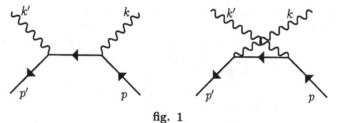

fig. 1

$$\mathcal{M} = e^2 \varepsilon_\mu'^* \varepsilon_\nu \bar{u}(\mathbf{p}') \left\{ \gamma^\mu \frac{1}{\not{p} + \not{k} - m} \gamma^\nu + \gamma^\nu \frac{1}{\not{p} - \not{k}' - m} \gamma^\mu \right\} u(\mathbf{p}) \tag{2}$$

Gauge invariance requires that the substitutions

$$\varepsilon_\mu \to \varepsilon_\mu + \alpha \, k_\mu \qquad \varepsilon_\mu' \to \varepsilon_\mu' + \beta \, k_\mu' \tag{3}$$

leave the amplitude invariant. Thus we expect

$$\delta\mathcal{M} = \bar{u}(\mathbf{p}') \left\{ \not{k}' \frac{1}{\not{p} + \not{k}' - m} \gamma^\nu + \gamma^\nu \frac{1}{\not{p} - \not{k}' - m} \not{k}' \right\} u(\mathbf{p}) \varepsilon_\nu = 0 \tag{4}$$

and similarly upon contraction with k_ν. Substituting $\not{k}'$ in this expression through the identities

$$\begin{aligned} \not{k}' &= \not{k}' + \not{p} - m + m - \not{p} \\ \not{k}' &= \not{k}' + m - \not{p} + (\not{p} - m) \end{aligned} \tag{5}$$

and enforcing the Dirac equation

$$\bar{u}(\mathbf{p}')(\not{p}' - m) = 0 \qquad (\not{p} - m)u(\mathbf{p}) = 0 \tag{6}$$

we obtain indeed

$$\delta\mathcal{M} = \bar{u}(\mathbf{p}') \left\{ \gamma^\nu - \gamma^\nu \right\} u(\mathbf{p}) \varepsilon_\nu = 0 \tag{7}$$

b) In the standard (Pauli) representation of the γ matrices we have

$$u(\mathbf{p}) = \sqrt{E + m} \begin{pmatrix} \mathrm{w} \\ \dfrac{\boldsymbol{\sigma} \cdot \mathbf{p}}{E + m} \mathrm{w} \end{pmatrix} \tag{8}$$

w is the two-component spinor describing the electron in the Foldy representation. The generic form of the amplitude is

$$u^\dagger(\mathbf{p}') \, \mathcal{O} \, u(\mathbf{p}) \tag{9}$$

The operator $\mathcal{O}$, being a combination of γ matrices, can be written as

$$\begin{pmatrix} \alpha_{11} + \boldsymbol{\beta}_{11} \cdot \boldsymbol{\sigma} & \alpha_{12} + \boldsymbol{\beta}_{12} \cdot \boldsymbol{\sigma} \\ \alpha_{21} + \boldsymbol{\beta}_{21} \cdot \boldsymbol{\sigma} & \alpha_{22} + \boldsymbol{\beta}_{22} \cdot \boldsymbol{\sigma} \end{pmatrix} \tag{10}$$

From (8) and (10) the most general expression for $\mathcal{M}$ is

$$\mathcal{M} = e^2 \mathrm{w}'^\dagger \left\{ A + \mathbf{B} \cdot \boldsymbol{\sigma} \right\} \mathrm{w} \tag{11}$$

In the amplitude of Eq.(1) $\mathcal{O}$ equals

$$\mathcal{O} = \gamma^0 \left\{ \not{\epsilon}^* \frac{1}{\not{p} + \not{k} - m} \not{\epsilon} + \not{\epsilon} \frac{1}{\not{p} - \not{k}' - m} \not{\epsilon}^* \right\}$$

$$= \gamma_0 \left\{ \frac{1}{s - m^2} \not{\epsilon}^* (\not{p} + \not{k} + m) \not{\epsilon} + \frac{1}{u - m^2} \not{\epsilon} (\not{p} - \not{k}' + m) \not{\epsilon}^* \right\} \quad (12)$$

Going to the laboratory frame, and choosing the transverse gauge in which ϵ and ϵ' have only spatial components, the following relations hold

$$p_\mu = (m, \mathbf{0}) \qquad \{\not{p}, \not{\epsilon}\}_+ = \{\not{p}, \not{\epsilon}'\}_+ = 0 \qquad (\not{p} - m)u(\mathbf{p}) = 0$$

$$\mathcal{O} = \gamma^0 \left\{ \frac{1}{s - m^2} \not{\epsilon}^* \not{k} \not{\epsilon} - \frac{1}{u - m^2} \not{\epsilon} \not{k}' \not{\epsilon}^* \right\} \quad (13)$$

Separating the space and time components of $\not{k}$ we obtain

$$\mathcal{O} = \frac{|\mathbf{k}'|}{u - m^2} \not{\epsilon} \not{\epsilon}^* - \frac{|\mathbf{k}|}{s - m^2} \not{\epsilon}^* \not{\epsilon}$$

$$- \frac{\gamma_0}{s - m^2} \not{\epsilon}^* \, i(\mathbf{k} \wedge \boldsymbol{\epsilon}) \cdot \boldsymbol{\Sigma} + \frac{\gamma_0}{u - m^2} \not{\epsilon} \, i(\mathbf{k}' \wedge \boldsymbol{\epsilon}'^*) \cdot \boldsymbol{\Sigma} \quad (14)$$

$$\boldsymbol{\Sigma} = \begin{pmatrix} \boldsymbol{\sigma} & 0 \\ 0 & \boldsymbol{\sigma} \end{pmatrix}$$

$$\mathcal{O} = -\frac{\gamma_0}{s - m^2} \not{\epsilon}' \, i(\mathbf{k} \wedge \boldsymbol{\epsilon}) \cdot \boldsymbol{\Sigma} + \frac{\gamma_0}{u - m^2} \not{\epsilon} \, i(\mathbf{k}' \wedge \boldsymbol{\epsilon}'^*) \cdot \boldsymbol{\Sigma}$$

$$+ \frac{|\mathbf{k}'|}{u - m^2} \{ -\epsilon \epsilon'^* - i(\boldsymbol{\epsilon} \wedge \boldsymbol{\epsilon}'^*) \cdot \boldsymbol{\Sigma} \} - \frac{|\mathbf{k}|}{s - m^2} \{ -\epsilon \epsilon'^* - i(\boldsymbol{\epsilon}'^* \wedge \boldsymbol{\epsilon}) \cdot \boldsymbol{\Sigma} \} \quad (15)$$

Therefore

$$\alpha_{11} = \alpha_{22} = \epsilon \epsilon'^* \left[\frac{|\mathbf{k}|}{s - m^2} - \frac{|\mathbf{k}'|}{u - m^2} \right]$$

$$\beta_{11} = \beta_{22} = -i\,(\boldsymbol{\epsilon} \wedge \boldsymbol{\epsilon}'^*) \left[\frac{|\mathbf{k}'|}{u - m^2} + \frac{|\mathbf{k}|}{s - m^2} \right]$$

$$\alpha_{12} = \alpha_{21} = i \left[\frac{\mathbf{k} \cdot (\boldsymbol{\epsilon} \wedge \boldsymbol{\epsilon}'^*)}{s - m^2} + \frac{\mathbf{k}' \cdot (\boldsymbol{\epsilon} \wedge \boldsymbol{\epsilon}'^*)}{u - m^2} \right]$$

$$\beta_{12} = \beta_{21} = -\frac{\boldsymbol{\epsilon}'^* \wedge (\mathbf{k} \wedge \boldsymbol{\epsilon})}{s - m^2} + \frac{\boldsymbol{\epsilon} \wedge (\mathbf{k}' \wedge \boldsymbol{\epsilon}'^*)}{u - m^2} \quad (16)$$

The form of the spinors for $\mathbf{p} = 0$ allows us to write

$$A + \mathbf{B} \cdot \boldsymbol{\sigma} = \sqrt{2m(E' + m)}(\alpha_{11} + \beta_{11} \cdot \boldsymbol{\sigma})$$

$$+ \sqrt{\frac{2m}{E' + m}} (\alpha_{12} \boldsymbol{\sigma} \cdot \mathbf{p}' + \beta_{12} \mathbf{p}' - i(\beta_{12} \wedge \mathbf{p}') \cdot \boldsymbol{\sigma}) \quad (17)$$

Finally

$$A = \sqrt{2m(E'+m)} \left[\frac{|\mathbf{k}|}{s-m^2} - \frac{|\mathbf{k}'|}{u-m^2} \right] \boldsymbol{\varepsilon}'^* \boldsymbol{\varepsilon} \tag{18a}$$

$$+ \sqrt{\frac{2m}{E'+m}} \mathbf{p}' \cdot \left[\frac{\boldsymbol{\varepsilon} \wedge (\mathbf{k}' \wedge \boldsymbol{\varepsilon}'^*)}{u-m^2} - \frac{\boldsymbol{\varepsilon}'^* \wedge (\mathbf{k} \wedge \boldsymbol{\varepsilon})}{s-m^2} \right]$$

$$\mathbf{B} = -\sqrt{2m(E'+m)} \, \mathrm{i} \, (\boldsymbol{\varepsilon} \wedge \boldsymbol{\varepsilon}'^*) \left[\frac{|\mathbf{k}'|}{u-m^2} + \frac{|\mathbf{k}|}{s-m^2} \right] \tag{18b}$$

$$+ \mathrm{i} \sqrt{\frac{2m}{E'+m}} \mathbf{p}' \left[\frac{\mathbf{k} \cdot (\boldsymbol{\varepsilon} \wedge \boldsymbol{\varepsilon}'^*)}{s-m^2} + \frac{\mathbf{k}' \cdot (\boldsymbol{\varepsilon} \wedge \boldsymbol{\varepsilon}'^*)}{u-m^2} \right]$$

$$+ \mathrm{i} \sqrt{\frac{2m}{E'+m}} \mathbf{p}' \wedge \left[\frac{\boldsymbol{\varepsilon} \wedge (\mathbf{k}' \wedge \boldsymbol{\varepsilon}'^*)}{u-m^2} - \frac{\boldsymbol{\varepsilon}'^* \wedge (\mathbf{k} \wedge \boldsymbol{\varepsilon})}{s-m^2} \right]$$

c) In the laboratory frame we have

$$s = m^2 + 2m\,|\mathbf{k}| \qquad u = m^2 - 2m\,|\mathbf{k}'| \tag{19}$$

$$\frac{|\mathbf{k}'|}{u-m^2} = -\frac{1}{2m} \qquad \frac{|\mathbf{k}|}{s-m^2} = \frac{1}{2m} \qquad |\mathbf{k}'| \simeq |\mathbf{k}| \left\{ 1 + \mathcal{O}\left(\frac{|\mathbf{k}|}{m}\right) \right\}$$

Setting $\mathbf{n} = \mathbf{k}/|\mathbf{k}|$, $\mathbf{n}' = \mathbf{k}'/|\mathbf{k}'|$, we obtain

$$A = 2\sqrt{\frac{E'+m}{2m}} \boldsymbol{\varepsilon}'^* \boldsymbol{\varepsilon} - \frac{\mathbf{p}'}{\sqrt{2m(E'+m)}} \cdot \{ \boldsymbol{\varepsilon} \wedge (\mathbf{n}' \wedge \boldsymbol{\varepsilon}'^*) + \boldsymbol{\varepsilon}'^* \wedge (\mathbf{n} \wedge \boldsymbol{\varepsilon}) \} \tag{20a}$$

$$\mathbf{B} = \frac{\mathrm{i}\,\mathbf{p}'}{\sqrt{2m(E'+m)}} [\mathbf{n} \cdot (\boldsymbol{\varepsilon} \wedge \boldsymbol{\varepsilon}'^*) - \mathbf{n}' \cdot (\boldsymbol{\varepsilon} \wedge \boldsymbol{\varepsilon}'^*)] \tag{20b}$$

$$- \frac{\mathrm{i}\,\mathbf{p}'}{\sqrt{2m(E'+m)}} \wedge [\boldsymbol{\varepsilon} \wedge (\mathbf{n}' \wedge \boldsymbol{\varepsilon}'^*) + \boldsymbol{\varepsilon}'^* \wedge (\mathbf{n} \wedge \boldsymbol{\varepsilon})]$$

To zeroth order in the frequency, $\mathbf{p}' \simeq 0$, $E' = m$, we find

$$A^{(0)} = 2(\boldsymbol{\varepsilon}'^* \boldsymbol{\varepsilon}) \qquad \mathbf{B}^{(0)} = 0 \tag{21}$$

To first order, $\mathbf{p}' = \mathbf{k} - \mathbf{k}' \simeq |\mathbf{k}|(\mathbf{n} - \mathbf{n}')$. Using the relations $\boldsymbol{\varepsilon}\mathbf{n} = 0 = \boldsymbol{\varepsilon}'\mathbf{n}'$ we are led to

$$A^{(1)} = |\mathbf{k}|(\mathbf{n} - \mathbf{n}') \{ (\boldsymbol{\varepsilon}\boldsymbol{\varepsilon}'^*)\mathbf{n}' - (\boldsymbol{\varepsilon}\mathbf{n}')\boldsymbol{\varepsilon}'^* + (\boldsymbol{\varepsilon}\boldsymbol{\varepsilon}'^*) \cdot \mathbf{n} - (\boldsymbol{\varepsilon}'^*\mathbf{n})\boldsymbol{\varepsilon} \} = 0 \tag{22a}$$

$$\mathbf{B}^{(1)} = \frac{\mathrm{i}\,|\mathbf{k}|}{2m}(\mathbf{n} - \mathbf{n}') [\boldsymbol{\varepsilon}'^* \cdot (\mathbf{n} \wedge \boldsymbol{\varepsilon}) + \boldsymbol{\varepsilon} \cdot (\mathbf{n}' \wedge \boldsymbol{\varepsilon}'^*)] \tag{22b}$$

$$- \frac{\mathrm{i}\,|\mathbf{k}|}{2m}(\mathbf{n} - \mathbf{n}') \wedge [\boldsymbol{\varepsilon} \wedge (\mathbf{n}' \wedge \boldsymbol{\varepsilon}'^*) + \boldsymbol{\varepsilon}'^* \wedge (\mathbf{n} \wedge \boldsymbol{\varepsilon})]$$

$$= \frac{i|\mathbf{k}|}{2m}[\mathbf{n}(\varepsilon'^* \cdot (\mathbf{n} \wedge \varepsilon)) + \mathbf{n}(\varepsilon \cdot (\mathbf{n}' \wedge \varepsilon'^*)) - \mathbf{n}'(\varepsilon'^* \cdot (\mathbf{n} \wedge \varepsilon)) - \mathbf{n}'(\varepsilon \cdot (\mathbf{n}' \wedge \varepsilon'^*))$$

$$-\varepsilon(\mathbf{n} \cdot (\mathbf{n}' \wedge \varepsilon'^*)) - (\mathbf{n}' \wedge \varepsilon'^*)(\mathbf{n}'\varepsilon) + \varepsilon'^*(\mathbf{n}' \cdot (\mathbf{n} \wedge \varepsilon)) + (\mathbf{n} \wedge \varepsilon)(\mathbf{n}\varepsilon'^*))]$$

$$= \frac{i|\mathbf{k}|}{2m}[2(\mathbf{n}' \wedge \varepsilon'^*) \wedge (\mathbf{n} \wedge \varepsilon) + (\mathbf{n} \wedge \varepsilon)(\mathbf{n}\varepsilon'^*) + \mathbf{n}(\varepsilon'^* \cdot (\mathbf{n} \wedge \varepsilon))$$

$$-(\mathbf{n}' \wedge \varepsilon'^*)(\mathbf{n}'\varepsilon) - \mathbf{n}'(\varepsilon \cdot (\mathbf{n}' \wedge \varepsilon'^*))]$$

d) The matrix element squared, summed over final polarizations, equals

$$\sum_{\varepsilon'} |w'^\dagger \{A + \mathbf{B} \cdot \boldsymbol{\sigma}\} w|^2 = \sum_{\varepsilon'} w^\dagger \{A^* + \mathbf{B}^* \cdot \boldsymbol{\sigma}\} \{A + \mathbf{B} \cdot \boldsymbol{\sigma}\} w$$

$$= \sum_{\varepsilon'} w^\dagger \{A^*A + \mathbf{B}^*\mathbf{B} + (A\mathbf{B}^* + A^*\mathbf{B}) \cdot \boldsymbol{\sigma} + i(\mathbf{B}^* \wedge \mathbf{B}) \cdot \boldsymbol{\sigma}\} w \qquad (23)$$

We sum over ε noticing that

$$\sum \varepsilon_i' \varepsilon_j'^* = \delta_{ij} - n_i' n_j' \qquad (24)$$

Using the following notation for A and $\mathbf{B}$

$$A = A_i \varepsilon_i'^* = \mathbf{A}\varepsilon'^* \qquad (25a)$$

$$B_i = B_{ij} \varepsilon_j'^* \qquad (25b)$$

and averaging over ε', we find

$$\langle AA^* \rangle = \mathbf{A}\mathbf{A}^* - (\mathbf{n}'\mathbf{A})(\mathbf{n}'\mathbf{A}^*) \qquad (26a)$$

$$\langle AB_i^* \rangle = A_j B_{ij}^* - (n_j' A_j)(n_r' B_{ir}^*) \qquad (26b)$$

$$\langle B_i B_j^* \rangle = B_{it} B_{jt}^* - (n_k' B_{ik})(n_t' B_{jt}^*) \qquad (26c)$$

To first order in $|\mathbf{k}|/m$, these averages give

$$\sum_{pol} A^*A = 4\left(1 - |\mathbf{n}'\varepsilon|^2\right) \qquad (27a)$$

$$\sum_{pol} A^*\mathbf{B} = i\frac{|\mathbf{k}|}{m}\{(\mathbf{n} - \mathbf{n}')[(\mathbf{n} - \mathbf{n}') \cdot (\varepsilon \wedge \varepsilon^*)] + (\varepsilon^*\mathbf{t})\mathbf{t} \wedge \varepsilon - (\mathbf{t}\varepsilon)\mathbf{t} \wedge \varepsilon^* \qquad (27b)$$

$$-\varepsilon[\varepsilon^* \cdot (\mathbf{n} \wedge \mathbf{n}')] - \varepsilon^*[\varepsilon \cdot (\mathbf{n} \wedge \mathbf{n}')]$$

$$-(\mathbf{n}\mathbf{n}')[(\mathbf{n}'\varepsilon)\mathbf{n} \wedge \varepsilon^* + (\mathbf{n}'\varepsilon^*)\mathbf{n} \wedge \varepsilon]\}$$

with $\mathbf{t} = \mathbf{n}' - (\mathbf{n}\mathbf{n}')\mathbf{n}$. In (27b) only the first three terms contribute to the sum $A^*\mathbf{B} + A\mathbf{B}^*$. In terms of the electron polarization $\mathbf{P}$, Eq.(23) becomes, to first order in $|\mathbf{k}|/m$

$$\text{Tr}\left\{\frac{1}{2}(1 + \mathbf{P} \cdot \boldsymbol{\sigma})(A^* + \mathbf{B}^* \cdot \boldsymbol{\sigma})(A + \mathbf{B} \cdot \boldsymbol{\sigma})\right\} = AA^* + \mathbf{P} \cdot (A\mathbf{B}^* + A^*\mathbf{B}) \qquad (28)$$

and, using eqs.(27)

$$\overline{|\mathcal{M}|^2} = 4\left(1 - |\mathbf{n}'\boldsymbol{\varepsilon}|^2\right) +$$

$$2\mathrm{i}\,\frac{|\mathbf{k}|}{m}\,\mathbf{P}\cdot\left\{(\mathbf{n} - \mathbf{n}')\left[(\mathbf{n} - \mathbf{n}')\cdot(\boldsymbol{\varepsilon}\wedge\boldsymbol{\varepsilon}^*)\right] + \left[(\boldsymbol{\varepsilon}^*\mathbf{t})\mathbf{t}\wedge\boldsymbol{\varepsilon} - (\mathbf{t}\boldsymbol{\varepsilon})\mathbf{t}\wedge\boldsymbol{\varepsilon}^*\right]\right\} \qquad (29)$$

In the nonrelativistic limit, the cross section summed over final polarizations is given by

$$\frac{d\sigma}{d\Omega} = \frac{1}{4}\left(\frac{\alpha}{m}\right)^2\overline{|\mathcal{M}|^2} \qquad (30)$$

In particular, for photon polarization $+$ or $-$, and electron polarization P along the beam

$$\frac{d\sigma^{(\pm)}}{d\Omega} = \left(\frac{\alpha}{m}\right)^2\left[\frac{1 + \cos^2\theta}{2} \mp P\frac{|\mathbf{k}|}{m}\cos\theta(1 - \cos\theta)\right]$$

$$\equiv \frac{d\sigma_0}{d\Omega} \mp P\frac{d\sigma_1}{d\Omega} \qquad (31)$$

The scattering angle θ is defined by $\cos\theta = \mathbf{nn}'$. Use has been made of the equality $\mathrm{i}\left[\mathbf{n}\cdot(\boldsymbol{\varepsilon}\wedge\boldsymbol{\varepsilon}^*)\right] = 1$, valid for right-handed photons (polarization $+1$); similarly, $\mathrm{i}\left[\mathbf{n}\cdot(\boldsymbol{\varepsilon}\wedge\boldsymbol{\varepsilon}^*)\right] = -1$, for left-handed photons. The total cross section is

$$\sigma^{(\pm)} = 4\pi\left(\frac{\alpha}{m}\right)^2\left[\frac{2}{3} \pm \frac{P}{3}\frac{|\mathbf{k}|}{m}\right] \qquad (32)$$

The electron polarization can be measured through the asymmetry A_{+-} in scattering

$$A_{+-} = \frac{\sigma^{(+)} - \sigma^{(-)}}{\sigma^{(+)} + \sigma^{(-)}} = \frac{P|\mathbf{k}|}{2m} \qquad (33)$$

Conversely, when the γ's have a partial right-handed polarization C, while the electrons are polarized along the beam, Eq.(31) implies

$$\sigma = C(\sigma_0 - P\,\sigma_1) + (1 - C)(\sigma_0 + P\,\sigma_1) = \sigma_0 + P(1 - 2C)\sigma_1 \qquad (34)$$

Thus, by measuring the cross section one can deduce the degree of photon polarization.

Problem 78.   Production of heavy lepton pairs from e^+e^-

Consider the process

$$e^+ + e^- \to L + \bar{L} \qquad (1)$$

where L is a charged particle with spin 1/2 and mass $m_L \simeq 1.8 \, \text{GeV}$, and $\bar{L}$ is the corresponding antiparticle.

As a consequence of the Coulomb interaction in the final state, the process is resonant on hydrogenoid bound states of the outgoing particles; even in the continuum spectrum there is a correction to the cross section.

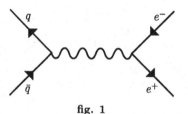

fig. 1

a) Identify the bound states giving rise to a resonance, treating the annihilation to first perturbative order (single photon intermediate state).

b) Calculate the decay width of such states into an $e^+ e^-$ pair.

c) Calculate the contribution of bound states to the amplitude of $e^+ e^- \rightarrow e^+ e^-$ scattering, using a Breit-Wigner parameterization.

Solution

a) The quantum numbers of a spin-1/2 particle-antiparticle system are

$$P = (-)^{L+1} \qquad\qquad C = (-)^{L+S} \tag{2}$$

where S is the eigenvalue of the total spin

$$\mathbf{S} = \mathbf{S}_L + \mathbf{S}_{\bar{L}} \tag{3}$$

If the interaction proceeds through a single photon, the intermediate state has definite charge conjugation $C = -1$ and parity $P = -1$, and total angular momentum $J = 1$. Then, parity invariance implies that L is even, while CP invariance requires $S = 1$. Therefore, the bound states which give a resonant cross section are

$$^3S_1 \qquad\qquad ^3D_1 \tag{4}$$

b) A hydrogenoid state $L\bar{L}$ is described by a superposition of states of the form

$$|\psi\rangle = \int \frac{\mathrm{d}^3 k}{(2\pi)^3} \, \tilde{\psi}(\mathbf{k}) |\frac{\mathbf{P}}{2} + \mathbf{k}\rangle |\frac{\mathbf{P}}{2} - \mathbf{k}\rangle \tag{5}$$

$\mathbf{P}$ and $\mathbf{k}$ stand for the total and relative momenta, respectively, of the heavy leptons; $\tilde{\psi}(\mathbf{k})$ is the Fourier transform of the Schrödinger wave function. In the nonrelativistic approximation, the state is normalized as follows

$$\langle \psi | \psi' \rangle = 4 E_L E_{\bar{L}} (2\pi)^3 \delta^3(\mathbf{P} - \mathbf{P}') \int \frac{\mathrm{d}^3 k}{(2\pi)^3} \left| \tilde{\psi}(\mathbf{k}) \right|^2 \simeq 4 m_L^2 (2\pi)^3 \delta^3(\mathbf{P} - \mathbf{P}') \tag{6}$$

Spin variables have been left out for the time being.

The width of the decay of the state $|\psi\rangle$ into a lepton pair $\ell^- \ell^+$ is given by

$$\Gamma = \frac{1}{4E_L E_{\bar{L}}} \sum_{pol.\ell} \int d\Phi^{(2)} \left| \langle \ell^+ \ell^- | J_\mu | 0 \rangle \frac{1}{q^2} \langle 0 | J^\mu | \psi \rangle \right|^2 \tag{7}$$

Here, q is the four-momentum of the virtual photon and J_μ is the electromagnetic current; the normalization is compatible with Eq.(6). Substituting the expression (5) for $|\psi\rangle$ we obtain

$$\begin{aligned}
\Gamma &= \frac{1}{4E_L E_{\bar{L}}} \sum_{pol.\ell} \int d\Phi^{(2)} \left| \langle \ell^+ \ell^- | J_\mu | 0 \rangle \frac{1}{q^2} \int \langle 0 | J^\mu | \mathbf{k}, -\mathbf{k} \rangle \frac{d^3 \mathbf{k}}{(2\pi)^3} \tilde{\psi}(\mathbf{k}) \right|^2 \\
&= \int d\Phi^{(2)} \frac{1}{4E_L E_{\bar{L}}} \left| \int \frac{d^3 \mathbf{k}}{(2\pi)^3} \mathcal{M}(\mathbf{k}) \tilde{\psi}(\mathbf{k}) \right|^2
\end{aligned} \tag{8}$$

$\mathcal{M}(\mathbf{k})$ is the amplitude of the process $L\bar{L} \to \ell^+ \ell^-$. Ignoring the $\mathbf{k}$-dependence in the amplitude, since the momenta involved in $\tilde{\psi}(\mathbf{k})$ are much smaller than the scales appearing in $\mathcal{M}(\mathbf{k})$, we find

$$\Gamma = |\psi(0)|^2 \int d\Phi^{(2)} \frac{1}{4E_L E_{\bar{L}}} |\mathcal{M}(0)|^2 \tag{9}$$

We have made use of the relation

$$\psi(0) = \int \frac{d^3 \mathbf{k}}{(2\pi)^3} \tilde{\psi}(\mathbf{k}) \tag{10}$$

By Eq.(9), only the S-wave amplitude is different from zero.

For a given lepton momentum $\mathbf{k}$, the annihilation cross section $L\bar{L} \to \ell^+ \ell^-$ can be written as

$$d\bar{\sigma}(\mathbf{k}) = \frac{1}{4E_L E_{\bar{L}} v_R} \frac{1}{4} \sum_{pol.L} \sum_{pol.\ell} \left| \langle \ell^+ \ell^- | J_\mu | 0 \rangle \frac{1}{q^2} \langle 0 | J^\mu | L\bar{L} \rangle \right|^2 d\Phi^{(2)} \tag{11}$$

v_R is the relative velocity of the heavy leptons. Denoting by $\mathcal{M}_T$ and $\mathcal{M}_S$ the triplet and singlet amplitudes, we have

$$\bar{\sigma} = \int d\Phi^{(2)} \frac{1}{4E_L E_{\bar{L}} v_R} \frac{1}{4} \left(3|\mathcal{M}_T|^2 + |\mathcal{M}_S|^2 \right) \tag{12}$$

We showed in part a) that only the triplet state contributes; consequently,

$$|\mathcal{M}_T|^2 \propto \frac{4}{3} \bar{\sigma} \tag{13}$$

The width now takes the form

$$\Gamma = \frac{4}{3} |\psi(0)|^2 \left(v_R \, \bar{\sigma}_{\ell^+ \ell^-} \right)_{v_R \to 0} \tag{14}$$

We recall that the wave function appearing above is

$$\psi_n(0) = \frac{1}{\sqrt{\pi}} \left(\frac{\alpha\mu}{n}\right)^{\frac{3}{2}} \tag{15}$$

$\mu = \frac{m}{2}$ is the reduced mass and $\frac{n}{\mu\alpha} = r_n$ is the Bohr radius of the nth level.

Eq.(9) gives zero for the D-wave, because $\psi \simeq r^2$ as $r \to 0$; one must consider instead successive terms in the k-expansion of $\mathcal{M}(k)$

$$\mathcal{M}(k) \simeq \mathcal{M}(0)(1 + c\frac{k^2}{m_L^2}) \tag{16}$$

The resulting width is

$$\Gamma_D \simeq \int d\Phi^{(2)} \frac{1}{4E_L E_{\overline{L}}} |\mathcal{M}(0)|^2 \left|\frac{c}{m_L^2} \nabla^2 \psi(0)\right|^2 \tag{17}$$

Given that

$$\nabla^2 \psi \simeq \frac{1}{r_B^2} \psi \tag{18}$$

where r_B is the system's Bohr radius, we conclude that

$$\Gamma_D \simeq \Gamma_S \left(\frac{1}{r_B m_L}\right)^4 \simeq \Gamma_S \alpha^4 \tag{19}$$

Let us now derive explicitly the width of the S-states. This entails calculating

$$\sigma(L\overline{L} \to e^+e^-) = \frac{1}{2E_L \, 2E_{\overline{L}} \, v_R} \overline{|\mathcal{M}|^2} \, d\Phi^{(2)} \tag{20}$$

In standard notation, $s = 4E_L E_{\overline{L}}$ and

$$\mathcal{M} = \frac{e^2}{s} \bar{u}(\mathbf{p}_{e^-})\gamma^\mu v(\mathbf{p}_{e^+}) \, \bar{v}(\mathbf{p}_{\overline{L}})\gamma_\mu u(\mathbf{p}_L) \tag{21}$$

The modulus squared of $\mathcal{M}$ can be expressed as

$$\overline{|\mathcal{M}|^2} = \frac{e^4}{(q^2)^2} \frac{1}{4} E_{\mu\nu} Q^{\mu\nu} \tag{22}$$

with

$$
\begin{aligned}
E_{\mu\nu} &= 2\left[q_\mu q_\nu - q^2 g_{\mu\nu} - r_\mu r_\nu\right] \\
Q_{\mu\nu} &= 2\left[q_\mu q_\nu - q^2 g_{\mu\nu} - l_\mu l_\nu\right]
\end{aligned}
$$

$$q = (p_{e^-} + p_{e^+}) = p_L + p_{\overline{L}} \qquad r = p_{e^-} - p_{e^+} \qquad l = p_L - p_{\overline{L}} \tag{23}$$

We now average the tensor $E_{\mu\nu}$ over final state phase space, with the result

$$\langle E_{\mu\nu} \rangle = \frac{4}{3} \left(1 + \frac{2m_e^2}{q^2} \right) (q_\mu \, q_\nu - q^2 g_{\mu\nu}) \tag{24}$$

Eq.(22) now becomes

$$\langle \overline{|M|^2} \rangle = \frac{4}{3} e^4 \left(1 + \frac{2m_e^2}{q^2} \right) \left(1 + \frac{2m_L^2}{q^2} \right) \tag{25}$$

and the cross section reads

$$\sigma_{L\overline{L} \to e^+e^-} = \frac{4\pi}{3} \frac{\alpha^2}{q^2} \left(1 + \frac{2m_L^2}{q^2} \right) \left(1 + \frac{2m_e^2}{q^2} \right) \frac{\sqrt{1 - \dfrac{4m_e^2}{q^2}}}{\sqrt{1 - \dfrac{4m_L^2}{q^2}}} \tag{26}$$

The expression for the relative velocity

$$v_R = 2\sqrt{1 - \frac{4m_L^2}{q^2}} \tag{27}$$

in the limit $v_R \to 0$, leads to $q^2 \to 4m_L^2$ and

$$v_R \, \sigma_{L\overline{L} \to e^+e^-} \to \frac{8\pi}{3} \frac{\alpha^2}{4m_L^2} \left(1 + \frac{2m_L^2}{4m_L^2} \right) = \frac{\pi\alpha^2}{m_L^2} \tag{28}$$

In conclusion, the decay width of a state $n \, {}^3S_1$ into e^+e^- is

$$\Gamma_{e^+e^-} = \frac{\alpha^5}{6} \frac{m_L}{n^3} \tag{29}$$

c)   Let us note at first that, in general, the total decay width of $L\overline{L}$ states includes a term corresponding to radiative transitions into bound states with lower energy

$$\Gamma_n = \Gamma_{e^+e^-} + \Gamma_{\mu^+\mu^-} + \Gamma_{\text{rad}} \tag{30}$$

Γ_{rad} can be estimated in the dipole approximation: A bound state $|n\rangle$ decays into a state $|m\rangle$ with width

$$\Gamma_{\text{rad}} = \frac{1}{3\pi} |\langle m|\mathbf{d}|n\rangle|^2 \omega_{mn}^3 \tag{31}$$

where $\mathbf{d}$ is the electric dipole operator and $\omega_{mn} = E_n - E_m$ is the transition frequency. Summing over states $|m\rangle$, and using the relations $\dot{\mathbf{d}} = i\,\omega\,\mathbf{d}$, $\ddot{\mathbf{d}} = -\omega^2\mathbf{d}$, we write

$$\Gamma_{\text{rad}} \simeq \sum_m \Gamma_{\text{rad}}(n \to m) = \frac{1}{3\pi} \langle n|i\,\dot{\mathbf{d}}\,\ddot{\mathbf{d}}|n\rangle \tag{32}$$

The above may now be estimated classically, setting $\mathbf{d} = e\mathbf{r}$ and noting the following relations, valid in the nS-level,

$$r = \frac{n^2}{m\alpha} \qquad |\dot{r}| = \omega r \qquad |\ddot{r}| = \omega^2 r = \frac{\alpha}{m\,r^2} \tag{33}$$

We are thus led to

$$\omega \simeq m\alpha^2 n^{-3} \qquad \Gamma_{\text{rad}} \simeq \frac{4\alpha^5 m}{3n^5} \ll \frac{1}{6}\frac{\alpha^5 m}{n^3} = \Gamma_{e^+e^-} \tag{34}$$

for large n.

Furthermore, one has

$$\frac{\Gamma_{\mu^+\mu^-}}{\Gamma_{e^+e^-}} \simeq \frac{\left(1 + \dfrac{2m_\mu^2}{m_L^2}\right)\left(1 - \dfrac{4m_\mu^2}{m_L^2}\right)^{\frac{1}{2}}}{\left(1 + \dfrac{2m_e^2}{q^2}\right)\left(1 - \dfrac{4m_e^2}{m_L^2}\right)^{\frac{1}{2}}} \simeq 1 - \frac{1}{8}\left(\frac{m_\mu}{m_L}\right)^4 \tag{35}$$

leading to

$$\Gamma(n) \simeq 2\Gamma_{e^+e^-} = \frac{\alpha^5 m}{3n^3} \tag{36}$$

In the region where the cross section is dominated by bound state resonances, we may write

$$\sigma = \frac{4\pi}{q^2}\frac{(2J+1)}{(2S_L+1)^2}|\mathcal{M}|^2 = \frac{3\pi}{q^2}|\mathcal{M}|^2 \tag{37}$$

The amplitude $\mathcal{M}$ is obtained summing over resonances

$$\mathcal{M} = \sum_n \frac{\Gamma_{e^+e^-}(n)}{\left(E - 2m_L + \dfrac{m_L\alpha^2}{4n^2}\right) + i\dfrac{\Gamma_T(n)}{2}} \tag{38}$$

The relative phase between nearby resonances has been neglected, because the level separation is large compared to the width; indeed, the level separation is

$$\Delta E \simeq \frac{dE}{dn} \simeq \frac{m_L\alpha^2}{2n^3} \tag{39}$$

while the width equals

$$\Gamma_T = \frac{\alpha^5}{3}\frac{m_L}{n^3} \tag{40}$$

Consequently,

$$\frac{\Gamma_T}{\Delta E} \simeq \frac{2}{3}\alpha^3 \ll 1 \tag{41}$$

The sum which appears in $\mathcal{M}$ can be easily performed when $E - 2m_L \gg m_L \alpha^2$, that is, far above the resonances. In that case, binding energies and widths are negligible and $\mathcal{M}$ becomes

$$\mathcal{M} = \frac{\alpha^5}{6} \frac{m_L}{(E - 2m_L)} \sum_1^\infty \frac{1}{n^3} = \zeta(3) \frac{\alpha^5}{6} \frac{m_L}{(E - 2m_L)} \tag{42}$$

Problem 79. Scattering $\gamma - \gamma$

The elastic scattering amplitude $\gamma - \gamma$ is described, to lowest perturbative order, by the diagram

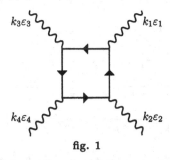

$k_3 \varepsilon_3 \qquad k_1 \varepsilon_1$

$k_4 \varepsilon_4 \qquad k_2 \varepsilon_2$

fig. 1

together with all diagrams obtained from the above by permuting the photons among themselves.

a) Show that the sum of these diagrams is gauge invariant.

b) What is the dimension of the invariant amplitude for this process, in natural units?

c) By virtue of gauge invariance and by the superposition principle, the amplitude must be quadrilinear in the electromagnetic tensors $F_{\mu\nu}^i$ ($i = 1, 2, 3, 4$) of the four photons. Write down the most general quadrilinear in $F_{\mu\nu}^i$, which is Lorentz invariant, parity invariant and symmetric with respect to particle interchange. Approximate the invariant functions contained in this expression by appropriate powers of α and of the electron mass.

d) How does the cross section at low energies grow, as a function of the center of mass energy? What is its order of magnitude?

e) Show that a static electric or magnetic field generate a rectilinear birefringence in the vacuum; calculate its order of magnitude.

Solution

a) We write the invariant amplitude in the form

$$\mathcal{M}_{fi} = e^4 \, \varepsilon_1^\mu \varepsilon_2^\nu \varepsilon_3^{*\rho} \varepsilon_4^{*\sigma} \int \frac{d^4 p}{(2\pi)^4}$$

$$\text{Tr}\left\{\gamma^\mu \frac{1}{\not{p}+\not{k}_1-m}\gamma^\nu\frac{1}{\not{p}+\not{k}_1+\not{k}_2-m}\gamma^\rho\frac{1}{\not{p}+\not{k}_1+\not{k}_2+\not{k}_3-m}\gamma^\sigma\frac{1}{\not{p}-m}\right\}$$
$$+(1342)+(1432)+(1324)+(1423)+(1243) \tag{1}$$

For the sake of demonstrating gauge invariance we substitute ε_1^μ by k_1^μ; using the equality

$$\frac{1}{\not{p}-m}\not{k}_1\frac{1}{\not{p}+\not{k}_1-m}=\frac{1}{\not{p}-m}-\frac{1}{\not{p}+\not{k}_1-m} \tag{2}$$

we obtain the following expression

$$\begin{aligned}
\text{Tr}\Bigg\{ &\frac{1}{\not{p}-m}\not{q}_2\frac{1}{\not{p}+\not{k}_1+\not{k}_2-m}\not{q}_3^*\frac{1}{\not{p}+\not{k}_1+\not{k}_2+\not{k}_3-m}\not{q}_4^* \\
&-\frac{1}{\not{p}+\not{k}_1-m}\not{q}_2\frac{1}{\not{p}+\not{k}_1+\not{k}_2-m}\not{q}_3^*\frac{1}{\not{p}+\not{k}_1+\not{k}_2+\not{k}_3-m}\not{q}_4^* \\
&+\frac{1}{\not{p}-\not{k}_2-m}\not{q}_2\frac{1}{\not{p}-m}\not{q}_3^*\frac{1}{\not{p}+\not{k}_1+\not{k}_3-m}\not{q}_4^* \\
&-\frac{1}{\not{p}-\not{k}_2-m}\not{q}_2\frac{1}{\not{p}+\not{k}_1-m}\not{q}_3^*\frac{1}{\not{p}+\not{k}_1+\not{k}_3-m}\not{q}_4^* \\
&+\frac{1}{\not{p}-\not{k}_2-\not{k}_3-m}\not{q}_2\frac{1}{\not{p}-\not{k}_3-m}\not{q}_3^*\frac{1}{\not{p}-m}\not{q}_4^* \\
&-\frac{1}{\not{p}-\not{k}_2-\not{k}_3-m}\not{q}_2\frac{1}{\not{p}-\not{k}_3-m}\not{q}_3^*\frac{1}{\not{p}+\not{k}_1-m}\not{q}_4^* \\
&+\frac{1}{\not{p}-m}\not{q}_2\frac{1}{\not{p}+\not{k}_1+\not{k}_2-m}\not{q}_4^*\frac{1}{\not{p}-\not{k}_3-m}\not{q}_3^* \\
&-\frac{1}{\not{p}+\not{k}_1-m}\not{q}_2\frac{1}{\not{p}+\not{k}_1+\not{k}_2-m}\not{q}_4^*\frac{1}{\not{p}-\not{k}_3-m}\not{q}_3^* \\
&+\frac{1}{\not{p}+\not{k}_1+\not{k}_3-m}\not{q}_2\frac{1}{\not{p}+\not{k}_1+\not{k}_2+\not{k}_3-m}\not{q}_4^*\frac{1}{\not{p}-m}\not{q}_3^* \\
&-\frac{1}{\not{p}+\not{k}_1+\not{k}_3-m}\not{q}_2\frac{1}{\not{p}+\not{k}_1+\not{k}_2+\not{k}_3-m}\not{q}_4^*\frac{1}{\not{p}+\not{k}_1-m}\not{q}_3^* \\
&+\frac{1}{\not{p}-\not{k}_2-m}\not{q}_2\frac{1}{\not{p}-m}\not{q}_4^*\frac{1}{\not{p}-\not{k}_2-\not{k}_3-m}\not{q}_3^* \\
&-\frac{1}{\not{p}-\not{k}_2-m}\not{q}_2\frac{1}{\not{p}+\not{k}_1-m}\not{q}_4^*\frac{1}{\not{p}-\not{k}_2-\not{k}_3-m}\not{q}_3^*\Bigg\}
\end{aligned} \tag{3}$$

To show that this expression vanishes identically, it suffices to shift the integration variable p, in each of the twelve summands, by a corresponding quantity given below

$$(0,0,0,k_2,k_1+k_2+k_3,k_3,0,0,-k_3,-k_1-k_2-k_3,0,k_2) \tag{4}$$

b) The invariant amplitude has dimension zero; this is true of any scattering process with two-body initial and final states.

c)   There exist only two quadrilinear forms in $F^i_{\mu\nu}$, which are Lorentz and parity invariant

$$L_1 = F^1_{\mu\nu}\, F^{2\,\mu\nu}\, F^{3\,\rho\sigma} F^4_{\rho\sigma} + \text{permutations} \tag{5a}$$

$$L_2 = \frac{1}{2}\varepsilon^{\mu\nu\rho\sigma}\, F^1_{\mu\nu}\, F^2_{\rho\sigma}\, \frac{1}{2}\varepsilon^{\alpha\beta\gamma\delta}\, F^3_{\alpha\beta}\, F^4_{\gamma\delta} + \text{permutations} \tag{5b}$$

It is easy to prove this result, noting that tensorial indices can only be contracted with $g_{\mu\nu}$ or with an even number of tensors $\varepsilon_{\mu\nu\rho\sigma}$, because of parity; in fact, the second case need not be considered explicitly, since a pair of tensors $\varepsilon_{\mu\nu\rho\sigma}\varepsilon_{\alpha\beta\gamma\delta}$ can always be decomposed into a sum of products of $g_{\mu\nu}$. Using also $F^\mu_{\ \mu} = 0$, the only possible contractions turn out to be L_1 and

$$F^1_{\mu\nu}\, F^2_{\nu\rho}\, F^3_{\rho\sigma}\, F^4_{\sigma\mu} + \text{perm} \equiv \frac{1}{2}\left(L_1 - 4\, L_2\right) \tag{6}$$

as can be explicitly verified.

The physical dimension of L_i is $[m^8]$ and the number of photon vertices involved is four; thus, for photon energies $k^0_i \ll m_e$, the amplitude can be parameterized as

$$\mathcal{L}_I = \frac{\alpha^2}{m^4_e}\left[C_1\, L_1 + C_2\, L_2\right] \tag{7}$$

where C_1 and C_2 are dimensionless functions of the kinematic invariants.

d)   Starting from

$$F^{(i)}_{\mu\nu}\, F^{(j)\mu\nu} = 2\left(\mathbf{E}^{(i)} \cdot \mathbf{E}^{(j)} - \mathbf{H}^{(i)} \cdot \mathbf{H}^{(j)}\right) \tag{8a}$$

$$\frac{1}{2}\varepsilon_{\mu\nu\rho\sigma}F^{(i)\mu\nu}\, F^{(j)\rho\sigma} = 2\left[\mathbf{E}^{(i)} \cdot \mathbf{H}^{(j)} + \mathbf{E}^{(j)} \cdot \mathbf{H}^{(i)}\right] \tag{8b}$$

and

$$\mathbf{E} = -i\,\omega\,\boldsymbol{\varepsilon}\, e^{-ikx} \tag{9a}$$

$$\mathbf{H} = -i\,\omega\,\hat{\mathbf{n}} \wedge \boldsymbol{\varepsilon}\, e^{-ikx} \tag{9b}$$

and noting that all photons have the same energy in the center of mass frame, we conclude that the invariants L_i are proportional to ω^4.

Using an order-of-magnitude value of 1 for the constants C_i, as suggested in the text, we find

$$\mathcal{M}_{fi} \sim \frac{\alpha^2}{m^4_e}\omega^4 \tag{10}$$

and

$$\sigma = \frac{1}{8\omega^2}|\mathcal{M}_{fi}|^2\frac{1}{8\pi} \simeq \frac{\alpha^4}{m^2}\left(\frac{\omega}{m}\right)^6 \tag{11}$$

For $\omega \ll m_e$, the dimensionless invariants can indeed be considered constant. On the other hand, at large energies $\omega \gg m_e$ the result cannot depend on m_e; consequently, apart from logarithms, we expect

$$\sigma \sim \frac{\alpha^4}{\omega^2} \tag{12}$$

for dimensional reasons.

e) Carrying through the calculation of the above amplitude, we end up with the following one-loop effective Lagrangian of electrodynamics

$$
\begin{aligned}
\mathcal{L} &= \mathcal{L}_0 + \mathcal{L}_I \\
\mathcal{L}_0 &= \frac{1}{2}(\mathbf{E}^2 - \mathbf{H}^2) \\
\mathcal{L}_I &= \frac{2\alpha^2}{45\,m_e^4}\left[\left(\mathbf{E}^2 - \mathbf{H}^2\right)^2 + 7\left(\mathbf{E} \cdot \mathbf{H}\right)^2\right]
\end{aligned}
\tag{13}
$$

In the above, we have also provided the computed coefficients of $\mathcal{L}_I$ (not required by the problem).

In order to study the propagation of electromagnetic waves in the system described by (10), we separate the field into an external (classical) component plus a part describing quantum fluctuations

$$A_\mu \rightarrow A_\mu + a_\mu \tag{14}$$

The propagator of electromagnetic field quanta can be derived from the part of the Lagrangian quadratic in a_μ. In particular, from the interaction part we have

$$\mathcal{L}_I^{(2)} = \frac{1}{2}a_\mu \left.\frac{\delta^2 \mathcal{L}_I}{\delta A_\mu \delta A_\nu}\right|_{a_\mu = 0} a_\nu \tag{15}$$

Calling $\mathbf{e}$ and $\mathbf{h}$ the electric and magnetic fields constructed out of the radiation field a_μ, the effective quadratic Lagrangian governing wave propagation is given by

$$
\begin{aligned}
\mathcal{L}^{(2)} &= \mathcal{L}_0(a) + \mathcal{L}_I^{(2)} = \mathcal{L}_0(a) + \\
&+ \frac{1}{2}e_i'^* \frac{\delta^2 \mathcal{L}_1}{\delta E_i \delta E_j}e_j + \frac{1}{2}e_i'^* \frac{\delta^2 \mathcal{L}_I}{\delta E_i \delta H_j}h_j + \frac{1}{2}h_i'^* \frac{\delta^2 \mathcal{L}_I}{\delta H_i \delta E_j}e_j + \frac{1}{2}h_i'^* \frac{\delta^2 \mathcal{L}_1}{\delta H_i \delta H_j}h_j \\
&= \frac{1}{2}\left(e_i'^* e_i - h_i'^* h_i\right) + \\
&+ \frac{\alpha^2}{45\,m_e^4}\left\{ e_i'^*\left[8E_i E_j + 4\delta_{ij}(E^2 - H^2) + 14 H_i H_j\right]e_j \right. \\
&\qquad\qquad + e_i'^*\left[-8H_j E_i + 14 E_j H_i + 14\delta_{ij}(\mathbf{E} \cdot \mathbf{H})\right]h_j \\
&\qquad\qquad + h_i'^*\left[-8H_i E_j + 14 E_i H_j + 14\delta_{ij}(\mathbf{E} \cdot \mathbf{H})\right]e_j \\
&\qquad\qquad \left. + h_i'^*\left[8H_i H_j + 4\delta_{ij}(E^2 - H^2) + 14 E_i E_j\right]h_j \right\}
\end{aligned}
\tag{16}
$$

The refractive index can be deduced from the form of the plane wave solutions

$$\boldsymbol{\varepsilon}\, e^{i(n k x - \omega t)} \qquad (17)$$

of the equations stemming from the above Langrangian. Let us first consider the case $\mathbf{H} = 0$. Eq.(16) simplifies as follows

$$\mathcal{L}^{(2)} = \frac{1}{2}\left(e_i'^* e_i - h_i'^* h_i\right) +$$
$$\frac{\alpha^2}{45\, m_e^4}\left[E_i E_j \left(8 e_i'^* e_j + 14 h_i'^* h_j\right) + 4\, E^2 (\mathbf{e}'^* \,\mathbf{e} - \mathbf{h}'^* \,\mathbf{h})\right] \qquad (18)$$

The equations of motion now follow immediately; going to the Coulomb gauge, and performing a Fourier transform, with the standard substitutions

$$\mathbf{e} = i\omega\,\boldsymbol{\varepsilon} \qquad \mathbf{h} = i\,n\omega\,\hat{\mathbf{k}}\wedge\boldsymbol{\varepsilon} \qquad (19)$$

we obtain

$$\frac{1}{2}\left(\omega^2 - n^2\mathbf{k}^2\right)\varepsilon_i + \frac{\alpha^2\omega^2}{45\, m_e^4}\left[8\, E_i E_j + 14(\hat{\mathbf{n}}\wedge\mathbf{E})_i(\hat{\mathbf{n}}\wedge\mathbf{E})_j\right]\varepsilon_j = 0 \qquad (20)$$

This equation does not alter the refractive index of a wave parallel to the electric field. For waves perpendicular to $\mathbf{E}$, on the other hand, the refractive index assumes two different values; expanding in the parameter $\alpha E^2/m^4$, we find

$$n = 1 + \frac{8\alpha^2 E^2}{45\, m_e^4} \qquad \boldsymbol{\varepsilon} \parallel \mathbf{E} \qquad (21\text{a})$$

$$n = 1 + \frac{14\alpha^2 E^2}{45\, m_e^4} \qquad \boldsymbol{\varepsilon} \perp \mathbf{E} \qquad (21\text{b})$$

This system thus behaves like a crystal with rectilinear birefringence. An electric field with intensity $E = 10^4\,\text{Volt/cm}$ leads to

$$n - 1 \simeq 10^{-28} \qquad (22)$$

The case with $\mathbf{E}=0$ is exactly analogous. The corresponding quadratic Lagrangian is

$$\mathcal{L}^{(2)} = \frac{1}{2}\left(e_i'^* e_i - h_i'^* h_i\right) +$$
$$\frac{\alpha^2\omega^2}{45\, m_e^4}\left[H_i H_j \left(14 e_i'^* e_j + 8 h_i'^* h_j\right) + 4\, H^2(\mathbf{h}'^* \,\mathbf{h} - \mathbf{e}'^* \,\mathbf{e})\right] \qquad (23)$$

with the resulting equations of motion

$$\frac{1}{2}\left(\omega^2 - n^2\mathbf{k}^2\right)\varepsilon_i + \frac{\alpha^2\omega^2}{45\, m_e^4}\left[14\, H_i H_j + 8(\hat{\mathbf{n}}\wedge\mathbf{H})_i(\hat{\mathbf{n}}\wedge\mathbf{H})_j\right]\varepsilon_j = 0 \qquad (24)$$

Problem 80.   The electromagnetic form factor of K^0 through electron scattering

In order to measure the scattering cross section of the process

$$K^0 + e^- \rightarrow K^0 + e^- \tag{1}$$

a beam of K^0 mesons with energy 10 GeV is sent in on a target.

a)   What is the total energy in the process, in the center-of-mass frame? What is the maximum value of the momentum transfer q^2?

b)   Suppose that the scattering proceeds through a single photon exchange, as shown in fig. 1.

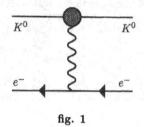

fig. 1

What is the most general form of the vertex $K^0 - \gamma - K^0$, consistent with Lorentz and gauge invariance? K^0 is neutral: How is this property reflected in the form factor?

c)   Calculate the K^0 form factor, representing the particle as a hydrogenoid bound state with radius 10^{-13} cm, made out of a quark and an antiquark with charge $\pm 1/3$ and mass $m_s = 450$ MeV, $m_d = 50$ MeV.

How can this form factor be approximated, at the accessible values of momentum transfer?

d)   Calculate the angular distribution in the laboratory frame and the total cross section, within the model of part c).

Solution

a)   In the laboratory frame we have

$$\begin{aligned}
p_K^\mu &= (E_K, \mathbf{p}_K) \\
p_{e^-}^\mu &= (m_e, \mathbf{0})
\end{aligned} \tag{2}$$

implying

$$E_{\text{cm}} = \sqrt{(p_K + p_{e^-})^2} = \sqrt{m_K^2 + m_e^2 + 2\, m_e E_K} \simeq m_K + \frac{E_K m_e}{m_K} \tag{3}$$

For a 10 Gev kaon beam, $E_{cm} \simeq m_K + 20 m_e$. The K_0 is thus nonrelativistic in the center-of-mass frame, while the electron is relativistic.

By energy conservation

$$\sqrt{p_{cm}^2 + m_e^2} + \sqrt{p_{cm}^2 + m_K^2} = E_{cm} \tag{4}$$

we obtain

$$p_{cm} \simeq E_{cm} - m_k \simeq \frac{E_K m_e}{m_K} \tag{5}$$

The momentum transfer is given by

$$q^2 = (p_K - p_K')^2 = -4 p_{cm}^2 \sin^2 \frac{\theta_{cm}}{2} \tag{6}$$

where θ_{cm} is the scattering angle in the center of mass. q_{max} corresponds to backward scattering $(\theta = \pi)$; it equals

$$q_{max} = 2 p_{cm} = \frac{2 E_K m_e}{m_k} \simeq 20 \frac{MeV}{c} \tag{7}$$

b) The most general form of the vertex is

$$
\begin{aligned}
\Gamma &= \varepsilon_\mu J^\mu(P, q) \\
P &= p_K + p_K' \\
q &= p_K - p_K'
\end{aligned}
\tag{8}
$$

Gauge invariance requires

$$q_\mu J^\mu = 0 \tag{9}$$

that is,

$$J_\mu = J(q^2) \cdot P_\mu \tag{10}$$

The Fourier transform of the time component of the current, at $q_\mu = 0$, is proportional to the charge of the particle. Since the charge is zero in this case, we find

$$J(0) = 0 \tag{11}$$

c) $J(q^2)$ can be evaluated explicitly within the hydrogenoid model suggested in the text. The electromagnetic vertex (8) is defined by the matrix element

$$\int d^4x \, \langle \mathbf{p}_K' | j_\mu(x) | \mathbf{p}_K \rangle \langle \mathbf{q} | A^\mu(x) | 0 \rangle = (2\pi)^4 \delta^{(4)}(p_K - p_K' - q) \varepsilon^\mu J_\mu(P, q) \tag{12}$$

Translational invariance has allowed us to factorize $\delta^{(4)}$. Eqs. (12) and (10) now lead to

$$(2\pi)^4 \delta^{(4)}(p_K - p_K' - q)\left(p_K^0 + p_{K'}^0\right) J(q^2) = \langle \mathbf{p}_K' | \int d^4x \, e^{iqx} j_0(x) | \mathbf{p}_K \rangle \tag{13}$$

The invariant function $J(q^2)$ can be easily evaluated in the center of mass or, more generally, in any reference frame in which K is nonrelativistic. In that case, the charge density operator in the Schrödinger representation has the form

$$j_0(0, \mathbf{x}) = e_1 \, \delta^{(3)}(\mathbf{x} - \mathbf{x}_1) + e_2 \, \delta^{(3)}(\mathbf{x} - \mathbf{x}_2) \tag{14}$$

Here, $\mathbf{x}_1$ and $\mathbf{x}_2$ stand for the coordinates of the components of K, and e_1, e_2 are the corresponding charges. In terms of the center-of-gravity coordinate $\mathbf{X}$ and the relative coordinate $\boldsymbol{\xi}$ we have

$$\mathbf{x}_1 = \mathbf{X} + \frac{\mu}{m_1} \boldsymbol{\xi} \qquad \mathbf{x}_2 = \mathbf{X} - \frac{\mu}{m_2} \boldsymbol{\xi} \tag{15}$$

where μ is the reduced mass. The state of K is characterized by its momentum, and by the internal hydrogenoid state $|s\rangle$. Making use of Eqs. (14) and (15), and of the nonrelativistic relation $p_K'^0 \simeq p_K^0 \simeq m_K$, Eq. (13) becomes

$$(2\pi)^4 \delta^{(4)}(p_K' + q - p_K) 2 m_K J(-\mathbf{q}^2) =$$
$$\int dt \, d^3\mathbf{X} \langle \mathbf{p}_K' | e^{iHt} \, e^{iq^0 t - i\mathbf{q}\mathbf{X}} \, e^{-iHt} | \mathbf{p}_K \rangle \, \langle s | e_1 \, e^{-i\mathbf{q}\frac{\mu}{m_1}\boldsymbol{\xi}} + e_2 \, e^{i\mathbf{q}\frac{\mu}{m_2}\boldsymbol{\xi}} | s \rangle \tag{16}$$

With the normalization adopted for momentum eigenstates, the first matrix element above reproduces the δ-function of momentum conservation and the factor of $2m_K$. Thus, denoting by $\psi_s(\boldsymbol{\xi})$ the wave function corresponding to state $|s\rangle$, we find

$$J(-\mathbf{q}^2) = \int d^3\boldsymbol{\xi} \, |\psi_s(\boldsymbol{\xi})|^2 \left(e_1 \, e^{-i\mathbf{q}\frac{\mu}{m_1}\boldsymbol{\xi}} + e_2 \, e^{i\mathbf{q}\frac{\mu}{m_2}\boldsymbol{\xi}} \right) \tag{17}$$

For the case at hand, $m_1 \gg m_2$ and $e_1 = e/3 = -e_2$, so that (17) reduces to

$$J(-\mathbf{q}^2) = \frac{e}{3} \left(1 - \int d^3\boldsymbol{\xi} \, |\psi_s(\boldsymbol{\xi})|^2 \, e^{i\mathbf{q}\boldsymbol{\xi}} \right) \tag{18}$$

This is simply the Fourier transform at momentum $\mathbf{q}$ of the following charge density

$$\rho = \delta^{(3)}(\boldsymbol{\xi}) - |\psi_s(\boldsymbol{\xi})|^2 \tag{19}$$

The ground state wave function of the hydrogenoid model is

$$\psi_s(\boldsymbol{\xi}) = \frac{1}{\sqrt{\pi a^3}} e^{-|\boldsymbol{\xi}|/a} \tag{20}$$

a is the Bohr radius, $a \sim 10^{-13}$ cm. We conclude that

$$J(-\mathbf{q}^2) = \frac{e}{3} \left[1 - \frac{1}{\left(1 + \mathbf{q}^2 a^2/4\right)^2} \right] = \frac{e}{3} \frac{\mathbf{q}^2 a^2}{4} \frac{2 + \mathbf{q}^2 a^2/4}{\left(1 + \mathbf{q}^2 a^2/4\right)^2} \tag{21}$$

Since $q_{max}a \ll 1$ (cf. Eq.(7)), we write

$$J\left(-\mathbf{q}^2\right) \simeq \frac{e}{6}\mathbf{q}^2 a^2 \tag{22}$$

in agreement with (11).

d)   The expression for the invariant cross section is

$$d\sigma = \frac{1}{v_r}\frac{1}{2E_K}\frac{1}{2E_e}|\mathcal{M}|^2 d\Phi^{(2)} \tag{23}$$

The relative velocity v_R is given by

$$v_R = \frac{\sqrt{(p_K \cdot p_{e-})^2 - m_K^2 m_e^2}}{E_K E_e} \tag{24}$$

and the transition amplitude is

$$\mathcal{M} = e\,\bar{u}(\mathbf{p}'_e)\gamma_\mu u(\mathbf{p}_e)\,P^\mu\,\frac{J(q^2)}{q^2} \tag{25}$$

We take the modulus squared of the amplitude, summing over final polarizations and averaging over initial ones, with the result

$$\overline{|\mathcal{M}|^2} = \left|\frac{e\,J(q^2)}{q^2}\right|^2 P^\mu P^\nu \left[q^2 g_{\mu\nu} - q_\mu q_\nu + r_\mu r_\nu\right] \tag{26}$$

where

$$
\begin{aligned}
r_\mu &= (p_{e-} + p'_{e-})_\mu \\
q_\mu &= (p'_{e-} - p_{e-})_\mu \\
q \cdot r &= 0
\end{aligned}
\tag{27}
$$

Introducing the standard kinematic invariants

$$
\begin{aligned}
s &= (p_K + p_{e-})^2 = m_K^2 + m_e^2 + 2p_K p_{e-} \\
t &= (p'_{e-} - p_{e-})^2 = m_K^2 + m_e^2 - 2p'_{e-} p_{e-} \\
u &= (p'_{e-} - p_K)^2 = m_K^2 + m_e^2 - 2p_K p'_{e-} \\
s + t + u &= 2m_K^2 + 2m_e^2
\end{aligned}
\tag{28}
$$

we simplify further Eq.(26)

$$
\begin{aligned}
\overline{|\mathcal{M}|^2} &= \frac{e^4}{9}a^4\left[q^2 m_K^2 + (q \cdot p_K)^2 + (r \cdot p_K)^2\right] \\
&= \frac{e^4}{9}a^4\left[\left(s - m_K^2 - m_e^2\right)^2 + t(s - m_e^2)\right]
\end{aligned}
\tag{29}
$$

It is convenient to cast Eq.(24) in the form

$$d\sigma = \frac{1}{64\pi} |\mathcal{M}|^2 \frac{dt}{(p_K \cdot p_{e^-})^2 - m_K^2 m_e^2} \tag{30}$$

In order to calculate the angular distribution in the laboratory frame, let us express the kinematic variables in terms of the angle θ_{lab} of emission of the electron. By momentum conservation, the final energy of the electron becomes

$$E'_e = m_e \frac{(E_K + m_e)^2 + \mathbf{p}_K^2 \cos^2 \theta_{\text{lab}}}{(E_K + m_e)^2 - \mathbf{p}_K^2 \cos^2 \theta_{\text{lab}}} \tag{31}$$

and the invariants read

$$\begin{aligned}
s &= m_K^2 + m_e^2 + 2E_K m_e \\
t &= 2m_e^2 - 2E'_e m_e = -4m_e^2 \frac{\mathbf{p}_K^2 \cos^2 \theta_{\text{lab}}}{(E_K + m_e)^2 - \mathbf{p}_K^2 \cos^2 \theta_{\text{lab}}}
\end{aligned} \tag{32}$$

The kinematic limits on t

$$-\frac{4m_e^2 \mathbf{p}_K^2}{m_K^2 + m_e^2 + 2m_e E_K} \leq t \leq 0 \tag{33}$$

correspond to the fact that the angle of emission of the electron in the laboratory frame is always less than 90°. Eqs. (29) and (32) now yield

$$\overline{|\mathcal{M}|}^2 = \frac{e^4}{9} a^4 \, 4m_e^2 \left[E_K^2 - \frac{\mathbf{p}_K^2 \cos^2 \theta_{\text{lab}}}{(E_K + m_e)^2 - \mathbf{p}_K^2 \cos^2 \theta_{\text{lab}}} \left(m_K^2 + 2E_K m_e \right) \right] \tag{34}$$

and

$$dt = \frac{8m_e^2 (E_K + m_e)^2 \, \mathbf{p}_K^2 \cos \theta_{\text{lab}}}{\left[(E_K + m_e)^2 - \mathbf{p}_K^2 \cos^2 \theta_{\text{lab}} \right]^2} \frac{d\Omega_{\text{lab}}}{2\pi} \tag{35}$$

In deriving the above, use was made of the azimuthal symmetry of the problem. Finally, inserting (34) and (35) in (30) we obtain the angular distribution in the laboratory frame

$$\begin{aligned}
d\sigma &= \frac{1}{64\pi} \frac{e^4}{9} a^4 \, 4m_e^2 \left[E_K^2 - \frac{\mathbf{p}_K^2 \cos^2 \theta_{\text{lab}}}{(E_K + m_e)^2 - \mathbf{p}_K^2 \cos^2 \theta_{\text{lab}}} \left(m_K^2 + 2E_K m_e \right) \right] \\
&\quad \cdot \frac{8m_e^2 (E_K + m_e)^2 \, \mathbf{p}_K^2 \cos \theta_{\text{lab}}}{\left[(E_K + m_e)^2 - \mathbf{p}_K^2 \cos^2 \theta_{\text{lab}} \right]^2} \frac{d\Omega_{\text{lab}}}{2\pi} \frac{1}{m_e^2 \mathbf{p}_K^2} \\
&= \frac{4\alpha^2 m_e^2 a^4}{9} \left[E_K^2 - \frac{\mathbf{p}_K^2 \cos^2 \theta_{\text{lab}}}{(E_K + m_e)^2 - \mathbf{p}_K^2 \cos^2 \theta_{\text{lab}}} \left(m_K^2 + 2E_K m_e \right) \right] \\
&\quad \cdot \frac{(E_K + m_e)^2 \cos \theta_{\text{lab}}}{\left[(E_K + m_e)^2 - \mathbf{p}_K^2 \cos^2 \theta_{\text{lab}} \right]^2} \, d\Omega_{\text{lab}}
\end{aligned} \tag{36}$$

To obtain the total cross section, we integrate (30) within the kinematic limits (33); we find

$$\sigma = \frac{\pi}{9} \frac{\alpha^2 a^4}{s} \left[\left(s - m_K^2 - m_e^2 \right)^2 - 2 \frac{m_e^2}{s} \mathbf{p}_K^2 \left(s - m_e^2 \right) \right] \tag{37}$$

At the given beam energy

$$\sigma \simeq 10^{-33} \, \text{cm}^2 \tag{38}$$

Problem 81. A general parameterization of the Compton amplitude

We would like to derive the most general form for the amplitude of the Compton effect on a spin-1/2 particle, compatible with
a) The superposition principle
b) Invariance under the Poincaré group and gauge invariance
c) Invariance under the discrete symmetries P and T.

Show that the most general amplitude satisfying the above properties depends on 6 real functions of the two independent kinematic invariants present. We denote by k (k') and p (p') the initial (final) four-momenta of the photon and of the fermion, respectively; it is suggested to make use of the following independent four-vectors

$$
\begin{aligned}
K_\mu &= k_\mu + k'_\mu \\
P_\mu &= p_\mu + p'_\mu - K_\mu \frac{(p + p') \cdot K}{K^2} \\
Q_\mu &= k'_\mu - k_\mu \\
N_\mu &= \varepsilon_{\mu\nu\rho\sigma} P^\nu Q^\rho K^\sigma
\end{aligned}
\tag{1}
$$

Check your result using the amplitudes for states of definite helicity and their transformation under the symmetries listed above.

Solution

a) The superposition principle implies a linear dependence of the amplitude on the initial and final photon polarizations and on the initial and final fermion wave functions.

$$\mathcal{M}_{fi} = \varepsilon'^*_\alpha \, \mathcal{M}_{\alpha\beta} \, \varepsilon_\beta \tag{2}$$

($\mathcal{M}_{\alpha\beta}$ is a bilinear in fermion wave functions).
b) Poincaré invariance implies that $\mathcal{M}_{\alpha\beta}$ must be a tensor, constructed out of the (four-)vectors at our disposal. Indeed, any tensor can be written as a superposition of products of the 4 independent vectors listed above (with Lorentz scalar coefficients).

The conditions

$$\varepsilon' k' = \varepsilon k = 0 \tag{3}$$

together with gauge invariance

$$k'_\alpha \mathcal{M}_{\alpha\beta} = \mathcal{M}_{\alpha\beta} k_\beta = 0 \tag{4}$$

exclude superpositions involving K_μ and Q_μ in $\mathcal{M}_{\alpha\beta}$. Thus, the most general gauge invariant form is

$$
\begin{aligned}
\mathcal{M}_{\alpha\beta} &= G_0 \left(\frac{P_\alpha P_\beta}{-P^2} + \frac{N_\alpha N_\beta}{-N^2} \right) + G_1 \frac{P_\alpha N_\beta + N_\alpha P_\beta}{\sqrt{P^2 N^2}} \\
&+ G_2 \frac{P_\alpha N_\beta - N_\alpha P_\beta}{\sqrt{P^2 N^2}} + G_3 \left(\frac{P_\alpha P_\beta}{-P^2} - \frac{N_\alpha N_\beta}{-N^2} \right)
\end{aligned} \tag{5}
$$

where the G's are given by

$$G_i = \bar{u}(\mathbf{p}') \, Q_i \, u(\mathbf{p}) \tag{6}$$

and Q_i are matrices to be determined.

c)   We want $\mathcal{M}$ to be invariant under parity; since P_μ is a polar vector while N_μ is an axial vector, G_0 and G_3 must be scalars, and G_1, G_2 pseudoscalars. Making use of the equations of motion to eliminate linearly dependent terms, we find [12]

$$
\begin{aligned}
Q_0 &= f_1 + f_2 \slashed{K} \\
Q_1 &= \gamma^5 \left(f_3 + f_4 \slashed{K} \right) \\
Q_2 &= \gamma^5 \left(f_5 + f_6 \slashed{K} \right) \\
Q_3 &= f_7 + f_8 \slashed{K}
\end{aligned} \tag{7}
$$

f_i are invariant amplitudes and can depend only on the kinematic invariants s, t, u.

Time reversal invariance requires

$$S_{fi} = S_{T_i T_f} \tag{8}$$

$S_{T_i T_f}$ is obtained from S_{fi} with the substitutions

$$
\begin{aligned}
(k^0, \mathbf{k}) &\leftrightarrow (k'_0, -\mathbf{k}') \\
(p^0, \mathbf{p}) &\leftrightarrow (p'_0, -\mathbf{p}') \\
(\varepsilon_0, \boldsymbol{\varepsilon}) &\leftrightarrow (\varepsilon_0'^*, -\boldsymbol{\varepsilon}'^*)
\end{aligned} \tag{9}
$$

s, t and u stay invariant. Thus, T invariance amounts to the requirement

$$
\begin{aligned}
T : \quad \mathcal{M}_{00} &\rightarrow \mathcal{M}_{00} \\
T : \quad \mathcal{M}_{i0} &\rightarrow -\mathcal{M}_{0i} \\
T : \quad \mathcal{M}_{ik} &\rightarrow \mathcal{M}_{ki}
\end{aligned} \tag{10}
$$

---

[12]In general, Q_i are polynomials in $\slashed{P}$, $\slashed{Q}$, $\slashed{K}$, multiplied possibly by γ_5. ($\slashed{N}$ can be eliminated by $\slashed{N} = \gamma_5 \slashed{P} \slashed{Q} \slashed{K}$.) Rewriting Q_i in terms of $\slashed{p}$, $\slashed{p}'$, $\slashed{K}$, we further eliminate $\slashed{p}$, $\slashed{p}'$ using the equations of motion.

Under transformation (9) we have

$$
\begin{aligned}
(K^0, \mathbf{K}) &\rightarrow (K^0, -\mathbf{K}) \\
(Q^0, \mathbf{Q}) &\rightarrow (-Q^0, \mathbf{Q}) \\
(P^0, \mathbf{P}) &\rightarrow (P^0, -\mathbf{P}) \\
(N^0, \mathbf{N}) &\rightarrow (N^0, -\mathbf{N})
\end{aligned}
\tag{11}
$$

Consequently, G_i must transform as follows

$$
G_0 \rightarrow G_0 \, ; \, G_1 \rightarrow G_1 \, ; \, G_3 \rightarrow G_3 \, ; \, G_2 \rightarrow -G_2
\tag{12}
$$

Comparing the above to the transformation of spinor bilinears under T,

$$
\begin{aligned}
\bar{u}' u &\rightarrow \bar{u}' u & \bar{u}' \gamma^5 u &\rightarrow -\bar{u}' \gamma^5 u \\
\bar{u}' \slashed{K} u &\rightarrow \bar{u}' \slashed{K} u & \bar{u}' \gamma^5 \slashed{K} u &\rightarrow \bar{u}' \gamma^5 \slashed{K} u
\end{aligned}
\tag{13}
$$

we conclude that $f_3 = 0$, $f_6 = 0$ and

$$
\begin{aligned}
\mathcal{M}_{\alpha\beta} = {} & \bar{u}'(f_1 + f_2 \slashed{K})u\,(\hat{P}_\alpha \hat{P}_\beta + \hat{N}_\alpha \hat{N}_\beta) \\
& + \bar{u}'(f_7 + f_8 \slashed{K})u\,(\hat{P}_\alpha \hat{P}_\beta - \hat{N}_\alpha \hat{N}_\beta) \\
& + \bar{u}' f_4 \gamma_5 \slashed{K} u\,(\hat{P}_\alpha \hat{N}_\beta + \hat{N}_\alpha \hat{P}_\beta) \\
& + \bar{u}' f_5 \gamma_5 u\,(\hat{P}_\alpha \hat{N}_\beta - \hat{N}_\alpha \hat{P}_\beta)
\end{aligned}
\tag{14}
$$

Indeed, the amplitude contains 6 independent functions of the kinematic invariants. In (14) we used the notation $\hat{P} = P/\sqrt{-P^2}$ and $\hat{N} = N/\sqrt{-N^2}$.

Let us now turn to helicity amplitudes. By virtue of P invariance we have

$$
\langle \lambda_c \lambda_d | S | \lambda_a \lambda_b \rangle = \langle -\lambda_c - \lambda_d | S | -\lambda_a - \lambda_b \rangle
\tag{15}
$$

while T invariance implies

$$
\langle \lambda_c \lambda_d | S | \lambda_a \lambda_b \rangle = \langle \lambda_a \lambda_b | S | \lambda_c \lambda_d \rangle
\tag{16}
$$

The first constraint reduces the number of independent amplitudes from 16 to 8, in agreement with (7). T invariance on its own would reduce the number of amplitudes to 10; together with P, it leads to 6 independent amplitudes: By Eqs. (15) and (16) we find

$$
\begin{aligned}
\langle + + | + + \rangle &\underset{P}{=} \langle - - | - - \rangle \\
\langle + - | + - \rangle &\underset{P}{=} \langle - + | - + \rangle \\
\langle + + | + - \rangle &\underset{T}{=} \langle + - | + + \rangle \underset{P}{=} \langle - + | - - \rangle \underset{T}{=} \langle - - | - + \rangle \\
\langle + + | - + \rangle &\underset{T}{=} \langle - + | + + \rangle \underset{P}{=} \langle + - | - - \rangle \underset{T}{=} \langle - - | + - \rangle \\
\langle + + | - - \rangle &\underset{P,T}{=} \langle - - | + + \rangle \\
\langle + - | - + \rangle &\underset{P,T}{=} \langle - + | + - \rangle
\end{aligned}
\tag{17}
$$

Problem 82.   Resonances in $e^+e^- \rightarrow e^+e^-$ at low energy

The cross section of $e^+e^- \rightarrow e^+e^-$ has a peak about the ρ^0 mass, at $E_{cm} = 768$ MeV, with width 151.5 ± 1.2 MeV. The total cross section at the peak equals $\sigma_{max} = (50 \pm 10)$ pb; ρ has spin 1.

a)   Calculate the mean life of ρ^0 and the branching ratio

$$BR \left(\frac{\rho^0 \rightarrow e^+e^-}{\rho^0 \rightarrow \pi^+\pi^-} \right) \tag{1}$$

knowing that practically 100% of the ρ^0 width is due to the $\pi^+\pi^-$ mode.

b)   Derive the expression for $\sigma_{e^+e^- \rightarrow \pi^+\pi^-}$ about the resonance, as a function of energy.

c)   Calculate the cross section of $\pi^+\pi^- \rightarrow e^+e^-$.

d)   At approximately the same energy as above (the mass of ω), the total cross section $\sigma_{e^+e^-}$ presents a peak with width 10 MeV; at the peak, $\sigma_{e^+e^- \rightarrow 3\pi} = (1.7 \pm 0.1)\mu b$. Find the superpositions of ρ and ω corresponding to physical states with definite mass and mean life, making use of

$$BR \left(\frac{\omega \rightarrow \pi^+\pi^-}{\omega \rightarrow \text{all}} \right) = (2.21 \pm 0.3) \ 10^{-2} \tag{2}$$

Solution

a)   The relativistic expression of the resonant cross section $e^+e^- \rightarrow e^+e^-$ is

$$\sigma = \frac{4\pi}{q^2} \left(\Gamma_{\rho \rightarrow e^+e^-} \right)^2 \frac{(2J+1)}{(2s_1+1)(2s_2+1)} \frac{4M_\rho^2}{\left(q^2 - M_\rho^2 \right)^2 + M_\rho^2 \Gamma^2} \tag{3}$$

where J is the spin of ρ and $s_1 = s_2 = \frac{1}{2}$ is the spin of e^+ and e^-. In the limit $q^2 \rightarrow M_\rho^2$ one obtains the nonrelativistic expression

$$\sigma = \frac{4\pi}{q^2} \left(\Gamma_{\rho \rightarrow e^+e^-} \right)^2 \frac{(2J+1)}{(2s_1+1)(2s_2+1)} \frac{1}{(\sqrt{q^2} - M_\rho)^2 + \Gamma^2/4} \tag{4}$$

At the peak, the total cross section becomes

$$\sigma_{max} = \frac{12\pi}{M_\rho^2} \left(\frac{\Gamma_{\rho \rightarrow e^+e^-}}{\Gamma} \right)^2 \tag{5}$$

Since the $\pi^+\pi^-$ decay channel accounts for almost 100% of the ρ width, we have

$$\frac{\Gamma_{\rho \rightarrow e^+e^-}}{\Gamma_{\rho \rightarrow \pi^+\pi^-}} \simeq \frac{\Gamma_{\rho \rightarrow e^+e^-}}{\Gamma} = \sqrt{\frac{M_\rho^2 \sigma_{max}}{12\pi}} \tag{6}$$

Substituting numerical values, we find

$$\frac{\Gamma_{\rho \rightarrow e^+e^-}}{\Gamma} = (0.43 \pm 0.04)\, 10^{-4} \tag{7}$$

b) By the previous analysis, there follows

$$\sigma_{e^+e^- \rightarrow \pi^+\pi^-} = \frac{3\pi}{q^2}\, \Gamma_{\rho \rightarrow e^+e^-}\, \Gamma\, \frac{4M_\rho^2}{\left(q^2 - M_\rho^2\right)^2 + M_\rho^2\, \Gamma^2} \tag{8}$$

c) The Breit-Wigner formula can be used once more in the case of the inverse process; it yields

$$\sigma_{\pi^+\pi^- \rightarrow e^+e^-} = \frac{12\pi}{q^2}\, \Gamma\, \Gamma_{\rho \rightarrow e^+e^-}\, \frac{4M_\rho^2}{\left(q^2 - M_\rho^2\right)^2 + M_\rho^2\, \Gamma^2} = 4\sigma_{e^+e^- \rightarrow \pi^+\pi^-} \tag{9}$$

The last equality reflects the principle of detailed balance.

d) Similar reasoning leads to

$$\sigma_{e^+e^- \rightarrow 3\pi} = \frac{3\pi}{q^2}\, \Gamma_{\omega \rightarrow e^+e^-}\, \Gamma_{\omega \rightarrow 3\pi}\, \frac{4M_\omega^2}{\left(q^2 - M_\omega\right)^2 + M_\omega^2\, \Gamma_\omega^2} \tag{10}$$

At the peak,

$$\sigma_{e^+e^- \rightarrow 3\pi}^{\text{max}} = \frac{12\pi}{M_\omega^2}\, \frac{\Gamma_{\omega \rightarrow e^+e^-}}{\Gamma_\omega}\, \frac{\Gamma_{\omega \rightarrow 3\pi}}{\Gamma_\omega} \tag{11}$$

or, equivalently,

$$\Gamma_{\omega \rightarrow e^+e^-} = \frac{M_\omega^2}{12\pi}\, \sigma_{e^+e^- \rightarrow 3\pi}^{\text{max}}\, \frac{\Gamma_\omega^2}{\Gamma_{\omega \rightarrow 3\pi}} \tag{12}$$

Using the numerical values

$$\begin{aligned}
\Gamma_\omega &= (8.43 \pm 0.1)\ \text{MeV} \\
\frac{\Gamma_{\omega \rightarrow 3\pi}}{\Gamma_\omega} &= (88.8 \pm 0.6)\, 10^{-2}
\end{aligned} \tag{13}$$

we conclude that $\Gamma_{\omega \rightarrow e^+e^-}/\Gamma_\omega = (0.79 \pm 0.05)\, 10^{-4}$.

The mixing between the ρ and ω states is due to the existence of decay channels, common to the two particles. The values of the branching ratios

$$\begin{aligned}
BR\,(\rho^0 \rightarrow e^+e^-) &= (4.44 \pm 0.21)\, 10^{-5} \\
BR\,(\omega \rightarrow e^+e^-) &= (7.15 \pm 0.19)\, 10^{-5} \\
BR\,(\rho^0 \rightarrow \pi^+\pi^-) &= 1 & \text{(strong)} \\
BR\,(\omega \rightarrow \pi^+\pi^-) &= (2.21 \pm 0.3)\, 10^{-2} & \text{(electromagnetic)}
\end{aligned} \tag{14}$$

show that the dominant process is

$$\rho \leftrightarrow \pi^+ \pi^- \leftrightarrow \omega \tag{15}$$

Let us consider the ρ and ω masses to be equal; then, in order to find the physical states, it suffices to diagonalize the imaginary part of the mass matrix. By the Breit-Wigner formula, the latter is easily shown to be

$$\Gamma = \begin{pmatrix} \Gamma_\rho & \sqrt{\Gamma_{\rho\to\pi\pi}\,\Gamma_{\omega\to\pi\pi}} \\ \sqrt{\Gamma_{\rho\to\pi\pi}\,\Gamma_{\omega\to\pi\pi}} & \Gamma_\omega \end{pmatrix} \tag{16}$$

Given that

$$\frac{\Gamma_{\rho\to\pi\pi}\,\Gamma_{\omega\to\pi\pi}}{(\Gamma_\omega\,\Gamma_\rho)} \simeq (2.21 \pm 0.3)\,10^{-2} \ll 1 \tag{17}$$

we can carry out the diagonalization to first order. We find the eigenvalues

$$\Gamma_\rho^{\text{phys}} = \Gamma_\rho + \frac{\Gamma_{\rho\to\pi\pi}\,\Gamma_{\omega\to\pi\pi}}{\Gamma_\rho - \Gamma_\omega} \qquad \Gamma_\omega^{\text{phys}} = \Gamma_\omega + \frac{\Gamma_{\rho\to\pi\pi}\,\Gamma_{\omega\to\pi\pi}}{\Gamma_\omega - \Gamma_\rho} \tag{18}$$

and the orthogonal eigenstates

$$|\rho^{\text{phys}}\rangle = |\rho\rangle + \frac{\sqrt{\Gamma_{\rho\to\pi\pi}\,\Gamma_{\omega\to\pi\pi}}}{\Gamma_\rho - \Gamma_\omega}|\omega\rangle \tag{19a}$$

$$|\omega^{\text{phys}}\rangle = |\omega\rangle + \frac{\sqrt{\Gamma_{\rho\to\pi\pi}\,\Gamma_{\omega\to\pi\pi}}}{\Gamma_\omega - \Gamma_\rho}|\rho\rangle \tag{19b}$$

The mixing coefficient equals $\dfrac{\sqrt{\Gamma_{\rho\to\pi\pi}\,\Gamma_{\omega\to\pi\pi}}}{\Gamma_\rho - \Gamma_\omega} \simeq (3.6 \pm 0.2)\,10^{-2}$

Problem 83. Electroweak asymmetry in $e^+ e^- \to \mu^+ \mu^-$

Electrons and muons interact with photons as well as with the neutral vector meson Z_μ having mass $M = (91.173 \pm 0.02)$ GeV. The interaction Lagrangian is

$$\begin{aligned}
\mathcal{L}_I &= e\,A_\lambda \left\{ \bar{\psi}_e\,\gamma^\lambda\,\psi_e + \bar{\psi}_\mu\,\gamma^\lambda\,\psi_\mu \right\} \\
&\quad + e'\,Z_\lambda \left\{ \bar{\psi}_e \left(a + b\gamma^5 \right) \gamma^\lambda\,\psi_e + \bar{\psi}_\mu \left(a + b\gamma^5 \right) \gamma^\lambda\,\psi_\mu \right\}
\end{aligned} \tag{1}$$

ψ_μ, ψ_e are the quantum fields of the muon and electron.
a) Derive the amplitude of the process $e^+ e^- \to \mu^+ \mu^-$, to lowest order in e and e'.
b) Calculate the cross section $\sigma(e^+ e^- \to \mu^+ \mu^-)$ as a function of the energy.
c) Calculate the forward-backward asymmetry of μ^- at a center-of-mass energy $E = 30$ Gev. Assume vanishing polarization of the incoming e^+ and e^-.

Solution

a) Two diagrams contribute to the amplitude to lowest order; they are shown in fig. 1.

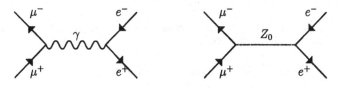

fig. 1

The two diagrams correspond to the exchange of a photon and of a Z_0, as prescribed by the interaction Lagrangian. The amplitude takes the form

$$\mathcal{M}_{fi} = ie^2 \left\{ \bar{v}(\mathbf{p}_{e^+})\gamma^\mu u(\mathbf{p}_{e^-})\frac{1}{q^2}\,\bar{u}(\mathbf{p}_{\mu^-})\gamma_\mu v(\mathbf{p}_{\mu^+}) \right\} + ie'^2 \left(g_{\mu\nu} - \frac{q_\mu q_\nu}{M_Z^2} \right)$$

$$\cdot \left\{ \bar{v}(\mathbf{p}_{e^+})\gamma^\mu(a + b\gamma^5)u(\mathbf{p}_{e^-})\frac{1}{q^2 - M_Z^2}\,\bar{u}(\mathbf{p}_{\mu^-})\gamma^\nu(a + b\gamma^5)v(\mathbf{p}_{\mu^+}) \right\}$$

with

$$q_\mu = (p_{e^+} + p_{e^-})_\mu = (p_{\mu^+} + p_{\mu^-})_\mu \tag{2}$$

We define vector and axial currents as follows

$$j^\lambda_{(e)} = \bar{v}_e(\mathbf{p}_{e^+})\gamma^\lambda u_e(\mathbf{p}_{e^-}) \qquad j^{5\lambda}_{(e)} = \bar{v}_e(\mathbf{p}_{e^+})\gamma^\lambda\gamma^5 u_e(\mathbf{p}_{e^-}) \tag{3a}$$

$$j^\lambda_{(\mu)} = \bar{u}_\mu(\mathbf{p}_{\mu^-})\gamma^\lambda v_\mu(\mathbf{p}_{\mu^+}) \qquad j^{5\lambda}_{(\mu)} = \bar{u}_\mu(\mathbf{p}_{\mu^-})\gamma^\lambda\gamma^5 v_\mu(\mathbf{p}_{\mu^+}) \tag{3b}$$

In the limit $m_e \simeq 0$, the transversality condition

$$q_\lambda j^\lambda_{(e)} = 0 \qquad q_\lambda j^\lambda_{(\mu)} = 0 \tag{4}$$

can be extended to the axial current of the electron

$$q_\lambda j^{5\lambda}_{(\mu)} \simeq 0 \tag{5}$$

This approximation consists in neglecting terms of order $\mathcal{O}(m_e^2/q^2)$.

We thus write in compact notation

$$\mathcal{M}_{fi} = i \left[A j^\lambda_{(e)} j_{(\mu)\lambda} + B \left(j^\lambda_{(e)} j^5_{(\mu)\lambda} + j^{5\lambda}_{(e)} j_{(\mu)\lambda} \right) + C j^{5\lambda}_{(e)} j^5_{(\mu)\lambda} \right] \tag{6}$$

where

$$A = \frac{e^2}{q^2} + \frac{a^2 e'^2}{q^2 - M_Z^2} \qquad B = \frac{a b e'^2}{q^2 - M_Z^2} \qquad C = \frac{b^2 e'^2}{q^2 - M_Z^2} \tag{7}$$

b) In order to calculate the cross section we must evaluate the modulus squared of the amplitude, averaged over initial spins

$$\frac{1}{4} \sum |\mathcal{M}_{fi}|^2 \tag{8}$$

At sufficiently high energies, the muon mass can also be neglected; the calculation is then simplified in terms of the tensors

$$\langle j_\mu j_\nu \rangle = \langle j_\mu^5 j_\nu^5 \rangle = \mathrm{Tr}\left\{ \gamma_\mu \not{p}_1 \gamma_\nu \not{p}_2 \right\}$$
$$\langle j_\mu j_\nu^5 \rangle = \langle j_\mu^5 j_\nu \rangle = \mathrm{Tr}\left\{ \gamma_\mu \not{p}_1 \gamma_\nu \gamma^5 \not{p}_2 \right\} \tag{9}$$

We see that the first tensor is symmetric under the exchange $p_1 \leftrightarrow p_2$ while the second one is antisymmetric. Carrying out the traces, we find

$$\langle j_\mu j_\nu \rangle = 4\left[p_{1\mu} p_{2\nu} + p_{1\nu} p_{2\mu} - g_{\mu\nu} (p_1 \cdot p_2) \right]$$
$$\langle j_\mu j_\nu^5 \rangle = -4i\, \varepsilon_{\mu\nu\alpha\beta}\, p_1^\alpha p_2^\beta \tag{10}$$

Exploiting the symmetry of these tensors, we obtain for the square of the amplitude

$$\begin{aligned}
\langle \overline{|\mathcal{M}|^2} \rangle = \frac{1}{4}\Big\{ &A^2 \langle j_\mu j_\nu \rangle_{e^+e^-} \langle j^\mu j^\nu \rangle_{\mu^+\mu^-} \\
&+ B^2 \left[\langle j_\mu j_\nu \rangle_{e^+e^-} \langle j^{5\mu} j^{5\nu} \rangle_{\mu^+\mu^-} + \langle j_\mu^5 j_\nu^5 \rangle_{e^+e^-} \langle j^\mu j^\nu \rangle_{\mu^+\mu^-} \right] \\
&+ C^2 \langle j_\mu^5 j_\nu^5 \rangle_{e^+e^-} \langle j^{5\mu} j^{5\nu} \rangle_{\mu^+\mu^-} \\
&+ B^2 \left[\langle j_\mu j_\nu^5 \rangle_{e^+e^-} \langle j^{5\mu} j^\nu \rangle_{\mu^+\mu^-} + \langle j_\mu^5 j_\nu \rangle_{e^+e^-} \langle j^\mu j^{5\nu} \rangle_{\mu^+\mu^-} \right] \\
&+ AC \left[\langle j_\mu j_\nu^5 \rangle_{e^+e^-} \langle j^\mu j^{5\nu} \rangle_{\mu^+\mu^-} + \langle j_\mu^5 j_\nu \rangle_{e^+e^-} \langle j^{5\mu} j^\nu \rangle_{\mu^+\mu^-} \right] \Big\}
\end{aligned} \tag{11}$$

Eqs.(9) further reduce the above to

$$\begin{aligned}
\langle \overline{|\mathcal{M}|^2} \rangle = &\frac{1}{4}(A^2 + 2B^2 + C^2)\, \langle j_\mu j_\nu \rangle_{e^+e^-} \langle j^\mu j^\nu \rangle_{\mu^+\mu^-} \\
&+ \frac{1}{2}(AC + B^2)\, \langle j_\mu j_\nu^5 \rangle_{e^+e^-} \langle j^\mu j^{5\nu} \rangle_{\mu^+\mu^-}
\end{aligned} \tag{12}$$

Substituting the explicit expressions (10) we obtain

$$\begin{aligned}
\langle \overline{|\mathcal{M}|^2} \rangle = &8\,(A^2 + 2B^2 + C^2) \left[(p_{e^+} \cdot p_{\mu^+})(p_{e^-} \cdot p_{\mu^-}) + (p_{e^+} \cdot p_{\mu^-})(p_{e^-} \cdot p_{\mu^+}) \right] \\
&+ 16\,(AC + B^2) \left[(p_{e^+} \cdot p_{\mu^-})(p_{e^-} \cdot p_{\mu^+}) - (p_{e^+} \cdot p_{\mu^+})(p_{e^-} \cdot p_{\mu^-}) \right]
\end{aligned} \tag{13}$$

In the center-of-mass frame the differential cross section equals

$$\frac{d\sigma}{d\Omega} = \frac{1}{64\pi^2 q^2}\left\{ \left(\bar{A}^2 + 2\bar{B}^2 + \bar{C}^2 \right)\left(1 + \cos^2\theta \right) + 4\left(\bar{A}\bar{C} + \bar{B}^2 \right)\cos\theta \right\} \tag{14}$$

where we have defined the dimensionless quantities

$$\bar{A} = e^2 + \frac{a^2 e'^2}{1 - M_Z^2/q^2} \qquad \bar{B} = \frac{ab\,e'^2}{1 - M_Z^2/q^2} \qquad \bar{C} = \frac{b^2 e'^2}{1 - M_Z^2/q^2} \tag{15}$$

c) The forward-backward asymmetry is, by definition,

$$A = \frac{\sigma_+ - \sigma_-}{\sigma_+ + \sigma_-} \tag{16}$$

with

$$\sigma_+ = +2\pi \int_0^{+1} d(\cos\theta)\, \frac{d\sigma}{d\Omega} \qquad \sigma_- = 2\pi \int_{-1}^0 d(\cos\theta)\, \frac{d\sigma}{d\Omega} \tag{17}$$

Eq.(14) then leads to

$$A = \frac{3\left(\bar{A}\,\bar{C} + \bar{B}^2\right)}{2\left(\bar{A}^2 + 2\bar{B}^2 + \bar{C}^2\right)} \tag{18}$$

Using the explicit values of the standard model

$$a = -\frac{1}{4} + \sin^2\theta_W \qquad b = \frac{1}{4} \qquad e' = \frac{e}{\sin\theta_W\,\cos\theta_W} \tag{19}$$

$$\sin^2\theta_W = 0.2325 \pm 0.0008$$

we find, at 30 GeV,

$$A \simeq 0.06 \tag{20}$$

The asymmetry increases as one approaches the mass of Z_0. In this case, however, one must also introduce an imaginary part to the Z_0 propagator.

Problem 84. π^0 production via the Primakov effect

The effective Lagrangian of π^0 decay is

$$\mathcal{L}_I = \frac{\alpha}{8\pi}\,\frac{1}{f_\pi}\,\varepsilon_{\mu\nu\rho\sigma}\,F^{\mu\nu} F^{\rho\sigma}\,\phi_{\pi^0} \tag{1}$$

$F_{\mu\nu}$ is the electromagnetic tensor operator, α is the fine structure constant and f_π is the pion decay constant $f_\pi = 93$ MeV.

Calculate the cross section of the Primakov effect, i.e. of π^0 production by γ in the Coulomb field of a nucleus

$$\gamma + Z \rightarrow \pi^0 + Z \tag{2}$$

Calculate the angular distribution of π^0 and exhibit the dependence of the cross section on the energy.

Solution

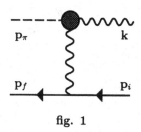

p_π k

p_f p_i

fig. 1

The process in question is depicted in the diagram of fig. 1; the corresponding amplitude is

$$\mathcal{M} = \frac{\alpha}{\pi f_\pi} \, \varepsilon_{\mu\nu\rho\sigma} \, k_\mu \, \varepsilon_\nu \, q_\rho \, A_\sigma(\mathbf{q}) \qquad (3)$$

where

$$q_\mu = (p_\pi - k)_\mu = (p_i - p_f)_\mu \qquad (4)$$

$A_\mu(q)$ is the Fourier transform of the Coulomb field. A factor of 2 corresponds to the two possibilities of associating the incident photon to $F_{\mu\nu}$; two further factors of 2 stem from the definition of $F_{\mu\nu}$. In the rest frame of the nucleus we have

$$
\begin{aligned}
\mathcal{M} &= \frac{\alpha}{\pi f_\pi} \varepsilon_{ijk} \varepsilon^i \, k^j \, p_\pi^k \, A_0(q) \\
&= \frac{\alpha}{\pi f_\pi} \, Z \, e \, \frac{\boldsymbol{\varepsilon} \cdot (\mathbf{k} \wedge \mathbf{p}_\pi)}{|\mathbf{p}_\pi - \mathbf{k}|^2}
\end{aligned}
\qquad (5)
$$

There follows, summing over initial polarizations,

$$\overline{|\mathcal{M}|^2} = \frac{2Z^2\alpha^3}{\pi f_\pi^2} \frac{|\mathbf{k} \wedge \mathbf{p}_\pi|^2}{|\mathbf{p}_\pi - \mathbf{k}|^4} \qquad (6)$$

Neglecting the recoil energy of the nucleus, we obtain the cross section

$$\mathrm{d}\sigma = \frac{\overline{|\mathcal{M}|^2}}{2|\mathbf{k}|} 2\pi \delta(k - E_\pi) \frac{\mathrm{d}^3 \mathbf{p}_\pi}{2E_\pi (2\pi)^3} \qquad (7)$$

or, equivalently,

$$\frac{\mathrm{d}\sigma}{\mathrm{d}\Omega} = \frac{1}{16\pi^2} \frac{|\mathbf{p}_\pi|}{|\mathbf{k}|} \overline{|\mathcal{M}|^2} = \frac{Z^2\alpha^3}{8\pi^3 f_\pi^2} \frac{|\mathbf{p}_\pi|}{|\mathbf{k}|} \frac{|\mathbf{k} \wedge \mathbf{p}_\pi|^2}{|\mathbf{p}_\pi - \mathbf{k}|^4} \qquad (8)$$

We define for convenience $\omega = |\mathbf{k}|$ and the scattering angle

$$\cos\theta = \hat{\mathbf{k}}\hat{\mathbf{p}}_\pi \qquad (9)$$

Then the angular distribution of the pion takes the form

$$\frac{\mathrm{d}\sigma}{\mathrm{d}\Omega} = \frac{Z^2\alpha^3}{8\pi^3 f_\pi^2} v_\pi \left[\frac{\sin\theta}{v_\pi + \dfrac{1}{v_\pi} - 2\cos\theta} \right]^2 \qquad (10)$$

with

$$v_\pi = \frac{|\mathbf{P}_\pi|}{\omega} = \sqrt{1 - \frac{m_\pi^2}{\omega^2}} \qquad (11)$$

To obtain the total cross section we perform the angular integration, by means of the formula

$$\int_{-1}^{1} \mathrm{d}\cos\theta \left[\frac{\sin\theta}{A - \cos\theta}\right]^2 = 2A \ln\frac{A+1}{A-1} - 4 \qquad (12)$$

As a result,

$$\sigma = \frac{Z^2\alpha^3}{32\pi^3 f_\pi^2} \, 4\pi v_\pi \left[\left(v_\pi + \frac{1}{v_\pi}\right) \ln\frac{1+v_\pi}{1-v_\pi} - 2\right] \qquad (13)$$

In the limit $\omega \to m_\pi$ ($v_\pi \to 0$), we find

$$\sigma \simeq \frac{Z^2\alpha^3}{3\pi^2 f_\pi^2} v_\pi^3 \qquad (14)$$

Problem 85.   The Compton effect on spin 0 particles

Consider a free scalar field, with Lagrangian $L_0 = \partial_\mu\phi^*\partial^\mu\phi - m^2\phi^*\phi$. Its interaction with the electromagnetic field is given by minimal coupling, i.e. by the prescription: $\partial_\mu\phi \to (\partial_\mu - ieA_\mu)\phi$.
a)   Derive the amplitude for the process $\gamma + \phi \to \gamma + \phi$, to order e^2 in perturbation theory.
b)   Calculate the energy distribution of ϕ particles in the laboratory frame, after they have undergone Compton scattering. Show that this distribution differs from that of a spin-1/2 particle, thus making it possible to deduce whether the target particle has spin 0 or 1/2.
c)   Compare the angular distribution of the processes

$$e^+e^- \to \mu^+\mu^- \qquad e^+e^- \to \phi\phi^* \qquad (1)$$

Solution

a)   The diagrams contributing to the amplitude are shown in fig.1.

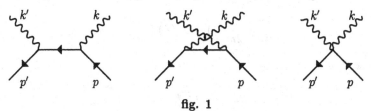

fig. 1

The amplitude is

$$
\begin{aligned}
\mathcal{M} &= e^2 \varepsilon_\mu \varepsilon_\nu'^* \left[\frac{(2p_\mu + k_\mu)(2p_\nu' + k_\nu')}{2p \cdot k + k^2} - \frac{(2p_\mu' - k_\mu)(2p_\nu - k_\nu')}{2p' \cdot k - k^2} - 2\, g_{\mu\nu} \right] \\
&= e^2 \varepsilon_\mu \varepsilon_\nu'^* \left[\frac{2p_\mu p_\nu'}{k \cdot p} - \frac{2p_\mu' p_\nu}{k \cdot p'} - 2\, g_{\mu\nu} \right]
\end{aligned}
\tag{2}
$$

In the last step we made use of

$$
k^2 = k'^2 = 0 \quad (\varepsilon \cdot k) = (\varepsilon'^* \cdot k') = 0
\tag{3}
$$

It is easy to verify the gauge invariance of this amplitude. The above expression simplifies considerably in the laboratory, with the choice of gauge $\varepsilon_\mu = (0, \boldsymbol{\varepsilon})$; we find

$$
\mathcal{M} = 2e^2 \boldsymbol{\varepsilon}\, \boldsymbol{\varepsilon'}^*
\tag{4}
$$

b) In order to calculate the spectrum of scattered particles we use the expression

$$
d\sigma = \frac{1}{4 m_\phi\, \omega} |\mathcal{M}|^2\, d\Phi^{(2)}
\tag{5}
$$

In the laboratory frame we have

$$
d\Phi^{(2)} = \frac{1}{16\pi^2} \frac{\omega'^2}{m_\phi \omega}\, d\Omega = \frac{1}{8\pi} \frac{d\omega'}{\omega}
\tag{6}
$$

and

$$
\frac{1}{\omega'} - \frac{1}{\omega} = \frac{1}{m_\phi} (1 - \cos\theta)
\tag{7}
$$

The amplitude squared, summed (averaged) over the polarizations of the outgoing (incoming) photon, becomes

$$
\langle |\mathcal{M}|^2 \rangle = 2e^2 \left(1 + \cos^2\theta \right)
\tag{8}
$$

leading to the angular distribution

$$
\frac{d\sigma}{d\Omega} = \frac{\alpha^2}{2 m_\phi^2} \left(1 + \cos^2\theta \right) \left(\frac{\omega'}{\omega} \right)^2
\tag{9}
$$

and the distribution in energy

$$
\frac{d\sigma}{d\omega'} = \frac{\pi \alpha^2}{m_\phi\, \omega^2} \left[1 + \left(\frac{m_\phi}{\omega'} - \frac{m_\phi}{\omega} - 1 \right)^2 \right]
\tag{10}
$$

For spin-1/2 particles the angular distribution equals

$$\frac{d\sigma}{d\Omega} = \frac{1}{2}\left(\frac{\alpha}{m}\right)^2\left(\frac{\omega'}{\omega}\right)^2\left(\frac{\omega}{\omega'} + \frac{\omega'}{\omega} - 1 + \cos^2\theta\right) \tag{11}$$

The difference between the two distributions (9) and (11) becomes more pronounced the more ω' and ω differ from each other, i.e. at scattering angles much different from $\theta = 0$.

In the low energy limit the two distributions coincide, reducing to the form of the Thompson cross section

$$\frac{d\sigma}{d\Omega} = \frac{\alpha^2}{2m^2}\left(1 + \cos^2\theta\right)\left[1 - \frac{2\omega}{m}\left(1 - \cos\theta\right)\right] \tag{12}$$

Terms of order $\mathcal{O}(\omega^2/m^2)$ have been left out in (12).

c)   The process $e^+e^- \rightarrow \phi\phi'$ is described by the diagram shown in fig.2.

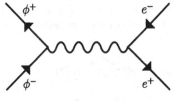

fig. 2

The corresponding amplitude reads

$$\mathcal{M} = e^2\bar{v}(\mathbf{p}_{e^+})\gamma_\mu u(\mathbf{p}_{e^-})\frac{1}{q^2}\left(p_+'^\mu - p_-'^\mu\right) \tag{13}$$

In the limit $m_e \rightarrow 0$, the modulus squared of the amplitude, averaged over initial spins, equals

$$\langle|\mathcal{M}|^2\rangle = \frac{e^4}{2}\left(1 - \cos^2\theta\right)\left(1 - \frac{4m_\phi^2}{q^2}\right) \tag{14}$$

and the cross section is

$$\frac{d\sigma}{d\Omega} = \frac{\alpha^2}{8q^2}\left(1 - \cos^2\theta\right)\left(1 - \frac{4m_\phi^2}{q^2}\right)^{\frac{3}{2}} \tag{15}$$

For comparison, we recall that the cross section for $e^+e^- \rightarrow \mu^+\mu^-$ is, in the limit $m_e = 0$,

$$\frac{d\sigma}{d\Omega} = \frac{\alpha^2}{4q^2}\left[1 + \cos^2\theta + \frac{4m_\mu^2}{q^2}\left(1 - \cos^2\theta\right)\right]\left(1 - \frac{4m_\mu^2}{q^2}\right)^{\frac{1}{2}} \tag{16}$$

At sufficiently large energies ($q^2 \gg m_\phi^2$, m_μ^2), we find

$$\frac{d\sigma}{d\Omega}(e^+e^- \to \phi^+\phi^-) \simeq \frac{\alpha^2}{8q^2}\left(1 - \cos^2\theta\right) \tag{17}$$

$$\frac{d\sigma}{d\Omega}(e^+e^- \to \mu^+\mu^-) \simeq \frac{\alpha^2}{8q^2}\left(1 + \cos^2\theta\right) \tag{18}$$

From Eqs. (17) and (18) we obtain the total cross sections

$$\sigma_{e^+e^- \to \phi^+\phi^-} = \frac{1}{4}\,\sigma_{e^+e^- \to \mu^+\mu^-} = \frac{\pi\alpha^2}{3q^2} \tag{19}$$

Problem 86.   The Compton effect on neutrons

a)   Consider the Compton effect on neutrons, induced by low energy photons; derive the expression for the amplitude in the limit of nonrelativistic neutrons. The effective interaction may be taken to be

$$H_I(x) = e\,A_\mu(x)\,\bar\psi_n(x)\,\frac{\mathrm{i}\,\sigma^{\mu\nu}\,q_\nu}{2B}\,\psi_n(x); \qquad \sigma_{\mu\nu} = \frac{1}{2\mathrm{i}}\,[\gamma_\mu,\,\gamma_\nu] \tag{1}$$

b)   Determine the parameter B, knowing that the magnetic moment of the neutron is

$$\mu_n = -1.91\,\frac{e\,\hbar}{2m_n\,c} \tag{2}$$

c)   Show that left- and right-handed γ's lead to the same cross section when scattering off unpolarized neutrons. Calculate the percentage difference in the cross section of photons and neutrons with parallel/antiparallel spins.

Solution

a)   The effective photon-neutron interaction gives rise to the diagrams shown in fig.1, in lowest order of perturbation theory

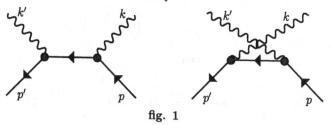

fig. 1

The corresponding amplitude is

$$\mathcal{M} = \frac{e^2}{4B^2} \varepsilon_\mu \varepsilon_{\mu'}^{'*} \bar{u}_n(\mathbf{p}') \left[\sigma_{\mu'\nu'} k_{\nu'}' \frac{\not{p} + \not{k} + m}{2p \cdot k} \sigma_{\mu\nu} k_\nu \right.$$

$$\left. - \sigma_{\mu\nu} k_\nu \frac{\not{p} - \not{k}' + m}{2p \cdot k'} \sigma_{\mu'\nu'} k_{\nu'}' \right] u_n(\mathbf{p}) \tag{3}$$

The relationship

$$(\varepsilon \cdot k) = 0 \tag{4}$$

implying also

$$i \sigma_{\mu\nu} \varepsilon^\nu k^\nu = \not{\varepsilon} \not{k} = -\not{k} \not{\varepsilon} \tag{5}$$

allows us to put the amplitude in the form

$$\mathcal{M} = \frac{e^2}{4B^2} \bar{u}_n(\mathbf{p}') \left[\not{\varepsilon}'^* \not{k}' \frac{\not{p} + m}{2p \cdot k} \not{\varepsilon} \not{k} - \not{\varepsilon} \not{k} \frac{\not{p} + m}{2p' \cdot k} \not{\varepsilon}'^* \not{k}' \right] u_n(\mathbf{p}) \tag{6}$$

It is easy to obtain the nonrelativistic limit of Eq.(6): In this limit

$$u_n(\mathbf{p}) \simeq \sqrt{2m} \begin{pmatrix} \mathbf{w} \\ 0 \end{pmatrix} \qquad \bar{u}_n(p') \simeq \sqrt{2m} \, (\mathbf{w}'^*, 0) \tag{7}$$

where w and w' are Pauli spinors and

$$\frac{\not{p} + m}{2p \cdot k} = \frac{1}{\omega} \begin{pmatrix} 1 & 0 \\ 0 & 0 \end{pmatrix} \qquad \frac{\not{p} + m}{2p' \cdot k} = \frac{1}{\omega'} \begin{pmatrix} 1 & 0 \\ 0 & 0 \end{pmatrix} \tag{8}$$

We see that the operators in (8) project onto the "large" components of the operators $\not{\varepsilon} \not{k}$ and $\not{\varepsilon}'^* \not{k}'$.

We thus arrive at

$$\mathcal{M} = -\frac{m e^2 \omega}{B^2} i \, \mathbf{w}'^* \boldsymbol{\sigma} \, \mathbf{w} \cdot [(\mathbf{n}' \wedge \boldsymbol{\varepsilon}') \wedge (\mathbf{n} \wedge \boldsymbol{\varepsilon})] \tag{9}$$

b) The parameter m/B in the interaction Hamiltonian plays the role of the anomalous magnetic moment of the neutron, in units of $e\hbar/2mc$; this can be verified in the nonrelativistic limit, where

$$\begin{aligned} H_I &= \frac{e}{2m} \left(\frac{m}{B} \right) A_i \, q_j \, \bar{\psi}_n \, i \, \sigma_{ij} \, \psi_n \\ &= \frac{e}{2m} \frac{m}{B} \mathbf{H} \cdot \mathbf{w}'^* \boldsymbol{\sigma} \, \mathbf{w} \end{aligned} \tag{10}$$

Therefore

$$\frac{m}{B} = -1.91 \qquad B = -\frac{m}{1.91} \tag{11}$$

c)   The cross section in the nonrelativistic limit is given by the formula

$$\frac{d\sigma}{d\Omega} = \frac{1}{64\pi^2 \, m^2} \, \overline{|\mathcal{M}|^2} \tag{12}$$

One must sum $|\mathcal{M}|^2$ over the final photon and nucleon polarizations, since these remain unobserved. (9) leads to

$$\overline{|\mathcal{M}|^2} = \left(\frac{e^2 m \omega}{B^2}\right)^2 \mathrm{Tr}\left\{\sigma_i \frac{1 + \mathbf{P}\,\boldsymbol{\sigma}}{2}\, \sigma_j\right\} \overline{A_i A_j^*} \tag{13}$$

Here, $\mathbf{P}$ is the polarization of the incoming neutron and $\mathbf{A}$ equals

$$\mathbf{A} = (\mathbf{n}' \wedge \boldsymbol{\varepsilon}'^*) \wedge (\mathbf{n} \wedge \boldsymbol{\varepsilon}) \tag{14}$$

Summing over final polarizations and averaging over phase space with use of

$$\sum_{pol} \varepsilon_i' \varepsilon_j'^* = \delta_{ij} - n_i' n_j'$$

$$\langle n_i' n_j' \rangle = \frac{1}{3}\delta_{ij} \tag{15}$$

we obtain

$$\overline{A_i A_j^*} = \frac{2}{3}\left[(\mathbf{n} \wedge \boldsymbol{\varepsilon}) \cdot (\mathbf{n} \wedge \boldsymbol{\varepsilon}^*)\, \delta_{ij} - (\mathbf{n} \wedge \boldsymbol{\varepsilon})_j \, (\mathbf{n} \wedge \boldsymbol{\varepsilon}^*)_i\right] \tag{16}$$

and

$$\mathrm{Tr}\left\{\sigma_i \frac{1 + \mathbf{P}\boldsymbol{\sigma}}{2}\, \sigma_j\right\} = \delta_{ij} - i\,\varepsilon_{ijk}\, P_k \tag{17}$$

Substituting into Eq.(12) yields

$$\sigma = \frac{1}{16\pi \, m^2} \left(\frac{e^2 m \omega}{B^2}\right)^2 \frac{2}{3} \left[2\,(\mathbf{n} \wedge \boldsymbol{\varepsilon}) \cdot (\mathbf{n} \wedge \boldsymbol{\varepsilon}^*) + i\,\mathbf{P} \cdot \{(\mathbf{n} \wedge \boldsymbol{\varepsilon}^*) \wedge (\mathbf{n} \wedge \boldsymbol{\varepsilon})\}\right]$$

or

$$\sigma = \frac{2\pi}{3}\, \alpha^2 \, \frac{\omega^2}{B^4}\left[2 - i\,(\mathbf{P}\,\mathbf{n})\,\{\mathbf{n} \cdot (\boldsymbol{\varepsilon} \wedge \boldsymbol{\varepsilon}^*)\}\right] \tag{18}$$

The two helicity states of the photon obey the relationship

$$i\,\boldsymbol{\varepsilon}_{(\pm)} \wedge \boldsymbol{\varepsilon}_{(\pm)}^* = \pm\mathbf{n} \tag{19}$$

As a result,

$$\sigma^{\pm} = \frac{2\pi\,\alpha^2}{3}\, \frac{\omega^2}{B^4}\, [2 \mp \mathbf{P}\,\hat{\mathbf{n}}] \tag{20}$$

If the neutrons are unpolarized, the cross section is independent of the photon polarization, and this is a consequence of parity invariance. If $|\mathbf{P}| = 1$, we have, in the case of parallel polarizations,

$$\sigma^{++} = \frac{2\pi}{3}\, \alpha^2 \, \frac{\omega^2}{B^4} \tag{21}$$

and for antiparallel polarizations

$$\sigma^{-+} = 2\pi \alpha^2 \frac{\omega^2}{B^4} \tag{22}$$

By Eq.(20), there follows also

$$\sigma^{--} = \sigma^{++} \qquad \sigma^{-+} = \sigma^{+-} \tag{23}$$

The above can again be deduced from parity invariance.

In conclusion, we have

$$\left| \frac{\sigma^{++} - \sigma^{-+}}{\sigma^{++} + \sigma^{-+}} \right| = \frac{1}{2} \tag{24}$$

Problem 87. $p\bar{p} \to e^+e^-$ at rest

We would like to calculate the cross section for the annihilation process $p\bar{p} \to e^+e^-$ at very low energies ($\sim 2m_p$) and to lowest order in the electromagnetic interaction.

a)   Provide the most general parameterization of the vertex $p\bar{p}\gamma$, compatible with the symmetries of the strong and electromagnetic interactions (Poincaré, C, P, T).

b)   Derive the cross section for $p\bar{p} \to e^+e^-$ and compare it with that of the inverse process $e^+e^- \to p\bar{p}$, at the same center-of-mass energy.

c)   Propose a minimum set of measurements, necessary to determine separately the two form factors obtained in part a).

Solution

a)   The coupling to photons has the general form $\mathcal{L}_I = e\, j_\mu\, A^\mu$, where

$$j_\mu(x) = \langle 0 | J_\mu^{\text{e.m.}}(x) | p\bar{p} \rangle = \langle 0 | J_\mu^{\text{e.m.}}(0) | p\bar{p} \rangle\, e^{i x \cdot (p + \bar{p})} \tag{1}$$

$$\langle 0 | J_\mu^{\text{e.m.}}(0) | \mathbf{p}\bar{\mathbf{p}} \rangle = \bar{v}(\bar{\mathbf{p}})\, \Gamma_\mu\, u(\mathbf{p}) \tag{2}$$

Translational invariance was invoked in (1). By Lorentz invariance, the quantity Γ_μ in (2) is the most general vector that can be made out of p_μ, $\bar{p}_\mu$ and the γ matrices. Finally, parity imposes that j_μ be a polar vector. Using the equations of motion we obtain the following relations between the possible vectors

$$\bar{v}(\bar{\mathbf{p}})\, (q^\mu + i\, \sigma^{\mu\nu} P_\nu)\, u(\mathbf{p}) = 0$$
$$\bar{v}(\bar{\mathbf{p}})\, (P^\mu + i\, \sigma^{\mu\nu} q_\nu - 2M\, \gamma^\mu)\, u(\mathbf{p}) = 0 \tag{3}$$

where

$$P_\mu = (p - \bar{p})_\mu \qquad q_\mu = (p + \bar{p})_\mu \tag{4}$$

Thus, j_μ assumes the form

$$j_\mu = \bar{v}(\bar{\mathbf{p}}) \left\{ A\,\gamma_\mu + B\,q_\mu + i\,C\,\sigma_{\mu\nu}q^\nu \right\} u(\mathbf{p}) \tag{5}$$

Gauge invariance now requires that $B = 0$, while the discrete symmetries C and T imply that the coefficients A and C are real. The most general current can thus be written as follows

$$j^\mu = \bar{v}(\bar{\mathbf{p}}) \left\{ F_1(q^2)\gamma^\mu + i\,\frac{F_2(q^2)}{2M}\sigma^{\mu\nu}q_\nu \right\} u(\mathbf{p}) \tag{6}$$

or, using Eq.(3),

$$j^\mu = \bar{v}(\bar{\mathbf{p}}) \left\{ (F_1 + F_2)\,\gamma^\mu - \frac{F_2}{2M}\,P^\mu \right\} u(\mathbf{p}) \tag{7}$$

b)   The diagram contributing to this process is shown in fig. 1.

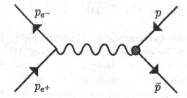

fig. 1

The corresponding amplitude is

$$\mathcal{M} = e^2\,\bar{v}(\bar{\mathbf{p}}) \left\{ (F_1 + F_2)\,\gamma_\mu - \frac{F_2}{2M}\,P_\mu \right\} u(\mathbf{p})\,\frac{1}{q^2}\,\bar{u}(\mathbf{p}_{e^-})\gamma^\mu v(\mathbf{p}_{e^+}) \tag{8}$$

Summing and averaging over polarizations we find

$$\overline{|\mathcal{M}|^2} = \left(\frac{e^2}{q^2}\right)^2 \frac{1}{4}\,L_{\mu\nu}\,T^{\mu\nu} \tag{9}$$

Here,

$$L_{\mu\nu} = \mathrm{Tr}\left\{ (\slashed{p}_{e^+} - m)\,\gamma_\mu\,(\slashed{p}_{e^-} + m)\,\gamma_\nu \right\} = 2\left[q^\mu q^\nu - q^2 g^{\mu\nu} - r^\mu r^\nu \right] \tag{10}$$

and

$$r_\mu = (p_{e^+} - p_{e^-})_\mu \tag{11}$$

The hadronic tensor equals

$$T_{\mu\nu} = \mathrm{Tr}\left\{ (\slashed{p} - M)\left[(F_1 + F_2)\,\gamma_\mu - \frac{F_2}{2M}\,P_\mu \right] (\slashed{p} + M)\left[(F_1 + F_2)\,\gamma_\nu - \frac{F_2}{2M}\,P_\nu \right] \right\}$$

$$= 2\,(F_1 + F_2)^2\left(q_\mu q_\nu - q^2 g_{\mu\nu} \right) - 2P_\mu P_\nu \left(F_1^2 - \frac{q^2}{4M^2}\,F_2^2 \right) \tag{12}$$

The phase space average of the tensor $L_{\mu\nu}$ is (see Appendix E)

$$\langle L_{\mu\nu}\rangle = \frac{4}{3}\left(1 + \frac{2m_e^2}{q^2}\right)\left(q_\mu q_\nu - q^2 g_{\mu\nu}\right) \tag{13}$$

Eq.(9) now becomes

$$\overline{|\mathcal{M}|^2} = \left(\frac{e^2}{q^2}\right)^2 \left(1 + \frac{2m_e^2}{q^2}\right)(q^2)^2$$
$$\cdot\left[2\left(F_1 + F_2\right)^2 + \frac{2}{3}\left(F_1^2 - \frac{q^2}{4M^2}F_2^2\right)\left(\frac{4M^2}{q^2} - 1\right)\right] \tag{14}$$

and the total cross section reads

$$\sigma = \frac{\pi\alpha^2}{q^2}\left[2\left(F_1 + F_2\right)^2 + \frac{2}{3}\left(F_1^2 - \frac{q^2}{4M^2}F_2^2\right)\left(\frac{4M^2}{q^2} - 1\right)\right]$$
$$\cdot\left(1 + \frac{2m_e^2}{q^2}\right)\frac{\sqrt{1 - 4m_e^2/q^2}}{\sqrt{1 - 4M^2/q^2}} \tag{15}$$

The cross section for the inverse process follows immediately from the above

$$\sigma_{inv} = \frac{\pi\alpha^2}{q^2}\left[2\left(F_1 + F_2\right)^2 + \frac{2}{3}\left(F_1^2 - \frac{q^2}{4M^2}F_2^2\right)\left(\frac{4M^2}{q^2} - 1\right)\right]$$
$$\cdot\left(1 + \frac{2m_e^2}{q^2}\right)\frac{\sqrt{1 - 4M^2/q^2}}{\sqrt{1 - 4m_e^2/q^2}} \tag{16}$$

The only difference in the two cross sections is due to an exchange between the phase space factor and the flux factor; these are both proportional to the corresponding center-of-mass velocities. In conclusion

$$\frac{\sigma(p\bar{p} \to e^+e^-)}{\sigma(e^+e^- \to p\bar{p})} = \frac{1 - \dfrac{4m_e^2}{q^2}}{1 - \dfrac{4M^2}{q^2}} \tag{17}$$

(Principle of detailed balance).
c)   A measurement of the total cross section alone is clearly not sufficient in order to deduce each of the two form factors separately, at a given energy. To this end, we measure the angular distribution which, in the limit $m_e \to 0$, equals

$$\frac{d\sigma}{d\Omega} = \frac{\alpha^2}{4q^2}\frac{1}{\sqrt{1 - 4M^2/q^2}}$$
$$\cdot\left[(1 + \cos^2\theta)\left(F_1 + F_2\right)^2 + \sin^2\theta\left(\frac{2M}{\sqrt{q^2}}F_1 + \frac{\sqrt{q^2}}{2M}F_2\right)^2\right] \tag{18}$$

Problem 88. W and Z production in $p\bar{p}$ collisions

During proton-antiproton collisions an antiquark ($\bar{q}$) of $\bar{p}$ may annihilate against a quark (q) of p, forming a weak-interaction boson W or Z.

a)   Knowing that the interaction Lagrangian is

$$\mathcal{L} = \frac{1}{\sqrt{2}} g \left(W_\mu^{(+)} J_\mu^+ + W_\mu^{(-)} J_\mu^{(-)} \right) + \frac{g}{\cos\theta_W} Z_\mu J_\mu^0 \tag{1}$$

and assuming that the W and Z are produced practically at rest, calculate their polarization state.

b)   How can this polarization state be determined by observing the angular distribution in the decay

$$W \to \mu\nu, \quad W \to e\nu, \quad Z \to e^+ e, \quad Z \to \mu^+ \mu^- \tag{2}$$

c)   From the result obtained above show that P and C are violated, while CP is conserved.

The currents appearing in Eq.(1) are defined by

$$
\begin{aligned}
J_\mu^{(-)} &= \bar{e}_L \gamma_\mu \nu_L + \bar{d}_L \gamma_\mu u_L \\
J_\mu^{(+)} &= \bar{\nu}_L \gamma_\mu e_L + \bar{u}_L \gamma_\mu d_L \\
J_\mu^0 &= \sum_{\text{particles}} \bar{\psi} \left(g_V \gamma^\mu - g_A \gamma^\mu \gamma^5 \right) \psi
\end{aligned}
\tag{3}
$$

The constants g_V and g_A in the expression for J^0 are given in the following table

	g_V	g_A	
ν	$\frac{1}{4}$	$\frac{1}{4}$	
e	$-\frac{1}{4} + \zeta$	$-\frac{1}{4}$	(4)
u	$\frac{1}{4} - \frac{2}{3}\zeta$	$\frac{1}{4}$	
d	$-\frac{1}{4} + \frac{1}{3}\zeta$	$-\frac{1}{4}$	

and

$$\zeta = \sin^2\theta_W = 0.2325 \pm 0.0008$$

$$\frac{g^2}{8M_W^2} = \frac{G_F}{\sqrt{2}} \qquad g^2 = \frac{e^2}{\sin^2\theta_W} \qquad M_Z = \frac{M_W}{\cos\theta_W}$$

$$M_W = (80.22 \pm 0.26) \text{ GeV} \tag{5}$$

Solution

a)   In order to determine the polarization of the W or Z boson, we must identify the quarks involved in its production and calculate the modulus squared of the transition amplitude from a quark-antiquark state to the bosonic state with a given polarization. Recalling that the constituents of p and $\bar{p}$ are

$$p : (u, u, d) \qquad \bar{p} : (\bar{u}, \bar{u}, \bar{d}) \tag{6}$$

we write for the production probabilities

$$
\begin{aligned}
P_{W^+} &\sim 2\,|\mathcal{M}(u\bar{d} \to W^+)|^2 \\
P_{W^-} &\sim 2\,|\mathcal{M}(d\bar{u} \to W^-)|^2 \\
P_{Z_0} &\sim 2\,|\mathcal{M}(u\bar{u} \to Z_0)|^2 + |\mathcal{M}(d\bar{d} \to Z_0)|^2
\end{aligned} \tag{7}
$$

The amplitudes appearing above are, apart from irrelevant overall factors,

$$
\begin{aligned}
\mathcal{M}(u\bar{d} \to W^+) &\sim \bar{d}\gamma^\mu \frac{(1-\gamma_5)}{2} u\, W_\mu^{(+)} \\
\mathcal{M}(d\bar{u} \to W^-) &\sim \bar{u}\gamma^\mu \frac{(1-\gamma_5)}{2} d\, W_\mu^{(-)} \\
\mathcal{M}(q\bar{q} \to Z_0) &\sim \bar{q}\left(g_V^q\,\gamma^\mu - g_A^q\,\gamma^\mu\gamma^5\right) q\, Z_\mu^0
\end{aligned} \tag{8}
$$

Consider now the limit in which quark masses are negligible and mesons are produced at rest; we have, calling $\hat{\mathbf{n}}$ the direction of the incoming proton in the center of mass

$$\mathbf{p}_q = \frac{M_V}{2}\,\hat{\mathbf{n}} \qquad \mathbf{p}_{\bar{q}} = -\mathbf{p}_q = -\frac{M_V}{2}\,\hat{\mathbf{n}} \tag{9}$$

Noticing that the wave function of a vector meson at rest has only space components, we calculate easily the probabilities (7) with the result

$$
\begin{aligned}
P_{W^+} &\sim 1 - (\mathbf{W}^+\,\hat{\mathbf{n}})\,(\mathbf{W}^{+*}\,\hat{\mathbf{n}}) - \mathrm{i}\,(\mathbf{W}^+ \wedge \mathbf{W}^{+*}) \cdot \hat{\mathbf{n}} \\
P_{W^-} &\sim 1 - (\mathbf{W}^-\,\hat{\mathbf{n}})\,(\mathbf{W}^{-*}\,\hat{\mathbf{n}}) - \mathrm{i}\,(\mathbf{W}^- \wedge \mathbf{W}^{-*}) \cdot \hat{\mathbf{n}} \\
P_{Z_0} &\sim [1 - (\mathbf{Z}\,\hat{\mathbf{n}})\,(\mathbf{Z}^*\,\hat{\mathbf{n}})]\left[2\left(g_V^{u\,2} + g_A^{u\,2}\right) + \left(g_V^{d\,2} + g_A^{d\,2}\right)\right] \\
&\quad -\mathrm{i}\,(\mathbf{Z} \wedge \mathbf{Z}^*) \cdot \hat{\mathbf{n}}\left(4g_V^u\,g_A^u + 2g_V^d\,g_A^d\right)
\end{aligned} \tag{10}
$$

By $\mathbf{W}^+$, $\mathbf{W}^-$, $\mathbf{Z}$, we have denoted the polarization vectors of the corresponding mesons. We substitute in (10) the values given in the text, finding

$$
\begin{aligned}
g_V^{u\,2} + g_A^{u\,2} &= \tfrac{1}{8} - \tfrac{\zeta}{3} + \tfrac{4}{9}\zeta^2 \;\; ; &\quad 2g_V^u\,g_A^u &= \tfrac{1}{8} - \tfrac{\zeta}{3} \\
g_V^{d\,2} + g_A^{d\,2} &= \tfrac{1}{8} - \tfrac{\zeta}{6} + \tfrac{1}{9}\zeta^2 \;\; ; &\quad 2g_V^d\,g_A^d &= \tfrac{1}{8} - \tfrac{\zeta}{6}
\end{aligned} \tag{11}
$$

Consequently,

$$P_{Z_0} \sim 1 - (\mathbf{Z}\,\hat{\mathbf{n}})\,(\mathbf{Z}^*\,\hat{\mathbf{n}}) + \mathrm{i}\,(\mathbf{Z} \wedge \mathbf{Z}^*) \cdot \hat{\mathbf{n}}\left(-1 + \frac{\zeta^2}{\dfrac{3}{8} - \dfrac{5}{6}\zeta + \zeta^2}\right) \tag{12}$$

Setting $\zeta = 0.2325$, we obtain

$$\frac{\zeta^2}{\frac{3}{8} - \frac{5}{6}\zeta + \zeta^2} \simeq 0.23 \tag{13}$$

States having spin 0, ± 1 along the beam axis will satisfy

$$\varepsilon^0 = \hat{\mathbf{n}}$$
$$i\,\varepsilon^+ \wedge \varepsilon^{+*} = \hat{\mathbf{n}}$$
$$i\,\varepsilon^- \wedge \varepsilon^{-*} = -\hat{\mathbf{n}} \tag{14}$$

By inspection of Eq.(10), we conclude that none of the particles can be produced in the state ε^0. We define the polarization through the probabilities P_+ and P_- of producing states with spin ± 1

$$P = \frac{P_+ - P_-}{P_+ + P_-} \tag{15}$$

and obtain

$$P(W^+) = P(W^-) = -1$$
$$P(Z^0) = -1 + \frac{\zeta^2}{\frac{3}{8} - \frac{5}{6}\zeta + \zeta^2} \simeq -0.77 \tag{16}$$

The polarization of the W's could have been predicted without any calculation: Indeed the $V - A$ interaction in the massless limit couples negative helicity fermions with positive helicity antifermions, leading to total angular momentum -1 along the direction of the quark. By angular momentum conservation, the vector meson produced at rest is completely polarized along the direction of $\bar{p}$.

b)   The weak leptonic decay of a vector meson is described by the Lagrangian (1); the corresponding decay amplitude is

$$\mathcal{M}_{fi} \simeq h\, V_\mu\, \bar{u}(\mathbf{p})\gamma^\mu \left(g_V - g_A\, \gamma^5\right) u(\mathbf{p}) \tag{17}$$

with $g_V = g_A = 1$ for charged currents, $g_V = -\frac{1}{4} + \zeta$, $g_A = -\frac{1}{4}$ for neutral currents, and

$$h = \frac{g}{\sqrt{2}} \quad \text{for } W^\pm \quad ; \quad h = \frac{g}{\cos\theta} \quad \text{for } Z_0 \tag{18}$$

Thus, the modulus squared of this amplitude reduces (in the massless limit) to the previous case; we find for $W^\pm$

$$|\mathcal{M}_{fi}|^2 = \frac{g^2}{2}\, M_W^2 \left\{ 1 - (\mathbf{W}\,\hat{\mathbf{n}}')^2 - i\,(\mathbf{W} \wedge \mathbf{W}^*)\cdot\hat{\mathbf{n}}' \right\} \tag{19}$$

where $\hat{\mathbf{n}}'$ is the direction of motion of the charged lepton.

For Z_0 we find

$$|\mathcal{M}_{fi}|^2 = \frac{2g^2}{\cos^2\theta_W} M_Z^2 \left\{ \left[1 - (\mathbf{Z}\,\hat{\mathbf{n}}')^2\right] \left(\frac{1}{4} - \zeta + 2\zeta^2\right) \right.$$
$$\left. - \mathrm{i}\,(\mathbf{Z}\wedge\mathbf{Z}^*)\cdot\hat{\mathbf{n}}'\left(\frac{1}{4} - \zeta\right) \right\} \tag{20}$$

The angular distribution is given by the formula

$$\frac{d\Gamma}{d\Omega} = \frac{1}{64\pi^2 M}|\mathcal{M}_{fi}|^2 \tag{21}$$

The initial state in $\mathcal{M}_{fi}$ is prescribed by the final density matrix of the production process. In the case of the W this simply amounts to using the state with polarization -1, yielding

$$\frac{d\Gamma}{d\Omega} = \frac{g^2 M_W}{128\pi^2} \left[\frac{1}{2}(1 + \cos^2\theta) + \cos\theta\right]$$
$$= \frac{g^2 M_W}{256\pi^2}(1 + \cos\theta)^2 \tag{22}$$

In the case of Z^0 we write instead the density matrix in the form

$$\rho = \frac{P_+\rho^+ + P_-\rho^-}{P_+ + P_-} \tag{23}$$

Here, $\rho^\pm$ are projectors to the states with helicity $\pm$, and P_+, P_- the relative probabilities. Eq.(21) now becomes

$$\frac{d\Gamma}{d\Omega} = \frac{g^2 M_Z}{32\pi^2\cos^2\theta_W}\left\{\frac{1}{2}\left(\frac{1}{4} - \zeta + 2\zeta^2\right)(1 + \cos^2\theta) - P\cos\theta\left(\frac{1}{4} - \zeta\right)\right\} \tag{24}$$

P is the quantity calculated in (16).

c)   Let us discuss the case of $W^\pm$ production.

Parity exchanges the momenta of the incoming p and $\bar{p}$, while polarizations stay the same. A subsequent rotation by 180° brings momenta back to the initial configuration, while the longitudinal polarization changes sign. Therefore, the presence of the state with $P = -1$ alone violates parity (maximal violation).

Charge conjugation exchanges p and $\bar{p}$, as well as W^+ and W^-. A 180° rotation again takes us to the initial configuration, whereas the oppositely charged vector meson is produced with opposite polarization $P = 1$. Thus C is also violated.

A CP transformation leaves the initial state unchanged, but exchanges W^+ and W^-, with the same polarization. We conclude that CP is conserved.

Problem 89.   Electroweak effects in μe scattering

Consider the scattering

$$\mu^- + e^- \to \mu^- + e^- \tag{1}$$

at a fixed angle θ in the center-of-mass frame.

a)   Show that parity invariance implies equality of the cross sections $\sigma_+(\theta)$ and $\sigma_-(\theta)$, corresponding to initial muons with helicity $+$ or $-$ and unpolarized electrons, summed over final spins.

b)   Process (1) receives contributions from the normal electromagnetic interaction

$$\mathcal{L}_{\rm em} = e\, A_\nu(x) \left[\bar{\psi}_e\, \gamma^\nu\, \psi_e + \bar{\psi}_\mu\, \gamma^\nu\, \psi_\mu \right] \tag{2}$$

and the interaction with the Z_0 meson, having mass $M = 91.17\,{\rm GeV}$,

$$\mathcal{L}_{Z_0} = e\, Z_0^\rho \left[\bar{\psi}_e\, \gamma_\rho \left(a - b\gamma^5 \right) \psi_e + \bar{\psi}_\mu\, \gamma_\rho \left(a - b\gamma^5 \right) \psi_\mu \right] \tag{3}$$

Show that (3) violates parity if both a and b are nonzero.

c)   Calculate, to lowest perturbative order, the ratio

$$A(\theta) = \frac{\sigma_+(\theta) - \sigma_-(\theta)}{\sigma_+(\theta) + \sigma_-(\theta)} \tag{4}$$

in terms of the parameters a and b.

Solution

a)   Helicity changes sign under parity. A parity transformation, accompanied by a 180° rotation on the scattering plane, leaves all momenta unchanged and flips all polarization components on the scattering plane. This configuration, if parity is conserved, must have the same transition probability as the original configuration. In particular, summing over final spins and averaging over the spin of the incoming electron, there follows immediately

$$\sigma_+(\theta) = \sigma_-(\theta) \tag{5}$$

Observing $\sigma_+ \neq \sigma_-$ is a clear sign of parity violation.

b)   Given the simultaneous presence of a vector and an axial vector current coupled to Z_0 (Eq.(3)), the Lagrangian will violate parity.

c)   Let us consider the limit in which all lepton masses are negligible. It is then convenient to express the interaction Lagrangian in terms of helicity components

$$\psi_L = \frac{1 - \gamma_5}{2}\, \psi \qquad \psi_R = \frac{1 + \gamma_5}{2}\, \psi \tag{6}$$

For each lepton we obtain

$$
\mathcal{L}_{\text{em}} + \mathcal{L}_{Z_0} = e \left[A^{\rho} \left(\bar{\psi}_L \gamma_{\rho} \psi_L + \bar{\psi}_R \gamma_{\rho} \psi_R \right) \right.
$$
$$
\left. + Z_0^{\rho} \left(\bar{\psi}_L \gamma_{\rho} \psi_L (a + b) + \bar{\psi}_R \gamma_{\rho} \psi_R (a - b) \right) \right] \tag{7}
$$

The amplitudes for processes initiated by left- and right-handed muons can now be decomposed in the form

$$
\mathcal{M}^L = e^2 J_L^{\mu} \left\{ \left(\mathcal{D}_0 + (a+b)^2 \mathcal{D}_M \right) j_{\mu}^L + \left(\mathcal{D}_0 + (a^2 - b^2) \mathcal{D}_M \right) j_{\mu}^R \right\}
$$
$$
\mathcal{M}^R = e^2 J_R^{\mu} \left\{ \left(\mathcal{D}_0 + (a-b)^2 \mathcal{D}_M \right) j_{\mu}^R + \left(\mathcal{D}_0 + (a^2 - b^2) \mathcal{D}_M \right) j_{\mu}^L \right\} \tag{8}
$$

with

$$
J_{\sigma}^{L,R} = \bar{\psi}_{L,R}^{(\mu)} \gamma_{\sigma} \psi_{L,R}^{(\mu)} \qquad j_{\sigma}^{L,R} = \bar{\psi}_{L,R}^{(e)} \gamma_{\sigma} \psi_{L,R}^{(e)} \tag{9}
$$

$$
\mathcal{D}_0 = \frac{1}{q^2} \qquad \mathcal{D}_M = \frac{1}{q^2 - M_Z^2} \tag{10}
$$

We have also used the fact that, in the chiral limit (vanishing masses),

$$
\partial^{\mu} J_{\mu}^{L,R} = \partial^{\mu} j_{\mu}^{L,R} = 0 \tag{11}
$$

This has allowed us to eliminate the term $q_{\mu} q_{\nu} / M^2$ from the Z_0 propagator.

The modulus squared of the amplitudes, summed over electron spins, contains no interference between opposite helicity electrons; we find

$$
|\mathcal{M}^L|^2 = e^4 M_L^{\mu\nu} \left\{ \left(\mathcal{D}_0 + (a+b)^2 \mathcal{D}_M \right)^2 E_{\mu\nu}^L + \left(\mathcal{D}_0 - (a^2 - b^2) \mathcal{D}_M \right)^2 E_{\mu\nu}^R \right\}
$$
$$
|\mathcal{M}^R|^2 = e^4 M_R^{\mu\nu} \left\{ \left(\mathcal{D}_0 + (a-b)^2 \mathcal{D}_M \right)^2 E_{\mu\nu}^R + \left(\mathcal{D}_0 - (a^2 - b^2) \mathcal{D}_M \right)^2 E_{\mu\nu}^L \right\} \tag{12}
$$

where

$$
M_{\mu\nu}^{L,R} = \overline{J_{\mu}^{L,R} J_{\nu}^{L,R}} = S_{\mu\nu} \pm A_{\mu\nu}
$$
$$
E_{\mu,\nu}^{R,L} = \overline{j_{\mu}^{R,L} j_{\nu}^{R,L}} = s_{\mu\nu} \pm a_{\mu\nu} \tag{13}
$$

The plus sign above corresponds to R and the minus sign to L. Denoting by p and p' the initial and final muon momentum, we have

$$
S_{\mu\nu} = 2 \left[p'_{\mu} p_{\nu} + p_{\mu} p'_{\nu} - (p \cdot p') g_{\mu\nu} \right]
$$
$$
A_{\mu\nu} = 2i \, \varepsilon_{\mu\nu\rho\sigma} \, p'^{\rho} p^{\sigma} \tag{14}
$$

$s_{\mu\nu}$ and $a_{\mu\nu}$ have a similar expression in terms of the electron momentum.

The amplitudes now read, in compact form,

$$
\begin{aligned}
|\mathcal{M}_L|^2 &= e^4 \, S_{\mu\nu} \, s^{\mu\nu} \left[\left(\mathcal{D}_0 + (a+b)^2 \mathcal{D}_M \right)^2 + \left(\mathcal{D}_0 + (a^2 - b^2)\mathcal{D}_M \right)^2 \right] \\
&\quad + e^4 \, A_{\mu\nu} \, a^{\mu\nu} \left[\left(\mathcal{D}_0 + (a+b)^2 \mathcal{D}_M \right)^2 - \left(\mathcal{D}_0 + (a^2 - b^2)\mathcal{D}_M \right)^2 \right] \\
|\mathcal{M}_R|^2 &= e^4 \, S_{\mu\nu} \, s^{\mu\nu} \left[\left(\mathcal{D}_0 + (a-b)^2 \mathcal{D}_M \right)^2 + \left(\mathcal{D}_0 + (a^2 - b^2)\mathcal{D}_M \right)^2 \right] \\
&\quad + e^4 \, A_{\mu\nu} \, a^{\mu\nu} \left[\left(\mathcal{D}_0 + (a-b)^2 \mathcal{D}_M \right)^2 - \left(\mathcal{D}_0 + (a^2 - b^2)\mathcal{D}_M \right)^2 \right] \qquad (15)
\end{aligned}
$$

Calculating the asymmetry

$$
A(\theta) = \frac{|\mathcal{M}_R|^2 - |\mathcal{M}_L|^2}{|\mathcal{M}_L|^2 + |\mathcal{M}_R|^2} \qquad (16)
$$

we are led to the expression

$$
A(\theta) = \frac{-\left[S_{\mu\nu} \, s^{\mu\nu} + A_{\mu\nu} \, a^{\mu\nu} \right] 2ab \, \mathcal{D}_M \left(\mathcal{D}_0 + (a^2 + b^2)\mathcal{D}_M \right)}{S_{\mu\nu} \, s^{\mu\nu} \left[\mathcal{D}_0^2 + 2a^2 \mathcal{D}_0 \mathcal{D}_M + (a^2 + b^2)^2 \mathcal{D}_M^2 \right] + A_{\mu\nu} \, a^{\mu\nu} \left[2b^2 \mathcal{D}_M \left(\mathcal{D}_0 + 2a^2 \mathcal{D}_M \right) \right]} \qquad (17)
$$

We must now evaluate explicitly $S_{\mu\nu} \, s^{\mu\nu}$ and $A_{\mu\nu} \, a^{\mu\nu}$. A straightforward calculation gives

$$
\begin{aligned}
S_{\mu\nu} \, s^{\mu\nu} &= 2 \left[s^2 + (s + q^2)^2 \right] \\
A_{\mu\nu} \, a^{\mu\nu} &= 2 \left[s^2 - (s + q^2)^2 \right]
\end{aligned} \qquad (18)
$$

with

$$
s = (p_{(e)} + p_{(\mu)})^2 = (p'_{(e)} + p'_{(\mu)})^2 \qquad q^2 = (p_{(e)} - p'_{(e)})^2 = (p'_{(\mu)} - p_{(\mu)})^2 \qquad (19)
$$

All angular dependence is contained in

$$
q^2 = -s \, \sin^2 \frac{\theta}{2} \qquad (20)
$$

A limit case of interest is $|q^2| \ll M_Z^2$; there, $A(\theta)$ becomes

$$
A(\theta) = -\frac{4ab \, \sin^2 \theta/2}{1 + \cos^4 \theta/2} \frac{s}{M_Z^2} \qquad (21)
$$

Problem 90. Quark model description and vector resonances in $e^+ e^- \to$ hadrons at low energy

The cross section for $e^+ e^- \to$ hadrons, below 3 GeV, can be attributed to the production of quark-antiquark pairs (u, d, s). In terms of quark fields, the electromagnetic current has the form

$$j_\mu = :\bar\psi \, \gamma_\mu \, Q \, \psi: \qquad Q = \frac{Y}{2} + T_3 \qquad (1)$$

a) Calculate the percentage of isospin singlet and triplet in the final state.

b) Let us assume that the cross section is saturated at low energies by the resonances ρ_0, ω, Φ; the states of ω and Φ can be written as

$$
\begin{aligned}
|\omega\rangle &= \cos\theta \, |1\rangle + \sin\theta \, |8\rangle \\
|\Phi\rangle &= -\sin\theta \, |1\rangle + \cos\theta \, |8\rangle
\end{aligned} \qquad (2)
$$

$|1\rangle$ is an $SU(3)$ singlet; $|8\rangle$ is a triplet state with $Y = 0$ and $T = 0$.

Calculate the ratio

$$R_1 = \frac{m_\rho \, \Gamma(\rho \to e^+e^-)}{m_\omega \, \Gamma(\omega \to e^+e^-) + m_\Phi \, \Gamma(\Phi \to e^+e^-)}) \qquad (3)$$

and compare with experimental data given below.

c) Calculate the mixing angle θ, using the answer to part b) and the observed ratio

$$R_2 = \frac{m_\rho^4 \, \Gamma_\rho^2 \, \sigma_{e^+e^- \to \rho_0 \to e^+e^-}}{m_\omega^4 \, \Gamma_\omega^2 \, \sigma_{e^+e^- \to \omega \to e^+e^-} + m_\Phi^4 \, \Gamma_\Phi^2 \, \sigma_{e^+e^- \to \Phi \to e^+e^-}} \simeq 14.5 \pm 2.5 \qquad (4)$$

In (4), σ is the cross section at the peak of the resonance. Show that the mixing angle θ obtained in this manner is compatible with the states

$$
\left.
\begin{aligned}
|\Phi\rangle &= |s\bar s\rangle \\
|\omega\rangle &= |u\bar u + d\bar d\rangle \tfrac{1}{\sqrt 2}
\end{aligned}
\right\} \text{ (Ideal mixing)} \qquad (5)
$$

Some relevant experimental data:

$$
\begin{array}{lll}
m_\rho = 768.1 \pm 0.5 \text{ MeV} & \Gamma_\rho = 151.5 \pm 1.2 \text{ MeV} & BR(e^+e^-) = (4.44 \pm 0.21)\cdot 10^{-5} \\
m_\omega = 781.9 \pm 0.1 \text{ MeV} & \Gamma_\omega = 8.43 \pm 0.1 \text{ MeV} & BR(e^+e^-) = (7.15 \pm 0.19)\cdot 10^{-5} \\
m_\Phi = 1019.413 \pm 0.008 \text{ MeV} & \Gamma_\Phi = 4.43 \pm 0.06 \text{ MeV} & BR(e^+e^-) = (3.09 \pm 0.07)\cdot 10^{-4}
\end{array}
$$

Solution

a) The total cross section for annihilation of e^+e^- into fermion-antifermion pairs is

$$\sigma_{Tot} = \frac{4\pi}{3s} \alpha^2 \sum_i Q_i^2 \qquad (6)$$

In standard notation, we write

$$Q = \frac{Y}{2} + T_3 \qquad (7)$$

$$Q^2 = \frac{Y^2}{4} + T_3^2 + YT_3$$

$$\sum Q_i^2 = \sum \frac{Y^2}{4} + \sum T_3^2$$

$$= \text{(Isosinglet)} + \text{(Isotriplet)} \tag{8}$$

The quantum numbers of the u, d and s quarks are

$$
\begin{array}{lll}
u & Y = \frac{1}{3} & T_3 = \frac{1}{2} \\
d & Y = \frac{1}{3} & T_3 = -\frac{1}{2} \\
s & Y = -\frac{2}{3} & T_3 = 0
\end{array}
\tag{9}
$$

Furthermore, there are 3 colors per quark species.

Thus the contribution of the triplet is $(4\pi\alpha^2/s)\,1/2$, whereas the singlet contributes $(4\pi\alpha^2/s)\,1/6$, leading to a total cross section of $(4\pi\alpha^2/s)\,2/3$. Finally, the triplet content equals

$$\frac{\sigma(Triplet)}{\sigma(Total)} = \frac{3}{4} \tag{10}$$

b) Let us start from the general expression

$$m_V\,\Gamma(V \to e^+e^-) = \frac{1}{2}|M|^2 \frac{1}{8\pi} \tag{11}$$

We denote by M_3 and M_8 the amplitudes corresponding to T_3 and Y; given that ρ is an isospin triplet

$$m_\rho\,\Gamma_{\rho \to e^+e^-} = \frac{1}{16\pi}|M_3|^2 \tag{12}$$

Now, state $|1\rangle$ decouples and therefore, by Eq.(2)

$$m_\omega\,\Gamma_{\omega \to e^+e^-} = \frac{1}{16\pi}|M_8|^2 \sin^2\theta$$

$$m_\Phi\,\Gamma_{\Phi \to e^+e^-} = \frac{1}{16\pi}|M_8|^2 \cos^2\theta \tag{13}$$

The answer to part a) allows us to conclude that

$$R_1 = \frac{|M_3|^2}{|M_8|^2} = 3 \tag{14}$$

in agreement with experimental data

$$R_1^{exp} = 2.8 \pm 0.2 \tag{15}$$

c) At the resonant peak we have

$$\sigma_{el} \sim \frac{1}{m^2}\frac{\Gamma_{e^+e^-}}{\Gamma^2} \tag{16}$$

Consequently, R_2 of Eq.(4) becomes

$$R_2 = \frac{\left(m_\rho \, \Gamma_{\rho \to e^+ e^-}\right)^2}{\left(m_\omega \, \Gamma_{\omega \to e^+ e^-}\right)^2 + \left(m_\phi \, \Gamma_{\Phi \to e^+ e^-}\right)^2} \tag{17}$$

Eqs. (12) and (13) allow us to write

$$R_2 = \frac{|M_3|^4}{|M_8|^4 \left(\sin^4 \theta + \cos^4 \theta\right)} = \frac{R_1^2}{\left(\sin^4 \theta + \cos^4 \theta\right)} \tag{18}$$

Using the values of R_1 and R_2 we end up with

$$\left(\sin^4 \theta + \cos^4 \theta\right)^{-1} = \frac{1}{9} \left(14.5 \pm 2.5\right) = 1.6 \pm 0.3 \tag{19}$$

Ideal mixing predicts

$$\cos\theta \,|1\rangle + \sin\theta \,|8\rangle \;=\; |\frac{u\bar{u} + d\bar{d}}{\sqrt{2}}\rangle$$
$$\cos\theta \,|8\rangle - \sin\theta \,|1\rangle \;=\; |s\bar{s}\rangle \tag{20}$$

$$|1\rangle \;=\; |\frac{u\bar{u} + d\bar{d} + s\bar{s}}{\sqrt{3}}\rangle$$
$$|8\rangle \;=\; |\frac{u\bar{u} + d\bar{d} - 2s\bar{s}}{\sqrt{6}}\rangle \tag{21}$$

which results in

$$\cos\theta = \sqrt{\frac{2}{3}} \qquad \sin\theta = \sqrt{\frac{1}{3}} \qquad \cos\theta^4 + \sin\theta^4 = \frac{5}{9} \tag{22}$$

and

$$\left(\sin^4 \theta + \cos^4 \theta\right)^{-1} = 1.8 \tag{23}$$

This value is compatible with the value of Eq.(19).

Problem 91.   The coupling of a charged vector boson to the electromagnetic field

a)   Derive the most general effective coupling between a charged massive vector field (vector meson) and the electromagnetic field, compatible with

   i) Lorentz symmetry

   ii) Invariance under P and C

   iii) Invariance under T

iv) Gauge invariance

b) Consider this coupling in the limit of vanishing momentum transfer, $q \to 0$; show that, to order q^2, it is determined by three parameters: Charge, magnetic moment, and electric quadrupole. Express these parameters in terms of the form factors introduced in part a).

c) Suppose the meson is also coupled to a massless scalar field through the interaction

$$g \, \phi \, W_\mu^+ \, W_\mu \tag{1}$$

Determine (to lowest perturbative order) a value for g such that two equally charged mesons at a distance r from each other do not interact. Discuss also the case of oppositely charged mesons.

Solution

a) The most general coupling to the electromagnetic field has the form

$$\mathcal{M} = e \, A_\mu \, J^\mu \tag{2}$$

where

$$J^\mu = \langle V' | j_{\text{em}}^\mu | V \rangle \tag{3}$$

is the matrix element of the current between states $|V\rangle$ and $|V'\rangle$ of the vector boson.

By the superposition principle J_μ must be bilinear in the wave functions of incoming and outgoing vector bosons. Lorentz invariance requires that J_μ be a four-vector, made out of the vectors at our disposal. Consequently,

$$J^\mu = \varepsilon'^{* \, \alpha} \, \varepsilon^\beta \, T_{\alpha\beta}^\mu \, (p, p') \tag{4}$$

where $T_{\alpha\beta}^\mu$ is a tensor under Lorentz transformations. By parity invariance J^μ must be a polar vector, thus excluding expressions containing $\varepsilon_{\mu\nu\rho\sigma}$.

The symmetries invoked so far lead to the following most general form for the current

$$
\begin{aligned}
J^\mu = \ & A(q^2) \, P^\mu \, (\varepsilon'^* \cdot \varepsilon) + B(q^2) \, (\varepsilon'^* \cdot \varepsilon) \, q^\mu \\
& + C(q^2) \, (\varepsilon'^* \cdot q) \, (\varepsilon \cdot q) \, P^\mu + D(q^2) \, (\varepsilon'^* \cdot q) \, (\varepsilon \cdot q) \, q^\mu \\
& + E(q^2) \, (\varepsilon'^* \cdot q) \, \varepsilon^\mu + F(q^2) \, (\varepsilon \cdot q) \, \varepsilon'^{* \, \mu}
\end{aligned}
\tag{5}
$$

with

$$P^\mu = (p + p')^\mu \qquad q^\mu = (p - p')^\mu \tag{6}$$

We have exploited the transversality of the vector wave functions

$$\varepsilon'^* \cdot p' = \varepsilon \cdot p = 0 \tag{7}$$

which implies

$$(\varepsilon \cdot P) = -(\varepsilon \cdot q) \qquad (\varepsilon' \cdot P) = (\varepsilon' \cdot q) \tag{8}$$

If T invariance hold, we can take the form factors in (5) to be real. Gauge invariance places a further condition: $q_\mu J^\mu = 0$. As a consequence

$$
\begin{aligned}
J_{\text{em}}^\mu &= a(q^2)\,(\varepsilon'^* \cdot \varepsilon)\,P^\mu + b(q^2)\,(\varepsilon'^* \cdot q)\,(\varepsilon \cdot q)\,P^\mu \\
&\quad + i\,c(q^2)\,[(\varepsilon'^* \cdot q)\,\varepsilon^\mu - (\varepsilon \cdot q)\,\varepsilon'^{*\,\mu}] \\
&\quad + d(q^2)\,\Big[q^2\,(\varepsilon'^* \cdot q)\,\varepsilon^\mu + q^2\,(\varepsilon \cdot q)\,\varepsilon'^{*\,\mu} - 2(\varepsilon \cdot q)\,(\varepsilon'^* \cdot q)\,q^\mu\Big]
\end{aligned} \tag{9}
$$

We note that the presence of 6 form factors is consistent with the number of independent helicity amplitudes for the processes (see appendix E)

$$
\begin{aligned}
\text{spin } 1 &\longrightarrow \text{ spin } 1 + \text{spin } 1 \\
\text{spin } 1 &\longrightarrow \text{ spin } 1 + \text{spin } 0
\end{aligned} \tag{10}
$$

Spin 0 is the scalar component of the current. Making now use of current conservation we eliminate the scalar component; we are left with 4 indipendent amplitudes.

b)   In the limit of small momentum transfer, the previous expression becomes

$$
\begin{aligned}
J_{\text{em}}^\mu &\simeq (\varepsilon'^* \cdot \varepsilon)\left[a_0 + \frac{a_1}{2}q^2\right]P^\mu + b_0\,(\varepsilon'^* \cdot q)\,(\varepsilon \cdot q)\,P^\mu \\
&\quad + c_0\,[(\varepsilon'^* \cdot q)\,\varepsilon^\mu - (\varepsilon \cdot q)\,\varepsilon'^{*\,\mu}]
\end{aligned} \tag{11}
$$

In the same limit, if the particle is initially at rest, the following relations hold

$$
\begin{aligned}
\varepsilon_0' &\simeq \frac{\mathbf{q}\,\boldsymbol{\varepsilon}'}{M} \\
q^0 &\simeq \frac{\mathbf{q}^2}{2M}
\end{aligned} \tag{12}
$$

Thus, to order $\mathbf{q}^2$ the current may be approximated as follows

$$
\begin{aligned}
J_{\text{em}}^\mu &\simeq -(\boldsymbol{\varepsilon}'^*\,\boldsymbol{\varepsilon})\left[a_0 - \frac{a_1}{2}\mathbf{q}^2\right]P_\mu + b_0\,(\boldsymbol{\varepsilon}'^*\,\mathbf{q})\,(\boldsymbol{\varepsilon}\,\mathbf{q})\,P_\mu \\
&\quad - c_0\left[(\boldsymbol{\varepsilon}'^*\,\mathbf{q})\,\varepsilon_\mu - (\boldsymbol{\varepsilon}\,\mathbf{q})\,\varepsilon_\mu'^*\right]
\end{aligned} \tag{13}
$$

Within the same approximation $\boldsymbol{\varepsilon}'$ is the polarization of a particle at rest.

For a physical interpretation of the coefficients in J_{em}, let us consider the case $\varepsilon' = \varepsilon$: in this case, J_{em}^μ becomes the expectation value of j^μ in a state with polarization ε. We substitute the product $\varepsilon_i\,\varepsilon_j'^*$ by the density matrix in the initial state ρ_{ij}, with the parametrization

$$
\rho_{ij} = \frac{1}{3}\delta_{ij} + \frac{1}{2}\varepsilon_{ijk}\,P^k + Q_{ij} \tag{14}
$$

$\mathbf{P}$ is the polarization of the initial state and Q_{ij} its alignment. Setting $p^\mu \simeq M\delta^{\mu 0}$ in the above limit, we find

$$
\begin{aligned}
\frac{1}{2M}\,J_{\text{em}}^0 &\simeq a_0 - \frac{1}{2}a_1\,\mathbf{q}^2 + \frac{b_0}{3}\,\mathbf{q}^2 + b_0\,q^i\,q^j\,Q_{ij} \\
\frac{1}{2M}\,\mathbf{J}_{\text{em}} &= \frac{c_0}{2M}\,\mathbf{q}\wedge\mathbf{P}
\end{aligned} \tag{15}
$$

The parameters in this expression now have a direct interpretation:

a_0 is the electric charge of the vector boson, in units of e;

$2b_0 - 3a_1$ is the square of the charge radius

b_0 is the electric quadrupole

c_0 is the magnetic moment.

For a pure state, the tensor Q_{ij} is completely determined by the Wigner-Eckart theorem; hence, one parameter suffices to characterize the coupling.

c) We turn to the case of two vector bosons at rest, noninteracting due to cancellation of the forces. This case corresponds to the limit $\mathbf{q} = 0$. Therefore, the effective coupling to photons is described completely by the electric monopole term

$$\mathcal{L}_I = e\, A_\mu\, (\varepsilon'^* \cdot \varepsilon)\, P^\mu \simeq 2M\, e\, A_0\, (\varepsilon'^* \cdot \varepsilon) \tag{16}$$

The interaction between two mesons is then described by the exchange of a Coulomb photon; the amplitude equals

$$-\mathrm{i}\,\mathcal{M} = -\mathrm{i}\,\frac{4M^2 e^2}{\mathbf{q}^2}\, (\varepsilon_1'^* \cdot \varepsilon_1)\,(\varepsilon_2'^* \cdot \varepsilon_2) \tag{17}$$

For the scalar interaction instead we have

$$-\mathrm{i}\,\mathcal{M} = (-\mathrm{i})^2\, g^2\, \frac{(-\mathrm{i})}{\mathbf{q}^2}\, (\varepsilon_1'^* \cdot \varepsilon_1)\,(\varepsilon_2'^* \cdot \varepsilon_2) = \mathrm{i}\,g^2\,\frac{1}{\mathbf{q}^2}\, (\varepsilon_1'^* \cdot \varepsilon_1)\,(\varepsilon_2'^* \cdot \varepsilon_2) \tag{18}$$

The Coulomb interaction between particles of equal charge is repulsive, whereas the scalar interaction is always attractive. The total amplitude reads

$$-\mathrm{i}\,\mathcal{M} = (\varepsilon_1'^* \cdot \varepsilon_1)\,(\varepsilon_2'^* \cdot \varepsilon_2)\,\frac{\mathrm{i}}{\mathbf{q}^2}\left(g^2 - 4M^2 e^2\right) \tag{19}$$

We conclude that, when $g = \pm 2Me$, vector bosons at rest with equal charge do not interact. By the same token, bosons of opposite charge will attract each other (at the same value of g) with twice the Coulomb force.

Problem 92.   Electrons in a radial electric field: Transformation of longitudinal into transverse polarization.

An electron with momentum $\simeq \mathbf{p}$, described by a localized wave packet, enters a quarter-cylinder shaped capacitor, as shown in fig. 1. After following a circular arc of radius r, it exits with momentum $\mathbf{p}' \perp \mathbf{p}$.

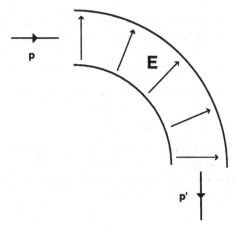

fig. 1

a) Write down the classical relativistic equation of motion and determine the value of the electric field necessary to deviate the electron.

b) Consider a nonrelativistic electron: Show that its spin is a constant of the motion. In particular, an electron entering with longitudinal polarization will exit transversely polarized.

c) In the relativistic case, let Γ^μ be the polarization four-vector, defined in an instantaneous rest frame by

$$\Gamma^\mu = (0, \mathbf{s}) \tag{1}$$

Starting from the equation of motion[13] for Γ^μ

$$\frac{d\Gamma^\mu}{d\tau} = g\mu_B F^{\mu\nu}\Gamma_\nu - (g-2)\mu_B p_\mu F^{\nu\rho}p_\nu\Gamma_\rho \tag{2}$$

with $\mu_B = e\hbar/2mc$, show that the spin magnetic moment is $\frac{g}{2}\mu_B\hbar\boldsymbol{\sigma}$.

d) Calculate the spin precession in the relativistic case, assuming for simplicity that the anomalous component of the magnetic moment is zero.

Solution

a) The classical relativistic trajectory in an electric field is described by the equation of motion

$$\frac{d\mathbf{p}}{dt} = e\,\mathbf{E} \qquad \mathbf{p} = \frac{m\,\mathbf{v}}{\sqrt{1-v^2}} \tag{3}$$

The constraint of a circular orbit with radius r implies

$$m\,\gamma|\mathbf{v}_\perp| = m\,\gamma\frac{v^2}{r} = |e\,\mathbf{E}| \tag{4}$$

---

[13]See problem 98 for a discussion of this equation.

Given that the electron charge is negative, the electric field points outward. In terms of the energy and momentum of the particle

$$v = \frac{p}{\varepsilon} \qquad \varepsilon = m\gamma \tag{5}$$

we conclude that the magnitude of the electric field, required to deflect the electron, is

$$|E| = \frac{p^2}{e r \varepsilon} \tag{6}$$

b) The magnetic moment couples to the magnetic field H. In the instantaneous rest frame of the electron the magnetic field is of the order of $\frac{v}{c}E$; consequently, in the nonrelativistic approximation $\boldsymbol{\mu} = g\,\mu_B\,\mathbf{s}$ is a constant of the motion.

c) Equation (2) may be written in the form

$$\frac{d\Gamma^\mu}{d\tau} = g\mu_B F^{\mu\nu}\Gamma_\nu \tag{7}$$

In the nonrelativistic limit, Γ^μ reads

$$\Gamma^\mu = (0, \mathbf{s}) \tag{8}$$

and Eq.(7) reduces to

$$\left(\frac{d\boldsymbol{\Gamma}}{dt}\right)^i = g\mu_B\, F^{ij}\Gamma_j \tag{9}$$

The explicit expression for $F^{\mu\nu}$

$$F^{\mu\nu} = \begin{pmatrix} 0 & -E_1 & -E_2 & -E_3 \\ E_1 & 0 & -B_3 & B_2 \\ E_2 & B_3 & 0 & -B_1 \\ E_3 & -B_2 & -B_1 & 0 \end{pmatrix} \tag{10}$$

allows us to rewrite (9) as

$$\frac{d\boldsymbol{\Gamma}}{dt} = g\mu_B\,(\boldsymbol{\Gamma} \wedge \mathbf{B}) \tag{11}$$

From this equation we infer that the magnetic moment is $g\mu_B$. For the electron $g - 2 = \alpha/\pi \ll 1$.

d) By its definition, Γ_μ observes the property

$$\Gamma_\mu u^\mu = 0 \tag{12}$$

Temporal evolution does not spoil this property: Indeed, the equation of motion of the particle

$$m\frac{du^\mu}{d\tau} = e\,F^{\mu\nu}u_\nu \tag{13}$$

leads to

$$\frac{\mathrm{d}}{\mathrm{d}\tau}(u^\mu\Gamma_\mu) = 2\mu_B F^{\mu\nu}u_\nu\Gamma_\mu + u_\mu 2\mu_B F^{\mu\nu}\Gamma_\nu = 0 \tag{14}$$

We have assumed $g = 2$. In the laboratory frame the components of Γ^μ are

$$\Gamma^0 = \frac{|\mathbf{p}|}{m}\zeta_\parallel \qquad \Gamma_\perp = \zeta_\perp \qquad \Gamma_\parallel = \frac{\varepsilon}{m}\zeta_\parallel \tag{15}$$

We have decomposed the polarization in the rest frame into the components $\zeta_\parallel$ and $\zeta_\perp$, parallel and perpendicular to the direction of motion.

In our case, only $F^{0i} = E^i$ differs from zero and therefore

$$\frac{\mathrm{d}\mathbf{\Gamma}}{\mathrm{d}\tau} = \frac{e}{mc}\Gamma_0\mathbf{E} \qquad \frac{\mathrm{d}\Gamma^0}{\mathrm{d}\tau} = \frac{e}{mc}\mathbf{\Gamma}\,\mathbf{E} \tag{16}$$

The initial condition is $\zeta_\perp = 0$; $\zeta_\parallel = 1$. The first of Eqs. (16) implies that the component of $\mathbf{\Gamma}$ perpendicular to the plane of the orbit is constantly zero, so that $\zeta_\perp$ points in the direction of the electric field.

Since $|\mathbf{p}|$ is constant and $\mathbf{E}$ is transverse, the second of Eqs. (16) becomes

$$\frac{|\mathbf{p}|}{m}\frac{\mathrm{d}\zeta_\parallel}{\mathrm{d}\tau} = \frac{e}{m}\mathbf{E}\,\zeta_\perp \tag{17}$$

The initial condition $\zeta_\parallel(0) = 1$, together with $\zeta_\parallel^2 + \zeta_\perp^2 = 1$, suggest that $\dot\zeta_\parallel \leq 0$; thus, $\zeta_\perp$ is anti-parallel to $\mathbf{E}$ and we have

$$v\gamma^2\frac{\mathrm{d}\zeta_\parallel}{\mathrm{d}\tau} = -\frac{eE}{m}\sqrt{1-\zeta_\parallel^2} \tag{18}$$

Use of Eq.(4) gives

$$\frac{\mathrm{d}\zeta_\parallel}{\mathrm{d}\tau} = -\frac{v}{\gamma r}\sqrt{1-\zeta_\parallel^2} \tag{19}$$

whose solution is

$$\zeta_\parallel(t) = \cos\left(\frac{vt}{r\gamma}\right) \equiv \cos\frac{\theta(t)}{\gamma} \tag{20}$$

Complete transverse polarization will be attained when $\frac{\theta(t)}{\gamma} = \frac{\pi}{2}$, that is, when

$$\theta = \frac{\pi\gamma}{2} \simeq \frac{\pi}{2}\left(1+\frac{v^2}{2}\right) \qquad \theta = \frac{\pi}{2} + \Delta\theta \qquad \Delta\theta = \frac{\pi}{4}\frac{v^2}{c^2} \tag{21}$$

Problem 93.   The Čerenkov effect

A charged particle moves with velocity **v** in a medium with a frequency-dependent index of refraction $n(\omega)$.

a)   Write down the wave function of a monochromatic photon propagating in the medium.

b)   The vertex describing photon emission (see fig. 1), is not sensitive to the presence of matter.

Calculate the probability per unit time for the particle to emit a γ with frequency in the interval $(\omega, \omega+d\omega)$; discuss the condition which the particle velocity must satisfy in order for the emission to take place. Make use of the appropriate phase space for a photon propagating in a medium.

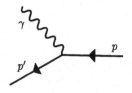

fig. 1

c)   Derive the angular distribution of the emitted photons and the energy per unit length. The emission considered here is known as the Čerenkov effect.

Solution

a)   The equation of electromagnetic waves in a medium is

$$\Delta \mathbf{A} - n^2 \frac{\partial^2 \mathbf{A}}{\partial t^2} = 0, \qquad \Delta \phi - n^2 \frac{\partial^2 \phi}{\partial t^2} = 0 \tag{1}$$

having imposed the generalized Lorentz gauge

$$\operatorname{div}\mathbf{A} + n^2 \frac{\partial \phi}{\partial t} = 0 \tag{2}$$

In general, a monochromatic photon with given momentum **k**, moving in a medium with index of refraction $n(\omega)$, is described by the wave function

$$A_\mu(\mathbf{x}, t) = \varepsilon_\mu \, e^{i\mathbf{k}\mathbf{x} - i\omega t} \tag{3}$$

The dispersion law and gauge condition have the form

$$|\mathbf{k}| = n(\omega)\,\omega \tag{4}$$

$$\mathbf{k}\,\boldsymbol{\varepsilon} - n^2(\omega)\,\omega \varepsilon_0 = 0 \tag{5}$$

We shall adopt the classical wave function defined in Eq.(3) as an effective wave function for the photon inside matter. According to (4), the corresponding momentum has modulus $n(\omega)\omega$; the photon energy is ω.

To calculate the transition amplitude, we will use the standard vertex of photon emission; phase space, on the other hand, will be modified, substituting the four-momentum of the photon by the effective four-momentum

$$k^\mu = (\omega\,,\mathbf{k}(\omega)) \qquad |\mathbf{k}(\omega)| = n(\omega)\,\omega \tag{6}$$

b) The emission probability per unit time is given by the formula

$$d\Gamma = \frac{1}{2E}\,|M_{fi}|^2\,d\Phi^{(2)} \tag{7}$$

$d\Phi^{(2)}$ denotes two-body phase space

$$
\begin{aligned}
d\Phi^{(2)} &= \frac{d^3\mathbf{p}'}{2p_0'(2\pi)^3}\,\frac{d^4k}{(2\pi)^4}\,2\pi\delta(\mathbf{k}^2 - n^2\omega^2)\,(2\pi)^3\delta^3(\mathbf{p}' + \mathbf{k} - \mathbf{p})\,2\pi\delta(\omega - \mathbf{k}\,\mathbf{v}) \\
&= \frac{1}{2p_0'}\,\frac{d\omega\,d^3\mathbf{k}}{(2\pi)^2}\,\delta(\mathbf{k}^2 - n^2\omega^2)\,\delta(\omega - \mathbf{k}\,\mathbf{v})
\end{aligned} \tag{8}
$$

In the above, the energy loss in single-photon emission was taken to be small with respect to the energy of the charged particle

$$E(\mathbf{p}) - E(\mathbf{p}') \simeq \frac{\partial E}{\partial \mathbf{p}}\cdot\mathbf{k} = \mathbf{v}\,\mathbf{k} \tag{9}$$

The angle of émission is related to the frequency by $\omega = \mathbf{k}\,\mathbf{v} \equiv |\mathbf{k}|v\cos\theta$, or, using (4),

$$\cos\theta = \frac{1}{nv} \tag{10}$$

We remark that $n = n(\omega)$, so that a definite angle corresponds to each frequency. By (10), emission will take place provided that

$$nv \geq 1 \qquad \text{i.e.} \qquad v \geq \frac{1}{n(\omega)} \tag{11}$$

The transition amplitude is given by

$$M_{fi} = e\,\varepsilon_\mu^*\,j^\mu \tag{12}$$

If the momentum transfer is small compared to the particle's four-momentum we can make the approximation

$$j^\mu = \bar{u}(\mathbf{p}')\gamma^\mu u(\mathbf{p}) \simeq \bar{u}(\mathbf{p})\gamma^\mu u(\mathbf{p}) = 2p^\mu \tag{13}$$

Eq.(5) allows us to express ε_0 in terms of ε, with the result

$$\mathcal{M}_{fi} = 2e\,\varepsilon\cdot\left[\frac{\mathbf{k}\,p^0}{n^2\omega} - \mathbf{p}\right] \tag{14}$$

Summing over polarizations, we have

$$\langle |\mathcal{M}_{fi}|^2 \rangle = 4e^2 \left[\frac{k_i \, E}{n^2 \omega} - p_i \right] \left[\frac{k_j \, E}{n^2 \omega} - p_j \right] \left(\delta_{ij} - \frac{k_i k_j}{k^2} \right) = 4e^2 \left(\mathbf{p}^2 - \frac{(\mathbf{p}\,\mathbf{k})^2}{k^2} \right) \tag{15}$$

This expression is simplified further through the relations

$$\mathbf{p}\,\mathbf{k} = \frac{p}{v}\,\omega \qquad k^2 = n^2 \omega^2 \tag{16}$$

leading to

$$\langle |\mathcal{M}_{fi}|^2 \rangle = 4e^2 p^2 \left(1 - \frac{1}{n^2 v^2} \right) \tag{17}$$

Finally, the frequency spectrum is

$$
\begin{aligned}
d\Gamma &= \frac{1}{2E} 4e^2 p^2 \left(1 - \frac{1}{n^2(\omega)\,v^2} \right) \frac{1}{8\pi} \frac{d\omega}{p} \\
&= \alpha v \left(1 - \frac{1}{n^2(\omega)\,v^2} \right) d\omega
\end{aligned}
\tag{18}
$$

c)   To obtain the angular distribution of emitted photons, we differentiate the relation

$$\cos\theta = \frac{1}{nv} \tag{19}$$

finding

$$
\begin{aligned}
d\cos\theta &= -\frac{1}{n^2 v} \frac{dn}{d\omega} d\omega \\
d\Gamma &= -\alpha \left(\frac{dn}{d\omega} \right)^{-1} \frac{\sin^2\theta}{\cos^2\theta} d\cos\theta
\end{aligned}
\tag{20}
$$

The energy loss of the particle equals

$$\frac{dE}{dx} = \frac{1}{v} \frac{dE}{dt} = \frac{1}{v} \int \omega \, d\Gamma = \alpha \int \left(1 - \frac{1}{n^2(\omega)\,v} \right) \omega \, d\omega \tag{21}$$

The above integral ranges over those frequencies for which emission can take place. The result is identical to the classical one. Thus, the quantum treatment of the Čerenkov effect simply amounts to introducing the correct effective phase space for the final photon.

Problem 94. Cross section for the Drell-Yan process in πP

Two hadrons, colliding at high energies, may be depicted by two collinear quark beams. Each quark is distributed with probability density $f_i(x_i)$ with respect to the variable $x_i = |\mathbf{p}_i|/|\mathbf{p}|$, where $\mathbf{p}_i$ is the momentum of the ith quark and $\mathbf{p}$ is the hadron momentum. One may consider the process $\pi + P \to \mu^+\mu^- + X$ as being due to the annihilation of $q\bar{q}$ pairs, with a quark coming from the proton and an antiquark coming from the pion.

a)   Derive the contribution of a quark-antiquark pair to the differential cross section at a given invariant mass M for the $\mu^+\mu^-$ system, in terms of the distribution $f_i(x)$. The quarks may be assumed to be massless.

b)   Show that the cross section has the scaling behaviour

$$\frac{\mathrm{d}\sigma}{\mathrm{d}M} = M^{-3}\,\Phi\left(\frac{s}{M^2}\right) \tag{1}$$

where s is the center-of-mass energy squared.

c)   Making use of the standard charge assignment of quarks, calculate the following ratio between inclusive cross sections

$$\sigma\left(\pi^- + P \to \mu^+\mu^- + X\right)/\sigma\left(\pi^+ + P \to \mu^+\mu^- + X\right) \tag{2}$$

Solution

a,b)   The inclusive production of $\mu^+\mu^-$ is due to the annihilation of the antiquark contained in π with one of the quarks inside the proton. Let us denote by p_P and p_π the proton and pion four-momenta in the center of mass; $q = x p_P$ and $\bar{q} = \bar{x} p_\pi$ stand for the quark four-momenta. The invariant mass of produced pairs is given by

$$M^2 = (q + \bar{q})^2 = 2x\bar{x}\,p_P \cdot p_\pi = x\bar{x}\,s \tag{3}$$

Here, $s = (p_\pi + p_P)^2$; by consistency with the approximation of massless quarks, we take hadronic masses to be zero as well. The matrix element for the process shown in the diagram of fig. 1 is

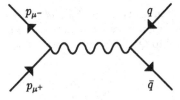

fig. 1

$$\mathcal{M} = e^2 Q_q\, \bar{v}(\bar{q})\gamma^\mu u(q)\, \frac{-\mathrm{i}}{M^2}\, \bar{u}(\mathbf{p}_{\mu^-})\gamma_\mu v(\mathbf{p}_{\mu^+}) \tag{4}$$

Q_q is the charge of the quark in units of e. We take the modulus squared and sum (average) over final (initial) spins; we further average over the direction of the muons with respect to their center of mass, with the result

$$\overline{|\mathcal{M}|^2} = \frac{1}{4} \frac{e^4 Q_q^2}{M^4} |\langle L_{\mu\nu} \rangle A^{\mu\nu}| \tag{5}$$

where

$$\langle L_{\mu\nu} \rangle = \frac{4}{3} \left(k_\mu k_\nu - k^2 g_{\mu\nu} \right) \tag{6}$$

$$k_\mu = (q + \bar{q})_\mu = (p_{\mu^+} + p_{\mu^-})_\mu \tag{7}$$

k_μ is the four-momentum of the virtual photon in the diagram of fig. 1. $A_{\mu\nu}$ equals

$$A_{\mu\nu} = 2 \left\{ k_\mu k_\nu - k^2 g_{\mu\nu} - (q - \bar{q})_\mu (q - \bar{q})_\nu \right\} \tag{8}$$

Substituting into Eq.(5) we find

$$\overline{|\mathcal{M}|^2} = \frac{4}{3} e^4 Q_q^2 \tag{9}$$

Introducing also phase space for two massless particles, $\Phi^{(2)} = 1/8\pi$, the expression for the inclusive cross section at fixed M^2 is readily obtained

$$d\sigma = \int dx\, d\bar{x}\, f_q^P(x) f_{\bar{q}}^\pi(\bar{x}) \frac{1}{2} \frac{1}{x\bar{x}\, 2(p_P \cdot p_\pi)} \overline{|\mathcal{M}|^2} \frac{1}{8\pi} \delta(M^2 - x\bar{x}s)\, dM^2 \tag{10}$$

or, equivalently,

$$\frac{d\sigma}{dM} = \frac{1}{M^3} \frac{8\pi\alpha^2}{3} Q_q^2 \, \Phi\left(\frac{s}{M^2}\right) \tag{11}$$

with

$$\Phi\left(\frac{s}{M^2}\right) = \int dx \int d\bar{x}\, f_q^P(x) f_{\bar{q}}^\pi(\bar{x})\, \delta(1 - x\bar{x}\frac{s}{M^2}) \tag{12}$$

c)   We recall that the proton is made out of a uud triplet, while π^- consists of a $\bar{u}d$ pair. Therefore, the annihilation cross section of $\pi^- P$ is proportional to $2Q_u^2$, while that of $\pi^+ P$ is proportional to Q_d^2. This leads to the ratio of cross sections

$$\frac{2Q_u^2}{Q_d^2} = 8 \tag{13}$$

We have assumed here that the u and d distributions inside the proton are equal, $f_u^P = f_d^P$. Antiquark distributions in π, as a consequence of isospin invariance, are the same in the two processes.

Problem 95.   The Bethe model for the Lamb shift

We propose to calculate[14] the radiative shift of atomic levels (Lamb shift). Let $H_I = -e\,\mathbf{v}\mathbf{A}$ be the interaction Hamiltonian of electrons in a radiation field. We denote by δE_n the radiative shift of an atomic level relative to the shift of a free electron ΔF:

$$\delta E_n = \Delta E_n - \Delta F \tag{1}$$

a)   Calculate the shift ΔE_n of states in the hydrogen atom.

b)   Calculate ΔF. Setting ΔE_n equal to ΔF above a certain cutoff frequency $\omega_{\max} \simeq mc^2$, show that

$$\delta E_n = \frac{2\alpha}{3\pi} \sum_n |\langle n|\mathbf{r}|s\rangle|^2 \, (E_n - E_s)^2 \, \ln\frac{mc^2}{(E_n - E_s)} \tag{2}$$

c)   Derive the shift due to $g - 2 = \alpha/\pi \neq 0$. You may consider the electron as nonrelativistic.

Solution

a)   The interaction $H_I = -e\,\mathbf{v}\mathbf{A}$ stems from the Hamiltonian

$$H = \frac{1}{2m}\,(\mathbf{p} - e\mathbf{A})^2 \tag{3}$$

to lowest order in the coupling constant e. $\mathbf{A}$ is the electromagnetic field operator in the Coulomb gauge.

The Hilbert space of the system is

$$H = H_A \otimes H_\gamma \tag{4}$$

where H_A is the space of states of the atom and H_γ is the Fock space of the electromagnetic field. Thus, for example, the atomic state $|n\rangle$ corresponds in this enlarged space to $|n\rangle \otimes |0\rangle$, having photon occupation number equal to zero.

We remark that the energy shift of the nth atomic level vanishes to first order in e

$$\langle 0|H_I|0\rangle = 0 \tag{5}$$

since the expectation value of $\mathbf{A}$ is zero.

To second order, we write

$$\Delta E_n = \sum_M \langle n, 0|H_I|M\rangle \, \frac{1}{E_n - E_M} \, \langle M|H_I|n, 0\rangle \tag{6}$$

where $|M\rangle$ is a state in the enlarged space H.

---

[14]H. A. Bethe, Phys. Rev. **72**, 339 (1947).

The expression for A,

$$\mathbf{A}(x) = \sum_i \int \frac{d^3k}{2\omega\,(2\pi)^3}\left[\boldsymbol{\varepsilon}_{(i)}\,a_i(\mathbf{k})\,e^{-ikx} + \boldsymbol{\varepsilon}_{(i)}^*\,a_i^\dagger(\mathbf{k})\,e^{ikx}\right] \tag{7}$$

implies that the sum in (6) will receive contributions only from states $|M\rangle$ with one photon

$$|M\rangle = |n\rangle \otimes |\mathbf{k}\,\boldsymbol{\varepsilon}\rangle \tag{8}$$

$\mathbf{k}$ is the photon momentum and and $\boldsymbol{\varepsilon}$ its polarization. In standard notation, we write

$$\sum_k |\mathbf{k}\rangle\langle\mathbf{k}| = \int \frac{d^3k}{(2\pi)^3\,(2\omega)}|\mathbf{k}\rangle\langle\mathbf{k}| \quad ; \quad \omega = |\mathbf{k}| \tag{9}$$

We calculate the matrix element of $\mathbf{A}$ and sum over photon polarizations, with the result

$$\Delta E_n = \sum_f \int \frac{2\alpha}{3\pi}\frac{\omega\,d\omega}{E_n - E_f - \omega}\,\langle n|\frac{\mathbf{P}}{m}|f\rangle\,\langle f|\frac{\mathbf{P}}{m}|n\rangle \tag{10}$$

Here, $|f\rangle$ stands for an atomic state; we have made use of the relation $E_M = \omega + E_f$. We note that the above expression diverges linearly in ω.

b)   The calculation for a free electron proceeds analogously. Given that energy eigenstates are also momentum eigenstates, we have $E_n = E_f$ and

$$\Delta E_F = \frac{2\alpha}{3\pi}\int \frac{\omega\,d\omega}{-\omega}\,\langle\mathbf{p},0|\frac{p^2}{m^2}|\mathbf{p},0\rangle \tag{11}$$

This energy shift amounts to a redefinition of the electron's rest energy. One can describe this effect in terms of an operator δH_0

$$\delta H_0 = -\frac{2\alpha}{3\pi}\int d\omega\,\frac{p^2}{m^2} \tag{12}$$

On a bound state $|n\rangle$, δH_0 gives

$$\langle n|\delta H_0|n\rangle = \frac{2\alpha}{3\pi}\int \frac{\omega\,d\omega}{-\omega}\sum_f |\langle n|\frac{\mathbf{P}}{m}|f\rangle|^2 \tag{13}$$

Consequently,

$$\begin{aligned}
\delta E_n &= \Delta E_n - \langle n|\delta H_0|n\rangle \\
&= \frac{2\alpha}{3\pi}\sum_f \int d\omega\,\frac{(E_n - E_f)}{(E_n - E_f - \omega)}\,|\langle f|\frac{\mathbf{P}}{m}|n\rangle|^2
\end{aligned} \tag{14}$$

Applying the Heisenberg equation

$$\frac{\mathbf{P}}{m} = i\,[H,\mathbf{x}] \tag{15}$$

to Eq.(14), we find

$$\delta E_n = \frac{2\alpha}{3\pi} \sum_f \int d\omega \, \frac{(E_n - E_f)^3}{(E_n - E_f - \omega)} |\langle f|\mathbf{x}|n\rangle|^2 \tag{16}$$

Eq.(14) may now be integrated up to the cutoff frequency $mc^2 \gg E_n$, yielding

$$\delta E_n = \frac{2\alpha}{3\pi} \sum_f (E_f - E_n) |\langle f|\frac{\mathbf{P}}{m}|n\rangle|^2 \ln \frac{mc^2}{E_n - E_f} \tag{17}$$

In order to evaluate further Eq.(17) we may ignore the E_n-dependence in the logarithm, and use the identity

$$\sum_f (E_f - E_n) \langle n|\frac{\mathbf{P}}{m}|f\rangle\langle f|\frac{\mathbf{P}}{m}|n\rangle = \frac{1}{i\,m^2} \langle n|\mathbf{p}\,\dot{\mathbf{p}}|n\rangle \tag{18}$$

$\dot{\mathbf{p}}$ can be eliminated by the equation of motion $\dot{\mathbf{p}} = -\nabla\Phi$, where Φ is the potential $-\alpha/r$; we obtain

$$\frac{1}{i}\langle n|\mathbf{p}\,\dot{\mathbf{p}}|n\rangle \; = \int d^3\mathbf{x}\,(-\nabla\psi_n^*)\,(-\nabla\Phi)\,\psi_n = -\frac{1}{2}\int |\psi_n|^2 \, \nabla^2\Phi$$

$$= 2\alpha\pi\,|\psi_n(0)|^2 = \frac{2\pi\alpha}{\pi\,n^3 a^3} = \frac{2\alpha}{n^3 a^3} \tag{19}$$

(a is the Bohr radius).

The approximation involved in (18) shifts only S states, by an amount

$$\delta E_n = \frac{4}{3\pi} \frac{m\alpha^5}{n^3} \ln\frac{mc^2}{\langle E - E_n\rangle} \tag{20}$$

where $\langle E - E_n\rangle$ indicates an average excitation energy of the atom.

c) An anomalous magnetic moment $g - 2$ corresponds to the following effective interaction with the electromagnetic field

$$H_I = -\frac{g - 2}{8m}\,e\,\bar{\psi}\sigma_{\mu\nu}F^{\mu\nu}\psi = -e\,\frac{g - 2}{4m}\,\psi^*\,(\mathbf{\Sigma}\,\mathbf{B} - i\,\boldsymbol{\alpha}\,\mathbf{E})\,\psi \tag{21}$$

with

$$\sigma_{\mu\nu} = -\frac{i}{2}\,[\gamma_\mu, \gamma_\nu] \qquad \boldsymbol{\alpha} = \gamma^0\boldsymbol{\gamma} \qquad \mathbf{\Sigma} = \begin{pmatrix} \boldsymbol{\sigma} & 0 \\ 0 & \boldsymbol{\sigma} \end{pmatrix} \tag{22}$$

in the standard representation. The field generated by a potential Φ due to the nucleus is

$$\mathbf{B} = 0 \qquad \mathbf{E} = -\frac{\mathbf{x}}{r}\frac{d\Phi}{dr} \tag{23}$$

We now set $\psi = \begin{pmatrix} \phi \\ \chi \end{pmatrix}$ and expand in powers of v/c, obtaining

$$\langle \phi | H_{\text{eff}} | \phi \rangle = \langle \phi | (g-2) \frac{1}{4m^2} e \, \boldsymbol{\sigma} \, \mathbf{L} \, \frac{1}{r} \frac{d\Phi}{dr} | \phi \rangle \tag{24}$$

To lowest nontrivial order the anomalous magnetic moment of the electron is

$$g - 2 = \frac{\alpha}{\pi} \tag{25}$$

Therefore,

$$H_{\text{eff}} = \frac{\alpha}{4\pi \, m^2} e \, \boldsymbol{\sigma} \, \mathbf{L} \, \frac{1}{r} \frac{d\Phi}{dr} \tag{26}$$

The effect of this term is a shift in levels with $L \neq 0$.

Eq.(20) predicts a shift in the $2S$ state of hydrogen by

$$\Delta E_S \simeq 1300 \text{MHz} \tag{27}$$

For P states, Eq.(26) gives

$$
\begin{array}{lll}
2P_{1/2} & \Delta E = \dfrac{\alpha}{4\pi m^2} \langle 2p | \dfrac{\alpha}{r^3} | 2p \rangle = \dfrac{m\alpha^5}{96\pi} = & 8.5 \,\text{MHz} \\[3mm]
2P_{3/2} & \Delta E = -\dfrac{2\alpha}{4\pi m^2} \langle 2p | \dfrac{\alpha}{r^3} | 2p \rangle = -\dfrac{m\alpha^5}{48\pi} = -17 \,\text{MHz}
\end{array} \tag{28}
$$

Problem 96. On the possibility of directing neutrino beams produced in accelerators onto distant laboratories

A proposal has been put forward to direct a neutrino beam, produced at CERN, onto a detector placed in the Italian underground laboratory of Gran Sasso (below the mountain peak bearing the same name). The distance between the two laboratories is $D \simeq 1000$ km.

a) Estimate the beam attenuation, assuming a uniform density $\delta \sim 3 \text{ gr/cm}^3$ of the earth near surface and an average isospin of $I = 0$; the cross section is

$$\sigma \simeq 0.3 \frac{G_F^2 \, s}{\pi} \qquad\qquad G_F \simeq 10^{-5} \, m_p^{-2} \tag{1}$$

b) The neutrinos are produced by pions with energy E, travelling in the same direction along a 1 Km trajectory in vacuum. Calculate the angular spread of the neutrino beam.

c) Suppose the detector below Gran Sasso has a sectional area of 20 m^2. How many neutrinos per initial pion will impinge on the detector as a function of energy?

Solution

a) The number of neutrinos absorbed per unit length is given by the formula

$$\frac{dN(x)}{dx} = -N(x)\,\rho\,\sigma_{\nu N} \tag{2}$$

At a distance D we have

$$N(D) = N_0\,e^{-\rho\sigma D} \tag{3}$$

Using $s \simeq m_p^2 + 2E_\nu\,m_p$, the cross section becomes

$$\sigma \simeq 0.3\,\frac{G_F^2 m_p^2}{\pi}\left(1 + \frac{2E_\nu}{m_p}\right) \tag{4}$$

and the quantity $\rho\,\sigma\,D$ equals

$$\rho\,\sigma\,D \simeq 0.7{\cdot}10^{-6}\left(1 + \frac{2E_\nu}{m_p}\right) \tag{5}$$

Thus, for neutrino energies $\leq 10^5\,\mathrm{GeV}$,

$$N(D) \simeq N_0\,(1 - \rho\,\sigma\,D) \tag{6}$$

The attenuation is

$$\frac{\Delta N}{N} \simeq \rho\,\sigma\,D \simeq 0.7{\cdot}10^{-6}\left(1 + \frac{2E_\nu}{m_p}\right) \tag{7}$$

b) The differential decay width for $\pi^- \to \mu^- \bar{\nu}_\mu$ is given by the formula

$$\begin{aligned}
d\Gamma &= \frac{1}{\gamma\,m_\pi}\left(G_F^2 m_\mu^2 f_\pi^2\right)\frac{m_\pi^2 - m_\mu^2}{8\pi\,m_\pi^2}\frac{d\Omega}{4\pi} \\
&\simeq \frac{1}{\gamma\,\tau_\pi}\frac{d\cos\theta_L}{2\gamma^2\left(1 - \beta\,\cos\theta_L\right)^2}
\end{aligned} \tag{8}$$

with

$$\beta = \frac{v_\pi}{c} \qquad \gamma = (1 - \beta^2)^{-1/2} \tag{9}$$

The mean life of π at rest equals

$$\begin{aligned}
\frac{1}{\tau_\pi} &= G_F^2 f_\pi^2\,\frac{m_\pi\,m_\mu^2}{8\pi}\left(1 - \frac{m_\mu^2}{m_\pi^2}\right)^2 \\
\tau_\pi &\simeq 2.6{\cdot}10^{-8}\,\mathrm{sec}
\end{aligned} \tag{10}$$

We define the angular spread as the angle containing half of the neutrinos. In the center of mass this angle is $\theta_c = \pi/2$; going over to the laboratory frame by means of

$$\cos\theta_c = \frac{\cos\theta_L - \beta}{1 - \beta\,\cos\theta_L} \tag{11}$$

we conclude that $\cos\theta_L = \beta$. Consequently, the angular spread is

$$\Delta\theta \simeq 2\left(1 - \sqrt{1 - \frac{m_\pi^2}{E^2}}\right) \underset{E \gg m_\pi}{\simeq} \frac{m_\pi^2}{E^2} \tag{12}$$

c)   As pions travel through a distance d, the number of neutrinos emitted per pion is

$$n = \frac{d}{\beta\gamma\tau_\pi} = \frac{d}{\tau_\pi}\frac{1}{\sqrt{\dfrac{E^2}{m_\pi^2} - 1}} \underset{E \gg m_\pi}{\simeq} \frac{m_\pi}{E}\frac{d}{\tau_\pi} \tag{13}$$

Setting $d = 1$ Km,

$$\frac{d}{c\tau_\pi} = 128 \tag{14}$$

For $\beta\gamma = \sqrt{\gamma^2 - 1} \lesssim 128$ practically all pions decay along the 1 Km track.

A sectional area of 20 m^2 at a distance of 1000 Km corresponds to a solid angle $\Delta\Omega \sim 2\cdot 10^{-11}$ sterad. The relation

$$2\pi(1 - \cos\theta^*) = \Delta\Omega \tag{15}$$

leads to

$$\cos\theta^* = 1 - 0.32\cdot 10^{-12} \qquad \text{or} \qquad \theta^* \simeq 0.8\cdot 10^{-6} \tag{16}$$

By Eq.(8), the fraction of neutrinos emitted within the angle θ^* is

$$\frac{\Delta\Gamma}{\Gamma} = \int_{\cos\theta^*}^{1} \frac{d\cos\theta_L}{2\gamma^2(1 - \beta\cos\theta_L)^2} = \frac{1}{2\beta\gamma^2}\frac{1}{1 - \beta\cos\theta_L}\Big|_{\cos\theta^*}^{1}$$

$$= \frac{1}{2\beta\gamma^2}\left(\frac{1}{1 - \beta} - \frac{1}{1 - \beta\cos\theta^*}\right) = \frac{1}{2\gamma^2}\frac{1 - \cos\theta^*}{(1 - \beta)(1 - \beta\cos\theta^*)} \tag{17}$$

In conclusion, the number of neutrinos $\bar{n}$ (per initial pion) arriving at the detector is, for sufficiently large γ,

$$\bar{n} = n\frac{\Delta\Gamma}{\Gamma} \tag{18}$$

Equivalently, by virtue of (13) and (17),

$$n = \frac{d}{\tau_\pi}\frac{\sqrt{1 - \beta^2}}{\beta}\frac{1 + \beta}{2}\frac{1 - \cos\theta^*}{1 - \beta\cos\theta^*} \simeq \frac{d}{\tau_\pi}\frac{2E}{m_\pi}\frac{\Delta\Omega}{2\pi} \simeq 0.82\cdot 10^{-9}\frac{E}{m_\pi} \tag{19}$$

In the numerical evaluation above we have assumed that

$$1 - \beta \gg 1 - \cos\theta^* \quad \text{or} \quad \gamma \ll (1 - \cos\theta^*)^{-1/2}$$

wich means

$$\frac{E}{m_\pi} \ll 10^6$$

Problem 97. Relativistic scattering of electrons from a given distribution of charge and magnetic moment

A charged particle with spin $1/2$ strikes a spinless nucleus with charge Z. The nucleus is taken to have infinite mass and vanishing spatial extent.

a) Derive the scattering amplitude and angular distribution in the Born approximation, assuming that the particle is pointlike.

b) Calculate the contribution of the particle's anomalous magnetic moment to the cross section.

c) Assume now the particle has an extended charge and magnetic moment distribution. Parameterize the amplitude and calculate the quantities of parts a) and b).

Solution

a) A relativistic pointlike particle with spin $1/2$ interacts with external electromagnetic fields through the interaction Lagrangian

$$\mathcal{L}_I = -e\,\bar{\psi}\gamma_\mu\psi\,A^\mu + \frac{e}{4m}\,\mu'\,\bar{\psi}\sigma_{\mu\nu}\psi\,F^{\mu\nu} \tag{1}$$

where μ' is the anomalous magnetic moment in units of $e/2m$, that is,

$$\mu' = \frac{g-2}{2} \quad \text{and} \quad \sigma_{\mu\nu} = -\frac{\mathrm{i}}{2}\,[\gamma_\mu, \gamma_\nu] \tag{2}$$

Introducing free-particle wave functions $\psi(x) = u(\mathbf{p})\,\mathrm{e}^{-\mathrm{i}px}$ we find, for a static external field,

$$\langle \mathbf{p}'|\int \mathcal{L}_I\,d^4x\,|\mathbf{p}\rangle = -e\,\bar{u}(\mathbf{p}')\left[\gamma^\mu - \frac{\mathrm{i}}{2m}\,\mu'\sigma_{\mu\nu}\,q^\nu\right]u(\mathbf{p})A^\mu(\mathbf{q})\,2\pi\delta(p_0' - p_0) \tag{3}$$

$q_\mu = p'_\mu - p_\mu$ is the four-momentum transfer. Use of the Gordon identity

$$\bar{u}(\mathbf{p}')\gamma_\mu u(\mathbf{p}) = \frac{1}{2m}\,\bar{u}(\mathbf{p}')\,[P_\mu - \mathrm{i}\,\sigma_{\mu\nu}\,q^\nu]\,u(\mathbf{p}) \tag{4}$$

with $P_\mu = p_\mu + p'_\mu$, gives the following expression for the matrix element of the electromagnetic current

$$j_\mu = \frac{1}{2m}\,\bar{u}(\mathbf{p}')\left[P_\mu - \mathrm{i}\,\frac{g}{2}\,\sigma_{\mu\nu}\,q^\nu\right]u(\mathbf{p}) \tag{5}$$

The interpretation of μ' as the anomalous magnetic moment is clear by inspection of the above expression.

The static limit of the field generated by a pointlike nucleus, having charge Z and infinite mass, is

$$A_0 = \frac{Z\,e}{4\pi\,r} \qquad \mathbf{E} = -\nabla A_0 = \frac{Z\,e\,\mathbf{r}}{4\pi\,r^3} \tag{6}$$

All other components vanish. The corresponding Fourier trasforms are

$$A_0(\mathbf{q}) = \frac{Z\,e}{\mathbf{q}^2} \qquad \mathbf{E}(\mathbf{q}) = i\mathbf{q}\,\frac{Z\,e}{\mathbf{q}^2} \tag{7}$$

The interaction Lagrangian is now reduced to the form

$$\mathcal{L}_I = -e\,\psi^\dagger\psi\,A_0 - i\,\frac{e}{2m}\,\mu'\,\psi^\dagger\gamma_i\psi\,E^i \tag{8}$$

and the amplitude of interest reads

$$\mathcal{M} = -\frac{e^2\,Z}{\mathbf{q}^2}\left[u^\dagger(\mathbf{p}')\,u(\mathbf{p}) + \frac{\mu'}{2m}\,u^\dagger(\mathbf{p}')\,\boldsymbol{\gamma}\cdot\mathbf{q}\,u(\mathbf{p})\right] \tag{9}$$

The cross section follows immediately from the above; it equals[15]

$$\frac{d\sigma}{d\Omega} = \frac{\alpha^2\,Z^2}{(\mathbf{q}^2)^2}\left|u'^\dagger u + \frac{\mu'}{2m}\,u'^\dagger\boldsymbol{\gamma}\cdot\mathbf{q}u\right|^2 \tag{10}$$

In the absence of an anomalous magnetic moment, taking a sum over final spins and an average over initial spins yields the Mott cross section

$$\frac{d\sigma}{d\Omega} = \frac{\alpha^2\,Z^2}{(\mathbf{q}^2)^2}\left(4p_0^2 - \mathbf{q}^2\right) \tag{11}$$

b,c) In order to avoid repeating many steps, we consider right away the case of an extended distribution of charge and anomalous magnetic moment. The current may now be parameterized as

$$j_\mu = \frac{1}{2m}\,\bar{u}(\mathbf{p}')\left[F(Q^2)\,P_\mu + i\,G(q^2)\,\sigma_{\mu\nu}\,q^\nu\right]u(\mathbf{p}) \tag{12}$$

with

$$F(0) = 1 \qquad\qquad G(0) = \frac{g}{2} \tag{13}$$

---

[15] We have made use of the fact that the flux factor, $1/2vp_0 = 1/2|\mathbf{p}|$, multiplied by phase space gives

$$\frac{1}{2|\mathbf{p}|}\,\frac{d^3\mathbf{p}'}{(2\pi)^3\,2p_0'}\,2\pi\delta(p_0 - p_0') = \frac{d\Omega'}{16\pi^2}$$

The Gordon identity allows us to rewrite j_μ as

$$j_\mu = \frac{1}{2m}\,\bar{u}(p')\left[(F - G)\,P_\mu + 2m\,G\,\gamma_\mu\right]u(p) \tag{14}$$

In our static case, we have

$$
\begin{aligned}
M &= -\frac{e^2\,Z}{\mathbf{q}^2}\,\bar{u}(\mathbf{p}')\left[G\,\gamma_0 + \frac{F - G}{2m}\,P_0\right]u(\mathbf{p}) \\[2mm]
\frac{d\sigma}{d\Omega} &= \frac{\alpha^2\,Z^2}{(\mathbf{q}^2)^2}\,\frac{1}{2}\,\mathrm{Tr}\Bigg\{G^2\left(\slashed{p}'\,\gamma_0\,\slashed{p}\,\gamma_0 + m^2\right) + \frac{F - G}{2m}\,P_0\,G\,(2m\slashed{p}\gamma_0 + 2m\slashed{p}'\gamma_0) \\[2mm]
&\quad + \left(\frac{F - G}{2m}\right)^2 P_0^2\left(\slashed{p}'\,\slashed{p} + m^2\right)\Bigg\}
\end{aligned}
\tag{15}
$$

Carrying out traces, we obtain

$$
\begin{aligned}
\frac{d\sigma}{d\Omega} &= \frac{\alpha^2\,Z^2}{(\mathbf{q}^2)^2}\,2\Bigg[G^2\left(2p_0\,p_0' + m^2 - p\cdot p'\right) + (F - G)\,G\,(p_0 + p_0')^2 \\[2mm]
&\quad + \left(\frac{F - G}{2m}\right)^2 (p_0 + p_0')^2\left(p\cdot p' + m^2\right)\Bigg]
\end{aligned}
\tag{16}
$$

Noting that $p\cdot p' = m^2 + q^2/2$ and $p_0' = p_0$, we conclude that

$$\frac{d\sigma}{d\Omega} = \frac{4\alpha^2\,Z^2}{(\mathbf{q}^2)^2}\,p_0^2\left[|F|^2 + \frac{\mathbf{q}^2}{4m^2}\,(F - G)^2 - \frac{q^2}{4p_0^2}\,G^2\right] \tag{17}$$

Eq.(17) is the expression sought.

The pointlike limit is now a special case of (17), with $F = 1$ and $G = g/2 = \mu' + 1$. The cross section in this case equals

$$\frac{d\sigma}{d\Omega} = \frac{4\alpha^2\,Z^2}{(\mathbf{q}^2)^2}\,p_0^2\left[1 + \frac{\mu'^2\mathbf{q}^2}{4m^2} - \frac{q^2}{4p_0^2}\,(\mu' + 1)^2\right] \tag{18}$$

This coincides with Eq.(11) for $\mu' = 0$.

Problem 98.   On the electric dipole moment of the neutron and its measurement

a) Show that the electric dipole moment **d** of the neutron, if it exists, must be parallel to the neutron spin. Show that $\mathbf{d} \neq 0$ signals a violation of P and T.

b) Consider a neutron moving with a given velocity in a uniform static field, electric or magnetic. In each of the two cases, calculate the evolution of the polarization vector in the rest frame, in terms of the electric and magnetic dipole moments. Express the polarization in the laboratory frame.

c) Propose a principle on which to base an experiment for the measurement of the electric dipole moment.

The present experimental limit on **d** is $|d| < 1.2 \cdot 10^{-25}$ e cm, with 95% confidence level.

Solution

a) The electric dipole operator is a vector under rotations; consequently, by the Wigner-Eckart theorem, its matrix elements between states of spin $1/2$ are proportional to those of the angular momentum operator. In other words, the matrix elements of **d** are proportional to the matrix elements of **S**, which amounts to stating that **d** is parallel to the spin.

The spin operator **S** is even under parity and odd under time reversal; precisely the opposite is true of **d**, leading to the conclusion that a nonzero expectation value for **d** breaks both P and T.

b) Let us write a relativistic equation for the polarization vector W_μ in the presence of an electromagnetic field. In the nonrelativistic limit,

$$W_\mu \simeq (0, \boldsymbol{\zeta}) \qquad \boldsymbol{\zeta} = 2\langle \mathbf{S} \rangle \qquad (1)$$

the spin equation of motion in a static magnetic field is

$$\frac{d\boldsymbol{\zeta}}{dt} = 2\mu\, \boldsymbol{\zeta} \wedge \mathbf{H} \qquad (2)$$

H is the magnetic field and μ is the magnetic moment: $\boldsymbol{\mu} = \mu\, \boldsymbol{\zeta}$. Since the force must be linear in the electromagnetic tensor $F_{\mu\nu}$ and it can only depend on the four-vector p_μ, the most general covariant equation, compatible with (2), is

$$\frac{dW^\mu}{d\tau} = 2\mu\, F^{\mu\nu} W_\nu + \frac{k}{m^2}\, p^\mu\, F^{\nu\rho} p_\nu\, W_\rho \qquad (3)$$

There are no other possible terms, by virtue of the relation

$$p_\mu\, W^\mu = 0 \qquad (4)$$

and by the antisymmetry of $F^{\mu\nu}$. To determine k, we note that Eq.(4) implies

$$\frac{dp^\mu}{d\tau}\, W_\mu + p_\mu\, \frac{dW^\mu}{d\tau} = 0 \qquad (5)$$

The proper time derivatives above can be made explicit using Eq.(3), as well as the Lorentz law

$$\frac{dp^\mu}{d\tau} = \frac{e}{m}\, F^{\mu\nu} p_\nu \qquad (6)$$

with the result

$$\frac{e}{m}\, F^{\mu\nu} p_\nu\, W_\mu + 2\mu\, p_\mu\, F^{\mu\nu} W_\nu + k\, F^{\nu\rho} p_\nu\, W_\rho = 0 \qquad (7)$$

There follows

$$k = -2\left(\mu - \frac{e}{2m}\right) \qquad (8)$$

Setting

$$\mu' = -\frac{k}{2} = \mu - \frac{e}{2m} \tag{9}$$

we have

$$\frac{dW^\mu}{d\tau} = 2\mu\, F^{\mu\nu}\, W_\nu - \frac{2\mu'}{m^2}\, p^\mu\, F^{\nu\rho}\, p_\nu\, W_\rho \tag{10}$$

μ' is the anomalous part of the magnetic dipole moment.

In the presence of an electric dipole moment a further term is introduced to the above equation; this term can be obtained from (10) by changing the electric and magnetic components of the electromagnetic tensor ($F^{\mu\nu} \rightarrow F^*_{\mu\nu} = \frac{1}{2}\,\varepsilon_{\mu\nu\rho\sigma}\, F^{\rho\sigma}$) and substituting the electric charge e in (8) by the (vanishing) magnetic charge. Thus, the term to add to the r.h.s. of (10) reads

$$F^{(e)\,\mu} = 2d \left[F^{*\,\mu\nu}\, W_\nu - \frac{p^\mu}{m^2}\, F^{*\,\nu\rho}\, p_\nu\, W_\rho \right] \tag{11}$$

One may also regard (11) as an outcome of the condition $p_\mu F^{(e)\,\mu} = 0$, which in turn follows from the absence of a term proportional to $F^*_{\mu\nu}$ in the Lorentz law.

Let us denote by $\mathbf{E}$ and $\mathbf{H}$ the electric and magnetic fields in the instantaneous rest frame of the neutron; these can be determined from the laboratory fields $\mathbf{E}'$ and $\mathbf{H}'$ by means of

$$\begin{aligned}
E'_\parallel &= E_\parallel & H'_\parallel &= H_\parallel \\
\mathbf{E}'_\perp &= (\mathbf{v} \wedge \mathbf{H}_\perp + \mathbf{E}_\perp)\,\gamma & \mathbf{H}'_\perp &= \gamma\,(\mathbf{H}_\perp - \mathbf{v} \wedge \mathbf{E}_\perp)
\end{aligned} \tag{12}$$

In this frame the polarization vector equals $W_\mu = (0, \boldsymbol{\zeta})$. A nonzero $\mathbf{H}$ leads to the precession equation

$$\frac{d\boldsymbol{\zeta}}{dt} = 2\mu\, \boldsymbol{\zeta} \wedge \mathbf{H} \tag{13}$$

Similarly a nonzero $\mathbf{E}$ leads to

$$\frac{d\boldsymbol{\zeta}}{dt} = 2d\, \boldsymbol{\zeta} \wedge \mathbf{E} \tag{14}$$

Consider now a pure magnetic field in the laboratory frame; the corresponding precession equation is, after some algebra and use of the equation of motion,

$$\frac{d\boldsymbol{\zeta}}{dt} = \frac{2m\mu + 2\mu'\,(\varepsilon - m)}{\varepsilon}\, \boldsymbol{\zeta} \wedge \mathbf{H}' + \frac{2\mu'\,\varepsilon}{\varepsilon + m}\,(\mathbf{v}\,\mathbf{H}')\,\mathbf{v} \wedge \boldsymbol{\zeta} + 2d\,\boldsymbol{\zeta} \wedge (\mathbf{v} \wedge \mathbf{H}') \tag{15}$$

In the case of the neutron we have $\mu = \mu'$, and (15) takes the form

$$\frac{d\boldsymbol{\zeta}}{dt} = 2\mu\, \boldsymbol{\zeta} \wedge \mathbf{H}' + \frac{2\mu\,\varepsilon}{\varepsilon + m}\,(\mathbf{v}\,\mathbf{H}')\,\mathbf{v} \wedge \boldsymbol{\zeta} + 2d\,\boldsymbol{\zeta} \wedge (\mathbf{v} \wedge \mathbf{H}') \tag{16}$$

If instead a pure electric field is present, the neutron precession equation becomes

$$\frac{d\boldsymbol{\zeta}}{dt} = 2d\,\boldsymbol{\zeta} \wedge \mathbf{E}' + \frac{2d\,\varepsilon}{\varepsilon + m}\,(\mathbf{v}\,\mathbf{E}')\,\mathbf{v} \wedge \boldsymbol{\zeta} - 2\mu\,\boldsymbol{\zeta} \wedge (\mathbf{v} \wedge \mathbf{E}') \qquad (17)$$

c) An experiment for determining the electric dipole moment of the neutron may be based on observation of the precession of the neutron spin, induced by the presence of an electric or magnetic field.

Problem 99.   Helicity states for spin 1/2 particles

a) Choosing a representation of the γ matrices, write down the wave function of a relativistic electron with momentum $\mathbf{p}$ and helicity $+1$ or -1.
b) For definiteness, let us take the electron momentum along the z-axis. Express the states with complete polarization in the x- and y-direction in terms of helicity states.
c) We now apply to the electron a rotation and a boost along the momentum axis. How do the states of part a) and b) transform?
d) How do the above states transform under parity?

Solution

a) Using the standard (Pauli) representation of the γ matrices, the wave functions in question are (with $\mathbf{p}$ pointing along the z-axis)

$$u_{\pm} = \begin{pmatrix} \sqrt{p^0 + m}\;\mathrm{w}_{\pm} \\ \dfrac{\boldsymbol{\sigma} \cdot \mathbf{p}}{\sqrt{p^0 + m}}\mathrm{w}_{\pm} \end{pmatrix} = \begin{pmatrix} \sqrt{p^0 + m}\;\mathrm{w}_{\pm} \\ \dfrac{\pm p}{\sqrt{p^0 + m}}\;\mathrm{w}_{\pm} \end{pmatrix} \qquad (1)$$

$$\mathrm{w}_{+} = \begin{pmatrix} 1 \\ 0 \end{pmatrix} \qquad \mathrm{w}_{-} = \begin{pmatrix} 0 \\ 1 \end{pmatrix}$$

b) The state with polarization along $+x$ is

$$\frac{1}{\sqrt{2}}\,(u_{+} + u_{-}) \qquad (2)$$

while the one with polarization along $+y$ is

$$\frac{1}{\sqrt{2}}\,(u_{+} + i\,u_{-}) \qquad (3)$$

c) Consider a rotation characterized by a vector $\boldsymbol{\alpha}$

$$R = \begin{pmatrix} e^{-i\boldsymbol{\alpha}\cdot\boldsymbol{\sigma}/2} & 0 \\ 0 & e^{-i\boldsymbol{\alpha}\cdot\boldsymbol{\sigma}/2} \end{pmatrix} \qquad (4)$$

The electron wave function transforms as

$$u_\pm \to u'_\pm = R u_\pm = \begin{pmatrix} \sqrt{p^0 + m} \; w'_\pm \\ \dfrac{\pm p}{\sqrt{p^0 + m}} \; w'_\pm \end{pmatrix} \tag{5}$$

with

$$w'_\pm = \cos \frac{|\boldsymbol{\alpha}|}{2} \, w_\pm - i \frac{\boldsymbol{\alpha} \cdot \boldsymbol{\sigma}}{|\boldsymbol{\alpha}|} \sin \frac{|\boldsymbol{\alpha}|}{2} \, w_\pm \tag{6}$$

A boost along the z-axis with rapidity β,

$$\Lambda = \begin{pmatrix} \cosh \frac{\beta}{2} & -\sigma_3 \sinh \frac{\beta}{2} \\ -\sigma_3 \sinh \frac{\beta}{2} & \cosh \frac{\beta}{2} \end{pmatrix} \tag{7}$$

transforms $u_\pm$ into

$$u_\pm \to u'_\pm = \Lambda u_\pm = \begin{pmatrix} \left[\cosh \frac{\beta}{2} \sqrt{p^0 + m} - \dfrac{p \sinh \frac{\beta}{2}}{\sqrt{p^0 + m}} \right] w_\pm \\ \mp \left[\sinh \frac{\beta}{2} \sqrt{p^0 + m} - \dfrac{p \cosh \frac{\beta}{2}}{\sqrt{p^0 + m}} \right] w_\pm \end{pmatrix} \tag{8}$$

d)   Under parity, we have

$$u_\pm \to \gamma_0 \, u_\pm \tag{9}$$

This is equivalent to

$$u_\pm (\mathbf{p}) \to u_\pm (-\mathbf{p}) \tag{10}$$

Problem 100.   Parity violating effects of the electroweak interaction in the hydrogen atom

a)   Write down the matrix element

$$J_\mu(x) = \langle \mathbf{p}' | \, j_\mu(x) \, | \mathbf{p} \rangle \tag{1}$$

of the current $j_\mu = \bar{\psi}(x) \, \gamma_\mu \, (g_V + g_A \gamma_5) \, \psi(x)$, between fermion states with momenta $\mathbf{p}$ and $\mathbf{p}'$. Derive the nonrelativistic limit of $J_\mu(x)$.

b)   Let $J_\mu(x)$ be coupled to a vector particle of mass M,

$$\left(\Box + M^2 \right) Z_\mu(x) = -J_\mu(x) \tag{2}$$

Calculate the potential $Z_\mu(x)$ generated by the fermion in the nonrelativistic limit.

c)   Suppose that the fermion described above is a proton and that Z_μ couples in a similar way to the electron; thus, an electron in the field Z_μ generated by the proton undergoes an interaction of the form $\int d^4x\, Z_\mu(x)\, j^e_\mu(x)$. Calculate the potential between the two fermions, generated by the interaction with Z, in the nonrelativistic limit.

d)   Show that the above potential violates parity. What is the mean life induced by it on the $2S$ state of the hydrogen atom? $(\Delta E(2S_{\frac{1}{2}}, 2P_{\frac{1}{2}}) = 1027\,\text{MHz}, M \simeq 91.173 \pm 0.020\,\text{GeV})$. For simplicity, you may neglect all dependence on the nuclear spin.

Solution

a)   The matrix element $J_\mu(x)$ is given by

$$J_\mu(x) = e^{i\,(p'-p)x}\,\bar{u}(p')\,\gamma_\mu\,(g_V + g_A\,\gamma_5)\,u(p) \tag{3}$$

The spinor wave function in the nonrelativistic limit becomes

$$u(p) \simeq \sqrt{2m}\begin{pmatrix} w \\ \dfrac{\sigma \cdot v}{2}\,w \end{pmatrix} + \mathcal{O}\left(\dfrac{v^2}{c^2}\right) \tag{4}$$

where $v = p/m$, and w is the Pauli spinor.

In the same representation, the γ matrices have the form

$$\gamma^0 = \begin{pmatrix} I & 0 \\ 0 & -I \end{pmatrix} \qquad \gamma = \begin{pmatrix} 0 & \sigma \\ -\sigma & 0 \end{pmatrix} \qquad \gamma_5 = \begin{pmatrix} 0 & I \\ I & 0 \end{pmatrix} \tag{5}$$

Dropping terms of order $\mathcal{O}(v^2/c^2)$ and setting $q_\mu = p'_\mu - p_\mu$ we find

$$2m\,V^\mu \equiv \bar{u}(p')\gamma^\mu u(p) = \begin{cases} 2m\,u'^\dagger u & \mu = 0 \\ 2m\,u'^\dagger\left\{\dfrac{v+v'}{2} + i\dfrac{\sigma \wedge q}{2m}\right\}u & \mu = 1,2,3 \end{cases}$$

$$2m\,A^\mu \equiv \bar{u}(p')\gamma^\mu\gamma_5 u(p) = \begin{cases} 2m\,u'^\dagger\sigma\dfrac{v+v'}{2}u & \mu = 0 \\ 2m\,u'^\dagger\sigma\, u & \mu = 1,2,3 \end{cases} \tag{6}$$

b)   The field Z_μ generated by the proton is given by

$$\left(M^2 - q^2\right) Z_\mu(q) = -J^{(p)}_\mu(q)$$

$$Z_\mu(x) = \int e^{iqx}\,\frac{1}{q^2 - M^2}\,2m\left[g^p_V\,V^{(p)}_\mu(q) + g^p_A\,A^{(p)}_\mu(q)\right]\frac{d^4q}{(2\pi)^4} \tag{7}$$

c)   In the nonrelativistic limit, the $e - p$ scattering amplitude reads

$$\mathcal{M} = \frac{2m_e\,2m_p}{q^2 - M^2}\left[g^e_V\,V^{(e)}_\mu + g^e_A\,A^{(e)}_\mu\right]\left[g^p_V\,V^\mu_{(p)} + g^p_A\,A^\mu_{(p)}\right] \tag{8}$$

We note that $q_0 = p'_0 - p_0 \simeq \mathcal{O}(v^2/c^2)$; switching to nonrelativistic normalization (dividing by $2m_e$, $2m_p$), we obtain

$$\mathcal{M} = -\frac{1}{\mathbf{q}^2 + M^2} \left[g_V^e V_\mu^{(e)} + g_A^e A_\mu^{(e)} \right] \left[g_V^p V_{(p)}^\mu + g_A^p A_{(p)}^\mu \right] \tag{9}$$

Neglecting the proton velocity we can write the above product of four-vectors as a matrix element of the following operator

$$\left[g_V^e V_\mu^{(e)} + g_A^e A_\mu^{(e)} \right] \left[g_V^p V_{(p)}^\mu + g_A^p A_{(p)}^\mu \right] \simeq \Big\langle g_V^e g_V^p + g_A^e g_A^p \, \boldsymbol{\sigma}_e \frac{\mathbf{v}_e + \mathbf{v}'_e}{2}$$
$$- g_A^p g_V^e \, \boldsymbol{\sigma}_p \left(\frac{\mathbf{v}_e + \mathbf{v}'_e}{2} + i \frac{\boldsymbol{\sigma}_e \wedge \mathbf{q}}{2m_e} \right) - g_A^p g_A^e \, \boldsymbol{\sigma}_p \cdot \boldsymbol{\sigma}_e \Big\rangle \tag{10}$$

To obtain the potential giving rise to this amplitude, we express the amplitude in terms of $\mathbf{v}_e$ and $\mathbf{q}$, and take the Fourier transform with respect to $\mathbf{q}$. The resulting potential is velocity-dependent.

Setting $\dfrac{\mathbf{v}_e + \mathbf{v}'_e}{2} = \mathbf{v}_e + \dfrac{\mathbf{q}}{2m_e}$, we write

$$V(\mathbf{x}, \mathbf{v}_e) = \int \frac{d^3 q}{(2\pi)^3} e^{i \mathbf{q} \mathbf{x}} \mathcal{M}(\mathbf{q}, \mathbf{v}_e) \tag{11}$$

Use of the formulae

$$\int \frac{d^3 q}{(2\pi)^3} \frac{e^{i \mathbf{q} \mathbf{x}}}{\mathbf{q}^2 + M^2} = \frac{e^{-Mr}}{4\pi r} \underset{M \to \infty}{\simeq} \frac{1}{M^2} \delta^3(\mathbf{x}) \tag{12}$$

$$\mathbf{v}_e = \frac{\mathbf{P}_e}{m_e} = \frac{1}{i\, m_e} \boldsymbol{\nabla} \tag{13}$$

in the Fourier transform of Eq.(11) leads to

$$V(\mathbf{x}, \mathbf{v}_e) = -\frac{1}{M^2} \Big\{ g_V^e g_V^p \delta^3(x) + g_A^e g_A^p \left(\delta^3(x) \boldsymbol{\sigma}_e \mathbf{v}_e + \frac{\boldsymbol{\sigma}_e}{2i\, m_e} \cdot \boldsymbol{\nabla} \delta^3(x) \right)$$
$$- g_A^e g_A^p \, \boldsymbol{\sigma}_p \boldsymbol{\sigma}_e \delta^3(x) - g_V^e g_A^p \, \boldsymbol{\sigma}_p \left(\delta^3(x) \mathbf{v}_e + \frac{1}{2i\, m_e} \boldsymbol{\nabla} \delta^3(x) + \frac{\boldsymbol{\sigma}_e}{2i\, m_e} \wedge \boldsymbol{\nabla} \delta^3(x) \right) \Big\} \tag{14}$$

Using (13) and neglecting nuclear spin, we finally obtain

$$V(x) = -\frac{1}{M^2} \Big\{ g_V^e g_V^p \, \delta^3(x) + \frac{g_A^e g_V^p \, \boldsymbol{\sigma}_e}{2i\, m_e} \left[\delta^3(\mathbf{x}) \boldsymbol{\nabla} + \boldsymbol{\nabla} \, \delta^3(x) \right] \Big\} \tag{15}$$

d) The potential $V(x)$ is not invariant under parity. It mixes the $2S$ and $2P$ states of the hydrogen atom, with mixing amplitude

$$\varepsilon = \frac{\langle 2P_{\frac{1}{2}} | V | 2S_{\frac{1}{2}} \rangle}{E_{2P} - E_{2S}} \tag{16}$$

If Γ is the mean life of the $2P$ state, then the mean life of $2S$ induced by the parity-violating mixing is

$$\gamma = \varepsilon^2 \, \Gamma \tag{17}$$

To determine ε we must calculate the matrix element

$$\langle 2P_{\frac{1}{2}} | V | 2S_{\frac{1}{2}} \rangle \tag{18}$$

The spatial wave functions are (r_B is the Bohr radius)

$$\psi_{2S} = \frac{1}{\sqrt{8\pi}} \frac{e^{r/2r_B}}{(r_B)^{3/2}} \left(1 - \frac{r}{r_B} \right)$$

$$\psi_{2P,m} = Y_1^m(\hat{\mathbf{r}}) \frac{1}{2\sqrt{6}(r_B)^{3/2}} \frac{r}{r_B} e^{-r/2r_B} \tag{19}$$

Since ε does not depend on the orientation of the angular momentum, we choose to calculate the matrix element between states $|2S_{1/2} \, J_Z = 1/2\rangle$ and $|2P_{1/2} \, J_Z = 1/2\rangle$. The corresponding wave functions are

$$\psi_{2S \, J_Z = 1/2} = \psi_{2S} \, u_\uparrow$$

$$\psi_{2P \, J_Z = 1/2} = -\frac{1}{\sqrt{3}} \psi_{2P,0} \, u_\uparrow + \sqrt{\frac{2}{3}} \, \psi_{2P,1} u_\downarrow \tag{20}$$

Here, $u_\uparrow$ and $u_\downarrow$ are spin states with $S_Z = 1/2$ and $S_Z = -1/2$.

Given that $\psi_{2P,\ell_z}(0) = 0$, the nonvanishing spatial matrix elements in Eq.(18) are

$$\langle \ell_Z | \mathbf{D} | 0 \rangle = \langle 2P \, \ell_Z | \left\{ \delta^3(\mathbf{x}) \, \boldsymbol{\nabla} + \boldsymbol{\nabla} \, \delta^3(x) \right\} | 2S \rangle$$

$$= -\psi_{2S}(0) \left[\boldsymbol{\nabla} \, \psi_{2P,\ell_z}^* \right]_{x=0} \tag{21}$$

$$\langle \ell_Z | \mathbf{D}' | 0 \rangle = \langle 2P \, \ell_Z | \left[\boldsymbol{\nabla} \, \delta^3(x) \right] | 2S \rangle$$

$$= -\int d^3x \, \delta^3(x) \, \boldsymbol{\nabla} \left(\psi_{2S}(x) \, \psi_{2P,\ell_z}^* \right) = \langle \ell_Z | \mathbf{D} | 0 \rangle \tag{22}$$

Substituting the explicit form of the wave functions, and keeping only terms which do not vanish as $r \to 0$, we find

$$\psi_{2S}(0) \, \boldsymbol{\nabla} \, \psi_{2P,\ell_z}^* \, |_{r=0} = \frac{1}{\sqrt{8\pi}} \frac{1}{2\sqrt{6}} \frac{1}{r_B^4} \frac{1}{\sqrt{4\pi}} \boldsymbol{\nabla} \begin{cases} -\dfrac{(x - i\,y)}{\sqrt{2}} \\[2mm] z \\[2mm] \dfrac{(x + i\,y)}{\sqrt{2}} \end{cases} \tag{23}$$

In spherical components we have

$$\langle 1|D^+|0\rangle = \frac{1}{\sqrt{2}}(D_x + i D_y) = \sqrt{\frac{2}{3}}\frac{1}{16\pi\, r_B^4}$$

$$\langle -1|D^-|0\rangle = \frac{1}{\sqrt{2}}(D_x - i D_y) = -\sqrt{\frac{2}{3}}\frac{1}{16\pi\, r_B^4}$$

$$\langle 0|D_3|0\rangle = D_Z = -\sqrt{\frac{1}{3}}\frac{1}{16\pi\, r_B^4} \tag{24}$$

We may now insert the above into Eq.(18), with the result

$$\langle 2P_{\frac{1}{2}},\tfrac{1}{2}|V|2S_{\frac{1}{2}},\tfrac{1}{2}\rangle = -\frac{g_A^e g_V^p}{2i\,m_e M^2}\left[\sigma_3 D_3 + \sigma_- D_+ + \sigma_+ D_-\right]$$

$$= \frac{i g_A^e g_V^p}{m_e M^2}\frac{1}{96\pi r_B^4}(1 + 2\sqrt{2}) \tag{25}$$

where $\sigma_\pm = (\sigma_x \pm i\sigma_y)/\sqrt{2}$. The mixing coefficient now follows immediately

$$\varepsilon = \frac{1}{\Delta E}\frac{i g_V^e g_V^p}{m_e M^2}\frac{1}{96\pi r_B^4}(1 + 2\sqrt{2}) \tag{26}$$

The couplings g_V and g_A of Z to the electron and quarks are, in units of $g/\cos\theta_W$

	V	A
e	$-\frac{1}{4} + \sin^2\theta_W$	$-\frac{1}{4}$
u	$\frac{1}{4} - \frac{2}{3}\sin^2\theta_W$	$\frac{1}{4}$
d	$-\frac{1}{4} + \frac{1}{3}\sin^2\theta_W$	$-\frac{1}{4}$

$$\tag{27}$$

where

$$\sin^2\theta_W = 0.2325 \pm 0.0008 \qquad g = \frac{e}{\sin\theta_W} \tag{28}$$

The coupling to the proton is obtained by summing over quark couplings

$$g_V^p = 2g_V^u + g_V^d = \left(\frac{1}{4} - \sin^2\theta_W\right)\frac{e}{\sin\theta_W\cos\theta_W}$$

$$g_A^p = 2g_A^u + g_A^d = \frac{1}{4}\frac{e}{\sin\theta_W\cos\theta_W} \tag{29}$$

The numerical value of ε thus turns out to be

$$\varepsilon \simeq 2.4\cdot 10^{-13} \tag{30}$$

leading to

$$\Gamma \simeq 6.2\cdot 10^{-26}\,\Gamma_{2P} \tag{31}$$

We remark that a mixing of the same order of magnitude as in Eq.(26) can be induced by an electric field $E \simeq 10^{-9}$ Volt/cm.

Problem 101. τ **polarization in the decay of** Z_0

The Z_0 meson can undergo the following decay

$$Z_0 \to \tau^+ + \tau^- \tag{1}$$

The mass of the τ lepton is $1784.1 \pm 3\,\text{MeV}$, that of Z_0 is $(91.173 \pm 0.020)\,\text{MeV}$.

This decay is described by the standard model Lagrangian, whose relevant term is

$$\mathcal{L}_I = -\frac{e}{\cos\theta_W \sin\theta_W}\, Z_\mu^0\, \bar{\psi}_\tau \gamma^\mu (a + b\gamma^5)\psi_\tau \tag{2}$$

with

$$a = -\frac{1}{4} + \sin^2\theta_W \qquad b = \frac{1}{4} \qquad \sin^2\theta_W = 0.2325 \pm .0003 \tag{3}$$

a)   Starting from nonpolarized Z_0's, calculate the polarization of τ^+ and τ^- in the final state; comparing this result to experimental data, show how to determine the Weinberg angle (θ_W).

b)   How can the τ polarization be measured?

Solution

a)   The amplitude for process (1) is

$$\mathcal{M} = \frac{e}{\cos\theta_W \sin\theta_W}\, \varepsilon^\mu \bar{u}(\mathbf{p}_{\tau^-})\gamma_\mu (a + b\gamma^5)v(\mathbf{p}_{\tau^+}) \tag{4}$$

If Z_0 is nonpolarized, the τ's can only be polarized along their direction of motion and must have opposite helicities.

The probability of decay into a state with τ^- helicity equal to h is proportional to

$$\varepsilon_\mu \varepsilon_\nu^* \,\text{Tr}\left[(\slashed{p}_{\tau^-} + m_\tau)\frac{1 + \gamma^5 \slashed{w} h}{2}\gamma^\mu(a + b\gamma^5)\,(\slashed{p}_{\tau^+} - m_\tau)\,\gamma^\nu(a^* + b^*\gamma^5)\right] \tag{5}$$

where

$$w = \left(\frac{|\mathbf{p}_\tau|}{m_\tau}, \frac{p_\tau^0}{m_\tau}\mathbf{n}\right) \tag{6}$$

and $\mathbf{n} = \mathbf{p}/|\mathbf{p}|$.

Dropping common factors between numerator and denominator, we write for the average helicity $\langle h \rangle$

$$\langle h \rangle = \frac{\varepsilon_\mu \varepsilon_\nu^* \,\text{Tr}\left[(\slashed{p}_{\tau^-} + m_\tau)\,\gamma^5 \slashed{w}\gamma^\mu(a + b\gamma^5)\,(\slashed{p}_{\tau^+} - m_\tau)\,\gamma^\nu(a^* + b^*\gamma^5)\right]}{\varepsilon_\mu \varepsilon_\nu^* \,\text{Tr}\left[(\slashed{p}_{\tau^-} + m_\tau)\,\gamma^\mu(a + b\gamma^5)\,(\slashed{p}_{\tau^+} - m_\tau)\,\gamma^\nu(a^* + b^*\gamma^5)\right]} \tag{7}$$

The average over Z_0 polarization brings in a factor

$$\sum_{pol} \varepsilon^\mu \varepsilon^{\nu*} = -\left(g^{\mu\nu} - \frac{p_Z^\mu p_Z^\nu}{M_Z^2}\right) \tag{8}$$

Carrying out the traces we obtain

$$\langle h \rangle = \frac{4m_\tau \operatorname{Re}(ab^*)(wp_{\tau+})}{M_Z^2(|a|^2 + |b|^2) + 2m_\tau^2(|a|^2 - 2|b|^2)} \tag{9}$$

We substitute for the product $(wp_{\tau+})$

$$wp_{\tau+} = \frac{M_Z^2}{2m_\tau}\sqrt{1 - \frac{4m_\tau^2}{M_Z^2}} \tag{10}$$

and leave out terms of order $\mathcal{O}(m_\tau^2/M_Z^2)$, finding

$$\langle h \rangle = \frac{2\operatorname{Re}(ab^*)}{|a|^2 + |b|^2} = -\frac{1 - 4\sin^2\theta_W}{1 - 4\sin^2\theta_W + 8\sin^4\theta_W} \tag{11}$$

From the known value of $\sin^2\theta_W$ we obtain a polarization of $\simeq 14\%$. This polarization, in turn, is a direct way of measuring the deviation of $\sin^2\theta_W$ from $1/4$.

b) τ^- can decay into $\nu_\tau + \mu^- + \bar\nu_\mu$ or $\nu_\tau + e^- + \bar\nu_e$. When a polarized τ decays, the charged lepton produced has an angular distribution of the form

$$\frac{\mathrm{d}P}{\cos\theta} \propto 1 - \frac{1}{3}\langle h \rangle \cos\theta \tag{12}$$

The polarization $\langle h \rangle$ can thus be determined from the angular distribution of the decay lepton (μ, e) in the center of mass of the τ.

Problem 102. Neutral currents and neutrino lepton scattering

The standard model couples leptons to the Z_0 meson through the interaction

$$\mathcal{L}^N = -\bar{g}Z_\mu^0 \sum_\ell \left\{ \frac{1}{2}\overline{\psi}_{\nu_\ell}\gamma^\mu(1-\gamma^5)\psi_{\nu_\ell} + \overline{\psi}_\ell\left[(-\frac{1}{2} + 2\sin^2\theta_W)\gamma^\mu + \frac{1}{2}\gamma^\mu\gamma^5\right]\psi_\ell \right\} \tag{1}$$

The index ℓ runs over leptons (e, μ, τ); $\bar{g}^2 = \sqrt{2}G_F M_Z^2$ and θ_W is the Weinberg angle.

The coupling to the charged current, giving rise to the usual Fermi interaction, is

$$\mathcal{L}^C = -\frac{g}{2\sqrt{2}} \sum_\ell \left\{ W_\mu\overline{\psi}_{\nu_\ell}\gamma^\mu(1-\gamma^5)\psi_\ell + \text{h.c.} \right\} \tag{2}$$

with $g^2 = 4\sqrt{2}\,G_F M_W^2$.

a) Calculate the differential and total cross section for the process $\nu_\mu + e^- \to \nu_\mu + e^-$, to order g^2.

b)   Show how one can determine the Weinberg angle from the ratio

$$R = \frac{\sigma(\nu_\mu + e^- \to \nu_\mu + e^-)}{\sigma(\bar{\nu}_\mu + e^- \to \bar{\nu}_\mu + e^-)} \tag{3}$$

Given that $\sin^2 \theta_W \simeq 1/4$, what is the experimental accuracy required of R, in order to determine $\sin^2 \theta_W$ to within 10^{-3}?

c)   The standard prediction of the Fermi theory for the cross sections

$$\bar{\nu}_e + e^- \to \bar{\nu}_e + e^-$$
$$\nu_e + e^- \to \nu_e + e^- \tag{4}$$

is modified by the presence of the neutral current. Calculate the magnitude of this effect. You may find useful the following Fierz identity

$$\overline{\psi}_{\nu_\ell} \gamma^\mu (1 - \gamma^5) \psi_\ell \, \overline{\psi}_\ell \gamma_\mu (1 - \gamma^5) \psi_{\nu_\ell} = \overline{\psi}_{\nu_\ell} \gamma^\mu (1 - \gamma^5) \psi_{\nu_\ell} \, \overline{\psi}_\ell \gamma_\mu (1 - \gamma^5) \psi_\ell \tag{5}$$

Solution

The amplitude for the process $\nu_\mu + e^- \to \nu_\mu + e^-$ has the form

$$\mathcal{M}(\nu) = -\frac{\bar{g}^2}{2} \, \bar{u}(\mathbf{p}'_{\nu_\mu}) \gamma^\mu (1 - \gamma^5) u(\mathbf{p}_{\nu_\mu}) \frac{1}{q^2 - M_Z^2} \, \bar{u}(\mathbf{p}'_e) \gamma_\mu (a + b\gamma^5) u(\mathbf{p}_e) \tag{6}$$

For brevity, we have set

$$a = -\frac{1}{2} + 2\sin^2 \theta_W \qquad b = \frac{1}{2} \tag{7}$$

and $q = p'_{\nu_\mu} - p_{\nu_\mu} = p_e - p'_e$.

Similarly, for the process $\bar{\nu}_\mu + e^- \to \bar{\nu}_\mu + e^-$ we have

$$\mathcal{M}(\bar{\nu}) = \frac{\bar{g}^2}{2} \, \bar{v}(\mathbf{p}_{\bar{\nu}_\mu}) \gamma^\mu (1 - \gamma^5) v(\mathbf{p}'_{\bar{\nu}_\mu}) \frac{1}{q^2 - M_Z^2} \, \bar{u}(\mathbf{p}'_e) \gamma_\mu (a + b\gamma^5) u(\mathbf{p}_e) \tag{8}$$

The kinematic invariant s is given by $s = m_e^2 + 2m_e E_\nu \simeq 2m_e E_\nu$, where E_ν is the neutrino energy in the laboratory frame. The maximum value of the momentum transfer is

$$q_{max}^2 \simeq -s \sim -2m_e E_\nu \tag{9}$$

Thus, for typical neutrino energies, the presence of q^2 in the Z_0 propagator is negligible compared to M_Z^2. The modulus squared of these amplitudes, summed over final spins and averaged over initial spins, equals

$$\overline{|\mathcal{M}(\nu)|^2} = \frac{G_F^2}{4} \operatorname{Tr} \left\{ \rlap{/}{p}_{\nu_\mu}' \gamma^\mu (1 - \gamma^5) \rlap{/}{p}_{\nu_\mu} \gamma^\nu (1 - \gamma^5) \right\} \tag{10a}$$

$$\operatorname{Tr} \left\{ (\rlap{/}{p}_e' + m_e) \gamma_\mu (a + b\gamma^5)(\rlap{/}{p}_e + m_e) \gamma_\nu (a + b\gamma^5) \right\}$$

$$\overline{|\mathcal{M}(\bar{\nu})|^2} = \frac{G_F^2}{4} \operatorname{Tr} \left\{ \rlap{/}{p}_{\bar{\nu}_\mu}' \gamma^\mu (1 - \gamma^5) \rlap{/}{p}_{\bar{\nu}_\mu} \gamma^\nu (1 - \gamma^5) \right\} \tag{10b}$$

$$\operatorname{Tr} \left\{ (\rlap{/}{p}_e' + m_e) \gamma_\mu (a + b\gamma^5)(\rlap{/}{p}_e + m_e) \gamma_\nu (a + b\gamma^5) \right\}$$

The trace containing electron momenta (leptonic tensor) evaluates to

$$E^{\mu\nu} = 4 \left\{ m^2(a^2 - b^2) g^{\mu\nu} + \left(a^2 + b^2 \right)(p_e'^\mu p_e^\nu + p_e'^\nu p_e^\mu - (p_e' \cdot p_e) g^{\mu\nu}) \right.$$
$$\left. + 2\mathrm{i}\, \varepsilon^{\mu\nu\alpha\beta}\, p_{e\alpha}' p_{e\beta} \right\} \tag{11}$$

while the neutrino tensor equals

$$8 \left\{ p_{\nu_\mu}'^\mu p_{\nu_\mu}^\nu + p_{\nu_\mu}'^\nu p_{\nu_\mu}^\mu - (p_{\nu_\mu}' \cdot p_{\nu_\mu}) g^{\mu\nu} \mp \mathrm{i}\varepsilon^{\mu\nu\alpha\beta}\, p_{\nu_\mu}'^\alpha p_{\nu_\mu}^\beta \right\} \tag{12}$$

where $\mp$ refers to ν_μ and $\bar{\nu}_\mu$ respectively. Taking the product of the two tensors we finally obtain

$$\overline{|\mathcal{M}|^2} = 8G_F^2 \left\{ (a^2 + b^2) \left[\frac{(s - m_e^2)^2}{2} + \frac{(u - m_e^2)^2}{2} \right] \right.$$
$$\left. + m_e^2\, t\, (a^2 - b^2) \mp 2ab \left[\frac{(s - m_e^2)^2}{2} - \frac{(u - m_e^2)^2}{2} \right] \right\} \tag{13}$$

Neglecting the electron mass we have

$$u = -\frac{s}{2}(1 + \cos\theta) \tag{14}$$

where θ is the scattering angle in the center of mass. The differential cross section thus becomes

$$\frac{d\sigma}{d\Omega} = \frac{\overline{|\mathcal{M}|^2}}{2s} \frac{1}{32\pi^2} = \frac{s\, G_F^2}{16\pi^2} \left[(a \mp b)^2 + \frac{(1 + \cos\theta)^2}{4} (a \pm b)^2 \right] \tag{15}$$

For the total cross section σ_{Tot} we find

$$\sigma_{Tot} = \frac{s\, G_F^2}{4\pi} \left[(a \mp b)^2 + \frac{1}{3}(a \pm b)^2 \right] \tag{16}$$

The ratio $R = \sigma(\nu_\mu e)/\sigma(\bar{\nu}_\mu e)$ now follows immediately

$$R = \frac{(a - b)^2 + \frac{1}{3}(a + b)^2}{(a + b)^2 + \frac{1}{3}(a - b)^2} \tag{17}$$

In terms of the values (7) for a and b, we obtain

$$R = \frac{\sigma(\nu_\mu e)}{\sigma(\bar{\nu}_\mu e)} = \frac{3 - 12 \sin^2 \theta_W + 16 \sin^4 \theta_W}{1 - 4 \sin^2 \theta_W + 16 \sin^4 \theta_W} \tag{18}$$

Setting $\sin^2 \theta_W = \frac{1}{4} - \varepsilon$, $\varepsilon \sim 0.02$, leads to

$$R = \frac{1 + 4\varepsilon + 16\varepsilon^2}{1 - 4\varepsilon + 16\varepsilon^2} \simeq 1 + 8\varepsilon \tag{19}$$

A precision of 10^{-3} in $\sin^2 \theta$ requires $\mathcal{O}(10^{-2})$ precision in ε and hence $\mathcal{O}(10^{-2})$ in R. This imposes a statistics of $\sim 10^4$ events.

The scattering matrix generated by interaction (2) to second order is given by

$$i \int d^4 x \int d^4 y \, \frac{g^2}{8} \, J^\mu(x) \mathcal{D}(x - y) J_\mu^\dagger(y) \tag{20}$$

where $\mathcal{D}(x - y)$ is the Fourier transform of $1/(q^2 - M_W^2)$. For small momentum transfer, we have

$$D(x - y) \simeq -\frac{1}{M_W^2} \delta^{(4)}(x - y) \tag{21}$$

and the S-matrix element may be considered as the first order matrix element of the effective Lagrangian operator $i\mathcal{L}_F$

$$\mathcal{L}_F = -\frac{G_F}{\sqrt{2}} \sum_\ell \overline{\psi}_{\nu_\ell} \gamma^\mu (1 - \gamma^5) \psi_\ell \, \overline{\psi}_\ell \gamma_\mu (1 - \gamma^5) \psi_{\nu_\ell} \tag{22}$$

This is the usual Fermi interaction. Use was made of the relation $g^2/8M_W^2 = G_F/\sqrt{2}$.

Similarly, the neutral current interaction (1) in the $\nu_\ell - \ell$ sector gives rise to

$$\mathcal{L}^N = -\frac{G_F}{\sqrt{2}} \sum_\ell \overline{\psi}_{\nu_\ell} \gamma^\mu (1 - \gamma^5) \psi_{\nu_\ell} \, \overline{\psi}_\ell (a\gamma_\mu + b\gamma_\mu \gamma^5) \psi_\ell \tag{23}$$

with a and b as defined in (7).

Applying the Fierz identity (5) to (22) we arrive at

$$\mathcal{L}_F = -\frac{G_F}{\sqrt{2}} \sum_\ell \overline{\psi}_{\nu_\ell} \gamma^\mu (1 - \gamma^5) \psi_{\nu_\ell} \, \overline{\psi}_\ell \gamma_\mu (1 - \gamma^5) \psi_\ell \tag{24}$$

This expression is still in the form of (23), with $a = 1$ and $b = -1$. Consequently, the cross sections calculated with the standard Fermi interaction are given by Eq.(15) with $a = 1$, $b = -1$. The effect of the neutral current is to modify a and b by the contributions of (7),

$$a \;\rightarrow\; 1 + (-\frac{1}{2} + 2\sin^2 \theta) = \frac{1}{2} + 2\sin^2 \theta \tag{25a}$$

$$b \;\rightarrow\; -1 + (\frac{1}{2}) = -\frac{1}{2} \tag{25b}$$

Problem 103. Neutron polarization by scattering on a magnetized sheet

A neutron beam is directed perpendicularly onto an iron sheet of thickness δ. The sheet is magnetized parallel to its surface.

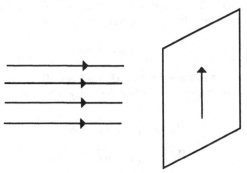

fig. 1

a) Rotating the sheet about itself, we observe that the fraction of neutrons going through it (transmission coefficient) does not change, i.e., it is independent of the direction of magnetization on the plane of the sheet; what can one infer from this measurement on the polarization of incoming neutrons? Write down the most general density matrix for the spin of incoming neutrons, compatible with this observation.

b) Let α_t (α_s) and r_t (r_s) be the scattering length and effective range of the interaction between neutrons and iron nuclei (assumed to have spin 1/2) in the triplet (singlet) channel. Calculate the transmission coefficient of the sheet as a function of the energy of incoming neutrons.

You may use the values $r_s = r_t = 2\,\mathrm{fm}$, $\alpha_t = -\alpha_s = 10\,\mathrm{fm}$; the density of iron and the thickness of the sheet are

$$\rho = 7.5\,\mathrm{gr/cm^3} \qquad \delta = 10^{-2}\,\mathrm{cm} \tag{1}$$

c) Calculate the polarization of scattered neutrons. The incoming density matrix is as given in part a).

Solution

a) The amplitude of neutron-iron scattering is spin-dependent; given the measurement of part a), the density matrix of the initial spin state must be invariant under rotations about the beam axis (denoted hereafter as the 3-axis). Hence, the most general form of the density matrix is

$$\rho = \frac{1}{2}(1 + P\sigma_3) \qquad P:\ \mathrm{real} \qquad |P| \le 1 \tag{2}$$

b) For small thickness δ, the fraction C of absorbed neutrons equals

$$C = n\,\delta\,\sigma \qquad (3)$$

Here, n is the number of nuclei per unit volume and σ is the total cross section. The transmission coefficient is $T = 1 - C$. For a state with given initial spins, the total cross section is the average value on that state of the quantity

$$\tilde{\sigma} = \Pi_t\,\sigma_t + \Pi_s\,\sigma_s \qquad (4)$$

where σ_t and σ_s are cross sections for each channel, and Π_t, Π_s are projectors to the triplet and singlet state, respectively. More explicitly, we have

$$\begin{aligned}
\Pi_t &= \frac{1}{4}\left[3 + \boldsymbol{\sigma}_n \cdot \boldsymbol{\sigma}_{Fe}\right] \\
\Pi_s &= \frac{1}{4}\left[1 - \boldsymbol{\sigma}_n \cdot \boldsymbol{\sigma}_{Fe}\right]
\end{aligned} \qquad (5)$$

The polarization of the sheet may be taken to be along the 1- axis; the density matrix of iron nuclei is then

$$\rho_{Fe} = \frac{1}{2}\left[1 + \sigma_{Fe}^1\right] \qquad (6)$$

Using the density matrix (2) for incoming neutrons, we compute

$$\begin{aligned}
\sigma = \langle\tilde{\sigma}\rangle &= \mathrm{Tr}\left\{\frac{1}{2}(1 + P\,\sigma_n^3)(1 + \sigma_{Fe}^1)[\Pi_t\,\sigma_t + \Pi_s\,\sigma_s]\right\} \\
&= \frac{1}{4}\sigma_s + \frac{3}{4}\sigma_t
\end{aligned} \qquad (7)$$

We see that the total cross section is the same as the one obtained in the absence of polarization.

In the effective range approximation, we write

$$\sigma_s = \frac{4\pi\alpha_s^2}{1 + \alpha_s^2\,k^2\left(1 - \dfrac{r_s}{\alpha_s}\right)} \qquad\qquad \sigma_t = \frac{4\pi\alpha_t^2}{1 + \alpha_t^2\,k^2\left(1 - \dfrac{r_t}{\alpha_t}\right)} \qquad (8)$$

We now use the specific values given in the problem $(-\alpha_s = \alpha_t \equiv \alpha,\ r_s = r_t \equiv r)$ and set

$$E = \frac{k^2\hbar^2}{2m_n} \quad,\quad \bar{E} = \frac{\hbar^2}{2m_n\alpha^2} \qquad (9)$$

yielding $\bar{E} = 200\,\mathrm{KeV}$ and

$$k^2\alpha^2 = \frac{E}{\bar{E}} \qquad (10)$$

Thus, if $E < 50$ KeV, $k^2\alpha^2 \ll 1$, leading to

$$
\begin{aligned}
\sigma &= 4\pi\,\alpha^2\left\{1 - \frac{E}{\overline{E}}\left[\frac{3}{4}\left(1 - \frac{r_t}{\alpha_t}\right) + \frac{1}{4}\left(1 - \frac{r_s}{\alpha_s}\right)\right]\right\} \\
&= 4\pi\,\alpha^2\left\{1 - \frac{E}{\overline{E}}\left(1 - \frac{1}{2}\frac{r}{\alpha}\right)\right\}
\end{aligned}
\tag{11}
$$

Introducing in (3) the numerical factors

$$
\begin{aligned}
n &= 0.73\cdot10^{23}\,\text{cm}^{-3} \\
\delta &= 10^{-2}\,\text{cm}
\end{aligned}
\tag{12}
$$

we obtain

$$
C = 9\cdot10^{-3}\left\{1 - 0.9\frac{E}{\overline{E}}\right\}
\tag{13}
$$

c)   The scattering amplitude may be expressed in the form

$$
A = A_t\Pi_t + A_s\Pi_s
\tag{14}
$$

Ignoring effective ranges, we write

$$
A_s = \frac{1}{-\dfrac{1}{\alpha_s} - ik} = \frac{1}{\dfrac{1}{\alpha} - ik}
\tag{15}
$$

$$
A_t = \frac{1}{-\dfrac{1}{\alpha_t} - ik} = \frac{1}{-\dfrac{1}{\alpha} - ik}
\tag{16}
$$

Equivalently,

$$
\begin{aligned}
\operatorname{Re} A_s &= -\operatorname{Re} A_t \simeq \frac{\alpha}{1 + k^2\alpha^2} \\
\operatorname{Im} A_s &= \operatorname{Im} A_t \simeq k\frac{\alpha^2}{1 + k^2\alpha^2}
\end{aligned}
\tag{17}
$$

The density matrix of scattered neutrons is given by

$$
\rho_n^f = \operatorname{Tr}_{Fe}[A\,\frac{1}{2}(1 + P\sigma_3^n)\,\frac{1}{2}(1 + \sigma_1^{Fe})\,A^\dagger]\Big/\frac{1}{4}\operatorname{Tr}_{Fe}\operatorname{Tr}_n(AA^\dagger)
\tag{18}
$$

To evaluate this expression we use the explicit form of the projectors Π_t and Π_s, with the result

$$
\rho_n^f = \frac{1}{2}(1 + \mathbf{P}\cdot\boldsymbol{\sigma})
\tag{19}
$$

With the given choice of axes, the final state polarization **P** reads

$$\mathbf{P} = (P_1, P_2, P_3)$$

$$P_1 = \frac{2|A_t|^2 - 2\mathrm{Re}(A_s A_t^*)}{3|A_t|^2 + |A_s|^2} = \frac{1}{1 + k^2\alpha^2}$$

$$P_2 = -2\frac{\mathrm{Im}(A_s A_t^*)P}{3|A_t|^2 + |A_s|^2} = P\frac{k\alpha}{1 + k^2\alpha^2}$$

$$P_3 = P\frac{2|A_t|^2 + 2\mathrm{Re}(A_s A_t^*)}{3|A_t|^2 + |A_s|^2} = P\frac{k^2\alpha^2}{1 + k^2\alpha^2} \tag{20}$$

At low energies, one has $k\alpha \sim E/\bar{E} \ll 1$; consequently, deflected neutrons are polarized almost completely parallel to the polarization of the iron sheet.

Problem 104.   Reflection of neutrons from a sandwich of metal sheets

Consider a neutron beam with kinetic energy $E \simeq 100°$ K, striking a slab made out of metal layers of different types, as shown in fig.1: Two metal layers (type 1) surrounding one layer (type 2) with width $0.1\,\mu$. The neutron-nucleus scattering lengths are $a_1 = -50$ fm and $a_2 = 100$ fm for the two metals, respectively.

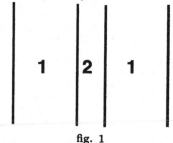

fig. 1

One observes the reflection off the middle layer as a function of energy. Calculate the reflection coefficient; show that it exhibits a pattern of peaks at certain values of the energy. Determine the position and intensity of these peaks.

Solution

In the low energy limit, the scattering amplitude reduces to the partial S-wave amplitude; this in turn, in the scattering length approximation, has the form

$$f_0 = \frac{1}{-\dfrac{1}{a} - i k} = -\frac{a}{1 + i k a} \tag{1}$$

There follows

$$
\begin{aligned}
\mathrm{Re} f &\simeq -\frac{a}{1+k^2 a^2} \simeq -a \\
\mathrm{Im} f &\simeq \frac{k\,a^2}{1+k^2 a^2} \simeq k\,a^2
\end{aligned}
\tag{2}
$$

A plane wave in a medium is attenuated according to the relation

$$
\frac{1}{\Phi}\frac{\mathrm{d}\Phi}{\mathrm{d}x} = -N\sigma_{Tot}
\tag{3}
$$

where N is the density of the medium. σ_{Tot} is given by the optical theorem

$$
\sigma_{Tot} = \frac{4\pi}{|\mathbf{k}|}\,\mathrm{Im}\,f
\tag{4}
$$

Applying (3) and (4) to the generic form of a plane wave in a medium

$$
\psi(x) = e^{i n \,\mathbf{k}\mathbf{x} - i\omega t}
\tag{5}
$$

we obtain

$$
2|\mathbf{k}|\,\mathrm{Im}\,n = N\,\sigma = \frac{4\pi}{|\mathbf{k}|}\,N\,\mathrm{Im}\,f
\tag{6}
$$

This implies

$$
\mathrm{Im}\,n = \frac{2\pi N}{\mathbf{k}^2}\,\mathrm{Im}\,f
\tag{7}
$$

and, by the forward dispersion relation,

$$
\mathrm{Re}\,n = 1 + \frac{2\pi N}{\mathbf{k}^2}\,\mathrm{Re}\,f
\tag{8}
$$

To calculate the reflection coefficient we match wave functions at the boundaries between the two metals, and impose a purely progressive wave in the rightmost layer. Setting $\mathbf{p}_1 = \mathbf{k}n_1$ and $\mathbf{p}_2 = \mathbf{k}n_2$, we write

$$
\begin{aligned}
\psi_{1L} &= A\,e^{i p_1 x} + B\,e^{-i p_1 x} \\
\psi_2 &= C\,e^{i p_2 x} + D\,e^{-i p_2 x} \\
\psi_{1R} &= e^{i p_1 x}
\end{aligned}
\tag{9}
$$

Matching ψ's at the two boundaries, $x = \pm d$, leads to

$$
\begin{aligned}
A\,e^{-i p_1 d} + B\,e^{i p_1 d} &= C\,e^{-i p_2 d} + D\,e^{i p_2 d} \\
p_1(A\,e^{-i p_1 d} - B\,e^{i p_1 d}) &= p_2(C\,e^{-i p_2 d} - D\,e^{i p_2 d}) \\
C\,e^{i p_2 d} + D\,e^{-i p_2 d} &= e^{i p_1 d} \\
p_2(C\,e^{i p_2 d} - D\,e^{-i p_2 d}) &= p_1\,e^{i p_1 d}
\end{aligned}
\tag{10}
$$

Solving the above, we find

$$2C\,\mathrm{e}^{\mathrm{i}p_2 d} = \mathrm{e}^{\mathrm{i}p_1 d}\left(1+\frac{p_1}{p_2}\right) \tag{11}$$

$$2D\,\mathrm{e}^{-\mathrm{i}p_2 d} = \mathrm{e}^{\mathrm{i}p_1 d}\left(1-\frac{p_1}{p_2}\right)$$

$$2A\,\mathrm{e}^{-\mathrm{i}p_1 d} = C\left(1+\frac{p_2}{p_1}\right)\mathrm{e}^{-\mathrm{i}p_2 d} + D\left(1-\frac{p_2}{p_1}\right)\mathrm{e}^{\mathrm{i}p_2 d}$$

$$2B\,\mathrm{e}^{\mathrm{i}p_1 d} = C\left(1-\frac{p_2}{p_1}\right)\mathrm{e}^{-\mathrm{i}p_2 d} + D\left(1+\frac{p_2}{p_1}\right)\mathrm{e}^{\mathrm{i}p_2 d}$$

and finally

$$A = \mathrm{e}^{2\mathrm{i}p_1 d}\left(\cos 2p_2 d - \mathrm{i}\,\frac{p_1^2+p_2^2}{2p_1 p_2}\sin 2p_2 d\right) \tag{12}$$

$$B = \mathrm{i}\,\mathrm{e}^{-2\mathrm{i}p_1 d}\sin 2p_2 d\,\frac{p_2^2-p_1^2}{2p_1 p_2} \tag{13}$$

We can check directly that flux conservation is satisfied

$$|A|^2 = 1 + |B|^2 \tag{14}$$

The transmission coefficient T is the ratio between the flux leaving the middle layer and the incident flux,

$$T = \frac{1}{|A|^2} = \frac{1}{1 + \dfrac{(p_1^2-p_2^2)^2}{4p_1^2 p_2^2}\sin^2 2p_2 d} \tag{15}$$

The reflection coefficient is the ratio between reflected and incident flux, that is,

$$R = \left|\frac{B}{A}\right|^2 \tag{16}$$

Indeed, Eq.(14) implies that

$$T + R = 1 \tag{17}$$

Peaks in the reflection coefficient R correspond to minima of T. Recalling that p_1 and p_2 are functions of $|\mathbf{k}|$, having the form

$$p_1 = |\mathbf{k}|\left(1 - \frac{2\pi N_1 a_1}{k^2}\right)$$

$$p_2 = |\mathbf{k}|\left(1 - \frac{2\pi N_2 a_2}{k^2}\right) \tag{18}$$

we conclude that the maxima of R correspond to $|\sin 2p_2 d| \simeq 1$, i.e., $p_2 \simeq n\pi/4d$. The peaks of the transmission coefficient, on the other hand, are solutions to $\sin 2p_2 d = 0$, giving $p_2 = n\pi/2d$; at the peaks we have $T = 1$.

Problem 105. A proton-nucleus scattering process at low energy

Consider protons scattering elastically off a nucleus $N(A, Z)$ with $T = 1/2$, $T_3 = 1/2$. The strong interaction cross section can be approximated by the formula

$$\frac{d\sigma}{d\Omega} = \frac{a^2}{1 + k^2 a^2} \tag{1}$$

for $(k\,a) \leq 1$, where a is the scattering length of the process. We would like to investigate the existence of a p–N bound state at low energy, by observing the interference with the Coulomb interaction at small angles.

a)   How can the existence of this state be decided by studying the above interference?

b)   Two further related processes are elastic neutron scattering and the charge-exchange reaction

$$n + (A, Z) \rightarrow p + (A, Z - 1) \tag{2}$$

What information does proton scattering give on these processes?

c)   Discuss the effect of final state Coulomb interaction on process (2).

Solution

a)   The parameterization (1) of the cross section for $(k\,a) \leq 1$ indicates that the scattering is S-wave. The S-wave scattering amplitude at low energies has the generic form

$$f_0 = \frac{1}{-\dfrac{1}{a} - i\,k} \tag{3}$$

which indeed leads to

$$\frac{d\sigma}{d\Omega} = \frac{a^2}{1 + (a\,k)^2} \tag{4}$$

f_0 exhibits a pole at $k = \dfrac{i}{a}$: As is well known, this pole corresponds to a bound or antibound state, according to whether $\operatorname{Im} k > 0$ (physical sheet) or $\operatorname{Im} k < 0$ (unphysical sheet). Thus, to ascertain whether a bound state exists, we must determine the sign of a. This can be done by studying the interference with Coulomb scattering. We note that for $ka \simeq 1$, $a \simeq 1\,\mathrm{fm}$, the relative velocity is

$$\frac{k}{m} \simeq \frac{1}{5} \tag{5}$$

For $Z\alpha \ll 1/5$ one may apply the Born approximation to the Coulomb interaction. In this approximation, we have

$$f_{\mathrm{Coul}} = -\frac{2m\,Z\,a}{q^2} \tag{6}$$

$\mathbf{k}$ and $\mathbf{k}'$ are the initial and final momenta, and $q = \mathbf{k}' - \mathbf{k}$, $q = |\mathbf{q}| = 2k\sin(\theta/2)$; θ is the scattering angle. The full cross section is

$$\frac{d\sigma}{d\Omega} = |f_0 + f_{\text{Coul}}|^2 = \left| \frac{1}{-\dfrac{1}{a} - i\,k} + \frac{m\,Z\,\alpha}{2k^2\sin^2\dfrac{\theta}{2}} \right|^2$$

$$\simeq a^2 \left[1 - \frac{m\,Z\,\alpha}{a k^2 \sin^2\dfrac{\theta}{2}} + \frac{m\,Z^2\,\alpha^2}{4a^2\,k^4\sin^4\dfrac{\theta}{2}} \right] \tag{7}$$

The sign of a can now be determined by studying the θ dependence of $\frac{d\sigma}{d\Omega}$.

b) Let us consider the isospin content of the nucleus and the nucleon. The nucleus (A, Z) is in a state $|T = \frac{1}{2}, T_3 = \frac{1}{2}\rangle$ and thus the proton-nucleon state is

$$\left|\frac{1}{2}, \frac{1}{2}\right\rangle \left|\frac{1}{2}, \frac{1}{2}\right\rangle = |1, 1\rangle \tag{8}$$

This is a pure triplet. In general, the nucleon-nucleus scattering amplitude is given by

$$f = a_s\,\Pi_s + a_t\,\Pi_t \tag{9}$$

where a_s (a_t) and Π_s (Π_t) is the scattering length and the projector to the isosinglet (isotriplet) state. In terms of eigenstates of T and T_3 we have

$$|p\rangle|A, Z\rangle = |1, 1\rangle \tag{10}$$

$$|n\rangle|A, Z\rangle = \frac{1}{\sqrt{2}}\{|1, 0\rangle - |0, 0\rangle\} \tag{11}$$

Consequently

$$f_p = a_t$$

$$f_n = \frac{1}{2}(a_t + a_s) \tag{12}$$

We see that p-nucleus scattering can give us only the triplet component of n-nucleus scattering. Let us now consider the charge exchange process (2): The initial state is given in (11), while the final state is

$$|p\rangle|A, Z - 1\rangle = \frac{1}{\sqrt{2}}\{|1, 0\rangle + |0, 0\rangle\} \tag{13}$$

Hence, the amplitude corresponding to (2) equals

$$f_{\text{ex}} = \frac{1}{2}(a_t - a_s) \tag{14}$$

Again, mere knowledge of σ_{el} of protons does not suffice to determine f_{ex}.

c) The charge exchange process

$$n + (A, Z) \rightarrow p + (A, Z - 1) \tag{15}$$

exhibits a Coulomb interaction in the final state. This entails the following correction
to the cross section

$$e^{-\frac{\pi}{k r_0}} \left| \Gamma \left(1 + \frac{i}{k r_0} \right) \right|^2 = \frac{\pi}{k r_0 (e^{-\frac{\pi}{k r_0}} - 1)} \simeq 1 - \frac{\pi}{2 k r_0} \tag{16}$$

In the above, we have set

$$r_0 = \frac{1}{\mu Z \alpha} \qquad \mu = \frac{m_p\, m_{(A,Z)}}{m_p + m_{(A,Z)}} \simeq m_p \tag{17}$$

The cross section becomes

$$\sigma \simeq \left(1 - \frac{\pi}{2 k r_0} \right) \sigma_{\text{strong}} \tag{18}$$

Problem 106.   Neutron interferometry

A neutron beam with energy corresponding to a temperature of 2°K is reflected N times
inside a cavity. The cavity walls are made out of spin-1/2 nuclei, polarized completely in
the direction perpendicular to the reflection plane.

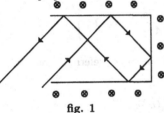

fig. 1

The nucleon-nucleus interaction can proceed
through a singlet and a triplet channel, with
scattering length

$$a_s = 2 \text{ fm}$$
$$a_t = 40 \text{ fm}$$

a) Using the scattering length approximation, determine the beam polarization after $N=5$
reflections.

b) How accurate is this approximation at the given beam energy?

c) The beam is sent into an interferometry
device as shown in fig.2. If the orientation
of the interferometer is changed from horizontal to vertical, what is the corresponding
change in interference phase induced by gravity? (Use the semiclassical approximation)

d) Consider the effect of part c) in the case
of an analogous experiment involving photons.

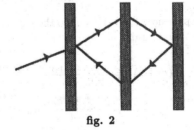

fig. 2

Solution

a) The scattering amplitude is

$$A = a_s \, \Pi_s + a_t \, \Pi_t \tag{1}$$

where Π_s and Π_t are singlet and triplet projection operators

$$\Pi_s = \frac{1 - \boldsymbol{\sigma}_n \cdot \boldsymbol{\sigma}_N}{4}$$

$$\Pi_t = \frac{3 + \boldsymbol{\sigma}_n \cdot \boldsymbol{\sigma}_N}{4}$$

After one scattering, the neutron density matrix ρ_n becomes

$$\rho_n' = \text{Tr}_N(A \, \rho_n \, \rho_N \, A^\dagger) \tag{2}$$

Tr_N denotes a trace over the nucleus degrees of freedom, and

$$\rho_N = \frac{1}{2}(1 + \mathbf{P}_N \cdot \boldsymbol{\sigma}_N)$$

is the density matrix of polarized nuclei. For the first reflection we have $\rho_n = \frac{1}{2}$ and

$$\rho_n'(1) = \frac{1}{2}\left[\frac{3\,a_t^2 + a_s^2}{4} + \frac{2\,a_t^2 - 2\,a_t\,a_s}{4}\,\boldsymbol{\sigma}_n \cdot \mathbf{P}_N\right] \tag{3}$$

The direction $\hat{\mathbf{P}}$ of the nucleus polarization is the only preferred direction in the problem; consequently, in the recursion relation

$$\rho_n'(m + 1) = \text{Tr}_N(A \, \rho_n'(m)\rho_N A^\dagger) \tag{4}$$

ρ has the form

$$\rho_n'(m) = \frac{1}{2}(1 + P_m\,\boldsymbol{\sigma}_n \cdot \hat{\mathbf{P}})F(m) \tag{5}$$

The presence of $F(m)$ is due to the fact that the density matrix (2) is not normalized. Eq.(4) now reads

$$\begin{aligned}\rho_n'(m + 1) = {}& \frac{1}{2}\Bigg\{\frac{3\,a_t^2 + a_s^2}{4} + \frac{a_t^2 - a_s^2}{4}P_m + \frac{2\,a_t^2 + 2\,a_s\,a_t}{4}\,P_m\boldsymbol{\sigma}_n \cdot \hat{\mathbf{P}} \\[2mm] & + \frac{2\,a_t^2 - 2\,a_t a_s}{4}\,\boldsymbol{\sigma}_n \cdot \hat{\mathbf{P}}\Bigg\}\,F(m) \equiv \frac{1}{2}[1 + P_{m+1}\boldsymbol{\sigma}_n \cdot \hat{\mathbf{P}}]F(m + 1)\end{aligned} \tag{6}$$

leading to the recursive equations

$$P_{m+1} = \frac{(2a_t^2 + 2a_s\,a_t)P_m + 2a_t^2 - 2a_s\,a_t}{(a_t^2 - a_s^2)P_m + 3a_t^2 + a_s^2}$$

$$F(m+1) = F(m)\left[\frac{3\,a_t^2 + a_s^2}{4} + \frac{a_t^2 - a_s^2}{4}P_m\right] \tag{7}$$

The first equation is easily solved with the substitution

$$P_m = 1 - \frac{1}{Q_m} \tag{8}$$

yielding

$$Q_{m+1} = \left(\frac{2a_t}{a_s + a_t}\right)^2 Q_m + \frac{a_s - a_t}{a_s + a_t} \tag{9}$$

The solution to the above is

$$Q_m = \frac{a_t + a_s}{3a_t + a_s} + \left[Q_0 - \frac{a_t + a_s}{3a_t + a_s}\right]\left(\frac{2a_t}{a_s + a_t}\right)^{2m} \tag{10}$$

Finally, for P_m we find

$$P_m = \frac{P_0 + (1 - P_0)\frac{2a_t}{3a_t + a_s}\left[1 - \left(\frac{a_s + a_t}{2a_t}\right)^{2m}\right]}{1 - (1 - P_0)\frac{a_s + a_t}{3a_t + a_s}\left[1 - \left(\frac{a_t + a_s}{2a_t}\right)^{2m}\right]}$$

$$= \frac{1 + \alpha\,P_0 - (1 - P_0)\alpha^{2m}}{1 + \alpha\,P_0 + (1 - P_0)\alpha^{2m+1}} \tag{11}$$

where

$$\alpha = \frac{a_t + a_s}{2\,a_t} \tag{12}$$

With the values of a_s, a_t at hand

$$P_m = \frac{P_0 + (1 - P_0)\left[1 - \left(\frac{21}{40}\right)^{2m}\right]\frac{40}{61}}{1 - (1 - P_0)\left[1 - \left(\frac{21}{40}\right)^{2m}\right]\frac{21}{61}} \tag{13}$$

If the neutron beam is initially nonpolarized

$$P_m = \frac{1 - \alpha^{2m}}{1 + \alpha^{2m+1}} \qquad \alpha = \frac{21}{40} \tag{14}$$

and for $m = 5$, $P_m \simeq 95\%$.

If, instead, the beam starts out completely polarized parallel to the nucleus polarization

$$P_m = P_0 = 1 \tag{15}$$

In all cases, $\alpha < 1$ leads to

$$P_m \to 1 \quad \text{as} \quad m \to \infty \tag{16}$$

b)   The energy corresponding to a temperature of $2°K$ is

$$\frac{3}{2} kT \; = 0.26 \cdot 10^{-3} \, \text{eV}$$

$$\frac{p^2}{2m} \; = 0.26 \cdot 10^{-3} \, \text{eV}$$

that is,

$$p \simeq 0.7 \cdot 10^3 \, \frac{\text{eV}}{c} \tag{17}$$

The scattering length approximation is applicable when

$$(p \, a) \ll 1 \tag{18}$$

A scattering length of $a_t = 40 \, \text{fm}$ gives $p \, a_t = 1.4 \cdot 10^{-4}$. $(p \, a_t)^2$ is a measure of the accuracy of the approximation.

c)   In the semiclassical approximation the phase is given by

$$\hbar \phi = \int p(z) dz \quad p = \sqrt{2m \left(E - V(z) \right)} \tag{19}$$

If the two paths in the interferometer (see fig.3) lie on a horizontal plane, the corresponding phases are

$$\hbar \phi_1 \; = \; p \left\{ \sqrt{D^2 + \ell^2} + \sqrt{\ell^2 + (D - x)^2} \right\}$$

$$\hbar \phi_2 \; = \; p \left\{ \sqrt{D^2 + \ell^2} + \sqrt{\ell^2 + (D + x)^2} \right\} \tag{20}$$

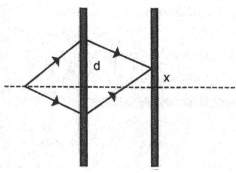

fig. 3

and, for $x \ll \ell, D$

$$\Delta \hbar \phi = p\sqrt{D^2 + \ell^2}\left(1 - \frac{Dx}{\ell^2 + D^2}\right) - p\sqrt{D^2 + \ell^2}\left(1 + \frac{Dx}{\ell^2 + D^2}\right)$$

$$\Delta \phi = \simeq -\frac{2p\,Dx}{\hbar\,\sqrt{\ell^2 + D^2}} = -\frac{2px}{\hbar}\sin\theta_0 \tag{21}$$

For vertical motion we have

$$U = m\,g\,h \qquad p(h) = \sqrt{p^2 - 2m^2\,g\,h} \tag{22}$$

The differential line elements along the two branches of the interferometer can be written as follows

$$dx_1 = dh\,\frac{\sqrt{D^2 + \ell^2}}{D} \qquad 0 \leq h \leq D$$

$$dx_1 = dh\,\frac{\sqrt{\ell^2 + (D - x)^2}}{x - D} \qquad D \geq h \geq x$$

$$dx_2 = dh\,\frac{\sqrt{D^2 + \ell^2}}{-D} \qquad 0 \geq h \geq -D$$

$$dx_2 = dh\,\frac{\sqrt{\ell^2 + (D + x)^2}}{D + x} \qquad -D \leq h \leq x \tag{23}$$

We thus find

$$\hbar\phi_1 = \frac{\sqrt{D^2 + \ell^2}}{D}\int_0^D \sqrt{p^2 - 2m^2 g\,h}\;dh\,+ \tag{24a}$$

$$+\frac{\sqrt{\ell^2 + (D - x)^2}}{x - D}\int_D^x \sqrt{p^2 - 2m^2 g\,h}\;dh$$

$$\hbar\phi_2 = -\frac{\sqrt{D^2 + \ell^2}}{D}\int_0^{-D} \sqrt{p^2 - 2m^2 g\,h}\;dh\,+ \tag{24b}$$

$$+\frac{\sqrt{\ell^2 + (D + x)^2}}{x + D}\int_{-D}^x \sqrt{p^2 - 2m^2 g\,h}\;dh$$

Evaluating the integral

$$\int^h \sqrt{p^2 - 2m^2 g\,h}\;dh = -\frac{1}{3\,m^2 g}(p^2 - 2\,m^2\,g\,h)^{\frac{3}{2}} \tag{25}$$

we write

$$\hbar\phi_1 = -\frac{\sqrt{D^2 + \ell^2}}{3\,m^2 g\,D}\left[(p^2 - 2\,m^2 g\,D)^{\frac{3}{2}} - p^3\right]$$

$$-\frac{\sqrt{\ell^2 + (D-x)^2}}{(x-D)3m^2g}\left[(p^2 - 2m^2gx)^{\frac{3}{2}} - (p^2 - 2m^2gD)^{\frac{3}{2}}\right]$$

$$\hbar\phi_2 = \frac{\sqrt{D^2 + \ell^2}}{3m^2gD}\left[(p^2 + 2m^2gD)^{\frac{3}{2}} - p^3\right]$$

$$-\frac{\sqrt{\ell^2 + (x+D)^2}}{(x+D)3m^2g}\left[(p^2 - 2m^2gx)^{\frac{3}{2}} - (p^2 + 2m^2gD)^{\frac{3}{2}}\right] \tag{26}$$

The quantity $\sqrt{2m^2gD} \simeq 7$ eV/c is much smaller than p of Eq.(17); thus, we can set

$$(p^2 - 2m^2gh)^{\frac{3}{2}} \simeq p^3 - 3m^2ghp + \frac{3}{2}\frac{m^4g^2h^2}{p} + \dots \tag{27}$$

For the phases we obtain

$$\hbar\phi_1 = \sqrt{D^2 + \ell^2}\left(p - \frac{1}{2}\frac{m^2gD}{p}\right) + \tag{28a}$$

$$+ \left(\sqrt{D^2 + \ell^2} - \frac{Dx}{\sqrt{D^2 + \ell^2}}\right)\left(p - \frac{1}{2}\frac{m^2g(x+D)}{p}\right)$$

$$\hbar\phi_2 = \sqrt{D^2 + \ell^2}\left(p + \frac{1}{2}\frac{m^2gD}{p}\right) + \tag{28b}$$

$$+ \left(\sqrt{D^2 + \ell^2} + \frac{Dx}{\sqrt{D^2 + \ell^2}}\right)\left(p + \frac{1}{2}\frac{m^2g(D-x)}{p}\right)$$

and

$$\hbar(\phi_1 - \phi_2) = -2\frac{m^2gD}{p}\sqrt{D^2 + \ell^2} - \frac{2Dx}{\sqrt{D^2 + \ell^2}}p + \mathcal{O}(x^2) \tag{29}$$

Comparing with Eq.(21) we see that

$$\phi_1 - \phi_2 = \Delta\phi - 2m^2g\frac{D\sqrt{D^2 + \ell^2}}{\hbar p} \tag{30}$$

Numerically, for $D = \ell \simeq 10$ cm

$$2m^2g\frac{D\sqrt{D^2 + \ell^2}}{p\hbar} \simeq 2\cdot10^5 \tag{31}$$

Therefore, the effect of gravity on interference fringes will be quite visible.
d) The gravitational mass of photons coincides with their frequency; thus,

$$p(h) = p(1 - \frac{gh}{c^2})$$

$$\int p(1 - \frac{gh}{c^2})\,dh = p\left(h - \frac{gh^2}{2c^2}\right) \tag{32}$$

From this, we obtain

$$\phi_1 = \frac{\sqrt{D^2+\ell^2}}{D}\frac{p}{\hbar}\left(D-\frac{gD^2}{2c^2}\right)+ \tag{33a}$$

$$+\frac{\sqrt{\ell^2+(D-x)^2}}{(x-D)}\frac{p}{\hbar}\left(x-D-\frac{gx^2}{2c^2}+\frac{gD^2}{2c^2}\right)$$

$$\phi_2 = \frac{\sqrt{D^2+\ell^2}}{-D}\frac{p}{\hbar}\left(-D-\frac{gD^2}{2c^2}\right)+ \tag{33b}$$

$$+\frac{\sqrt{\ell^2+(D+x)^2}}{(x+D)}\frac{p}{\hbar}\left(x+D-\frac{gx^2}{2c^2}+\frac{gD^2}{2c^2}\right)$$

$$\phi_1 - \phi_2 = -2\frac{p}{\hbar}\sqrt{D^2+\ell^2}\frac{gD}{c^2}-\frac{2Dxp}{\hbar\sqrt{\ell^2+D^2}} \tag{33c}$$

$$= \Delta\phi - 2gD\sqrt{D^2+\ell^2}\frac{p}{\hbar c^2}$$

Numerically, for $D=\ell\simeq10$ cm and the same wave number as before

$$2gD\sqrt{D^2+\ell^2}\cdot\frac{p}{\hbar c^2}\simeq10^{-7} \tag{34}$$

We see that this phase difference is 12 orders of magnitude smaller than in the case of neutrons.

Problem 107.   A model of the deuteron

The deuteron has spin 1 and isospin 0; it can be modelled by an S-wave p - n bound state in a potential well with radius a and depth V_0.
a)   Calculate V_0 and a, in terms of the binding energy ($E_B = 2.23$ MeV) and the p - n scattering length ($\alpha_T = 5.39$ fm, $J = 1$, $I = 0$). Determine the effective range, and compare it to the experimental value $r_0 = 1.70$ fm.
   Consider now a nonrelativistic charged particle interacting electromagnetically with the deuteron.
b)   Use form factors to parameterize the elastic collision.
c)   Calculate the photodisintegration cross section of the deuteron.

Solution

a)   The S-wave Schrödinger equation in a spherical well has the following bound state solutions

$$\psi(r) = \frac{1}{r}\frac{\sin k'r}{\sin k'a} \qquad r < a \qquad k' = \sqrt{2m_R(-E_B+V_0)}$$

$$\xi(r) = \frac{1}{r}\frac{e^{-kr}}{e^{-ka}} \qquad r > a \qquad k = \sqrt{2m_R E_B} \tag{1}$$

Matching logarithmic derivatives at the boundary, we find

$$k' \cot k'a = -k \tag{2}$$

equivalently,

$$\cot \sqrt{2m_R(-E_B + V_0)a^2} = -\sqrt{\frac{E_B}{V_0 - E_B}} \tag{3}$$

$m_R = 463.66$ MeV is the reduced mass. Since the binding energy E_B is known, this equation establishes a relation between V_0 and a.

The scattering solution is

$$\begin{aligned}
\psi(r) &= A \sin k'r & r < a \\
\psi(r) &= \sin(kr + \delta_0) & r > a
\end{aligned} \tag{4}$$

$$\frac{k'^2}{2m_R} + V_0 = \frac{k^2}{2m_R}$$

with the condition

$$\delta_0 = \operatorname{arccot}\left(\frac{k'}{k} \cot k'a\right) - ka \tag{5}$$

The scattering length thus equals

$$-\alpha_T = \frac{1}{k \cot \delta_0}\bigg|_{k=0} = a \left.\frac{1 - (k'a)\cot(k'a)}{(k'a)\cot(k'a)}\right|_{k=0} \tag{6}$$

Given α_T, the above relationship, together with (3), allows us to determine a and V_0. To do so, we introduce the dimensionless quantities

$$\begin{aligned}
x &= \sqrt{2m_R E_B a^2} \\
R &= \sqrt{2m_R V_0 a^2} \\
C &= \sqrt{2m_R E_B}\, \alpha_T = 1.25
\end{aligned} \tag{7}$$

In terms of these, a and V_0 are given by

$$\begin{cases}
a = \dfrac{x}{\sqrt{2\, m_R E_B}} \\[2mm]
V_0 = \dfrac{R^2}{x^2} E_B
\end{cases} \tag{8}$$

and the conditions (3), (6) become

$$\begin{cases}
\cot \sqrt{R^2 - x^2} = \dfrac{-x}{\sqrt{R^2 - x^2}} \\[3mm]
-C = x \left(\dfrac{tgR}{R} - 1\right)
\end{cases} \tag{9}$$

For $C = 1.25$ the numerical solution is

$$x = 0.48 \qquad R = 1.89 \qquad \sqrt{R^2 - x^2} = 1.828 \qquad \frac{R^2}{x^2} = 15.5 \qquad (10)$$

There follows

$$\begin{aligned} V_0 &= 34.57 \, \text{MeV} \\ a &= 2.07 \, \text{fm} \end{aligned} \qquad (11)$$

To calculate the effective range one must expand $k \cot \delta_0$ to order k^2, using Eq.(5). After some tedious algebra one arrives at

$$\begin{aligned} \rho &= a \frac{1 - 2g + \frac{2}{3}g^2 + g\frac{(1-g)}{R^2}}{(1-g)^2} \\ g &= R \cot R = -0.626 \\ \rho &= 0.84 \, a = 1.75 \, \text{fm} \end{aligned} \qquad (12)$$

This is compatible with the experimental value.

b) Let us derive the most general form for the matrix element of the electromagnetic current, between states of a spin-1 particle. The initial and final states are characterized by their four-momenta (p_μ and p'_μ) and polarizations (ε_μ and ε'_μ). The matrix element reads

$$\begin{aligned} \langle \mathbf{p}', \varepsilon' | J_\mu | \mathbf{p}, \varepsilon \rangle &= P_\mu \left\{ F_1(q^2)\,(\varepsilon'^* \cdot \varepsilon) + F_2(q^2)\,(\varepsilon'^* \cdot q)\,(\varepsilon \cdot q) \right\} \\ &\quad + F_3(q^2) \left\{ \varepsilon'^*_\mu\,(\varepsilon \cdot q) - \varepsilon_\mu\,(\varepsilon'^* \cdot q) \right\} \\ &\quad + F_4(q^2) \left\{ \varepsilon'^*_\mu\,(\varepsilon \cdot q) + \varepsilon_\mu\,(\varepsilon'^* \cdot q) \right\} \\ &\quad + q_\mu \left\{ F_5(q^2)\,(\varepsilon'^* \cdot \varepsilon) + F_6(q^2)\,(\varepsilon'^* \cdot q)\,(\varepsilon \cdot q) \right\} \end{aligned} \qquad (13)$$

Here, as usual, $P_\mu = p_\mu + p'_\mu$, $q_\mu = p_\mu - p'_\mu$. To arrive at (13), the following conditions were imposed

i) J_μ must be a polar four-vector, constructed from the available four-vectors: q_μ, P_μ, ε_μ, ε'_μ

ii) J_μ must be bilinear in ε_μ and ε'^*_μ

iii) ε and ε' satisfy

$$(\varepsilon \cdot P) = -(\varepsilon \cdot q) \qquad (\varepsilon' \cdot P) = (\varepsilon' \cdot q) \qquad (14)$$

Additional requirements reduce further these form factors:

i) Hermiticity of the current requires

$$\langle \mathbf{p}', \varepsilon' | J_\mu | \mathbf{p}, \varepsilon \rangle = \langle \mathbf{p}', \varepsilon' | J_\mu^\dagger | \mathbf{p}, \varepsilon \rangle = \langle \mathbf{p}, \varepsilon | J_\mu | \mathbf{p}', \varepsilon' \rangle^* \tag{15}$$

Consequently,

$$\begin{matrix} F_1 = & F_1^* & F_2 = & F_2^* \\ F_3 = & F_3^* & F_4 = -F_4^* \\ F_5 = -F_5^* & F_6 = -F_6^* \end{matrix} \tag{16}$$

ii) Imposing time reversal invariance

$$\begin{aligned} \langle \mathbf{p}, \varepsilon | g_{\mu\mu} J_\mu | \mathbf{p}', \varepsilon' \rangle &= \langle \mathbf{p}', \varepsilon' | T^\dagger J_\mu T | \mathbf{p}, \varepsilon \rangle = \\ \langle T(\mathbf{p}', \varepsilon') | J_\mu | T(\mathbf{p}, \varepsilon) \rangle^* &= \langle -\mathbf{p}, -\varepsilon | J_\mu | -\mathbf{p}', -\varepsilon' \rangle^* \end{aligned} \tag{17}$$

we find

$$\begin{matrix} F_1 = F_1^* & F_2 = F_2^* \\ F_3 = F_3^* & F_4 = F_4^* \\ F_5 = F_5^* & F_6 = F_6^* \end{matrix} \tag{18}$$

Together with (16), the above leads to

$$F_4 = F_5 = F_6 = 0 \tag{19}$$

iii) Current conservation is automatically satisfied, given Eq.(19).

Thus, the matrix element of the current takes the form

$$\begin{aligned} \langle \mathbf{p}', \varepsilon' | J_\mu | \mathbf{p}, \varepsilon \rangle &= P_\mu \left\{ F_1(q^2) (\varepsilon'^* \cdot \varepsilon) + F_2(q^2) (\varepsilon'^* \cdot q)(\varepsilon \cdot q) \right\} \\ &\quad + F_3(q^2) \left\{ \varepsilon_\mu'^* (\varepsilon \cdot q) - \varepsilon_\mu (\varepsilon'^* \cdot q) \right\} \end{aligned} \tag{20}$$

In the nonrelativistic limit, the energy transfer q^0 is negligible; denoting by j_μ the current matrix elements (between states normalized to 1), (20) becomes

$$j^0 = f_1(q^2) (\varepsilon'^* \varepsilon) + f_2(q^2) (\varepsilon'^* \mathbf{q})(\varepsilon \mathbf{q}) \tag{21}$$
$$\mathbf{j} = f_3(q^2) \mathbf{q} \wedge (\varepsilon'^* \wedge \varepsilon) \tag{22}$$

j^0 is the Fourier transform of the charge density at momentum $\mathbf{q}$

$$j^0(\mathbf{q}) = \int d^3x \, j^0(\mathbf{x}) \, e^{-i\mathbf{q}\mathbf{x}} \tag{23}$$

At small $\mathbf{q}$, (23) may be expanded to give

$$j^0(\mathbf{q}) \simeq 1 - \frac{1}{6} \langle \mathbf{r}^2 \rangle \mathbf{q}^2 - \frac{1}{6} q_i q_j Q_{ij} \tag{24}$$

where

$$1 = \int d^3x\, j^0(\mathbf{x}) \tag{25}$$

$$\langle \mathbf{r}^2 \rangle = \int d^3x\, j^0(\mathbf{x})\, \mathbf{r}^2 \tag{26}$$

$$Q_{ij} = \int d^3x\, j^0(\mathbf{x}) \left(3x_i x_j - \delta_{ij}\, \mathbf{x}^2 \right) \tag{27}$$

The above are, respectively, the total charge, the mean square radius of the charge distribution, and the quadrupole moment of the system. By virtue of the Wigner-Eckart theorem, we have

$$Q_{ij} = C \left(\varepsilon_i' \varepsilon_j + \varepsilon_i \varepsilon_j' - \frac{2}{3} \delta_{ij}\, \varepsilon' \varepsilon \right) \tag{28}$$

Substituting into Eq.(21) allows us to identify

$$f_1(0) = 1$$
$$f_2(0) = \frac{1}{4} Q_{zz} \tag{29}$$

At small momenta, the vector part of the current couples to external fields as follows

$$\mathbf{j}\,\mathbf{A} \simeq f_3(0)\, [\mathbf{q} \wedge (\boldsymbol{\varepsilon}'^* \wedge \boldsymbol{\varepsilon})] \cdot \mathbf{A} \tag{30}$$

As an example, let us take a state with polarization +1; then $\boldsymbol{\varepsilon} = -\mathrm{i}\dfrac{(\hat{\mathbf{x}} + \mathrm{i}\hat{\mathbf{y}})}{\sqrt{2}}$ and we have

$$\mathbf{j}\,\mathbf{A} \simeq -\mathrm{i} f_3(0)\, [\mathbf{q} \wedge \hat{\mathbf{z}}]\, \mathbf{A} \tag{31}$$

Since the magnetic field is $\mathbf{B} = -\mathrm{i}\mathbf{q} \wedge \mathbf{A}$, we identify the above as a magnetic dipole coupling, with

$$f_3(0) = \mu \tag{32}$$

c) We must calculate the probability of a transition from a bound proton-neutron state to a state in the continuous spectrum, triggered by the absorption of a photon (photoelectric effect). Let $\mathbf{k}$ be the initial photon momentum and $\mathbf{p}$ the (final) relative momentum of the proton-neutron pair. The cross section is given by the following nonrelativistic formula

$$d\sigma = \frac{2\pi}{2\omega}\, e^2\, |\mathcal{M}_{fi}|^2\, \delta(k_0 - E_B - E)\, \frac{d^3\mathbf{p}}{(2\pi)^3} \tag{33}$$

with

$$\mathcal{M}_{fi} = \boldsymbol{\varepsilon} \cdot \langle d|\mathbf{j}|n,p\rangle \tag{34}$$

The relation $p^2\, dp \simeq p\, m_R\, dE$ gives

$$\frac{d\sigma}{d\Omega} = \frac{\alpha p M}{4\pi\omega} |\mathcal{M}_{fi}|^2 \tag{35}$$

$M = 2m_R$ is the nucleon mass. Neglecting magnetic moment terms, $\mathbf{j}$ is the convective current of the proton

$$\mathbf{j} \simeq \frac{\mathbf{p}}{2m_R} = \frac{\mathbf{p}}{M} = \frac{\mathbf{v}}{2} = \mathbf{v}_p \tag{36}$$

$\mathbf{v}_p$ is the proton velocity. Averaging over photon polarizations, the photodisintegration cross section of the deuteron is found to be

$$\sigma = \frac{\alpha p}{4\pi\omega} \frac{4\pi}{3} |\mathbf{p}_{fi}|^2 \tag{37}$$

We must now calculate the matrix element of the electric dipole; it equals

$$\mathbf{p}_{fi} = \int d^3x \, \psi'^* \mathbf{p} \, \psi = \int d^3x \, e^{-i\mathbf{p}\mathbf{x}} \mathbf{p} \sqrt{\frac{\bar{k}}{2\pi}} \frac{e^{-\bar{k}r}}{r} \tag{38}$$

Here, the expression of the deuteron wave function for $r > a$ was taken to be approximately valid everywhere, with due normalization. An integration by parts yields

$$\mathbf{p}_{fi} = \sqrt{\frac{\bar{k}}{2\pi}} \frac{4\pi\mathbf{p}}{p^2 + \bar{k}^2} \tag{39}$$

We make use of

$$E_B = \frac{\bar{k}^2}{2m_R} \tag{40}$$

and neglect the nucleus recoil energy, $\omega^2/4m_R$, for $\omega \ll 4m_R$ with the result

$$\sigma = \frac{8\pi}{3} \frac{\alpha}{M} \sqrt{E_B} \frac{(\omega - E_B)^{\frac{3}{2}}}{\omega^3} \tag{41}$$

The cross section has its maximum at $\omega \simeq 2E_B$.

Problem 108.   Some properties of the deuteron

The deuteron has isospin 0, angular momentum 1 and well-defined parity.
a)   Which states are compatible with the above properties?
b)   Knowing the following values of magnetic moment

$$\begin{aligned}
\mu_p &= 2.7928 \frac{e\hbar}{2m_p c} && \text{(proton)} \\
\mu_n &= -1.913 \frac{e\hbar}{2m_p c} && \text{(neutron)} \\
\mu_D &= 0.8565 \frac{e\hbar}{2m_p c} && \text{(deuteron)}
\end{aligned} \tag{1}$$

deduce the state of the deuteron.
c)   Consider elastic scattering of electrons off deuterons polarized transversely to the relative momentum. What is the effect of the magnetic dipole and of the quadrupole moment on the angular distribution?

Solution

a) By the Pauli exclusion principle, the wave function of two-nucleon states is completely antisymmetric. Consequently, the spin S, orbital angular momentum L and isospin T must satisfy: $S + T + L = $ odd.

Isospin singlet states of the n-p system with total angular momentum 1 will thus have the following quantum numbers

$$^3S_1^+ \qquad ^1P_1^- \qquad ^3D_1^+ \tag{2}$$

Here we have used the notation $^{2S+1}L_J^P$. We see that the deuteron can be either a mixture of $^3S_1$ and $^3D_1$ or a pure $^1P_1$ state.

b) The magnetic moment operator for a system of two spin-1/2 particles, expressed in units of the nuclear magneton $e\hbar/2m_pc$, equals

$$\mathbf{M} = \mu_p\,\boldsymbol{\sigma}_p + \mu_n\,\boldsymbol{\sigma}_n + \frac{\mathbf{L}}{2} \tag{3}$$

The factor of 1/2 in the orbital term is due to the fact that only one of the two particles (the proton) is charged. The magnetic moment must be parallel to the angular momentum $\mathbf{J} = \mathbf{L} + \boldsymbol{\sigma}_p/2 + \boldsymbol{\sigma}_n/2$, by the Wigner- Eckart theorem. We define μ_D through the relationship

$$\mathbf{M} = \mu_D\mathbf{J} \tag{4}$$

equivalently

$$\mu_D = \frac{\mathbf{M}\cdot\mathbf{J}}{\mathbf{J}\cdot\mathbf{J}} \tag{5}$$

Writing $\mathbf{M}$ in the form

$$\mathbf{M} = (\mu_p + \mu_n)\,\mathbf{S} + \frac{\mathbf{L}}{2} + (\mu_p - \mu_n)\,(\mathbf{S}_p - \mathbf{S}_n) \tag{6}$$

and substituting into (5) we find

$$\mu_D = \frac{1}{2}\left[\mu_p + \mu_n + \frac{1}{2} + \left(\mu_p + \mu_n - \frac{1}{2}\right)\frac{\mathbf{S}^2 - \mathbf{L}^2}{2}\right] \tag{7}$$

The following scalar products have been used in deriving (7)

$$\mathbf{S}\cdot\mathbf{J} = \frac{\mathbf{J}^2 + \mathbf{S}^2 - \mathbf{L}^2}{2} \quad , \quad \mathbf{L}\cdot\mathbf{J} = \frac{\mathbf{J}^2 + \mathbf{L}^2 - \mathbf{S}^2}{2} \tag{8}$$

and

$$(\mathbf{S}_p - \mathbf{S}_n)\cdot\mathbf{J} = (\mathbf{S}_p - \mathbf{S}_n)\cdot\mathbf{L} = 0 \tag{9}$$

Eq.(9) is a consequence of the fact that the only nonzero matrix elements of the operator $\mathbf{S}_p - \mathbf{S}_n$ occur in the singlet-triplet transition.

The numerical values of μ_D can now be read off (7)

$$\mu_D\left({}^3S_1\right) = 0.8798 \tag{10}$$

$$\mu_D\left({}^3D_1\right) = 0.3101 \tag{11}$$

$$\mu_D\left({}^1P_1\right) = 0.5 \tag{12}$$

A comparison of the above with the experimental result excludes the state ${}^1P_1$, so that the deuteron must be a superposition of the following form

$$|d\rangle = \cos\alpha\,|{}^3S_1\rangle + \sin\alpha\,|{}^3D_1\rangle \tag{13}$$

The magnetic moment of this state is

$$\mu_D = \cos^2\alpha\,\mu_D({}^3S_1) + \sin^2\alpha\,\mu_D({}^3D_1) \tag{14}$$

The experimental value of μ_D now fixes α

$$\cos\alpha = 0.98 \qquad \sin\alpha = 0.2 \tag{15}$$

c) The expression for the electron-deuteron scattering amplitude is

$$\mathcal{M}_{fi} = \bar{u}(\mathbf{p}')\gamma_\mu u(\mathbf{p})\,\frac{e^2}{q^2}\,J_d^\mu(\mathbf{q}) \tag{16}$$

$J_d^\mu(\mathbf{q}) = \langle d_f|j^\mu(0)|d_i\rangle$ is the deuteron electromagnetic current.

In the nonrelativistic limit, the contribution of this current to the amplitude becomes

$$J_d^0(\mathbf{q}) = 2m_D\left[\boldsymbol{\varepsilon}'^*\boldsymbol{\varepsilon} + \frac{Q}{2}\left\{(\boldsymbol{\varepsilon}'^*\,\mathbf{q})\,(\boldsymbol{\varepsilon}\,\mathbf{q}) - \frac{\mathbf{q}^2}{2}\,(\boldsymbol{\varepsilon}'^*\,\boldsymbol{\varepsilon})\right\}\right] \tag{17}$$

$$\mathbf{J}_d = 2m_D\mathbf{q}\wedge\boldsymbol{\mu} \tag{18}$$

where $Q = Q_{zz}/2$ is the quadrupole moment of the deuteron and $\boldsymbol{\mu} = \mu\,(\boldsymbol{\varepsilon}'^*\wedge\boldsymbol{\varepsilon})$ is the magnetic moment; μ is given in units of $e\hbar/2m_Dc$.

To derive the angular distribution we evaluate

$$|\mathcal{M}_{fi}|^2 = \frac{e^4}{(q^2)^2}\,L_{\mu\nu}\,\overline{J_d^{\mu*}J_d^\nu} \tag{19}$$

where

$$L_{\mu\nu} = \frac{1}{2}\,\mathrm{Tr}\left[(\not{p} + m)\,\gamma_\mu\,(\not{p}' + m)\,\gamma_\nu\right] = q^2 g_{\mu\nu} - P_\mu P_\nu \tag{20}$$

$$\overline{J_d^{\mu*}J_d^\nu} = \sum_{\text{final pol.}} J_d^{\mu*}J_d^\nu$$

$$P_\mu = p_\mu + p'_\mu \qquad q_\mu = p_\mu - p'_\mu$$

In the limit of small momentum transfer we have

$$L_{00} = 4p_0^2 - \mathbf{q}^2 \qquad L_{ij} = \mathbf{q}^2 \delta_{ij} + 4p_i p_j \qquad L_{0i} = 4p_0 p_i \tag{21}$$

We perform the sum over final polarizations, λ, using

$$\sum_\lambda \varepsilon_i'(\lambda)\varepsilon_j'^*(\lambda) = \delta_{ij} \tag{22}$$

with the result

$$
\begin{aligned}
\overline{J_d^{0*} J_d^0} &= (2m_D)^2 \left\{ \left(1 - \frac{Q}{6}\mathbf{q}^2\right)^2 + |\boldsymbol{\varepsilon}\,\mathbf{q}|^2 Q^2 \left(1 + \frac{1}{12}Q^2\mathbf{q}^2\right) \right\} \\[4pt]
\overline{J_d^{0*}\mathbf{J}^*} &= (2m_D)^2 \Big\{ [\boldsymbol{\varepsilon}\,(\mathbf{q}\,\boldsymbol{\varepsilon}^*) - \boldsymbol{\varepsilon}^*\,(\mathbf{q}\,\boldsymbol{\varepsilon})]\left(1 - Q\frac{\mathbf{q}^2}{6}\right) \\[2pt]
&\qquad + \frac{1}{2}Q\left(|\mathbf{q}\,\boldsymbol{\varepsilon}|^2 \mathbf{q} - (\mathbf{q}\,\boldsymbol{\varepsilon})\boldsymbol{\varepsilon}^*\,\mathbf{q}^2\right) \Big\} \\[4pt]
\overline{J_i^* J_j} &= (2m_D)^2 \Big\{ \delta_{ij}|\mathbf{q}\,\boldsymbol{\varepsilon}|^2 + \varepsilon_i^* \varepsilon_j\,\mathbf{q}^2 - \varepsilon_i^* q_j (\mathbf{q}\,\boldsymbol{\varepsilon}) - \varepsilon_j q_i (\mathbf{q}\,\boldsymbol{\varepsilon})^* \Big\}
\end{aligned} \tag{23}
$$

Using (21) and (23), the squared amplitude reads

$$
\begin{aligned}
|\mathcal{M}_{fi}|^2 &= \frac{e^4}{q^4}(2m_D)^2 \Big[\left(4p_0^2 - \mathbf{q}^2\right)\left\{\left(1 - \frac{Q}{6}\mathbf{q}^2\right)^2 + |\boldsymbol{\varepsilon}\,\mathbf{q}|^2 Q^2 \left(1 + \frac{1}{12}Q^2\mathbf{q}^2\right)\right\} \\[4pt]
&\quad + 2\mu\,\mathbf{Re}\Big\{4p_0\left[(\mathbf{p}\,\boldsymbol{\varepsilon})(\mathbf{q}\,\boldsymbol{\varepsilon}^*) - (\mathbf{p}\,\boldsymbol{\varepsilon}^*)(\mathbf{q}\,\boldsymbol{\varepsilon})\right]\left(1 - Q\frac{\mathbf{q}^2}{6}\right) \\[4pt]
&\quad + 2p_0 Q\left[|\mathbf{q}\,\boldsymbol{\varepsilon}|^2(\mathbf{q}\,\mathbf{p}) - \mathbf{q}^2(\mathbf{q}\,\boldsymbol{\varepsilon})(\boldsymbol{\varepsilon}^*\,\mathbf{p})\right]\Big\} \\[4pt]
&\quad + \mu^2\Big\{\mathbf{q}^2\left(|\boldsymbol{\varepsilon}\,\mathbf{p}|^2 + \mathbf{q}^2\right) + 4(\mathbf{p}^2|\boldsymbol{\varepsilon}\,\mathbf{q}|^2 + \mathbf{q}^2|\boldsymbol{\varepsilon}\,\mathbf{p}|^2 \\[4pt]
&\quad - (\boldsymbol{\varepsilon}\,\mathbf{p})(\boldsymbol{\varepsilon}^*\,\mathbf{q})(\mathbf{p}\,\mathbf{q}) - (\boldsymbol{\varepsilon}\,\mathbf{q})(\boldsymbol{\varepsilon}^*\,\mathbf{p})(\mathbf{p}\,\mathbf{q}))\Big\}\Big]
\end{aligned} \tag{24}
$$

If the polarization is perpendicular to the incident momentum, $\boldsymbol{\varepsilon}\,\mathbf{p} = 0$, the above expression simplifies to

$$
\begin{aligned}
|\mathcal{M}_{fi}|^2 &= \frac{e^4}{q^4}(2m_D)^2 \Big[\left(4p_0^2 - \mathbf{q}^2\right)\left\{\left(1 - \frac{Q}{6}\mathbf{q}^2\right)^2 + |\boldsymbol{\varepsilon}\,\mathbf{q}|^2 Q^2 \left(1 + \frac{1}{12}Q^2\mathbf{q}^2\right)\right\} \\[4pt]
&\quad + \mu^2 \mathbf{q}^2\left(\mathbf{q}^2 + |\boldsymbol{\varepsilon}\,\mathbf{q}|^2\right)\Big]
\end{aligned} \tag{25}
$$

The magnetic dipole and the quadrupole moment introduce a dependence on the angle between momentum transfer and polarization. Setting the polarization along the x-axis, we write

$$
\begin{aligned}
\mathbf{p} &= (0,0,p) \\
\boldsymbol{\varepsilon} &= (1,0,0) \\
\mathbf{q} &= \mathbf{p}' - \mathbf{p} = p\,(\sin\theta\cos\phi,\ \sin\theta\sin\phi,\ \cos\theta - 1) \\
\boldsymbol{\varepsilon}\,\hat{\mathbf{q}} &= \cos\theta\cos\phi
\end{aligned} \tag{26}
$$

Inserting these expression into Eq.(24) we are led to the explicit dependence of the angular distribution on the azimuthal angle.

Problem 109.   Neutron polarization by reflection

Consider a beam of thermal, monochromatic neutrons striking a layer of solid hydrogen. The singlet and triplet scattering lengths are

$$\alpha_0 = 5.39 \cdot 10^{-13} \, \text{cm} \qquad \alpha_1 = -23.7 \cdot 10^{-13} \, \text{cm} \tag{1}$$

a)   Suppose that both incoming neutrons and hydrogen nuclei are completely polarized in the same direction. Calculate the polarization of scattered neutrons.

b)   Consider now the case in which incoming neutrons and/or hydrogen nuclei are nonpolarized. Again, calculate the polarization of scattered neutrons.

c)   Calculate the reflection coefficient for all the above cases.

Solution

Neutrons scatter off hydrogen nuclei with amplitude

$$A = -\alpha_0 \, \Pi_0 - \alpha_1 \, \Pi_1 \tag{2}$$

where

$$\Pi_0 = \frac{1}{4} - \mathbf{s}_n \cdot \mathbf{s}_p \qquad \Pi_1 = \frac{3}{4} + \mathbf{s}_n \cdot \mathbf{s}_p \tag{3}$$

are the projectors onto singlet and triplet states, respectively. The density matrix of initial states is

$$\rho = \rho_n \cdot \rho_p \qquad \begin{aligned} \rho_n &= \tfrac{1}{2}(1 + \mathbf{P}_n \cdot \boldsymbol{\sigma}_n) \\ \rho_p &= \tfrac{1}{2}(1 + \mathbf{P}_p \cdot \boldsymbol{\sigma}_p) \end{aligned} \tag{4}$$

whereas the final (non normalized) density matrix equals

$$\rho' = A \rho A^+ \tag{5}$$

In deriving the final density matrix of the neutron, final states of hydrogen remain unobserved and must be summed over. We obtain

$$\begin{aligned} \rho'_n = \ & \frac{1}{2}\left[\frac{3\alpha_1^2 + \alpha_0^2}{4} + \frac{\alpha_1^2 - \alpha_0^2}{4} \mathbf{P}_n \cdot \mathbf{P}_p + \frac{2\alpha_1^2 + 2\alpha_0\alpha_1}{4} \boldsymbol{\sigma}_n \cdot \mathbf{P}_n \right. \\ & \left. + \frac{2\alpha_1^2 - 2\alpha_0\alpha_1}{4} \boldsymbol{\sigma}_n \cdot \mathbf{P}_p \right] \end{aligned} \tag{6}$$

a) If both neutrons and hydrogen are polarized completely along the same direction $\hat{\mathbf{P}}$, we find

$$\rho_n'^{(1)} = \frac{\alpha_1^2}{2} \left(1 + \boldsymbol{\sigma}_n \cdot \hat{\mathbf{P}}\right) \tag{7}$$

Scattering takes place exclusively in the triplet channel.

b) If both hydrogen and neutrons are nonpolarized, we have

$$\rho_n'^{(2)} = \frac{1}{2} \frac{3\alpha_1^2 + \alpha_0^2}{4} \tag{8}$$

Polarized neutrons striking nonpolarized hydrogen give

$$\rho_n'^{(3)} = \frac{1}{2} \left[\frac{3\alpha_1^2 + \alpha_0^2}{4} + \frac{2\alpha_1(\alpha_1 + \alpha_0)}{4} \boldsymbol{\sigma}_n \mathbf{P}_n \right] \tag{9}$$

that is, outgoing neutrons will have a partial polarization

$$\mathbf{P}_n' = \frac{2\alpha_1(\alpha_1 + \alpha_0)}{3\alpha_1^2 + \alpha_0^2} \mathbf{P}_n \tag{10}$$

Finally, nonpolarized neutrons striking polarized hydrogen lead to

$$\rho_n'^{(4)} = \frac{1}{2} \left[\frac{3\alpha_1^2 + \alpha_0^2}{4} + \frac{2\alpha_1(\alpha_1 - \alpha_0)}{4} \boldsymbol{\sigma}_n \cdot \mathbf{P}_p \right] \tag{11}$$

Here, too, we find a partial polarization

$$\mathbf{P}_n' = \frac{2\alpha_1(\alpha_1 - \alpha_0)}{3\alpha_1^2 + \alpha_0^2} \mathbf{P}_p \tag{12}$$

Numerically, after one reflection on completely polarized hydrogen, we have

$$\mathbf{P}_n' = 0.804 \hat{\mathbf{P}}_p \qquad \hat{\mathbf{P}}_p = \frac{\mathbf{P}_p}{|\mathbf{P}_p|} \tag{13}$$

c) The total cross section in the case described in part a) equals

$$\sigma_T = 4\pi \, \alpha_1^2 \tag{14}$$

The cases in part b) have cross section

$$\sigma_T = \pi \left(3\alpha_1^2 + \alpha_0^2\right) \tag{15}$$

The reflectance of a layer with thickness ΔL is given by the formula

$$\frac{1}{2} \sigma_T \Delta L \, \rho \tag{16}$$

where ρ is the number of hydrogen atoms per unit volume. Eq.(16) results from the fact that the scattering takes place in the S-wave; consequently, half of the scattered particles proceed forward, the other half backward. The reflectance in the triplet channel is much greater than in the singlet.

Problem 110.   Muon catalyzed fusion

The $\mu^- - p$ system behaves like a hydrogen atom.

a)   Knowing that the mass of μ^- is 105.66 MeV calculate the binding energy and the Bohr radius of this system.

b)   When the muonic atom in its ground state strikes a nucleus, it is insensitive to Coulomb repulsion (since it is neutral), at least up to distances at which the atom remains bound. At small distances, the Coulomb field of the nucleus is strong enough to give rise to the photoelectric effect. In the presence of a field **E**, the scission probability per unit time is[1] (the charge e is in CGS units)

$$W = \frac{4m^3 e^9}{|\mathbf{E}|\,\hbar^7}\, e^{-2m^2 e^5/(3\,|\mathbf{E}|\hbar^4)} \tag{1}$$

Discuss up to what distance, r, scission time remains small compared to the time of crossing $\frac{r}{v}$. Consider in particular the case of a plasma of protons, at temperature $T = 10^6$ K.

c)   How does the fusion rate get modified, if the plasma contains N protons/cm^3 and n muons/cm^3 ?

Solution

a)   A hydrogenoid system with reduced mass μ has ground state energy

$$E_0 = -\frac{1}{2}\,\mu\alpha^2 \tag{2}$$

and Bohr radius

$$r_B = \frac{1}{\mu\alpha} \tag{3}$$

In the case at hand $\mu = \frac{m_\mu m_p}{m_\mu + m_p} = 94.96\,\text{MeV}$, so that

$$E = -2.53\,\text{KeV} \qquad\qquad r_B = 2.5{\cdot}10^{-11}\,\text{cm} \tag{4}$$

Compared to the usual hydrogen atom, the energy here is approximately 200 times larger, and the Bohr radius 200 times smaller.

b)   The probability (1) in our case reads

$$W = \frac{4}{|\mathbf{E}|}\,\frac{\mu^3 e^9}{\hbar^7}\,\exp\left(-\frac{2\mu^2 e^5}{3\,|\mathbf{E}|\,\hbar^4}\right) \tag{5}$$

In terms of the electric field generated by the proton, $|\mathbf{E}| = e/r^2$, the above becomes

$$W(r) = 4\alpha \left(\frac{r}{r_B}\right)^2 \frac{1}{r_B}\,\exp\left[-\frac{2}{3}\left(\frac{r}{r_B}\right)^2\right] \tag{6}$$

---

[1] L. D. Landau, E. M. Lifshitz, *Quantum Mechanics*, Pergamon Press

The screening produced by μ^- may catalyze the process of nuclear fusion, since it diminishes the Coulomb repulsion of nuclei. This mechanism becomes effective when the ionization probability of μ^- is negligible. To assess the effectiveness of this mechanism, one can compare the characteristic ionization time ($\tau = 1/W$) to the time of flight (r/v) of the nuclei. Note that $\tau = \tau(r)$ increases exponentially with distance. This condition takes the form

$$\frac{1}{W(\bar{r})} \simeq \frac{\bar{r}}{v} \tag{7}$$

with $W(r)$ as given in (6). Solving (7) numerically, we find $\frac{\bar{r}}{r_B} = 3.2$ at a temperature $T = 10^6$ K ($v = \sqrt{3kT/m}$).

The same calculation carried out for deuterium gives $\frac{\bar{r}}{r_B} = 3.4$.

As v grows (that is, as T rises) this ratio obviously diminishes. Thus, the projectile proton can reach distances of the order of $\bar{r}$ without ionizing the $\mu^- - p$ system: the screening is effective up to a distance $\bar{r}$.

c)   At a temperature $T \simeq 10^6$ K, we have

$$\frac{kT}{E_0} = \frac{kT}{\frac{1}{2}\mu\alpha^2} \sim 3\cdot10^{-2} \ll 1 \tag{8}$$

(E_0 is the ionization energy of the $\mu^- - p$ system); hence, practically all muons are bound to a proton.

Let us denote by σ the cross section for nucleus-nucleus fusion. If N nuclei are present per cm^3, the number of fusion events per unit time and volume, F_0, is

$$F_0 = \frac{1}{2}N^2\sigma v \tag{9}$$

v is the relative velocity of the two nuclei. Consider now a concentration n/N of $(\mu^- p)$; the fusion cross section in the presence of a μ^-–nucleus system is σ', and the $p - (\mu^- p)$ relative velocity is v'. The fusion rate now equals

$$F = \frac{1}{2}N^2\sigma v + \sigma'(N-n)nv' \simeq \frac{1}{2}N^2\sigma v \left(1 + 2\frac{\sigma'v' - \sigma v}{\sigma v}\frac{n}{N}\right) \tag{10}$$

We have assumed $n/N \ll 1$. The fusion probability will be enhanced if

$$\frac{n}{N}\frac{\sigma'v'}{\sigma v} \gg 1 \tag{11}$$

The fusion cross section σ can be written as $\sigma = \sigma_0 P$, where P is the probability of crossing the Coulomb barrier and σ_0 is the velocity in the absence of a barrier. The factor P is proportional to

$$\frac{1}{v}\exp\left(-2\int_{r_0}^{r^*} p\, dr\right) = \exp\left(-2\sqrt{2\mu_N}\int_{r_0}^{r^*}\sqrt{\frac{\alpha}{r} - \frac{1}{2}\mu v^2}\right) \tag{12}$$

Here, r^* is the classical inversion point $r^* = \frac{2\alpha}{\mu_N v^2}$, r_0 is the nuclear radius, and μ is the reduced mass. Considering, for definiteness, the hydrogen case

$$r^* = r_B \frac{\alpha^2}{v^2} 2 \left(\frac{\mu}{\mu_N}\right) \tag{13}$$

At $T \simeq 10^6$ K, we find $\frac{v}{c} \simeq 6 \cdot 10^{-4}$ and $r^* = 100 r_B$.

We obtain σ' from the relation $\sigma' = \sigma_0 P'$, where P' is the probability of barrier crossing at the point $\bar{r}$ where the system μ-p ionizes; P' is proportional to

$$\frac{1}{v'} \exp\left(-2 \int_{r_0}^{\bar{r}} p \, dr\right) \tag{14}$$

Consequently,

$$\begin{aligned}
\frac{\sigma' v'}{\sigma v} &= \exp 2 \int_{\bar{r}}^{r^*} p \, dr \\
&= \exp \frac{4\alpha}{v} \left[\arccos\left(\sqrt{\frac{\bar{r}}{2r_B}} \frac{v}{\alpha}\right) - \frac{v}{\alpha}\sqrt{\frac{\bar{r}}{2r_B}\left(1 - \frac{\bar{r}}{2r_B}\frac{v}{\alpha}\right)}\right]
\end{aligned} \tag{15}$$

The fact that $v \ll \alpha$ allows us to simplify

$$\frac{\sigma' v'}{\sigma v} \simeq \exp \frac{2\pi\alpha}{v} \tag{16}$$

Thus, catalysis will take place if the muon concentration satisfies

$$\frac{n}{N} \gtrsim \exp^{-2\pi\alpha/v} \tag{17}$$

Appendix A

Relativistic kinematics

A 1.   Definitions and notation

Einstein's special relativity requires that physical laws be invariant under Lorentz transformations and spacetime translations. The set of all these transformations forms a group, called the Poincaré group. Let us denote by $\delta x^\mu = x^\mu - y^\mu$ the difference in the coordinates of two events, where

$$x^\mu = (x^0, x^1, x^2, x^3) \tag{1}$$

($x^0 = t$ in natural units); then, the elements of the Poincaré group are those transformations which leave invariant the quadratic form

$$\delta s^2 = (\delta x^0)^2 - (\delta \mathbf{x})^2 \tag{2}$$

One subgroup is formed by translations, $x^\mu \to x^\mu + a^\mu$, while another one is the set of Lorentz transformations

$$x^\mu \overset{\Lambda}{\to} x'^\mu = \Lambda^\mu{}_\nu x^\nu \tag{3}$$

To define Λ, we first put (2) in the form

$$\delta s^2 = g_{\mu\nu} \delta x^\mu \delta x^\nu \tag{2'}$$

where $g_{\mu\nu}$ is the metric tensor

$$g_{\mu\nu} = \begin{pmatrix} 1 & 0 & 0 & 0 \\ 0 & -1 & 0 & 0 \\ 0 & 0 & -1 & 0 \\ 0 & 0 & 0 & -1 \end{pmatrix} \tag{4}$$

Requiring invariance of δs^2, we write

$$g_{\mu\nu} \delta x^\mu \delta x^\nu = g_{\rho\sigma} \delta x'^\rho \delta x'^\sigma = g_{\rho\sigma} \Lambda^\rho{}_\mu \Lambda^\sigma{}_\nu \delta x^\mu \delta x^\nu \tag{5}$$

Since the above must hold for arbitrary values of δx^μ, we conclude that

$$g_{\mu\nu} = g_{\rho\sigma} \Lambda^\rho_{\ \mu} \Lambda^\sigma_{\ \nu} \tag{6}$$

Transformations of the form (3), with Λ satisfying Eq. (6), are elements of the Lorentz group. An important subgroup of the Lorentz group is formed by those transformations which can be decomposed into a series of infinitesimal transformations (thus excluding, for instance, time reversal); this subgroup is called the connected Lorentz group.

Quantities transforming as x^μ under Lorentz transformations are called contravariant four-vectors.

We define a covariant four-vector x_μ by means of the formula

$$x_\mu = g_{\mu\nu} x^\nu \tag{7}$$

Eq. (6) shows that $g_{\mu\nu}$ is an invariant tensor. There exists another independent tensor, which is invariant under the connected Lorentz group: this is the completely antisymmetric Ricci tensor, $\varepsilon^{\mu\nu\alpha\beta}$, with $\varepsilon^{0123} = 1$.

The energy and momentum of a particle form a four-vector

$$p^\mu = (p^0, \mathbf{p}) \tag{8}$$

$p^2 = p^\mu p^\nu g_{\mu\nu} = p_0^2 - \mathbf{p}^2 = m^2$ is an invariant, equal to the mass squared of the particle.

A four-vector v^μ is called timelike if $v^\mu v_\mu > 0$, spacelike if $v^\mu v_\mu < 0$, or lightlike if $v^\mu v_\mu = 0$. Given a timelike vector, t^μ, the sign of t^0 is invariant under connected (proper) Lorentz transformations; further, it is always possible to find a reference frame in which t^μ has the form

$$t^\mu = (t^0, \mathbf{0}) \tag{9}$$

Such a frame, defined up to an arbitrary spatial rotation, is called a standard frame.

For a spacelike four-vector s^μ we can always find a reference frame (which we shall call standard) where

$$s^\mu = (0, \mathbf{s}) \tag{10}$$

The sign of s^0 can vary from one frame to another.

For a lightlike four-vector c^μ a standard frame is one in which

$$c^\mu = (c, 0, 0, c) \tag{11}$$

Such a frame always exists. As with timelike vectors, the sign of the time component c^0 of c^μ is frame-independent.

A nonzero four-vector will be spacelike if it is orthogonal to some timelike four-vector. Indeed, $t \cdot a = 0$ implies that, in the standard frame of t, we will have $a^0 = 0$, and hence $(a \cdot a) = -\mathbf{a}^2 < 0$.

A four-vector a orthogonal to a spacelike four-vector $(a \cdot s = 0)$ is either timelike (reducing to $a = (a^0, \mathbf{0})$ in the standard frame of s) or spacelike (becoming $(0, \mathbf{a})$ with

$\mathbf{s}\,\mathbf{a} = 0$ in the standard frame of s). a can also be a superposition of the two cases above.

Finally, suppose a is orthogonal to a lightlike vector ($a \cdot c = 0$); going to the standard frame of c, we deduce that a^μ will either be proportional to c^μ, or it will be a spacelike vector with only 1- and 2-components (or it will be a superposition of the above).

A 2.   Mandelstam variables

Consider a process in which the number of initial plus final (spinless) particles is N. The transition amplitude will in general be a function of $3N - 10$ independent variables: 3 variables describe the motion of each particle with fixed mass; of these, a total of 4 variables are fixed by energy and momentum conservation, while 6 variables are fixed by the 6 independent parameters of a Lorentz transformation.

In particular, for $N = 4$, that is for a process of the form

$$a + b \to c + d \tag{12}$$

or

$$a \to b + c + d \tag{13}$$

the number of independent variables is 2. It is customary to use the so-called Mandelstam variables for processes of this type. For the scattering process (12), these variables are defined as follows

$$
\begin{aligned}
s &\equiv (p_a + p_b)^2 = (p_c + p_d)^2 \\
t &\equiv (p_a - p_c)^2 = (p_b - p_d)^2 \\
u &\equiv (p_a - p_d)^2 = (p_b - p_c)^2
\end{aligned}
\tag{14}
$$

s, t and u obey a linear constraint

$$s + t + u = m_a^2 + m_b^2 + m_c^2 + m_d^2 \tag{15}$$

Similarly, for the decay process (13) we have

$$
\begin{aligned}
s &\equiv (p_a - p_b)^2 = (p_c + p_d)^2 \\
t &\equiv (p_a - p_c)^2 = (p_b + p_d)^2 \\
u &\equiv (p_a - p_d)^2 = (p_b + p_c)^2
\end{aligned}
\tag{16}
$$

The constraint (15) is valid here as well.

The same variables describe also the case of particles with spin, once the modulus squared of the amplitude is summed over final spins and averaged over initial spins.

Having chosen 2 out of the 3 Mandelstam variables, for example s and t, the set of all values that s and t may assume in a process is called the physical region for that process. Let us determine this region for the scattering (12). We write

$$s = (p_a + p_b)^2 = (p_c + p_d)^2 \tag{17}$$

leading to

$$s = m_a^2 + m_b^2 + 2(E_a E_b - \mathbf{p}_a \mathbf{p}_b) = m_c^2 + m_d^2 + 2(E_c E_d - \mathbf{p}_c \mathbf{p}_d) \tag{18}$$

s is always positive, since $E_a E_b \geq \mathbf{p}_a \mathbf{p}_b$; it can become zero only if $m_a = m_b = 0$, $m_c = m_d = 0$, and $\mathbf{p}_a \mathbf{p}_b = E_a E_b$, $\mathbf{p}_c \mathbf{p}_d = E_c E_d$, i.e. only in a process which involves two massless particles moving in the same direction. Apart from this case, the total four-momentum $(p_a + p_b)^\mu = (p_c + p_d)^\mu$ is timelike; thus there exists a frame, called the center of mass, in which it takes the form

$$(p_a + p_b)^\mu = (\sqrt{s}, \mathbf{0}) \tag{19}$$

For given particle masses, the minimum value of s is, by (18),

$$s_{\min} = \text{Max}\{(m_a + m_b)^2, (m_c + m_d)^2\} \tag{20}$$

For each value of s, let us find the kinematic limits of the momentum transfer t

$$t = (p_a - p_c)^2 = (p_b - p_d)^2 \tag{21}$$

Going to the center of mass

$$\begin{aligned} \mathbf{p}_a &= -\mathbf{p}_b \\ \mathbf{p}_c &= -\mathbf{p}_d \end{aligned} \tag{22}$$

and using

$$E_a + E_b = \sqrt{\mathbf{p}_a^2 + m_a^2} + \sqrt{\mathbf{p}_b^2 + m_b^2} = \sqrt{s} \tag{23}$$

we find

$$E_a = \frac{s + m_a^2 - m_b^2}{2\sqrt{s}} \tag{24a}$$

$$E_b = \frac{s + m_b^2 - m_a^2}{2\sqrt{s}} \tag{24b}$$

$$\mathbf{p}_a^2 = \mathbf{p}_b^2 = \frac{s^2 + (m_a^2 - m_b^2)^2 - 2s(m_a^2 + m_b^2)}{4s} \tag{24c}$$

Similarly,

$$E_c = \frac{s + m_c^2 - m_d^2}{2\sqrt{s}} \tag{25a}$$

$$E_d = \frac{s + m_d^2 - m_c^2}{2\sqrt{s}} \tag{25b}$$

$$\mathbf{p}_c^2 = \mathbf{p}_d^2 = \frac{s^2 + (m_c^2 - m_d^2)^2 - 2s(m_c^2 + m_d^2)}{4s} \tag{25c}$$

$$t = (p_a - p_c)^2 = m_a^2 + m_c^2 - 2p_a \cdot p_c = m_a^2 + m_c^2 - 2(E_a E_c - \mathbf{p}_a \mathbf{p}_c) \qquad (26)$$

The minimum value of the momentum transfer corresponds to forward scattering[1]
($\cos \theta_{ac} = 1$)

$$t_{\min} = m_a^2 + m_c^2 - 2(E_a E_c - |\mathbf{p}_a||\mathbf{p}_c|) \qquad (27)$$

and the maximum value corresponds to backward scattering

$$t_{\max} = m_a^2 + m_c^2 - 2(E_a E_c + |\mathbf{p}_a||\mathbf{p}_c|) \qquad (28)$$

where E_a, E_b, $|\mathbf{p}_a|$, $|\mathbf{p}_c|$ are as given in (24), (25). If all masses are equal, $m_a = m_b = m_c = m_d = m$, we find

$$t_{\min} = 0 \qquad t_{\max} = 4m^2 - s \qquad (29)$$

For the decay process, Eqs. (16) in the rest frame of a (center of mass) give

$$
\begin{align}
s &= m_a^2 + m_b^2 - 2E_b m_a \tag{30a}\\
t &= m_a^2 + m_c^2 - 2E_c m_a \tag{30b}\\
u &= m_a^2 + m_d^2 - 2E_d m_a \tag{30c}
\end{align}
$$

Summing up s, t and u one arrives immediately at (15), by virtue of $E_b + E_c + E_d = m_a$. The maximum of s corresponds to particle b being at rest ($E_b = m_b$); then, Eq. (30) leads to

$$s_{\max} = (m_a - m_b)^2 \qquad (31)$$

The minimum value is reached when $(p_c + p_d)^2$ is at a minimum; in the center of mass of the $c - d$ pair, the latter reads

$$m_c^2 + m_d^2 + 2(E_c E_d + |\mathbf{p}_c||\mathbf{p}_d|) \qquad (32)$$

The above is minimized at $\mathbf{p}_c = \mathbf{p}_d = 0$, that is,

$$s_{\min} = (m_c + m_d)^2 \qquad (33)$$

Having fixed s (equivalently, E_b) within $s_{\min} \leq s \leq s_{\max}$, the kinematic limits on t are determined as follows: momentum conservation $\mathbf{p}_d = -(\mathbf{p}_c + \mathbf{p}_b)$ gives

$$E_d = \sqrt{m_d^2 + (\mathbf{p}_c + \mathbf{p}_b)^2} \qquad (34)$$

Combined with energy conservation,

$$E_d = m_a - E_c - E_b \qquad (35)$$

E_d can be eliminated with the result

$$m_a - E_c - E_b = \sqrt{m_d^2 + (\mathbf{p}_c + \mathbf{p}_b)^2} \qquad (36)$$

---

[1]Since t is typically negative, $t_{\min}$ stands for the minimum value of $|t|$, and $t_{\max}$ is the maximum value of $|t|$.

Having fixed E_b, Eq. (36) says that E_c, and therefore t, depends only on the angle between $\mathbf{p}_c$ and $\mathbf{p}_b$. Thus, the extremal values of t correspond to $\cos\theta_{bc} = \pm 1$; they are given by the roots of the equation

$$m_a^2 - 2m_a(E_c + E_b) + 2E_cE_b - m_d^2 + m_b^2 + m_c^2 = \pm\sqrt{E_b^2 - m_b^2}\sqrt{E_c^2 - m_c^2} \qquad (37)$$

In terms of s and t,

$$E_c = \frac{m_a^2 + m_c^2 - t}{2m_a} \qquad E_b = \frac{m_a^2 + m_b^2 - s}{2m_a} \qquad (38)$$

At the boundary of the physical region, given in Eq. (37), two of the momenta become collinear; therefore, the triangle formed by $\mathbf{p}_a$, $\mathbf{p}_b$, $\mathbf{p}_c$ has zero area. In the special case $m_b = m_c = m_d = 0$ Eqs.(31) and (37) give

$$\begin{aligned} s_{\min} &= 0 \quad s_{\max} = m_a^2 \\ t_{\max} &= m_a^2 - s \qquad (39) \end{aligned}$$

The physical region in this case is the triangle shown in fig.1.

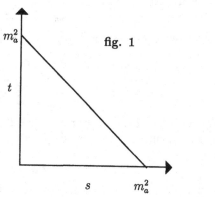

fig. 1

Another particular case of interest is $m_b = m_c = 0$; Eq(37) becomes

$$\begin{aligned} s_{\min} &= 0 \quad s_{\max} = m_a^2 \\ t_{\max} &= m_a^2 - s \qquad (40) \end{aligned}$$

$t_{\min}$ is the curve shown in fig. 2. If the final masses are all equal, $m_b = m_c = m_d = \mu$, we find

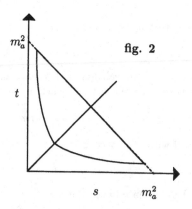

fig. 2

$$t_{\substack{\max \\ \min}} = \frac{1}{2}(M^2 - s + 3\mu^2) \pm \qquad (41)$$

$$\pm\sqrt{\frac{(M^2-\mu^2+s)^2}{4}-\frac{M^2s^2+\mu^2(M^2+\mu^2-s)^2-4\mu^4M^2}{s}}$$

Clearly, the same formula also gives $u_{\max}$ and $u_{\min}$ as a function of s.

When the masses are equal, it is convenient to plot events in triangular coordinates. We set

$$\sigma=\frac{s-m_b^2}{m_a^2}\qquad \tau=\frac{t-m_c^2}{m_a^2}\qquad v=\frac{u-m_d^2}{m_a^2}$$
$$\sigma+\tau+v=1 \tag{42}$$

σ, τ and v are the distances from the three sides of an equilateral triangle, with height 1. A point with Cartesian coordinates (x,y), with respect to the center of the triangle, has triangular coordinates

$$\sigma = y+\frac{1}{3}$$
$$\tau = \frac{\sqrt{3}}{2}x-\frac{y}{2}+\frac{1}{3} \tag{43}$$
$$v = -\frac{\sqrt{3}}{2}x-\frac{y}{2}+\frac{1}{3}$$

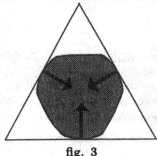

fig. 3

If the masses vanish, then $\sigma_{\min}=0$ and $\sigma_{\max}=1$; at a given value of σ we have $\tau_{\min}=0$ and $\tau_{\max}=1-\sigma$ ($v=0$). The physical region fills up the triangle in this case. More generally, for nonzero masses, the physical region is as shown in fig. 3.

A 3. Phase space

The phase space element of a set of n particles, with total four-momentum P_i^μ, equals

$$d\Phi^{(n)}=(2\pi)^4\delta^{(4)}(P_i-\sum_{k=1}^{n}p_k)\prod_{i=1}^{n}d\Omega_{\mathbf{p}_i} \tag{44}$$

$$d\Omega_{\mathbf{p}}=\frac{d^3\mathbf{p}}{2p^0(2\pi)^3}\qquad p^0=\sqrt{\mathbf{p}^2+m^2} \tag{45}$$

The cross section for the scattering process

$$a+b\to c_1+c_2+\ldots c_n \tag{46}$$

is expressible in terms of $d\Phi^{(n)}$ as follows

$$d\sigma=\frac{1}{2p_a^0 2p_b^0 v_r}|\mathcal{M}_{fi}|^2\,d\Phi^{(n)} \tag{47}$$

with $P_i = p_a + p_b$. $\mathcal{M}_{fi}$ is the matrix element of the scattering operator T between the initial and final state, with normalization

$$\langle \mathbf{p}, r | \mathbf{p}', r' \rangle = \delta_{rr'} 2p^0 (2\pi)^3 \delta^3(\mathbf{p} - \mathbf{p}') \tag{48}$$

v_r is the relative velocity of incoming particles, defined by

$$v_r = \sqrt{(\mathbf{v}_a - \mathbf{v}_b)^2 - (\mathbf{v}_a \wedge \mathbf{v}_b)^2} \tag{49}$$

Decay processes

$$a \to c_1 + c_2 + \ldots c_n \tag{50}$$

also involve the phase space element in the expression for the width

$$d\Gamma = \frac{1}{2p_a^0} |R_{fi}|^2 \, d\Phi^{(n)} \tag{51}$$

R_{fi} is the matrix element of the decay operator, R, between the (covariantly normalized) initial and final states.

Given that both the volume element $d\Omega_{\mathbf{p}_i}$ and the delta function $\delta^{(4)}(P_i - \sum p_k)$ are Lorentz invariant, the same is true of $d\Phi^{(n)}$.

The simplest case of interest is $n = 2$

$$d\Phi^{(2)} = (2\pi)^4 \delta^{(4)}(P_i - p_1 - p_2) \frac{d^3 \mathbf{p}_1}{2p_1^0 (2\pi)^3} \frac{d^3 \mathbf{p}_2}{2p_2^0 (2\pi)^3} \tag{52}$$

Integrating over $\mathbf{p}_2$ we eliminate $\delta^3(\mathbf{P} - \mathbf{p}_1 - \mathbf{p}_2)$ and set $\mathbf{p}_2 = \mathbf{P} - \mathbf{p}_1$. One further integration may be carried out to eliminate $\delta(P_i^0 - p_1^0 - p_2^0)$. Since $d\Phi^{(2)}$ is invariant, we can calculate it in any reference frame. In the center of mass, we have

$$\mathbf{P}_i = 0 \qquad \mathbf{p}_1 = \mathbf{p}_2$$
$$p_1^0 = \sqrt{\mathbf{p}_1^2 + m_1^2} \qquad p_2^0 = \sqrt{\mathbf{p}_1^2 + m_2^2} \qquad d^3 \mathbf{p}_1 = p_1^2 \, dp_1 \, d\Omega_1 \tag{53}$$

We can now integrate over p_1, making use of the relation

$$\frac{\partial}{\partial p_1}(p_1^0 + p_2^0) = \frac{p_1}{p_1^0} + \frac{p_2}{p_2^0} \tag{54}$$

The result is

$$d\Phi^{(2)} = \frac{1}{(2\pi)^2} \frac{p_1 \, d\Omega_1}{4(p_1^0 + p_2^0)} = \frac{d\Omega_1}{32\pi^2} \sqrt{1 + \frac{(m_1^2 - m_2^2)^2}{s^2} - 2\frac{m_1^2 + m_2^2}{s}} \tag{55}$$

In particular,

$$\text{For} \quad m_1 = m_2 = m \qquad d\Phi^{(2)} = \frac{d\Omega_1}{32\pi^2} \sqrt{1 - \frac{4m^2}{s}}$$

$$\text{For} \quad m_1 = m, \quad m_2 = 0 \qquad d\Phi^{(2)} = \frac{d\Omega_1}{32\pi^2} \left(1 - \frac{m^2}{s}\right) \tag{56}$$

Consider now a frame in which $\mathbf{P}_i \neq \mathbf{0}$; here, one may choose as independent variables the energy p_1^0 and the azimuthal angle φ of $\mathbf{p}_1$ relative to $\mathbf{P}_i$. With this choice, one has

$$d\Phi^{(2)} = \frac{1}{16\pi^2} \frac{dp_1^0 \, d\varphi}{|\mathbf{P}_i|} \tag{57}$$

Equivalently, choosing as independent variables the two polar angles of $\mathbf{p}_1$ relative to $\mathbf{P}_i$, one finds

$$d\Phi^{(2)} = \frac{1}{16\pi^2} \frac{p_1 \, d\Omega_1}{P_i^0 - \dfrac{|\mathbf{P}_i| p_1^0 \cos\theta}{p_1}} \tag{58}$$

p_1^0 and $p_1 \equiv |\mathbf{p}_1|$ are determined by energy and momentum conservation

$$p_1^0 = \frac{P_0(s + m_1^2 - m_2^2) + P\cos\theta\sqrt{(s + m_1^2 - m_2^2)^2 - 4m_1^2(s + |\mathbf{P}_i|^2 \sin^2\theta)}}{2(s + |\mathbf{P}_i|^2 \sin^2\theta)}$$

$$p_1 = \frac{p_1^0(s + m_1^2 - m_2^2)\cos\theta + P_0\sqrt{(s + m_1^2 - m_2^2)^2 - 4m_1^2(s + |\mathbf{P}_i|^2 \sin^2\theta)}}{2(s + |\mathbf{P}_i|^2 \sin^2\theta)} \tag{59}$$

Let us also examine 3-body phase space

$$d\Phi^{(3)} = (2\pi)^4 \delta^4(P_i - p_1 - p_2 - p_3) \frac{d^3\mathbf{p}_1}{2p_1^0(2\pi)^3} \frac{d^3\mathbf{p}_2}{2p_2^0(2\pi)^3} \frac{d^3\mathbf{p}_3}{2p_3^0(2\pi)^3} \tag{60}$$

Four variables can be eliminated by integrating out $\delta^4(P_i - p_1 - p_2 - p_3)$, leaving us with 5 variables. Actually, if the particles are spinless, or if one sums/averages over spins, then the matrix element squared is invariant under global rotations in the center of mass; consequently, 3 out the remaining 5 variables do not enter the dynamics, and can be easily integrated out.

Let us then integrate (60) over $d^3\mathbf{p}_3$, over the solid angle of $\mathbf{p}_1$ and over the azimuthal angle of $\mathbf{p}_2$ about $\mathbf{p}_1$; denoting by θ the angle between $\mathbf{p}_1$ and $\mathbf{p}_2$, we obtain

$$d\Phi^{(3)} = \frac{2}{(2\pi)^3} \delta(P_i^0 - p_1^0 - p_2^0 - p_3^0) \frac{p_1^2 \, dp_1 \, p_2^2 \, dp_2 \, d\cos\theta}{2p_1^0 2p_2^0 2p_3^0} \tag{61}$$

We can eliminate δ by integrating over $\cos\theta$: the only term in the argument of δ which depends on $\cos\theta$ is p_3^0

$$p_3^0 = \sqrt{m_3^2 + (\mathbf{p}_1 + \mathbf{p}_2)^2}$$

$$\frac{\partial p_3^0}{\partial\cos\theta} = \frac{p_1 p_2}{p_3^0} \tag{62}$$

Also making use of $p_1 \, dp_1 = p_1^0 \, dp_1^0$ and $p_2 \, dp_2 = p_2^0 \, dp_2^0$ we end up with

$$d\Phi^{(3)} = \frac{dp_1^0 \, dp_2^0}{4(2\pi)^3} \tag{63}$$

In terms of $s = (p_2 + p_3)^2 = (P - p_1)^2$ and $t = (p_1 + p_3)^2 = (P - p_2)^2$ we have

$$dp_1^0 \, dp_2^0 = \frac{ds \, dt}{4P_0^2} \tag{64}$$

and

$$d\Phi^{(3)} = \frac{ds \, dt}{16P_0^2(2\pi)^3} = \frac{ds \, du}{16P_0^2(2\pi)^3} \tag{65}$$

We see that phase space alone corresponds to a uniform event distribution in the physical region of the $s - u$ plane (Dalitz plot).

It is often useful to study the invariant mass distribution of a pair of outgoing particles, for instance when these tend to form a resonance; in doing so, it is convenient to introduce in $d\Phi^{(3)}$ the following decomposition of unity

$$1 = \int \frac{d\mu^2}{2\pi} \int \frac{d^4q}{(2\pi)^4} 2\pi\delta(q^2 - \mu^2)(2\pi)^4\delta^{(4)}(q - p_2 - p_3) \tag{66}$$

Eq. (60) becomes

$$
\begin{aligned}
d\Phi^{(3)} &= \int \frac{d\mu^2}{2\pi}(2\pi)^4\delta^{(4)}(P_i - p_1 - q)\frac{d^3\mathbf{p}_1}{2p_1^0(2\pi)^3}\frac{d^3\mathbf{q}}{2q^0(2\pi)^3} \\
&\quad (2\pi)^4\delta^{(4)}(q - p_2 - p_3)\frac{d^3\mathbf{p}_2}{2p_2^0(2\pi)^3}\frac{d^3\mathbf{p}_3}{2p_3^0(2\pi)^3}
\end{aligned}
\tag{67}
$$

that is,

$$d\Phi^{(3)} = \int \frac{d\mu^2}{2\pi} \, d\Phi^{(2)}(P_i; m_1, \mu; \mathbf{p}, \mathbf{q}) \, d\Phi^{(2)}(\mu; m_2, m_3; \mathbf{p}_2, \mathbf{p}_3) \tag{68}$$

Similarly, for a system of four particles coupled in pairs, we write

$$d\Phi^{(4)} = \int \frac{d\mu^2}{2\pi} \int \frac{d\mu'^2}{2\pi} \, d\Phi^{(2)}(P_i; \mu, \mu'; \mathbf{q}, \mathbf{q}') \tag{69}$$

$$d\Phi^{(2)}(\mu; m_1, m_2; \mathbf{p}_1, \mathbf{p}_2) \, d\Phi^{(2)}(\mu'; m_3, m_4; \mathbf{p}_3, \mathbf{p}_4) \tag{70}$$

More generally, an expression similar to (68) can be used as a recursive formula in order to calculate the complete integral over phase space

$$\Phi^{(n)} = \int \frac{d\mu^2}{2\pi} \, d\Phi^{(2)}(P; m_1, \mu; \mathbf{p}_1, \mathbf{q}) \, d\Phi^{(n-1)}(\mu; m_2 \dots m_n; \mathbf{p}_2 \dots \mathbf{p}_n) \tag{71}$$

Indeed, this equation allows us to calculate $\Phi^{(n)}(P)$ recursively.

Appendix B

The groups $SU(2)$ and $SU(3)$ and their representations

B 1. $SU(2)$

The rotation group coincides with the group $SU(2)$ of 2×2 unitary matrices with determinant 1.

A generic matrix in this group can be written as

$$U = e^{iM} \tag{1}$$

$$M = M^\dagger \qquad \mathrm{Tr}\, M = 0 \tag{2}$$

The most general 2×2 matrix satisfying (2) is

$$M = \frac{\alpha}{2}\sigma \tag{3}$$

α is a generic real vector; the factor of 2 has been introduced for convenience. By σ we denote the Pauli matrices

$$\sigma_1 = \begin{pmatrix} 0 & 1 \\ 1 & 0 \end{pmatrix} \qquad \sigma_2 = \begin{pmatrix} 0 & -i \\ i & 0 \end{pmatrix} \qquad \sigma_3 = \begin{pmatrix} 1 & 0 \\ 0 & -1 \end{pmatrix} \tag{4}$$

The matrices in (1) are entire functions of the parameters α: the group is a Lie group, and its algebra is that of angular momentum

$$S_i = \frac{\sigma_i}{2} \qquad [S_i, S_j] = i\varepsilon_{ijk}S_k \tag{5}$$

The parameters α range over the ball of radius 2π: the surface of the ball corresponds to the transformation $U = -I$. Thus the group is connected and also simply connected; its representations are those of the Lie algebra (5).

The 2-dimensional representation which was used to define the group is called the fundamental representation.

Consider now the group $SO(3)$ of 3×3 orthogonal matrices with unit determinant; this is none other but the group of proper rotations in 3-dimensional space. $SO(3)$ has the same Lie algebra (5) as $SU(2)$: indeed, a generic matrix of $SO(3)$ is real, so that $U^\dagger = U^T$, and orthogonal $U^T = U^{-1}$; thus, $U^\dagger = U^{-1}$, i.e. the matrix is unitary and can be written as

$$U = e^{iM} \tag{6}$$

$$M = M^\dagger \qquad \mathrm{Tr}\, M = 0 \tag{7}$$

In order for U to be real, M must be purely imaginary and antisymmetric, that is,

$$M_{ij} = -i \sum_{\ell=1}^{3} \varepsilon_{i\ell j} \alpha^\ell \tag{8}$$

$\varepsilon_{i\ell j}$ is the Levi-Civita symbol; it is completely antisymmetric, with $\varepsilon_{123} = 1$. Setting

$$\left(\Sigma^\ell\right)_{ij} = -i\varepsilon_{i\ell j} \tag{9}$$

we verify immediately that

$$\left[\Sigma^i, \Sigma^j\right] = i\varepsilon^{ijk}\Sigma^k \tag{10}$$

namely, the algebra of $SO(3)$ coincides with that of $SU(2)$.

A finite $SU(2)$ transformation in the fundamental representation reads

$$U = e^{\frac{i}{2}\boldsymbol{\alpha}\boldsymbol{\sigma}} = \cos\frac{\alpha}{2} + i\, \mathbf{n}\boldsymbol{\sigma}\sin\frac{\alpha}{2} \tag{11}$$

with $\mathbf{n} = \boldsymbol{\alpha}/|\boldsymbol{\alpha}|$.

Similarly, for a transformation (6) of $SO(3)$ we have

$$U_{ij} = \cos\alpha\, \delta_{ij} + \sin\alpha\, \varepsilon_{ikj}n_k + (1 - \cos\alpha)\, n_i n_j \tag{12}$$

The parameters of $SO(3)$ range over a sphere of radius π; diametrically opposite points correspond to the same group element. Thus the group is connected, but not simply connected, and not all representations of the algebra are also representations of the group. Rotations by 0 and 2π are distinct in $SU(2)$, but correspond to the same transformation in $SO(3)$. A representation of $SU(2)$ is also representation of $SO(3)$ only if the two rotations above are associated to the identity element; this is so for representations with integer angular momentum.

B 2.   The representations of $SU(2)$

$SU(2)$ is a compact group; therefore, its representations are completely reducible, equivalent to unitary representations with finite dimension. As is well known, they are labeled by the maximum eigenvalue J of one of the components of $\mathbf{J}$, say J_3. J can be integer or half-integer; the dimensionality of the representation is $2J + 1$.

Eq. (4) specifies the spin-1/2 representation, up to a unitary transformation. We note that $J_\pm = J_1 \pm iJ_2$ are represented by real matrices.

Let us denote by $|J, J_3\rangle$ the basis of the Jth representation. We shall omit, for simplicity, any other quantum numbers which characterize the state of our system.

Under a rotation R

$$|J, J_3\rangle \to U(R)|J, J_3\rangle = \mathcal{D}^J_{J'_3 J_3}(R)|J, J'_3\rangle \tag{13}$$

$\mathcal{D}^J$ is the unitary matrix corresponding to R in the given representation.

A tensor operator $T^J_{J_3}$ is one transforming as

$$U(R)\, T^J_{J_3}\, U^\dagger(R) = \mathcal{D}^J_{J'_3 J_3}(R)\, T^J_{J'_3} \tag{14}$$

As an example, the operator $T^1_{J_3}$, defined by

$$T^1_1 = J_+ \qquad T^1_0 = J_3 \qquad T^1_{-1} = J_- \tag{15}$$

is a tensor operator with $J = 1$.

The matrices $\mathcal{D}^J_{m_1 m_2}(\alpha, \beta, \gamma)$ obey the relationship

$$\int \frac{d\mu}{8\pi^2}\, \mathcal{D}^J_{m_1 m_2}(\alpha, \beta, \gamma)\mathcal{D}^{J'\,*}_{m'_1 m'_2}(\alpha, \beta, \gamma) = \delta_{JJ'}\delta_{m_1 m'_1}\delta_{m_2 m'_2}\frac{1}{2J+1} \tag{16}$$

where $d\mu = d\alpha\, d\gamma\, d\cos\beta$; α, β, γ are the Euler angles specifying the rotation.

Consider a rotationally invariant operator T^0_0; this operator will commute with $\mathbf{J}$, and will satisfy

$$U(R)T^0_0 U^\dagger(R) = T^0_0 \tag{17}$$

Consequently,

$$\begin{aligned}
\langle Jm|T^0_0|J'm'\rangle &= \langle Jm|U^\dagger U T^0_0 U^\dagger U|J'm'\rangle \\
&\quad \langle J\bar{m}|T^0_0|J'\bar{m}'\rangle \mathcal{D}^J_{\bar{m}'m'}(\alpha)\mathcal{D}^{J'\,*}_{\bar{m}m}(\alpha)
\end{aligned} \tag{18}$$

Integrating both sides over $d\mu/(8\pi^2)$ one finds

$$\langle Jm|T^0_0|J'm'\rangle = \delta_{mm'}\frac{\sum_{\bar{m}}\langle J\bar{m}|T^0_0|J'\bar{m}\rangle}{2J+1}\delta_{JJ'} \tag{19}$$

We see that T^0_0 has only diagonal matrix elements, all equal among themselves: it is proportional to the identity matrix of the Jth representation, if $J = J'$, and equals 0 otherwise (Schur's Lemma).

Let us now consider the matrix element

$$\langle Jm|T^k_q|J'm'\rangle \tag{20}$$

The vector $T_q^k|J'm'\rangle$ transforms under rotations as the direct product of the representations k and J'. In general, this product is reducible, and may be written as follows

$$T_q^k|J'm'\rangle = \sum_{\bar{J}} C_{m'\,q\,q+m'}^{J'\,k\,\bar{J}}\,|k, J'; \bar{J}, q + m'\rangle \tag{21}$$

The coefficients C are known as Clebsch-Gordan coefficients. Eq. (21) implies that

$$U(R)T_q^k|J'm'\rangle = \mathcal{D}_{\bar{q}q}^k \mathcal{D}_{\bar{m}m'}^{J'} T_{\bar{q}}^k|J'\bar{m}\rangle = \sum_{\bar{J}} C_{m'\,q\,q+m'}^{J'\,k\,\bar{J}} \mathcal{D}_{z,q+m'}^{\bar{J}}|k, J'; \bar{J}\,z\rangle \tag{22}$$

We may now repeat the argument which led to Eq. (19), with the result

$$
\begin{aligned}
\langle Jm|T_q^k|J'm'\rangle &= \int \frac{\mathrm{d}\mu}{8\pi^2}\langle Jm|U^\dagger\,UT_q^k|J'm'\rangle \\
&= C_{m'\,q\,q+m'}^{J'\,k\,J}\delta_{m,m'+q'}\frac{\displaystyle\sum_z \langle Jz|k, J', Jz\rangle}{2J+1}
\end{aligned}
\tag{23}
$$

The last factor in (23) is independent of m and m'; it is called reduced matrix element of the operator. Eq. (23) is the Wigner - Eckart theorem. In standard notation, we rewrite it as follows

$$\langle Jm|T_q^k|J'm'\rangle = C_{m'qm}^{J'kJ}\langle J\|T^k\|J'\rangle \tag{24}$$

Matrix elements of a tensor operator (in multiplets of given J) are thus proportional among themselves and proportional to the Clebsch - Gordan coefficients.

The standard definition of the coefficients C is given in the relation

$$|jm\rangle|j'm'\rangle = \sum_J C_{mm'm+m'}^{j\,j'\,J}|jj'; Jm + m'\rangle \tag{25}$$

In the above, states are normalized to 1. C's are the elements of the unitary matrix connecting two different representations: one in which j, j', m, m' are diagonal and one in which j, j', J, M are diagonal.

The modulus of the C's is fixed by Eq. (25), whereas their phase depends on the phase choices in the two representations. One may always choose phases in such a way that J^+ (and J^-) correspond to real matrices; this is always possible, thus leaving at most a global phase in each representation. Finally, such global phases may be chosen so that all Clebsch - Gordan coefficients become real.

The following symmetry property holds

$$C_{m_2 m_1 M}^{j_2\,j_1\,J} = (-1)^{J-j_1-j_2} C_{m_1 m_2 M}^{j_1\,j_2\,J} \tag{26a}$$

$$C_{-m_1 -m_2 -M}^{j_1\,j_2\,J} = (-1)^{J+j_1+j_2-2M} C_{m_1 m_2 M}^{j_1\,j_2\,J} \tag{26b}$$

Further, completeness reads

$$\sum_{m_1 m_2} C^{j_2\ j_1\ J'}_{m_2 m_1 M'}\, C^{j_2\ j_1\ J}_{m_2 m_1 M} = \delta_{JJ'}\delta_{MM'} \tag{27a}$$

$$\sum_{JM} C^{j_2\ j_1\ J}_{m_2 m_1 M}\, C^{j_2\ j_1\ J}_{m'_2 m'_1 M} = \delta_{m_1 m'_1}\delta_{m_2 m'_2} \tag{27b}$$

Particular values of C for $J = 0, 1/2, 1$ are

$$C^{j\ j'\ 0}_{m m' 0} = (-1)^{j-m}\frac{1}{\sqrt{2j+1}}\delta_{jj'}\delta_{m-m'}$$

$$C^{j+\frac{1}{2}\ j\ \frac{1}{2}}_{m\ -m-\frac{1}{2}-\frac{1}{2}} = (-1)^{j-m-\frac{1}{2}}\sqrt{\frac{2j-2m+1}{(2j+1)(2j+2)}}$$

$$C^{j\ j\ 1}_{m m 0} = \frac{2m\sqrt{3}}{(2j(2j+1)(2j+2))^{1/2}}$$

$$C^{j\ j\ 1}_{m m-1} = -\left[\frac{6(j-m)(j+m+1)}{2j(2j+1)(2j+2)}\right]^{1/2}$$

$$C^{j+1\ j\ 1}_{m\ m0} = \left[\frac{6(j+m+1)(j-m+1)}{(2j+1)(2j+2)(2j+3)}\right]^{1/2}$$

$$C^{j+1\ j\ 1}_{m\ m-1} = -\left[\frac{3(j-m)(j-m+1)}{(2j+1)(2j+2)(2j+3)}\right]^{1/2} \tag{28}$$

Coefficients not listed explicitly may be obtained by use of Eq. (26).

The choice of phases made above implies that the spherical harmonics Y_ℓ^m obey the relation

$$Y_\ell^{m*}(\hat{n}) = (-1)^{\ell-m}Y_\ell^{-m}(\hat{n}) \tag{29}$$

Up to $\ell = 2$, the functions $Y_\ell^m(\hat{n})$ are given by

$$Y_0^0 = \sqrt{\frac{1}{4\pi}}$$

$$Y_1^0 = i\sqrt{\frac{3}{4\pi}}\cos\theta \qquad Y_1^1 = -i\sqrt{\frac{3}{8\pi}}\sin\theta e^{i\varphi} \qquad Y_1^{-1} = i\sqrt{\frac{3}{8\pi}}\sin\theta e^{-i\varphi}$$

$$Y_2^2 = -\sqrt{\frac{15}{32\pi}}\sin^2\theta e^{2i\varphi} \qquad Y_2^1 = \sqrt{\frac{15}{8\pi}}\sin\theta\cos\theta e^{i\varphi}$$

$$Y_2^0 = -\sqrt{\frac{5}{16\pi}}(3\cos^2\theta - 1)$$

$$Y_2^{-1} = -\sqrt{\frac{15}{8\pi}}\sin\theta\cos\theta e^{-i\varphi} \qquad Y_2^{-2} = -\sqrt{\frac{15}{32\pi}}\sin^2\theta e^{-2i\varphi} \tag{30}$$

In terms of the Euler angles α, β, γ, one has $\mathcal{D}(\alpha,\beta,\gamma) = e^{im'\gamma}\, d^{(j)}_{m'm}(\beta)\, e^{im\alpha}$, and

$$\text{for } j = \frac{1}{2} \quad d^{(1/2)}_{m'm} = \begin{cases} \begin{array}{c|cc} {}_{m'}\!\!\diagdown^{\,m} & \frac{1}{2} & -\frac{1}{2} \\ \hline \frac{1}{2} & \cos\frac{\beta}{2} & \sin\frac{\beta}{2} \\ -\frac{1}{2} & -\sin\frac{\beta}{2} & \cos\frac{\beta}{2} \end{array} \end{cases}$$

$$\text{for } j = 1 \quad d^{(1)}_{m'm} = \begin{cases} \begin{array}{c|ccc} {}_{m'}\!\!\diagdown^{\,m} & 1 & 0 & -1 \\ \hline 1 & \frac{1}{2}(1+\cos\beta) & \frac{1}{\sqrt{2}}\sin\beta & \frac{1}{2}(1-\cos\beta) \\ 0 & -\frac{1}{\sqrt{2}}\sin\beta & \cos\beta & \frac{1}{\sqrt{2}}\sin\beta \\ -1 & \frac{1}{2}(1-\cos\beta) & -\frac{1}{\sqrt{2}}\sin\beta & \frac{1}{2}(1+\cos\beta) \end{array} \end{cases}$$

B 3. $SU(3)$

$SU(3)$ is the group of 3×3 unitary matrices with unit determinant.

$$U = e^{iM} \qquad M = M^\dagger \qquad \mathrm{Tr}\, M = 0 \tag{31}$$

We construct the most general matrix M in terms of the matrices A^i_k

$$\left(A^i_k\right)_{m,n} = \delta^i_m \delta^k_n \qquad A^k_i = \left(A^i_k\right)^\dagger \tag{32}$$

obeying the algebra

$$\left[A^i_k, A^l_m\right] = \delta^i_m A^k_l - \delta^l_k A^i_m \tag{33}$$

We now define

$$\begin{aligned} \lambda_1 &= A^1_2 + A^2_1 & \lambda_2 &= i(A^1_2 - A^2_1) \\ \lambda_3 &= A^1_1 - A^2_2 & \lambda_4 &= A^1_3 + A^3_1 & \lambda_5 &= i(A^1_3 - A^3_1) \\ \lambda_6 &= A^2_3 + A^3_2 & \lambda_7 &= i(A^2_3 - A^3_2) & \lambda_8 &= \frac{1}{\sqrt{3}}(A^1_1 + A^2_2 - 2A^3_3) \end{aligned} \tag{34}$$

The most general form for M, compatible with (31), is

$$M = \sum_{i=1}^{8} \alpha_i \frac{\lambda_i}{2} \tag{35}$$

α_i is a set of real coefficients. The commutation rules among λ matrices read

$$[\lambda_a, \lambda_b] = 2i\, f^{abc} \lambda_c \tag{36}$$

f^{abc} is a completely antisymmetric tensor; its elements are called the structure constants of the group. The nonzero elements of f^{abc} are easily found to be (cf. (33), (34))

$$
\begin{array}{lll}
f^{123} = 1 & f^{147} = \tfrac{1}{2} & f^{156} = -\tfrac{1}{2} \\[2mm]
f^{246} = \tfrac{1}{2} & f^{257} = \tfrac{1}{2} & f^{345} = \tfrac{1}{2} \\[2mm]
f^{367} = -\tfrac{1}{2} & f^{458} = \tfrac{\sqrt{3}}{2} & f^{678} = \tfrac{\sqrt{3}}{2}
\end{array}
\tag{37}
$$

A further property of the λ's is

$$
U^\dagger(\boldsymbol{\alpha})\lambda_i U(\boldsymbol{\alpha}) = \mathcal{U}(\boldsymbol{\alpha})_{ij}\lambda_j
\tag{38}
$$

This property follows from

$$
e^{iM}\lambda_i e^{-iM} = \lambda_i + i\,[M, \lambda_i] + \frac{i^2}{2}\,[M, [M, \lambda_i]] + \cdots
\tag{39}
$$

and the fact that $[M, \lambda_i]$ is a linear combination of the λ's themselves, by virtue of (35), (36). The group multiplication law, $\alpha \cdot \beta = \gamma$, leads to

$$
U^\dagger(\beta)U^\dagger(\alpha)\lambda_i U(\alpha)U(\beta) = U^\dagger(\gamma)\lambda_i U(\gamma)
\tag{40}
$$

and, by (38),

$$
\mathcal{U}(\alpha)_{ij}\mathcal{U}(\beta)_{jk} = \mathcal{U}_{ik}(\gamma)
\tag{41}
$$

Therefore, the matrices $\mathcal{U}$ constitute a representation, called the adjoint representation, whose basis is the Lie algebra of the group. The defining representation of the group, Eq. (31), is called the fundamental representation. Using (39), (36), we may write an infinitesimal transformation as

$$
\lambda_i \rightarrow \lambda_i - i\,\alpha_k f_{ikc}\lambda_c
\tag{42}
$$

Consequently, the generators in the adjoint representation are

$$
H^i_{ab} = -i f_{aib}
\tag{43}
$$

Three commonly encountered $SU(2)$ subgroups of $SU(3)$ are generated by $\{\lambda_1, \lambda_2, \lambda_3\}$ (isospin), $\{\lambda_4, \lambda_5, \lambda_3 + \frac{\lambda_8}{\sqrt{3}}\}$ (V - spin), and $\{\lambda_6, \lambda_7, \lambda_3 - \frac{\lambda_8}{\sqrt{3}}\}$ (U - spin).

In the $SU(3)$ flavour group of Gell Mann, $\frac{\lambda_3}{2}$ is the third component of isotopic spin and $\frac{\lambda_8}{\sqrt{3}}$ is the hypercharge $Y = N + S$. The electric charge is given by

$$
Q = T_3 + Y
\tag{44}
$$

In the fundamental representation, λ_3 and λ_8 are diagonal and the three states have charge $2/3, -1/3, -1/3$. On the Cartesian plane with axes Y, T_3, the states of this representation correspond to the vertices of an equilateral triangle; they are denoted by u, d, s.

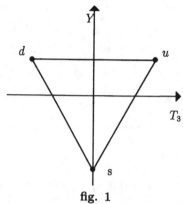

fig. 1

A further representation of the group is defined by the complex conjugates U^* of the matrices U appearing in (31), since complex conjugation preserves the group multiplication law. The corresponding generators are

$$\frac{\tilde{\lambda}_i}{2} = -\frac{\lambda_i^*}{2} \tag{45}$$

In particular, $\lambda_1, \lambda_3, \lambda_4, \lambda_6, \lambda_8$ change sign, while $\lambda_2, \lambda_5, \lambda_7$ remain the same. The eigenvalue spectrum of λ_8, $(\frac{1}{3}, -\frac{2}{3})$ differs from that of $-\lambda_8$. Thus, no unitary transformation can connect λ_8 to $-\lambda_8$, and we conclude that the two representations (U and U^*) are inequivalent. The states of this new representation are depicted in fig. 2.

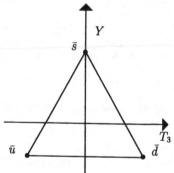

fig. 2

The states of representation (31) are called quarks; those of the complex conjugate representation are called antiquarks.

Given a triplet $\psi(x)$ of fields

$$\psi(x) = \begin{pmatrix} \psi_1(x) \\ \psi_2(x) \\ \psi_3(x) \end{pmatrix} \tag{46}$$

transforming according to the fundamental representation

$$\psi(x) \to U\psi(x) \tag{47}$$

the quantity $\bar{\psi}\psi$ is an invariant

$$\bar{\psi}\psi \to \bar{\psi}U^\dagger U\psi = \bar{\psi}\psi \tag{48}$$

Another important bilinear is

$$\bar{\psi}\lambda_i\psi \to \bar{\psi}U^\dagger \lambda_i U\psi \tag{49}$$

By Eq. (38),

$$\bar{\psi}U^\dagger \lambda_i U\psi = \mathcal{U}_{ij}\bar{\psi}\lambda_j\psi \tag{50}$$

so that this bilinear transforms as the 8-dimensional adjoint representation. A tensor quantity $T^{\alpha_1\alpha_2\ldots\alpha_n}_{\beta_1\beta_2\ldots\beta_n}$, transforming as the direct product of vectors $v^{\alpha_1}\ldots v^{\alpha_n}$ in the quark representation and $u_{\beta_1}\ldots u_{\beta_n}$ in the antiquark representation, forms a basis for an $SU(3)$ representation.

The Kronecker tensor is invariant

$$\delta^\alpha_\beta \to U^\alpha{}_{\alpha'}U^*{}^{\beta'}{}_\beta\delta^{\alpha'}_{\beta'} = \delta^\alpha_\beta \tag{51}$$

since U is unitary. Similarly, $\varepsilon^{\alpha_1\alpha_2\alpha_3}$ and $\varepsilon_{\beta_1\beta_2\beta_3}$ are invariant, because U and U^* have unit determinant.

It can be shown that irreducible representations are in a one-to-one correspondence with the tensors

$$T^{\alpha_1\alpha_2\ldots\alpha_m}_{\beta_1\beta_2\ldots\beta_n} \tag{52}$$

symmetric in the indices α and in the indices β, and traceless in any pair α_i, β_j. For any tensor V^α_β we can write

$$V^\alpha_\beta = \left(V^\alpha_\beta - \frac{1}{3}\delta^\alpha_\beta V^\gamma_\gamma\right) + \frac{1}{3}\delta^\alpha_\beta V^\gamma_\gamma \tag{53}$$

δ^α_β is invariant and corresponds to the trivial (singlet) representation; $V^\alpha_\beta - \frac{1}{3}\delta^\alpha_\beta V^\gamma_\gamma$ has 8 independent components.

The center of the group is made up of all 3×3 matrices proportional to the unit matrix; in order to have determinant 1 they must coincide with the cubic roots of unity, thus forming the discrete group Z_3. The quotient group $SU(3)/Z_3$ has the same algebra as $SU(3)$; however, representations of $SU(3)/Z_3$ are only those representations of $SU(3)$ which associate the group center to the identity. This will be the case whenever the number of α indices in (52) equals the number of β indices modulo 3. The singlet and the octet are indeed representations of $SU(3)/Z_3$, as is the dimension-10 representation $T^{\alpha_1\alpha_2\alpha_3}$ (completely symmetric in its 3 indices).

It is easy to verify that, in the $Y\,S$ plane, the octet representation has the form

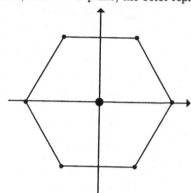

fig. 3

while the decuplet representation is given in fig.4. Representations of $SU(3)/Z_3$ have integer charge: indeed, this is true of any $q\bar{q}$ pair, as well as of any quark or antiquark triplet. Besides their $SU(3)$ flavour index, physical quarks also form a basis for a representation of the $SU(3)$ colour group. Since colour is confined, the only observable physical states are colour singlets, that is, $\bar{q}q$ pairs (δ_β^α), or triplets of quarks or antiquarks ($\varepsilon^{\alpha\beta\gamma}$, $\varepsilon_{\alpha\beta\gamma}$). Thus, colour confinement explains why only integer charges are observed in nature, and in particular only $SU(3)/Z_3$ flavour representations for particles made of light quarks.

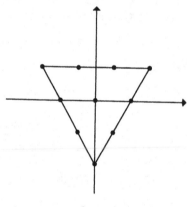

fig. 4

As an exercise, let us perform a rotation by an angle θ about the 7-axis on the isospin triplet

$$e^{i\theta\frac{\lambda_7}{2}}(\lambda_1 \pm i\lambda_2)e^{-i\theta\frac{\lambda_7}{2}} \tag{54}$$

Making use of the structure constants (37) we obtain

$$\cos\theta(\lambda_1 \pm i\lambda_2) + \sin\theta(\lambda_4 \pm i\lambda_5) \tag{55}$$

The above transformation is called a Cabibbo rotation.

Appendix C

Scattering and decays

C 1. The decay operator

The decay amplitude connecting an initial state $|i\rangle$ to a final state $|f\rangle$ is

$$\mathcal{M}_{fi} = \langle f|R(E)|i\rangle \tag{1}$$

where

$$R(E) = H_I + H_I Q \frac{1}{E - H + i\epsilon} Q H_I \tag{2}$$

$$H = H_0 + H_I$$

H_I is the interaction Hamiltonian inducing the decay. Q is a projector onto the final states in the Hilbert space. Eq. (1) may also be written as

$$\mathcal{M}_{fi} = \langle f_-|H_I|i\rangle \tag{3}$$

where $|f_-\rangle$ is the eigenstate of the final state Hamiltonian QHQ, whose asymptotic limit in the remote future is $|f\rangle$. The matrix $H_{ii'}^{\text{eff}}$ which describes the effective evolution of decaying states $\{|i\rangle, |i'\rangle, \dots\}$ is

$$\langle i|H_0 + R(E)|i'\rangle = M_{ii'} - i\Gamma_{ii'}$$

$$
\begin{aligned}
\Gamma_{ii'} &= \langle i|H_I Q\, 2\pi\delta(E - H)\, H_I|i'\rangle \\
M_{ii'} &= \langle i|H_0|i'\rangle + \Delta M_{ii'} \\
\Delta M_{ii'} &= \langle i|H_I Q \frac{1}{E - H} Q H_I|i'\rangle
\end{aligned} \tag{4}
$$

$H_{ii'}^{\text{eff}}$ is also called mass matrix. Its eigenstates have definite width Γ and mass M.

C 2.   The S matrix

The probability amplitude $\mathcal{A}_{fi}$ for a transition between free particle states is given by

$$\mathcal{A}_{fi} = \langle f|S|i\rangle \qquad (5)$$

The corresponding transition probability is

$$P_{fi} = \frac{|\langle f|S|i\rangle|^2}{\langle f|f\rangle\,\langle i|i\rangle} \qquad (6)$$

The operator S, also called S-matrix, is unitary

$$SS^\dagger = S^\dagger S = 1 \qquad (7)$$

If H_0 is the free Hamiltonian, describing incoming and outgoing particles, then

$$[S, H_0] = 0 \qquad (8)$$

Energy is thus conserved in a scattering process. Furthermore, S has all the symmetries of the Hamiltonian: it commutes with angular momentum if the system is rotationally invariant, it commutes with parity if the latter is not violated, and it is a Lorentz scalar. If time reversal is also a symmetry, then

$$U(T)SU^\dagger(T) = S^\dagger \qquad (9)$$

We set

$$H = H_0 + H_I \qquad (10)$$

If state $|i\rangle$ has a given energy E, we may write

$$\begin{aligned} S|i\rangle &= S(E)|i\rangle \\ S(E) &= 1 - (2\pi)^4 \mathrm{i}\,\delta(E - H_0)\,\mathcal{M}(E) \end{aligned} \qquad (11)$$

$$\mathcal{M}(E) = H_I + H_I \frac{1}{E - H + \mathrm{i}\varepsilon} H_I \qquad (12)$$

We now define $\mathcal{M}_{fi}$ by the relation

$$\langle f|\mathcal{M}(E)|i\rangle = \mathcal{M}_{fi}(2\pi)^3\delta^3(\mathbf{P}_i - \mathbf{P}_f) \qquad (13)$$

In terms of $\mathcal{M}_{fi}$, the cross section for a generic process $a + b \to f$ reads

$$\mathrm{d}\sigma = \frac{|\mathcal{M}_{fi}|^2}{2p_a^0\,2p_b^0\,v_R}\,\mathrm{d}\Phi^{(n)} \qquad (14)$$

Here, we have adopted an invariant normalization for single particle states

$$\langle \mathbf{p}_a, r|\mathbf{p}_a', s\rangle = \delta_{rs}\,(2\pi)^3 2p_a^0\,\delta^3(\mathbf{p}_a - \mathbf{p}_a') \qquad (15)$$

$d\Phi^{(n)}$ is the invariant phase space element, (44), and

$$v_R = \sqrt{(\mathbf{v}_a - \mathbf{v}_b)^2 - (\mathbf{v}_a \wedge \mathbf{v}_b)^2} \qquad (16)$$

is the relative velocity of the two particles. In the laboratory frame, where particle a is at rest, $v_R = |\mathbf{v}_b|$.

Obviously, the expression for the cross section must be independent of the normalization chosen for the states. The choice (15) is convenient because $d\Phi^{(n)}$, $|\mathcal{M}_{fi}|^2$ and the flux factors are all separately invariant.

Let us recall some essential consequences of unitarity, (7):

1) The optical theorem

$$2\,\mathrm{Im}\,\frac{\langle\alpha|\mathcal{M}|\alpha\rangle}{\langle\alpha|\alpha\rangle} = v_R\,\sigma_{Tot} \qquad (17)$$

equivalently

$$\mathrm{Im}\,f(0) = \frac{k}{4\pi}\sigma_{Tot} \qquad (18)$$

Here, k is the momentum of one of the particles in the center of mass and $f(0)$ is the forward elastic amplitude.

2) The S-matrix may be parameterized in terms of phases, in a basis in which it is diagonal. In particular, the amplitude for elastic scattering of spinless particles takes the form

$$f(\theta) = \sum_\ell (2\ell + 1)\frac{1}{2i\,k}\left(e^{2i\delta_\ell} - 1\right)P_\ell(\cos\theta) \qquad (19)$$

Near an elastic resonance with angular momentum ℓ, we have

$$e^{2i\delta_\ell} = \frac{E - E_R + i\Gamma/2}{E - E_R - i\Gamma/2}$$

$$e^{2i\delta_\ell} - 1 = \frac{i\,\Gamma}{E - E_R - i\Gamma/2}$$

and we can split the amplitude into a resonant part f_R plus a non resonant background produced by angular momenta different from ℓ

$$f_R(\theta) = \frac{1}{2i\,k}(2\ell + 1)P_\ell(\cos\theta)\frac{i\,\Gamma}{E - E_R - i\Gamma/2} \qquad (20)$$

C 3.   Scattering in a potential with spherical symmetry

In a spherically symmetric potential $V(r)$ the Schrödinger equation is separable in polar coordinates, $\psi(\mathbf{r}) = \sum Y_l{}^m(\hat{r})R_l(r)$ (l is the angular momentum and m its projection on the z-axis). Setting $\chi_l = rR_l(r)$, one has the following equation for χ

$$\chi_l''(r) + [k^2 - \frac{l(l+1)}{r^2} - U(r)]\chi_l(r) = 0 \qquad (21)$$

where $U(r) = V(r)\,2m/\hbar^2$, $k = p/\hbar$.

The physical solution of Eq. (21) is regular at the origin, defined by the condition $\chi_l(r{=}0) = 0$. Assuming that $V(r)$ is less singular than $1/r^2$ as $r \to 0$, such a solution exists and is unique. If $r^3V(r) \xrightarrow{r\to\infty} 0$, then at large distances the equation may be approximated by the free equation ($U(r) = 0$). Therefore, the solution will be a superposition of an outgoing wave e^{ikr} and an incoming wave e^{-ikr}; for convenience, we write it in the form

$$\chi_l(kr) \underset{r\to\infty}{\simeq} \frac{1}{2(ik)^{l+1}}\left[e^{ikr}\,\phi_l^+(k) - (-)^l\,e^{-ikr}\,\phi_l^-(-k)\right] \qquad (22)$$

In the free case, $\phi_l^+(k) = \phi_l^-(k) = 1$: indeed,

$$\chi_l^0(kr) = \frac{r\,j_l(kr)}{k^l} \underset{r\to\infty}{\simeq} \frac{1}{2(ik)^{l+1}}\left[e^{ikr} - (-)^l\,e^{-ikr}\right] = \frac{1}{k^{l+1}}\,\sin(kr - l\pi/2) \qquad (23)$$

$\phi_l^\pm(k)$ is called a Jost function. In the presence of an interaction, the asymptotic behaviour of χ is

$$\chi_l(kr) \underset{r\to\infty}{\simeq} (e^{ikr}\,e^{2i\delta_l} - (-)^l e^{-ikr}) \simeq \sin(kr - l\pi/2 + \delta_l) \qquad (24)$$

with $e^{2i\delta_l} = \phi_l^+(k)/\phi_l^-(-k)$. In terms of the Jost functions the S-matrix is completely specified. If the potential has the properties $r^2V(r)\xrightarrow{r\to0} 0$, $r^3V(r)\xrightarrow{r\to\infty} 0$, then the function $\phi_l^-(k)$ is analytic in the lower half of the complex k-plane, $\mathrm{Im}\,k \le 0$. All zeroes in this half-plane lie on the imaginary axis and correspond to bound states. For $\mathrm{Im}\,k \ge 0$, zeroes appear in symmetric pairs with respect to the imaginary axis; they correspond to resonances.

C 4.   The Goldberger theorem

This theorem applies to scattering in the presence of two interactions, U and V.

Let us denote by $|\psi_\pm\rangle$ the eigenstates of the Hamiltonian $H = H_0 + U + V$, whose asymptotic *in* and *out* conditions are described by the free state $|\psi\rangle$; then, by definition,

$$T_{fi} = \langle f_-|i_+\rangle = \langle f_-|(U+V)|i_+\rangle \qquad (25)$$

We now indicate by $|\phi_\pm\rangle$ the eigenstates of the Hamiltonian $H_0 + U$, corresponding to the presence of the single interaction U; the asymptotic conditions are described by the free state $|\phi\rangle$. The theorem states

$$\langle f_- | i_+ \rangle = \langle f_- | U | i_+ \rangle + \langle f_- | V | i_+ \rangle \tag{26}$$

The first term is the amplitude when only U is present. The second term would be the amplitude due exclusively to V, if U were negligible.

Let us take U to be the Coulomb interaction and V to be a short-range interaction. Coulomb scattering states behave at short distances like free waves, apart from an energy-dependent factor which tends to 1 at high energies. We write

$$(\mathbf{r} | k_-) \simeq e^{i \mathbf{k} \mathbf{r}} \, F(k) \tag{27}$$

$$F(k) = e^{-\pi/2 k r_0} \, \Gamma\left(1 - \frac{i}{k r_0}\right) \qquad \frac{1}{r_0} = Z_1 Z_2 \, \alpha m$$

where Z_1, Z_2 are the electric charges (in units of e) of the two particles in collision, $\alpha = 1/137$, and m is the reduced mass. There follows

$$|F(k)|^2 = \frac{2\pi}{k r_0 \left(e^{2\pi/k r_0} - 1\right)} \xrightarrow[k \gg 2\pi/r_0]{} 1 \tag{28}$$

Thus if U is the Coulomb potential, then, at sufficiently high energies, the complete amplitude is the sum of the amplitudes one would obtain in the presence of each interaction separately.

Appendix D

Relativistic equations

D 1. Spin 0 particles

The states of a spinless relativistic particle with mass m are positive energy solutions of the equation

$$\left(\Box + m^2\right)\varphi(x) = 0 \tag{1}$$

The most general solution of (1) has the form

$$\varphi(x) = \int \frac{\mathrm{d}^4 p}{(2\pi)^4}\, \mathrm{e}^{-ipx}\, \tilde{\varphi}(p)\, 2\pi\delta(p^2 - m^2) \tag{2}$$

where $\tilde{\varphi}(p)$ is an arbitrary function of p. Eq. (2) can also be written as follows

$$\varphi(x) = \int \mathrm{d}\Omega_{\mathbf{p}} \left\{ \mathrm{e}^{-ipx}\varphi_+(\mathbf{p}) + \mathrm{e}^{ipx}\varphi_-(\mathbf{p}) \right\} \tag{3}$$

The integration measure

$$\mathrm{d}\Omega_{\mathbf{p}} = \frac{\mathrm{d}^3 p}{(2\pi)^3 2p^0} \tag{4}$$

is Lorentz invariant. We have set

$$p^0 = \sqrt{\mathbf{p}^2 + m^2} \qquad \varphi_+(\mathbf{p}) = \tilde{\varphi}(p^0, \mathbf{p}) \qquad \varphi_-(\mathbf{p}) = \tilde{\varphi}(-p^0, -\mathbf{p}) \tag{5}$$

Positive energy solutions have $\varphi_-(\mathbf{p}) = 0$, and arbitrary $\varphi_+(\mathbf{p})$. The scalar product between two states $\varphi_a(x)$ and $\varphi_b(x)$ is

$$\langle b|a\rangle = \int \mathrm{d}^3 x\, \varphi_b^*(x^0, \mathbf{x}) i \overset{\leftrightarrow}{\partial}_0 \varphi_a(x^0, \mathbf{x}) = \int \mathrm{d}\Omega_{\mathbf{p}}\, \varphi_b^*(\mathbf{p})\varphi_a(\mathbf{p}) \tag{6}$$

This product is independent of x^0 and it is Lorentz invariant, as a consequence of the transformation property

$$\varphi(x) \overset{\Lambda,a}{\to} \varphi'(x) = \varphi(\Lambda^{-1}(x - a)) \tag{7}$$

384

For a state with given momentum $|\mathbf{p}\rangle$, the corresponding wave function is e^{-ipx}. Using the invariant scalar product (6), we have

$$\langle \mathbf{p}'|\mathbf{p}\rangle = (2\pi)^3 \, 2p^0 \, \delta^3(\mathbf{p} - \mathbf{p}') \tag{8}$$

The generators of the Poincaré group, given by

$$\begin{aligned}
P_\mu &= \mathrm{i}\frac{\partial}{\partial x^\mu} \\
M^{\mu\nu} &= \mathrm{i}\left(x^\mu \partial^\nu - x^\nu \partial^\mu\right)
\end{aligned} \tag{9}$$

in the x representation, act on the functions $\varphi_+(\mathbf{p})$ as follows

$$\begin{aligned}
P_\mu &= p_\mu \\
\mathbf{J} &= -\mathrm{i}\,\mathbf{p} \wedge \frac{\partial}{\partial \mathbf{p}} \\
\mathbf{K} &= \mathrm{i} p^0 \frac{\partial}{\partial \mathbf{p}}
\end{aligned} \tag{10}$$

D 2.   Spin 1/2 particles

Spin 1/2 particles obey the Dirac equation

$$(\mathrm{i}\slashed{\partial} - m)\,\psi = 0 \tag{11}$$

$\psi(x)$ has four components and

$$\slashed{\partial} = \partial_\mu \gamma^\mu \qquad \partial_\mu = \frac{\partial}{\partial x^\mu} \tag{12}$$

The matrices γ^μ obey the algebra

$$\{\gamma^\mu, \gamma^\nu\} = 2g^{\mu\nu} \tag{13}$$

The action of a Lorentz transformation $\Lambda = \exp(\mathrm{i}\,\omega_{\mu\nu} M^{\mu\nu}/2)$ is

$$\psi(x) \rightarrow S(\Lambda)\psi((\Lambda^{-1}x - a) \tag{14}$$

where S is a 4-dimensional representation of the Lorentz group. For Eq. (11) to be invariant we must have

$$S(\Lambda)\gamma^\mu S^{-1}(\Lambda) = \Lambda^\mu{}_\nu \gamma^\nu \tag{15}$$

The above is satisfied by setting

$$S(\Lambda) = e^{\frac{\mathrm{i}}{4}\omega^{\mu\nu}\sigma_{\mu\nu}} \tag{16}$$

where $\sigma_{\mu\nu} = \frac{1}{2i}[\gamma_\mu, \gamma_\nu]$.

Representations of the γ matrices which differ by a unitary transformation U

$$\psi \to U\psi \qquad \gamma_\mu \to \gamma'_\mu = U\gamma_\mu U^\dagger \tag{17}$$

are equivalent to all effects.

The most general solution of (11) reads

$$\psi(x) = \int d\Omega_{\mathbf{p}} \sum_r \left\{ e^{-ipx} u(r, \mathbf{p})\varphi^+(r, \mathbf{p}) + e^{ipx} v(r, \mathbf{p})\varphi^-(r, \mathbf{p}) \right\} \tag{18}$$

Here, $u(r, \mathbf{p})$ are two independent vectors obeying the equation

$$(\not{p} - m)\, u(r, \mathbf{p}) = 0 \qquad u^\dagger(r, \mathbf{p})u(r', \mathbf{p}) = C\delta_{rr'} \tag{19}$$

We choose the normalization $C = 2p^0$. Similarly,

$$(\not{p} + m)\, v(r, \mathbf{p}) = 0 \tag{20}$$

Particle states are precisely the positive frequency solutions ($\varphi^-(r, \mathbf{p}) = 0$).

The scalar product between two states a and b is

$$\langle b|a \rangle = \int d^3x\, \psi_b^\dagger(x^0, \mathbf{x})\psi_a(x^0, \mathbf{x}) = \int d\Omega_{\mathbf{p}} \sum_r \varphi_b^\dagger(r, \mathbf{p})\varphi_a(r, \mathbf{p}) \tag{21}$$

The most frequently used representations of the γ matrices are the Pauli and the Kramers (ultrarelativistic) representation. The first one is given by

$$\gamma^0 = \begin{pmatrix} 1 & 0 \\ 0 & -1 \end{pmatrix} \qquad \boldsymbol{\gamma} = \begin{pmatrix} 0 & \boldsymbol{\sigma} \\ -\boldsymbol{\sigma} & 0 \end{pmatrix} \tag{22}$$

In this representation, a rotation by an angle θ around the direction $\hat{\boldsymbol{\theta}}$ and a boost with rapidity y along the direction $\hat{\mathbf{y}}$ (th$y = v$) become

$$S(R(\boldsymbol{\theta})) = \cos\frac{\theta}{2} + i(\hat{\boldsymbol{\theta}}\boldsymbol{\Sigma})\sin\frac{\theta}{2} \tag{23a}$$

$$S(\Lambda(\mathbf{y})) = \mathrm{ch}\frac{y}{2} - (\hat{\mathbf{y}}\boldsymbol{\alpha})\,\mathrm{sh}\frac{y}{2} \tag{23b}$$

where $\boldsymbol{\Sigma}$ and $\boldsymbol{\alpha}$ are

$$\boldsymbol{\Sigma} = \begin{pmatrix} \boldsymbol{\sigma} & 0 \\ 0 & \boldsymbol{\sigma} \end{pmatrix} \qquad \boldsymbol{\alpha} = \gamma^0\boldsymbol{\gamma} = \begin{pmatrix} 0 & \boldsymbol{\sigma} \\ \boldsymbol{\sigma} & 0 \end{pmatrix} \tag{24}$$

The solutions to Eqs. (19) and (20) take the form

$$u(r, \mathbf{p}) = \begin{pmatrix} \sqrt{p^0 + m}\, \mathbf{w} \\ \sqrt{p^0 - m}\,(\mathbf{n}\cdot\boldsymbol{\sigma})\,\mathbf{w} \end{pmatrix} \tag{25a}$$

$$v(r, \mathbf{p}) = \begin{pmatrix} \sqrt{p^0 - m}\,(\mathbf{n}\cdot\boldsymbol{\sigma})\,\tilde{\mathbf{w}} \\ \sqrt{p^0 + m}\,\tilde{\mathbf{w}} \end{pmatrix} \tag{25b}$$

with $\mathbf{n} = \mathbf{p}/|\mathbf{p}|$. w and $\tilde{\text{w}}$ are Pauli (two component) spinors. For the spinors in the negative energy solution we choose the basis

$$\tilde{\text{w}}_r = -\mathrm{i}\,\sigma_2\,\text{w}_r \tag{26}$$

We may obtain the Kramers representation from the one of Pauli through a unitary transformation $\psi_K = U\,\psi_P$ with

$$U = \frac{1}{\sqrt{2}}\begin{pmatrix} 1 & 1 \\ 1 & -1 \end{pmatrix} \tag{27}$$

In this representation the γ matrices become

$$\gamma^0 = \begin{pmatrix} 0 & 1 \\ 1 & 0 \end{pmatrix} \qquad \gamma = \begin{pmatrix} 0 & -\boldsymbol{\sigma} \\ \boldsymbol{\sigma} & 0 \end{pmatrix} \tag{28}$$

while Σ and α equal

$$\Sigma = \begin{pmatrix} \boldsymbol{\sigma} & 0 \\ 0 & \boldsymbol{\sigma} \end{pmatrix} \qquad \alpha = \gamma^0\gamma = \begin{pmatrix} \boldsymbol{\sigma} & 0 \\ 0 & -\boldsymbol{\sigma} \end{pmatrix} \tag{29}$$

The transformation rules (23) remain unchanged.

We also define

$$\gamma^5 = \mathrm{i}\,\gamma^0\gamma^1\gamma^2\gamma^3 \tag{30}$$

with the property $(\gamma^5)^2 = 1$. In the Kramers representation, we have

$$\gamma^5 = \begin{pmatrix} 1 & 0 \\ 0 & -1 \end{pmatrix} \tag{31}$$

and in the Pauli representation

$$\gamma^5 = \begin{pmatrix} 0 & 1 \\ 1 & 0 \end{pmatrix} \tag{32}$$

The choice (30) implies

$$\mathrm{Tr}\{\gamma^5\gamma^\mu\gamma^\nu\gamma^\rho\gamma^\sigma\} = \mathrm{i}\,\varepsilon^{\mu\nu\rho\sigma} \qquad (\varepsilon^{0123} = 1) \tag{33}$$

Generators of the Poincaré group act on spinor wave functions as follows

$$\begin{aligned} P_\mu &= \mathrm{i}\sigma_{\mu\nu} \\ J_{\mu\nu} &= \frac{1}{2}\sigma_{\mu\nu} + \frac{1}{\mathrm{i}}\left(x_\mu\partial_\nu - x_\nu\partial_\mu\right) \end{aligned} \tag{34}$$

The solutions (25) satisfy the relations

$$u(r,\mathbf{p})\bar{u}(r,\mathbf{p}) \equiv \rho^{(+)} = (\not{p}+m)\frac{1+\gamma^5\not{W}}{2} \tag{35a}$$

$$v(r,\mathbf{p})\bar{v}(r,\mathbf{p}) \equiv \rho^{(-)} = (-\not{p}+m)\frac{1+\gamma^5\not{W}}{2} \tag{35b}$$

W_μ is the polarization four-vector.

In terms of the center-of-mass polarization vector $\boldsymbol{\zeta}$

$$W^0 = \frac{\boldsymbol{\zeta}\,\mathbf{p}}{m} \qquad \mathbf{W} = \boldsymbol{\zeta} + \frac{\mathbf{p}\,(\mathbf{p}\boldsymbol{\zeta})}{m(m+p^0)} \qquad (W \cdot p) = 0 \tag{36}$$

In particular, for helicity eigenstates we have $\boldsymbol{\zeta} = \lambda\hat{\mathbf{p}}$, $\lambda = \pm 1$ and

$$W_\mu = \lambda\left(\frac{|\mathbf{p}|}{m}, \hat{\mathbf{p}}\,\frac{p^0}{m}\right) \tag{37}$$

When $|\mathbf{p}| \gg m$, (35) becomes

$$\rho^{(\pm)} = \frac{1 \pm \lambda\gamma^5}{2}\,\not{p} \tag{38}$$

and, when $\mathbf{p} \to 0$,

$$\rho^{(+)} = \begin{pmatrix} m(1+\boldsymbol{\sigma}\boldsymbol{\zeta}) & 0 \\ 0 & 0 \end{pmatrix} \qquad \rho^{(-)} = \begin{pmatrix} 0 & 0 \\ 0 & m(1+\boldsymbol{\sigma}\boldsymbol{\zeta}) \end{pmatrix} \tag{39}$$

The above expressions are seen to coincide with the nonrelativistic spin density matrices

$$\rho = \frac{1}{2}(1 + \boldsymbol{\sigma}\boldsymbol{\zeta}) \tag{40}$$

For reference, we give here the Pauli spinors which are helicity eigenstates

$$\frac{1}{2}\frac{\mathbf{p}\boldsymbol{\sigma}}{|\mathbf{p}|}\,\mathrm{w}^{(\lambda)} = \lambda\mathrm{w}^{(\lambda)} \tag{41}$$

Denoting by φ and θ the polar angles of $\mathbf{p}$, we find

$$\mathrm{w}^{(\frac{1}{2})} = \begin{pmatrix} e^{-i\varphi/2}\cos\frac{\theta}{2} \\ e^{i\varphi/2}\sin\frac{\theta}{2} \end{pmatrix} \qquad \mathrm{w}^{(-\frac{1}{2})} = \begin{pmatrix} -e^{-i\varphi/2}\sin\frac{\theta}{2} \\ e^{i\varphi/2}\cos\frac{\theta}{2} \end{pmatrix} \tag{42}$$

The negative energy spinors can now be obtained using Eq. (26).

D 3. Spin 1 particles

The equations of motion for spin 1 particles are

$$\begin{align} \left(\Box + m^2\right)W^\mu(x) &= 0 \tag{43} \\ \partial_\mu W^\mu(x) &= 0 \tag{44} \end{align}$$

They have the general solution

$$W_\mu(x) = \sum_{r=1}^{3} \int d\Omega_{\mathbf{p}}\left\{ W(r,\mathbf{p})\varepsilon_\mu(r,\mathbf{p})\,e^{-ipx} + \tilde{W}(r,\mathbf{p})\varepsilon_\mu^*(r,\mathbf{p})e^{ipx} \right\} \tag{45}$$

Here, $\varepsilon_\mu(r, \mathbf{p})$ are three independent four-vectors subject to the constraint $\varepsilon^\mu p_\mu = 0$. The scalar product between two states reads

$$\langle a | b \rangle = -\mathrm{i} \int \mathrm{d}^3 \mathbf{x} W_\mu^{(a)*}(x) \overleftrightarrow{\partial}_0 W^{(b)\,\mu}(x) = - \int \mathrm{d}\Omega_\mathbf{p} W_\mu^{(a)*}(\mathbf{p}) W^{(b)\,\mu}(\mathbf{p}) \qquad (46)$$

For the sake of brevity, we have written $W_\mu^{(a)}(\mathbf{p}) = \sum_{r=1}^3 W^{(a)}(r, \mathbf{p}) \varepsilon_\mu(r, \mathbf{p})$. The polarization four-vectors $\varepsilon_\mu(r, \mathbf{p})$ satisfy a completeness relation

$$\sum_{r=1}^3 \varepsilon_\mu(r, \mathbf{p}) \varepsilon_\nu^*(r, \mathbf{p}) = -g_{\mu\nu} + \frac{p_\mu p_\nu}{m^2} \qquad (47)$$

and may be cast in the form

$$\varepsilon^0(r, \mathbf{p}) = \frac{\mathbf{p} \cdot \boldsymbol{\varepsilon}(r, 0)}{m}$$

$$\boldsymbol{\varepsilon}(r, \mathbf{p}) = \boldsymbol{\varepsilon}(r, 0) + \mathbf{p} \frac{\mathbf{p} \cdot \boldsymbol{\varepsilon}(r, 0)}{m(m + p^0)} \qquad (48)$$

Two possible bases for the vectors $\varepsilon^i(r, 0)$ are the canonical one, defined by $\varepsilon^i(r, 0) = \delta_r^i$, and the one made out of eigenstates of S_z

$$\boldsymbol{\varepsilon}(+1, 0) = -\frac{\mathrm{i}}{\sqrt{2}}(\mathbf{e}_x + \mathrm{i}\,\mathbf{e}_y)$$

$$\boldsymbol{\varepsilon}(0, 0) = \mathrm{i}\,\mathbf{e}_z \qquad (49)$$

$$\boldsymbol{\varepsilon}(-1, 0) = \frac{\mathrm{i}}{\sqrt{2}}(\mathbf{e}_x - \mathrm{i}\,\mathbf{e}_y)$$

$\mathbf{e}_x$, $\mathbf{e}_y$, $\mathbf{e}_z$, are unit vectors along the 3 axes.

Appendix E

Discrete symmetries

Parity, charge conjugation and time reversal act on the Fock space of free particles as follows

$$U(P)|a, r, \mathbf{p}\rangle = \eta_a^P |a, r, -\mathbf{p}\rangle \tag{1a}$$

$$U(C)|a, r, \mathbf{p}\rangle = \eta_a^C |\bar{a}, r, \mathbf{p}\rangle \tag{1b}$$

$$|U(T)(a, r, \mathbf{p})\rangle = \eta_a^{T*} S_{rr'} |a, -r', -\mathbf{p}\rangle \tag{1c}$$

a ($\bar{a}$) indicates a particle (antiparticle), r is the eigenvalue of a spin component (e.g. S_z) and $\mathbf{p}$ is the particle momentum. The operators $U(P)$ and $U(C)$ are unitary, whereas $U(T)$ is antiunitary. The phases η for particles and antiparticles are related to each other through

$$\eta_a^P = (-1)^{2s} \eta_{\bar{a}}^P \qquad \eta_a^C = \eta_{\bar{a}}^C \tag{2}$$

A fermion and its antifermion have opposite intrinsic parities. The matrix S in (1c) is proportional to a rotation by π about the y-axis, $S = \exp[i\pi J_y]$, (if z is the axis along which spin is quantized).

It is clear from (1) that charge conjugation eigenstates must correspond either to particles with $\bar{a} = a$, or to particle-antiparticle pairs.

One can rewrite Eqs. (1) as transformation rules on fields; these rules can then be adopted also in the case of interactive fields. In particular,

Spin 0

$$U^\dagger(P)\varphi(x^0, \mathbf{x})U(P) = \eta^P \varphi(x^0, -\mathbf{x})$$
$$U^\dagger(C)\varphi(x^0, \mathbf{x})U(C) = \eta^C \varphi^\dagger(x^0, \mathbf{x}) \tag{3}$$
$$U^\dagger(T)\varphi(x^0, \mathbf{x})U(T) = \eta^T \varphi(-x^0, \mathbf{x})$$

Spin $\frac{1}{2}$ (Pauli representation of γ matrices)

$$U^\dagger(P)\psi(x^0, \mathbf{x})U(P) = \eta^P \gamma^0 \psi(x^0, -\mathbf{x})$$

$$U^\dagger(C)\psi_\alpha(x^0,\mathbf{x})U(C) \;=\; \eta^C i\gamma^2_{\alpha\beta}\psi^\dagger_\beta(x^0,\mathbf{x}) \tag{4}$$
$$U^\dagger(T)\psi(x^0,\mathbf{x})U(T) \;=\; -\eta^T i\gamma^5\gamma^0\gamma^2\psi(-x^0,\mathbf{x})$$

Spin 1

$$U^\dagger(P)W_\mu(x^0,\mathbf{x})U(P) \;=\; \eta^P\gamma^0 W_\mu(x^0,-\mathbf{x})$$
$$U^\dagger(C)W_\mu(x^0,\mathbf{x})U(C) \;=\; \eta^C W^\dagger_\mu(x^0,\mathbf{x}) \tag{5}$$
$$U^\dagger(T)W_\mu(x^0,\mathbf{x})U(T) \;=\; \eta^T g_{\mu\mu}W_\mu(-x^0,\mathbf{x})$$

Particle-antiparticle pairs with spin 0   Given that $U^2(C)=1$, the possible eigenvalues of charge conjugation are ±1. Bose symmetry requires that the state be symmetric under particle exchange. If the angular momentum is ℓ, then an exchange leads to the condition $1=C(-1)^\ell$. As for parity, one has $P=(-1)^\ell$, since $\eta^P_a=\eta^P_{\bar a}$. Thus, the possible states of a and $\bar a$ have the following P and C assignments

$$P \;=\; (-1)^\ell \tag{6a}$$
$$C \;=\; (-1)^\ell \tag{6b}$$
$$CP \;=\; +1 \tag{6c}$$

Particle-antiparticle pairs with spin $\frac{1}{2}$ By the Pauli principle, the state $a-\bar a$ must be completely antisymmetric under exchange, if a has spin $1/2$. In a state with total spin $S=0,1$, a particle exchange introduces a factor of $(-1)^{S+1}$ to the spin wave function, and a factor of $(-1)^\ell$ to the orbital wave function; therefore, $-1=C(-1)^\ell(-1)^{S+1}$. Under parity, recalling that $\eta^P_a=-\eta^P_{\bar a}$ one finds $P=(-1)^{\ell+1}$. In conclusion,

$$P \;=\; (-1)^{\ell+1} \tag{7a}$$
$$C \;=\; (-1)^{\ell+S} \tag{7b}$$
$$CP \;=\; (-1)^{S+1} \tag{7c}$$

One obtains the same results (6) and (7) by considering a generic particle-antiparticle state with a given relative angular momentum ℓ and spin S:

$$|A\rangle = \int \frac{d^3\mathbf{p}_1}{(2\pi)^3}\frac{d^3\mathbf{p}_2}{(2\pi)^3}\varphi^\ell(\mathbf{p}_1,\mathbf{p}_2)\chi^S_{r_1r_2}b^\dagger(r_1,\mathbf{p}_1)d^\dagger(r_2,\mathbf{p}_2)|0\rangle \tag{8}$$

To this end, one must apply the transformation rules (1) and use commutation or anticommutation rules for $b^\dagger$ and $d^\dagger$, according to whether the particles are scalars or spinors.

E 1. Helicity amplitudes

The amplitude for a relativistic process, at given momentum values, can always be represented as a matrix in the spin quantum numbers of the particles involved. A natural basis in this space is provided by helicity states. Let us denote by g_i the number of helicity states of the ith particle. Massive spin-S particles have $g = 2S + 1$, while massless particles have $g = 1$ or $g = 2$, depending on the number of helicity states actually realized in nature. Neutrinos have $g = 1$; photons have $g = 2$.

Consider a process whose initial and final states contain n particles altogether; the number of possible helicity amplitudes is

$$N = \prod_{i=1}^{n} g_i \tag{9}$$

The helicity amplitudes have well-defined symmetry properties under P, C and T transformations; thus, if the process is invariant under some of these transformations, the number of independent amplitudes is appropriately reduced.

Let us illustrate with some examples the way in which invariances reduce the number of independent amplitudes, in two-body decays and in scattering processes.

We first consider two-body decays

$$a \to b + c \tag{10}$$

We denote by s_i, λ_i and η_i the spin, helicity and intrinsic parity of the ith particle, respectively. In the center-of-mass frame, choosing the direction of motion of b as the quantization axis for angular momentum, one finds

$$\lambda_a = \lambda_b - \lambda_c \tag{11}$$

by conservation of angular momentum. Parity invariance requires

$$\langle \lambda_b, \lambda_c | S | \lambda_a \rangle = \eta_a^P \eta_b^{P*} \eta_c^{P*} \langle -\lambda_b, -\lambda_c | S | -\lambda_a \rangle \tag{12}$$

If a coincides with its antiparticle, and if C invariance holds, one can relate the amplitudes of $a \to b + c$ to those of $a \to \bar{b} + \bar{c}$, as well as the amplitudes of $a \to b + \bar{b}$ among themselves. C invariance gives

$$\langle \lambda_b, \lambda_c | S | \lambda_a \rangle = \eta_a^C \eta_b^{C*} \eta_c^{C*} \langle \bar{\lambda}_b, \bar{\lambda}_c | S | \bar{\lambda}_a \rangle \tag{13}$$

Under CP,

$$\langle \lambda_b, \lambda_c | S | \lambda_a \rangle = \eta_a^{CP} \eta_b^{CP*} \eta_c^{CP*} \langle -\bar{\lambda}_b, -\bar{\lambda}_c | S | -\bar{\lambda}_a \rangle \tag{14}$$

By $\bar{\lambda}_a$ we denote the state of antiparticle with helicity λ_a.

Consider now two-body scattering processes

$$a + b \to c + d \tag{15}$$

and define a generalized reaction as the set of all processes obtained from (15) by trading a particle on one side of the reaction for its antiparticle on the other side. Thus, three-body decays $a \to \bar{b} + c + d$ will also be included in this set, whenever they are kinematically allowed.

Parity invariance, together with invariance under 180° rotations on the scattering plane, lead to the constraint

$$P: \qquad \langle \lambda_c, \lambda_d | S | \lambda_a, \lambda_b \rangle = \eta_a^P \eta_b^P \eta_c^{P*} \eta_d^{P*} \langle -\lambda_c, -\lambda_d | S | -\lambda_a, -\lambda_b \rangle \qquad (16)$$

If time reversal invariance holds, we have $U^\dagger(T) S U(T) = S^\dagger$; using the antiunitarity of $U(T)$ and applying again a 180° rotation on the scattering plane, we find

$$T: \qquad \langle \lambda_c, \lambda_d | S | \lambda_a, \lambda_b \rangle = \eta_a^{T*} \eta_b^{T*} \eta_c^T \eta_d^T \langle \lambda_a, \lambda_b | S | \lambda_c, \lambda_d \rangle \qquad (17)$$

Thus, T-invariance leads to constraints only for elastic processes.

Charge conjugation invariance connects an S-matrix element to one in which all particles turn into their antiparticles

$$C: \qquad \langle \lambda_c, \lambda_d | S | \lambda_a, \lambda_b \rangle = \eta_a^C \eta_b^C \eta_c^{C*} \eta_d^{C*} \langle \bar{\lambda}_c, \bar{\lambda}_d | S | \bar{\lambda}_a, \bar{\lambda}_b \rangle \qquad (18)$$

Further conditions based on symmetry properties can be obtained in the case of elastic scattering of identical particles, as well as in particle-antiparticle scattering.

The number of independent amplitudes is the same in all channels of a given generalized reaction.

Example 1   Consider the decay of a spin 0 particle into a fermion-antifermion pair. There are 4 possible amplitudes, by virtue of (9),

$$S_{++} \quad S_{--} \quad S_{+-} \quad S_{-+} \qquad (19)$$

where the two signs stand for the helicities of the two fermions. Angular momentum conservation, Eq. (11), reduces the number of possible amplitudes to 2

$$S_{++} \quad S_{--} \qquad (20)$$

This excludes, for example, the possibility of decaying into a neutrino-antineutrino pair. If parity is conserved, the initial state has well-defined parity, η^P, and (16) implies

$$S_{++} = -\eta^P S_{--} \qquad (21)$$

Therefore, only one independent amplitude exists. The only possible final state is

$$\frac{|\lambda_q \mathbf{p}_q, \bar{\lambda}_q \bar{\mathbf{p}}_q \rangle - \eta^P | - \lambda_q \mathbf{p}_q, -\bar{\lambda}_q \bar{\mathbf{p}}_q \rangle}{\sqrt{2}} \qquad (22)$$

For example, if $\eta^P = -1$, only the state with $\ell = 0$ contributes to the amplitude, in agreement with (7a).

Example 2   Consider π - nucleon scattering

$$\pi + N \rightarrow \pi + N \tag{23}$$

By (9), 4 amplitudes are possible. Parity invariance sets two constraints

$$S_{++} = S_{--} \qquad S_{+-} = S_{-+} \tag{24}$$

reducing to 2 the number of independent invariant amplitudes. Since the process is elastic, time reversal invariance introduces the constraint

$$S_{+-} = S_{-+} \tag{25}$$

E 2.   Invariant amplitudes

The use of helicity amplitudes enables us to enumerate in a simple way all invariant amplitudes. The explicit construction of independent amplitudes, once their number is known, can be performed using the covariant variables which describe particle states (momenta and polarizations), and respecting the superposition principle which imposes a linear dependence on the wave function of each particle.

In two-body decays all kinematic invariants are constants, whereas scattering processes and three-body decays exhibit an s- and t-dependence

$$\mathcal{M}_{fi} = \sum_n f_n(s,t)\, F_n \tag{26}$$

F_i are the independent invariants constructed out of the wave functions. A similar analysis can be applied to matrix elements of currents between states of a particle. Indeed, under rotations, the spatial part of a current transforms as a spin-1 representation, while the time component transforms as spin 0; hence, the classification of the matrix elements $\langle b|J_\mu|a\rangle$ is identical to that of the two processes $a \rightarrow (\text{spin } 1) + b$ and $a \rightarrow (\text{spin } 0) + b$. Imposing current conservation places a further constraint.

Example 1   Consider the matrix element of a current between two spin 0 states. The spin 0 part of the current obeys the kinematic constraint (11), whereas among the three helicity components of the spin 1 part $(\mathbf{J}_+, \mathbf{J}_-, \mathbf{J}_0)$ only the third one obeys the constraint. Thus, there are only two independent amplitudes. Since the wave functions of spinless particles are constant, we have

$$\langle \mathbf{p}_b|J_\mu|\mathbf{p}_a\rangle = f_1 P_\mu + f_2 q_\mu \tag{27}$$

where f_1 and f_2 are functions of the only invariant q^2 and

$$P^\mu = p_a^\mu + p_b^\mu \qquad q^\mu = p_b^\mu - p_a^\mu \tag{28}$$

If the current is conserved, a further constraint arises, reducing the number of independent amplitudes to one: setting $q^\mu J_\mu = 0$ we find

$$\langle \mathbf{p}_b | J_\mu | \mathbf{p}_a \rangle = f_1 \left[P_\mu - \frac{m_b^2 - m_a^2}{q^2} q_\mu \right] \tag{29}$$

Under parity, the components of a vector current transform as $(0^+, 1^-)$ and those of an axial current as $(0^-, 1^+)$. Suppose parity invariant holds; then, if the intrinsic parities of a and b are equal (opposite), only vector (axial) currents will have nonzero matrix elements.

Example 2 Consider the matrix elements of a current between two spin 1/2 states. The constraint (11) becomes $\lambda_b = \lambda_a - \lambda_J$, where λ_J is the helicity of the current. It is easy to see that only the 4 amplitudes indicated here at the left can be constructed out of the spin 1 part of the current. Similarly, only two amplitudes are possible for the spin 0 part: $\lambda_a = \lambda_b = \pm 1/2$. Thus, there is a total of 6 independent amplitudes. If parity is conserved, and if the current is purely vector or axial, the number of amplitudes reduces to 3. For example, given two states with the same relative parity, one can write

Spin1

λ_a	λ_b	λ_J
$\frac{1}{2}$	$\frac{1}{2}$	0
$-\frac{1}{2}$	$-\frac{1}{2}$	0
$\frac{1}{2}$	$-\frac{1}{2}$	-1
$-\frac{1}{2}$	$\frac{1}{2}$	1

$$\langle \mathbf{p}_b | V_\mu | \mathbf{p}_a \rangle = \bar{u}(\mathbf{p}_b) \left[f_1 \gamma_\mu - i f_2 \sigma_{\mu\nu} q^\nu + f_3 q_\mu \right] u(\mathbf{p}_a) \tag{30a}$$

$$\langle \mathbf{p}_b | A_\mu | \mathbf{p}_a \rangle = \bar{u}(\mathbf{p}_b) \gamma^5 \left[g_1 \gamma_\mu - i g_3 \sigma_{\mu\nu} q^\nu + g_2 q_\mu \right] u(\mathbf{p}_a) \tag{30b}$$

The form factors f_i and g_i are functions of q^2. Current conservation eliminates one form factor. In order to show that the expressions (30) exhaust the number of independent amplitudes, one may find useful the Gordon identity

$$\bar{u}(\mathbf{p}_b) i \sigma_{\mu\nu} q^\nu u(\mathbf{p}_a) = \bar{u}(\mathbf{p}_b)(P_\mu - (m_a + m_b)\gamma_\mu) u(\mathbf{p}_a) \tag{31}$$

with

$$\sigma_{\mu\nu} = -\frac{i}{2} [\gamma_\mu, \gamma_\nu] \tag{32}$$

Tables 1 and 2 give the number of helicity amplitudes involved in the most common scattering and decay processes. This number has been indicated as follows:

N : In the absence of any symmetry constraints
N_P : When parity is conserved
N_ν : In the presence of a massless spin 1/2 particle of given helicity
N_G : In the presence of a massless spin 1 particle of given helicity
N_T : When time reversal holds
The rightmost part of the tables refers to amplitudes of processes involving photons.

Two body decays

Spin states		N	$N_P(N_\nu)$		N_G	$N_{PG}(N_{VG})$
0	0	1	1			
0	1/2	2	1			
0	0	1	1		0	0
0	1	3	2		2	1
1	1/2	4	2		2	1
1	1	7	4	(1γ)	4	2
1	1			(2γ)	0	0

Table 1

Three-body decays and two-body scattering

Spin states			N	N_P	N_T	N_{PT}		N_G	N_{PG}	N_{TG}	N_{PTG}
0	0	0	1	1	1	1					
0	0	0	3	2	3	2					
0	0	1	9	5	6	4	(1γ)	2	1		
0	0	1					(2γ)	6	3		
1	1	1					(4γ)	4	2	3	2
0	0	1/2	4	2	3	2		10	6		5
0	1	1/2	12	6	9	5		8	4		
1/2	1/2	1/2	16	8	10	6					
1/2	ν	ν		4	3						
1	1	1/2					(2γ)	16	8	10	6

Table 2

Appendix F

Some useful formulae

Amplitudes involving spin 1/2 particles are often expressed in terms of tensors, made out of vector or axial currents of the form

$$
\begin{aligned}
V_\mu &= \bar{u}(\mathbf{p_1})\gamma_\mu u(\mathbf{p_2}) & \quad (1) \\
A_\mu &= \bar{u}(\mathbf{p_1})\gamma^5\gamma_\mu u(\mathbf{p_2}) & \quad (2)
\end{aligned}
$$

Frequently encountered cases are

$$
L_{\mu\nu}^{V,V} = \sum_{Pol} V_\mu V_\nu^\dagger \;;\; L_{\mu\nu}^{A,A} = \sum_{Pol} A_\mu A_\nu^\dagger \;;\; L_{\mu\nu}^{V,A} = \sum_{Pol} V_\mu A_\nu^\dagger \;;\; L_{\mu\nu}^{A,V} = \sum_{Pol} A_\mu V_\nu^\dagger \quad (3)
$$

Setting

$$
q = p_1 - p_2 \qquad r = p_1 + p_2 \quad (4)
$$

we have

$$
\begin{aligned}
L_{\mu\nu}^{V,V} &= \operatorname{Tr}\left[(\not{p}_1 + m_1)\gamma_\mu(\not{p}_2 + m_2)\gamma_\nu\right] \\
&= 2\left\{r_\mu r_\nu + q^2 g_{\mu\nu} - q_\mu q_\nu - (m_1 - m_2)^2 g_{\mu\nu}\right\} \\
L_{\mu\nu}^{A,A} &= -\operatorname{Tr}\left[(\not{p}_1 + m_1)\gamma_\mu\gamma^5(\not{p}_2 + m_2)\gamma^5\gamma_\nu\right] \\
&= 2\left\{r_\mu r_\nu + q^2 g_{\mu\nu} - q_\mu q_\nu - (m_1 + m_2)^2 g_{\mu\nu}\right\} \\
L_{\mu\nu}^{V,A} &= -\operatorname{Tr}\left[(\not{p}_1 + m_1)\gamma_\mu(\not{p}_2 + m_2)\gamma^5\gamma_\nu\right] \\
&= 4\mathrm{i}\,\varepsilon_{\mu\nu\alpha\beta}p_1^\alpha p_2^\beta \\
L_{\mu\nu}^{A,V} &= L_{\mu\nu}^{V,A} \quad (5)
\end{aligned}
$$

In the case of polarized particles, the corresponding formulae are obtained by means of the substitution

$$
(\not{p} + m) \rightarrow (\not{p} + m)\frac{1 + \gamma^5 \not{W}}{2} \quad (6)
$$

where W_μ is the polarization four-vector. For example, in the case of neutrino-lepton coupling we have

$$
\begin{aligned}
J_\mu &= \bar{u}_\ell(\mathbf{p}_1)\gamma_\mu(1-\gamma^5)u_{\nu_\ell}(\mathbf{p}_2) \\
\sum_{Pol} J_\mu J_\nu^\dagger &= \mathrm{Tr}\left[(\not{p}_1 + m)(1+\gamma^5\not{W})\gamma_\mu\not{p}_2\gamma_\nu(1-\gamma^5)\right] = \\
&= 4\left[L_\mu p_{2\nu} + L_\nu p_{2\mu} - (Lp_2)g_{\mu\nu} - i\,\varepsilon_{\mu\nu\alpha\beta}L^\alpha p_2^\beta\right]
\end{aligned}
\tag{7}
$$

with $L = p_1 - mW$; W_μ is the lepton polarization.

Summing over polarizations, (7) becomes

$$
\sum_{Pol} J_\mu J_\nu^\dagger = 4\left\{r_\mu r_\nu + q^2 g_{\mu\nu} - q_\mu q_\nu - m^2 g_{\mu\nu} - 2i\,\varepsilon_{\mu\nu\alpha\beta}q^\alpha r^\beta\right\}
\tag{8}
$$

The above formulae correspond to a case with a particle having four-momentum p_1 in the final state and one having four-momentum p_2 in the initial state. If, instead, a particle and an antiparticle are present in the final state, the formulae change by the substitution $p_2 \to -p_2$: therefore, in this channel we have

$$
q = p_1 + p_2 \qquad r = p_1 - p_2
\tag{9}
$$

A quantity of interest in several applications is the average of a tensor over the directions of the outgoing lepton. This amounts to averaging the tensor over phase space with invariant mass q^2

$$
\langle T_{\mu\nu}\rangle = \frac{\displaystyle\int T_{\mu\nu}\,d\Phi^{(2)}(q \to p_1 + p_2)}{\displaystyle\int d\Phi^{(2)}(q \to p_1 + p_2)}
\tag{10}
$$

The average value of $r_\mu r_\nu$ can be expressed in terms of $g_{\mu\nu}$ and $q_\mu q_\nu$, with the result

$$
\begin{aligned}
\langle L_{\mu\nu}^{V,V}\rangle &= \frac{4}{3}\left(q^2 g_{\mu\nu} - q_\mu q_\nu\right)\left[1 + \frac{m_1^2 + m_2^2}{q^2} - \frac{1}{2}\frac{(m_1^2 - m_2^2)^2}{q^4}\right] \\
&\quad -2(m_1 - m_2)^2\left[g_{\mu\nu} - \frac{q_\mu q_\nu}{q^4}(m_1 + m_2)^2\right]
\end{aligned}
\tag{11}
$$

$$
\begin{aligned}
\langle L_{\mu\nu}^{A,A}\rangle &= \frac{4}{3}\left(q^2 g_{\mu\nu} - q_\mu q_\nu\right)\left[1 + \frac{m_1^2 + m_2^2}{q^2} - \frac{1}{2}\frac{(m_1^2 - m_2^2)^2}{q^4}\right] \\
&\quad -2(m_1 + m_2)^2\left[g_{\mu\nu} - \frac{q_\mu q_\nu}{q^4}(m_1 - m_2)^2\right]
\end{aligned}
\tag{12}
$$

In the case $\ell\,\nu_\ell$, $(J_\mu = V_\mu - A_\mu)$, we obtain

$$
\langle L_{\mu\nu}\rangle = \frac{8}{3}\left(q^2 g_{\mu\nu} - q_\mu q_\nu\right)\left[1 + \frac{m^2}{q^2} - \frac{1}{2}\frac{m^4}{q^4}\right] - 4m^2\left[g_{\mu\nu} - \frac{q_\mu q_\nu}{q^4}m^2\right]
\tag{13}
$$